AF302227

Vibration of Structures and Machines

Third Edition

Springer Science+Business Media, LLC

Giancarlo Genta

Vibration of Structures and Machines

—— Practical Aspects

Third Edition

With 173 Illustrations

 Springer

Giancarlo Genta
Dipartimento di Meccanica
Politecnico di Torino
Corso Duca degli Abruzzi, 24
10129 Torino
Italy

Cover illustration: See Figure 4.34 for details.

Library of Congress Cataloging-in-Publication Data
Genta, G. (Giancarlo)
    Vibration of structures and machines : practical aspects /
Giancarlo Genta. — 3rd ed.
        p.   cm.
    Includes bibliographical references and index.
    ISBN 978-1-4612-7149-9   ISBN 978-1-4612-1450-2 (eBook)
    DOI 10.1007/978-1-4612-1450-2
    1. Vibration—Mathematical models.   2. Structural dynamics.
3. Machinery—Vibration.   I. Title.
TA355.G44   1998
621.8′11—dc21   98-48084

Printed on acid-free paper.

Production managed by Timothy Taylor; manufacturing supervised by Jacqui Ashri.
Camera-ready copy supplied by the author.

9 8 7 6 5 4 3 2 1

ISBN 978-1-4612-7149-9

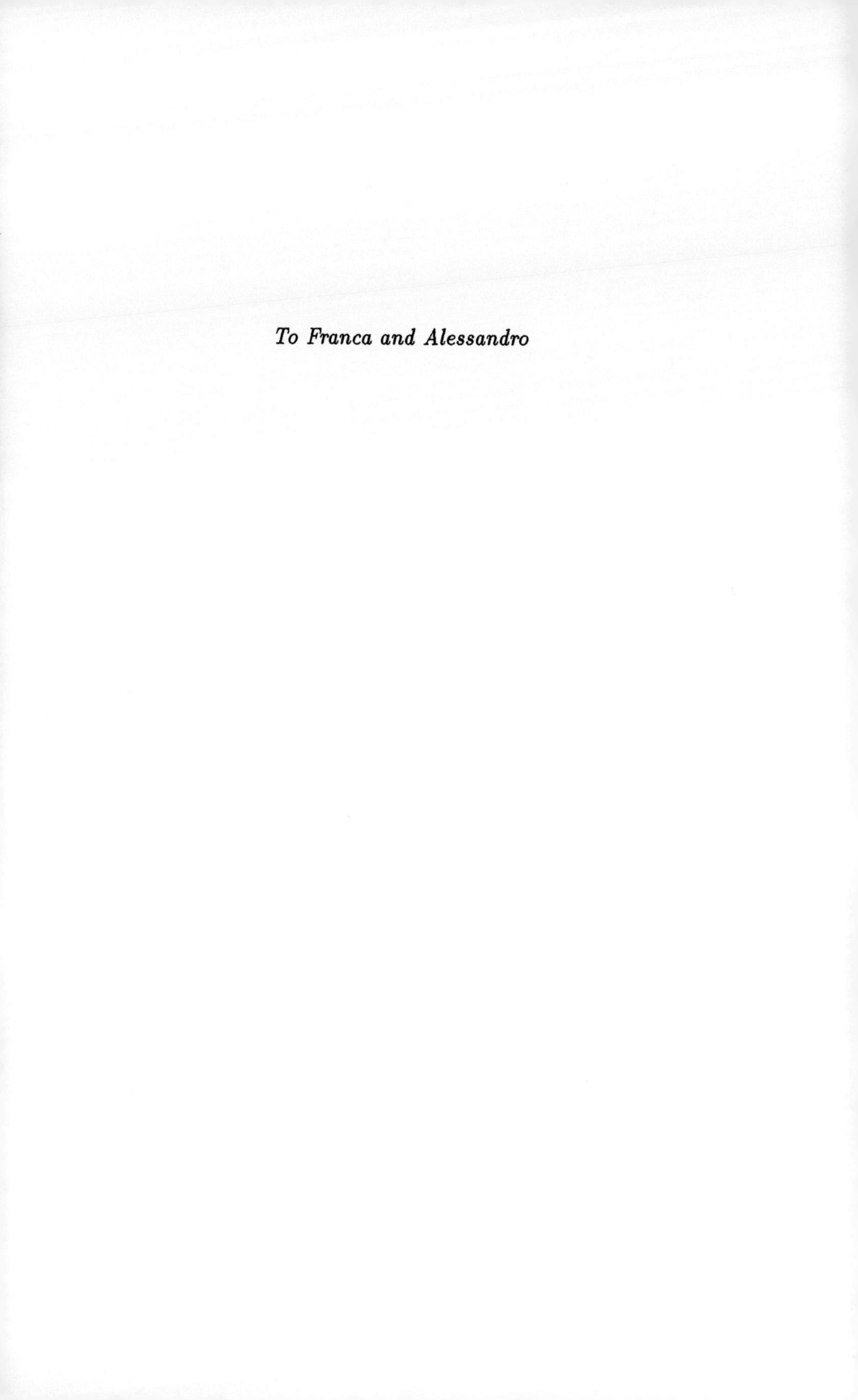

To Franca and Alessandro

Preface to the third edition

When asked to prepare a third edition of the current book, the author realized that come changes were needed.

The first change is a formal one; it regards notation for matrices and vectors. In earlier editions they were written using square brackets and braces, which are now dropped and replaced with bold characters. This type of notation is simpler and graphically cleaner, but it can be confusing to some students. The decision to do this was not an easy one, but it was made mostly because almost all recent books use this convention. This led to the need to change a few symbols to maintain a distinction between vectors, written with lower case letters, and matrices, for which capital letters are used.

More substantial changes deal with a greater integration on the classical approach to mechanics of vibrations with that typical of system dynamics and control theory. Altough this book has not the aim of dealing with control systems in any detailed way, the recent experience of the author in the new Mechatronics Laboratory of Politecnico di Torino led to the suggestion of looking at some aspects of vibration dynamics in a more interdisciplinary way.

G. Genta
Torino, September 1998

Preface to the second edition

This second edition follows the first by about two years. Apart from correcting a number of printing errors, few additions have been introduced.

In the chapter on the finite element method, a short summary on finite elements in time (four-dimensional finite elements) has been introduced. In the chapter on rotordynamics, two new graphical representations are used, namely, the roots locus plot for free whirling and a tridimensional plot for the forced response for which the designation of "orbital tubes" has been proposed. The former has been borrowed from controlled systems dynamics, where it is widely used, while the latter is made practical by the availability of powerful graphical tools for postprocessing numerical or experimental results.

As a last point, a more detailed description of the behaviour of magnetic bearings has been introduced in Chapter 6.

G. Genta
Torino, September 1994

Preface

The current book originates from the need felt by the author to give a systematic form to the contents of the lectures he gives to mechanical and aeronautical engineering students of the Technical University (Politecnico) of Torino, within the frames of the courses of Principles and Methodologies of Mechanical Design and Construction of Aircraft Engines. Its main aim is to summarize the fundamentals of mechanics of vibrations to give the needed theoretical background to the engineer who has to deal with vibration analysis and to show a number of design applications of the theory. Because the emphasis is mostly on the practical aspects, the theoretical aspects are not dealt with in detail, particularly in areas in which a long and complex study would be needed.

The book is structured in six chapters. The basic concepts of linear dynamics of discrete systems are summarized in Chapter 1. Following the lines just described, some specialized topics, such as random vibrations, are just touched on, more to remind the reader that they exist and to stimulate him to undertake a deeper study of these aspects than to supply detailed information.

The dynamics of continuous systems is the subject of Chapter 2. As the analysis of the dynamic behaviour of continuous systems is now mostly performed using discretization techniques, the stress is laid mostly on these, particularly the finite element method, with the aim of supplying the users of commercial computer codes with the theoretical background needed to build adequate mathematical models and critically evaluate the results obtained from the computer.

The behaviour of nonlinear systems is studied in Chapter 3, with the aim of stressing the aspects of this subject that are of interest to engineers more than to theoretical mechanicists. The recent advances in all fields of technology often result in an increased nonlinearity of machines and structural elements and design engineers must increasingly face nonlinear problems: This chapter is meant to be of help in this instance.

Chapters 4 and 5 are devoted to the study of the dynamics of rotating and reciprocating machines. They are meant as specific applications of the more general topics studied before, and they intend to be more application-oriented than the previous ones. However, methods and mathematical models that have not yet entered everyday design practice and are still regarded as research topics, are dealt with herein.

The last chapter constitutes an introduction to the dynamics of controlled structural systems, which are increasingly entering design practice and will unquestionably be used more often in the future.

The subjects studied here are usually considered different fields of applied mechanics or mechanical design. Specialists in rotor dynamics, torsional vibration, modal analysis, nonlinear mechanics and controlled systems often speak different languages, and it is difficult for students to be aware of the unifying ideas that are at the base of all these different specialized fields. The inconsistency of the symbols used in the different fields can be particularly confusing. In order to use a consistent symbol system throughout the book, some deviation from the common practice is unavoidable.

The author believes that it is possible to explain all the aspects related to mechanical vibrations (actually not only mechanical) using a unified approach. The current book is an effort in this direction.

S.I. units are used in the whole book, with few exceptions. The first exception is the measure of angles, for which in some cases the old unit degree is preferred to the S.I. unit radian, particularly where phase angles are concerned. Frequencies and angular velocities should be measured in rad/s. Sometimes the older units (Hz for frequencies and revolutions per minute [rpm]) are used, when the author feels that this makes things more intuitive or where normal engineering practice suggests it. In most formulas, at any rate, consistent units are used. In very few cases this rule is not followed, but the reader is expressly warned in the text.

For frequencies, no distinction is generally made between frequency in Hz and circular frequency in rad/s. Although the author is aware of the subtle differences between the two quantities (or better, between the two different ways of seeing the same quantity), which are subtended by the use of two different names, he chose to regard the two concepts as equivalent. A single symbol (λ) is used for both and the symbol f is never used for a frequency in Hz. The period is then always equal to $T = 2\pi/\lambda$ because consistent units (in this case, rad/s) must be used in all formulas. A similar rule holds for angular velocities, which are always referred to with the symbol ω. No different symbol is used for angular velocities in rpm, which in some texts are referred to by n. The use of λ instead of ω for frequencies is due to the need to avoid confusion between frequencies and angular velocities. In rotor dynamics, the speed at which the whirling motion takes place is regarded as a whirl frequency and not a whirl angular velocity (even if the expression *whirl speed* is sometimes used in opposition to spin speed), and symbols are used accordingly. It can be said that the concept of angular velocity is used only for the rotation of material objects, and the rotational speed of a vector in the Argand plane or of the deformed shape of a rotor (which does not involve actual rotation of a material object) is considered a frequency.

The author is grateful to colleagues and students in the Mechanics Department of the Politecnico di Torino for their suggestions, criticism, and general exchange of ideas and, in particular, to the postgraduate students working in the dynamics field at the department for reading the whole manuscript and checking most of the equations. Particular thanks are due to my wife, Franca, both for her encouragement and for doing the tedious work of revising the manuscript.

G. Genta
Torino, October 1992

Contents

Symbols

c	viscous damping coefficient, clearance.
c^*	complex viscous damping coefficient.
c_{cr}	critical value of c.
c_{eq}	equivalent damping coefficient.
d	distance (between axis of cylinder and center of crank).
e	base of natural logarithms.
$\mathbf{f}$	force vector.
$\overline{f}_i$	ith modal force.
f_0	amplitude of the force $F(t)$.
g	acceleration of gravity.
$g(t)$	response to a unit step input.
h	thickness of oil film, relaxation factor.
$h(t)$	response to a unit-impulse function.
i	imaginary unit $(i = \sqrt{-1})$.
k	stiffness, gain.
k^*	complex stiffness $(k^* = k' + ik'')$.
l	length, length of the connecting rod.
l_0	length in a reference condition.
m	mass, number of modes taken into account, number of outputs.
n	number of degrees of freedom.
p	pressure.

q_i	ith generalized coordinate.
$\mathbf{q}$	vector of the (complex) coordinates.
$\mathbf{q}_i$	ith eigenvector.
$q_i(xyz)$	ith eigenfunction.
r	radius, number of inputs.
$\mathbf{r}$	Ritz vector, vector of the complex coordinates (rotating frame), vector of the command inputs, vector of modal participation factors.
s	Laplace variable.
$\mathbf{s}$	state vector (transfer matrices method).
t	time, thickness.
u	displacement.
$\mathbf{u}$	vector of the inputs.
$u(t)$	unit-step function.
v	velocity.
v_s	velocity of sound.
$\mathbf{x}$	vector of the coordinates.
x_0	amplitude of $x(t)$.
x_m	maximum value of periodic law $x(t)$.
xyz	(fixed) reference frame.
$\mathbf{y}$	vector of the outputs.
z	complex coordinate ($z = x + iy$).
$\mathbf{z}$	state vector.
A	area of the cross section.
$\mathcal{A}$	dynamic matrix (state space approach).
$\mathcal{B}$	input gain matrix.
$\mathbf{C}$	damping matrix.
$\mathcal{C}$	output gain matrix.
$\overline{\mathbf{C}}$	modal damping matrix.
$\mathbf{D}$	dynamic matrix.
$\mathcal{D}$	direct link matrix.
E	Young's modulus.
F	force.
$\mathcal{F}$	Rayleigh disipation function.
G	shear modulus, balance-quality grade.
$G(s)$	transfer function.
$H(\lambda)$	frequency response.
$\mathbf{H}$	controllability matrix.
I	area moment of inertia.
$\mathbf{I}$	identity matrix.
J	mass moment of inertia.
$\mathcal{L}$	work.
$\mathbf{K}$	stiffness matrix.
$\mathbf{K}''$	imaginary part of the stiffness matrix.
$\overline{K}_i$	ith modal stiffness.
M	moment.
$\mathbf{M}$	mass matrix.
$\overline{M}_i$	ith modal mass.
$\mathbf{N}$	matrix of the shape functions.

O	load factor (Ocvirk number).
$\mathbf{O}$	observability matrix.
Q	quality factor.
Q_i	ith generalized force.
R	radius.
$\mathbf{R}$	rotation matrix.
S	Sommerfeld number.
$\mathbf{S}$	Jacobian matrix.
$S(\lambda)$	power spectral density.
T	period of the free oscillations.
$\mathcal{T}$	kinetic energy.
$\mathbf{T}$	transfer matrix, matrix linking the forces to the inputs.
$\mathcal{U}$	potential energy.
V	velocity, volume.
α	slenderness of a beam, phase of static unbalance, nondimensional parameter.
β	attitude angle, phase of couple unbalance, nondimensional parameter.
γ	shear strain.
δ	logarithmic decrement, phase in phase-angle diagrams.
$\delta(t)$	unit-impulse function (Dirac delta).
$\delta\mathcal{L}$	virtual work.
δx	virtual displacement.
ϵ	strain, eccentricity.
ζ	damping factor ($\zeta = c/c_{cr}$); nondimensional coordinate ($\zeta = z/l$), complex coordinate ($\zeta = \xi + i\eta$).
η	loss factor.
η	modal coordinates.
θ	angular coordinate.
λ	frequency, complex frequency, whirl speed.
λ'	whirl speed in the rotating frame.
$[\lambda^2]$	eigenvalue matrix.
λ_n	natural frequency of the undamped system.
λ_p	frequency of the resonance peak in damped systems.
μ	coefficient of the nonlinear term of stiffness, viscosity.
ν	Poisson's ratio, complex frequency.
$\xi\eta\zeta$	rotating reference frame.
ρ	density.
σ	decay rate, stress.
σ_y	yield stress.
τ	shear stress.
ϕ	angular displacement, complex coordinate ($\phi = \phi_y - i\phi_x$)
χ	shear factor, angular error for couple unbalance.
ψ	relative damping.

ω	angular velocity (spin speed).
ω_{cr}	critical speed.
$\mathbf{B}$	compliance matrix.
Φ	phase angle.
$\boldsymbol{\Phi}$	eigenvector matrix.
$\boldsymbol{\Phi}^*$	eigenvector matrix reduced to m modes.
Ω	angular velocity.
$\Im$	imaginary part.
$\Re$	real part.
—	complex conjugate ($\bar{a}$ is the conjugate of a)

SUBSCRIPTS

d	deviatoric.
m	mean.
n	nonrotating.
r	rotating.
I	imaginary part.
R	real part.

Introduction

Vibration is one of the most common aspects of life. Many natural phenomena, as well as manmade devices, involve periodic motion of some sort. Our own bodies include many organs that perform periodic motion, with a wide spectrum of frequencies, from the relatively slow motion of the lungs or heart, to the high-frequency vibration of the eardrums. When we shiver, hear or speak, even when we snore, we directly experience vibration. Vibration is often associated with dreadful events; indeed one of the most impressive and catastrophic natural phenomena is the earthquake, a manifestation of vibration. In manmade devices vibration is often less impressive, but it can be a symptom of malfunctioning and is often a signal of danger. When traveling by vehicle, particularly driving or flying, any increase of the vibration level makes us feel uncomfortable. Vibration is also what causes sound, from the most unpleasant noise to the most delightful music.

Vibration can be put to work for many useful purposes: Vibrating sieves, mixers, and tools are the most obvious examples. Vibrating machines also find applications in medicine, curing human diseases. Another useful aspect of vibration is that it conveys a quantity of useful information about the machine producing it.

Vibration produced by natural phenomena and, increasingly, by manmade devices is also a particular type of pollution, which can be heard as noise if the frequencies that characterize the phenomenon lie within the audible range (spanning from about 18 Hz to 20 kHz) or felt directly as vibration. This type of pollution can cause severe discomfort. The discomfort due to noise depends on the intensity of the noise and its frequency, but many other features are also of great importance. The sound of a bell and the noise from some machine can have the same intensity and frequency but can create very different sensations. Although even the psychological disposition of the subject can be important in assessing how much discomfort a certain sound creates, some standards must be assessed in order to evaluate the acceptability of noise sources.

Generally speaking, there is growing awareness of the problem and designers are asked, sometimes forced, to reduce the noise produced by all sorts of machinery. When vibration is transmitted to the human body by a solid surface, different effects are likely to be felt. Generally speaking, what causes discomfort is not the amplitude of the vibration but the peak value of the acceleration. The level of acceleration that causes discomfort depends on the frequency and the time of exposure, but other factors such as the position of the human body and the part that is in contact with the source are also important. Also, for this case, some standards have been stated. The maximum r.m.s. (root mean square) values of acceleration that cause reduced proficiency when applied for a stated time in a vertical direction to a sitting subject are plotted as a function of frequency in Figure 1. The figure, which is taken from the ISO 2631-1978 standard, deals with a field from 1 to 80 Hz and with daily exposure times from 1 minute to 24 hours.

The exposure limits can be obtained by multiplying the values reported in Figure 1 by 2, while the reduced comfort boundary is obtained by dividing the same values by 3.15 (i.e., by decreasing the r.m.s. value by 10 dB). From the plot, it is clear that the frequency field in which man is more affected by vibration lies between 4 and 8 Hz. Frequencies lower than 1 Hz produce sensations similar to motion sickness. They depend on many parameters other than acceleration and are variable from individual to individual. At frequencies greater than 80 Hz, the effect of vibration is too dependent on the part of the body involved and on the skin conditions to give general guidelines. An attempt to classify the effects of vibration with different frequencies on man is shown in Figure 2. Note that there are resonance fields at which some parts of the body vibrate with particularly large amplitudes.

As an example, the thorax-abdomen system has a resonant frequency of about 3 to 6 Hz, although all resonant frequency values are dependent on individual characteristics. The head-neck-shoulder system has a resonant frequency of about 20 to 30 Hz, and many other organs have more or less pronounced resonances at other frequencies (e.g., the eyeball at 60 to 90 Hz, the lower jaw-skull system at 100 to 220 Hz, etc.).

In English, as in many other languages, there are two terms used to designate oscillatory motion: oscillation and vibration. The two terms are used almost interchangeably; however, if a difference can be found, oscillation more often used to emphasize the kinematic aspects of the phenomenon (i.e., the time history of the motion in itself), while the vibration implies dynamic considerations (i.e., considerations on the relationships between the motion and the causes from which it originates). Actually, not all oscillatory motions can be considered vibrations: For a vibration to take place, it is necessary that a continuous exchange of energy between two different forms occurs. In mechanical systems, the particular forms of energy that are involved are kinetic energy and potential (elastic or gravitational) energy. Oscillations in electrical circuits are due to exchange of energy between the electrical and magnetic fields.

Many periodic motions taking place at low frequencies are thus oscillations but not vibrations, including the motion of the lungs. It is not, however, the slowness of the motion that is important but the lack of dynamic effects. To be subject to vibration, a system must be able to store energy in two different forms and allow energy to be transferred from one to the other. The simplest mechanical oscillators are the pendulum and the spring-mass system. The corresponding simplest electrical oscillator is the capacitor-inductor system. Their behaviour can be studied using the same linear second-order differential equation with constant coefficients, even if in the case of the pendulum the application of a simple linear model requires the assumption that the amplitude of the oscillation is small.

For centuries, the pendulum, and later the spring-mass system (later still the capacitor-inductor system), has been more than a model. It constituted a paradigm through which the oscillatory behaviour of actual systems has been interpreted. All oscillatory phe-

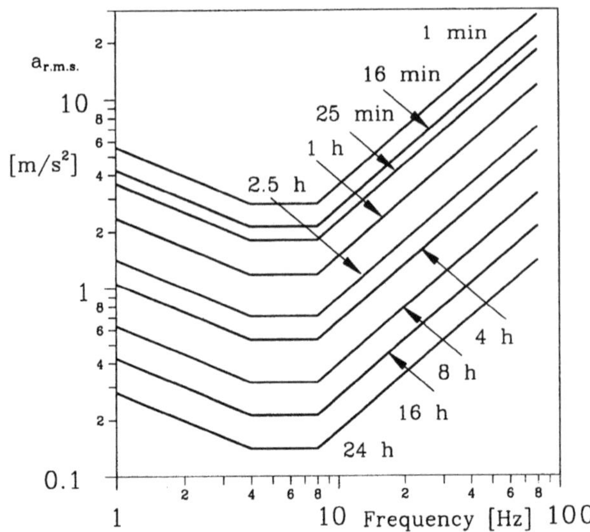

FIGURE 1. Vertical vibration exposure criteria curves defining the "fatigue-decreased proficiency boundary" (ISO 2631-1978 standard).

nomena in real life are more complex than that, at least for the presence of dissipative mechanisms owing to which at each vibration cycle, i.e., each time the energy is transformed forward and backward between the two main energy forms, some of the energy of the system is dissipated, usually being transformed into heat. This causes the vibration amplitude to decay in time until the system comes to rest, unless some form of excitation sustains the motion by providing the required energy.

The basic model can easily accommodate this fact, by simply adding some form of energy dissipator to the basic oscillator. The spring-mass-damper and the damped-pendulum models constitute a paradigm for mechanical oscillators, while the inductor-capacitor-resistor system is the basic damped electrical oscillator.

Although the very concept of periodic motion was well known, ancient natural philosophy failed to understand vibratory phenomena, with the exception of the study of sound and music. This is not surprising, as vibration could neither be predicted theoretically, owing to the lack of the concept of inertia, nor observed experimentally, as the wooden or stone structures were not prone to vibrate, and, above all, ancient machines were heavily damped owing to the very high friction. The beginnings of the theoretical study of vibrating systems is traced back to observations made by Galileo Galilei in

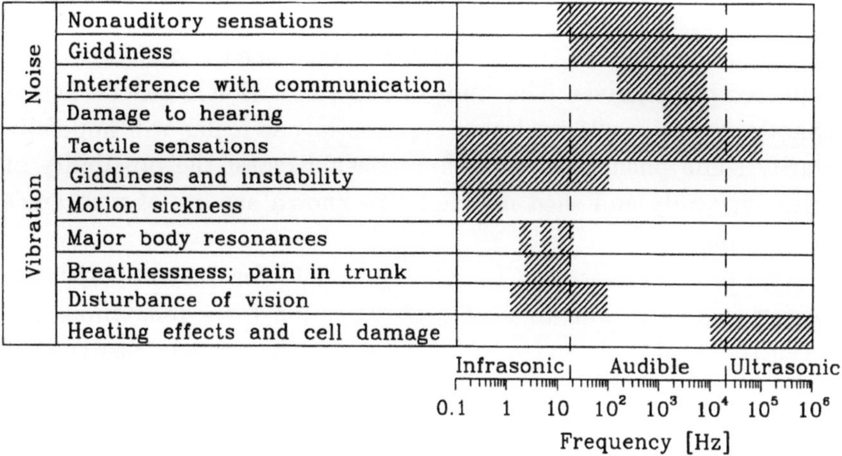

FIGURE 2. Effects of vibration and noise (intended as airborne vibration) on man as functions of frequency (R.E.D. Bishop, *Vibration*, Cambridge Univ. Press, Cambridge, 1979).

1583 regarding the motions of one of the lamps hanging from the ceiling of the cathedral of Pisa. It is said that he timed the period of oscillations using the beat of his heart as a time standard to conclude that the period of the oscillations is independent from the amplitude.

Whether or not this is true, he described in detail the motion of the pendulum in his *Dialogo sopra i due massimi sistemi del mondo*, published in 1638, and stated clearly that its oscillations are isochronous. It is not surprising that the beginning of the studies of vibratory mechanics occurred at the same time as the formulation of the law of inertia. The idea that a mechanical oscillator could be used to measure time, due to the property of moving with a fixed period, clearly stimulated the theoretical research in this field. While Galileo seems to have believed that the oscillations of a pendulum have a fixed period even if the amplitude is large (he quotes a displacement from the vertical as high as 50°), certainly Huygens knew that this is true only in linear systems and introduced in around 1656 a modified pendulum whose oscillations would have been truly isochronous even at large amplitudes. He published his results in his *Horologium Oscillatorium* in 1673.

The great development of theoretical mechanics in the eighteenth and nineteenth centuries gave the theory of vibration very deep and solid roots. When it seemed that theoretical mechanics could not offer anything new, the introduction of computers to theoretical sciences, with the possibility of performing very complex numerical experiments, revealed completely new phenomena and disclosed unexpected perspectives. The study of chaotic motion in general and of chaotic vibrations of nonlinear systems in particular will hopefully clarify some phenomena that have been beyond the possibility of scientific study and shed new light on known aspects of mechanics of vibration.

Mechanics of vibration is not just a field for theoretical study. Design engineers had to deal with vibration for a long time, but recently the current tendencies of technology have made the dynamic analysis of machines and structures more important.

The load conditions the designer has to take into account in the structural analysis of any member of a machine or a structure can be conventionally considered as static, quasi-static, or dynamic. A load condition belongs to the first category if it is constant and is applied to the structure for all or most of its life. A typical example is the self-weight of a building. The task of the structural analyst is usually limited to determining whether the stresses caused are within the allowable limits of the material, taking into account all possible environmental effects (creep, corrosion, etc.). Sometimes the analyst ensures that the deformations of the structure are consistent with the regular working of the machine. Also, loads that are repeatedly exerted on the structure, but that are applied and removed slowly and stay at a constant value for a long enough time, are assimilated to static loads. An example of these static load conditions is the pressure loading on the structure of the pressurized fuselage of an airliner and the thermal loading of many pressure vessels. In this case, the designer also has to take into account the fatigue phenomena that can be caused by repeated application of the load. Because the number of stress cycles is usually low, low-cycle fatigue usually is encountered.

Quasi-static load conditions are those conditions that, although due to dynamic phenomena, share with static loads the characteristics of being applied slowly and remaining for comparatively long periods at more or less constant values. Examples are the centrifugal loading of rotors and the loads on the structures of space vehicles due to inertia forces during launch or re-entry. Also, in this case, fatigue phenomena can be very importat in the structural analysis.

Dynamic load conditions are those in which the loads are rapidly varying and cause strong dynamic effects. The distinction is due mainly to the speed at which loads vary in time. Because it is necessary to state in some way a time scale to assess whether a certain load is applied slowly, it is possible to say that a load condition is static or quasi-static if the characteristic times of its variation are far longer than the longest period of the free vibrations of the structure.

A load can be considered static if it is applied to a structure whose first natural frequency is high or dynamic if it is applied to a structure that vibrates at low frequency. Dynamic loads can cause the structure to vibrate and can sometimes produce a resonant response. Causes of dynamic loading can be the motion of what supports the structure (as in the case of seismic loading of buildings or the stressing of the structure of ships due to wave motion), the motion of the structure (as in the case of ground vehicles moving on uneven roads), or the interaction of the two motions (as in the case of aircraft flying in gusty air). Other sources of dynamic loads are unbalanced rotating machinery and aero- or gas-dynamic phenomena in jet and rocket engines.

The task the structural analyst must perform in these cases is much more demanding. To check that the structure can withstand the dynamic loading for the required time and that the amplitude of the vibration does not affect the ability of the machine to perform its tasks, the analyst must acquire a knowledge of the dynamic behaviour, which is often quite detailed. The natural frequencies of the structure and the corresponding mode shapes must first be obtained, and then its motion under the action of the dynamic loads and the resulting stresses in the material must be computed. Fatigue must generally be taken into account, and often the methods based on fracture mechanics must be applied.

Fatigue is not necessarily due to vibration; it can be defined more generally as the possibility that a structural member fails under repeated loading at stress levels lower than those that could cause failure if applied only once. However, the most common way in which this repeated loading takes place is linked with vibration. If a part of a machine or structure vibrates, particularly if the frequency of the vibration is high, it can be called on to withstand a high number of stress cycles in a comparatively short time, and this is usually the mechanism triggering fatigue damage.

Another source of difficulty is the fact that, while static loads are usually defined in deterministic terms, often only a statistical knowledge of dynamic loads can be reached.

Progress causes machines to be lighter, faster, and, generally speaking, more sophisticated. All these trends make the tasks of the structural analyst more complex and demanding. Increasing the speed of machines is often a goal in itself, as in the transportation field. This is sometimes useful in increasing production and lowering costs (as in machine tools) or causing more power to be produced, transmitted, or converted (as in energy-related devices). Faster machines, however, are likely to be the cause of more intense vibrations, and, often, they are prone to suffer damages due to vibrations. Speed is just one of the aspects. Machines tend to be lighter, and materials with higher strength are constantly being developed. Better design procedures allow the exploitation of these characteristics with higher stress levels, and all these efforts often result in less stiff structures, which are more prone to vibrate. All these aspects compel designers to deal in more detail with the dynamic behaviour of machines.

Dynamic problems, which in the past were accounted for by simple overdesign of the relevant elements, must now be studied in detail, and dynamic design is increasingly the most important part of the design of many machines. Most of the methods used nowadays in dynamic structural analysis were first developed for nuclear or aerospace applications, where safety and lightness are of the utmost importance. These methods are spreading to other fields of industry, and the number of engineers working in the design area, particularly those involved in dynamic analysis, is growing. A good technical background in this field, at least enough to understand the existence and importance of these problems, is increasingly important for persons not directly involved in structural analysis, such as production engineers, managers, and users of machinery.

It is now almost commonplace to state that about half of the engineers working in mechanical industries, and particularly in the motor-vehicle industry, are employed in tasks directly related to design. A detailed analysis of the tasks in which engineers are engaged in an industrial group working in the field of energy systems is reported in Figure 3a. An increasing number of engineers are engaged in design and the relative economic weight of design activities on total production costs is rapidly increasing. An increase of 300% in the period from 1950 to 1990 has been recorded.

Within design activities, the relative importance of structural anal-

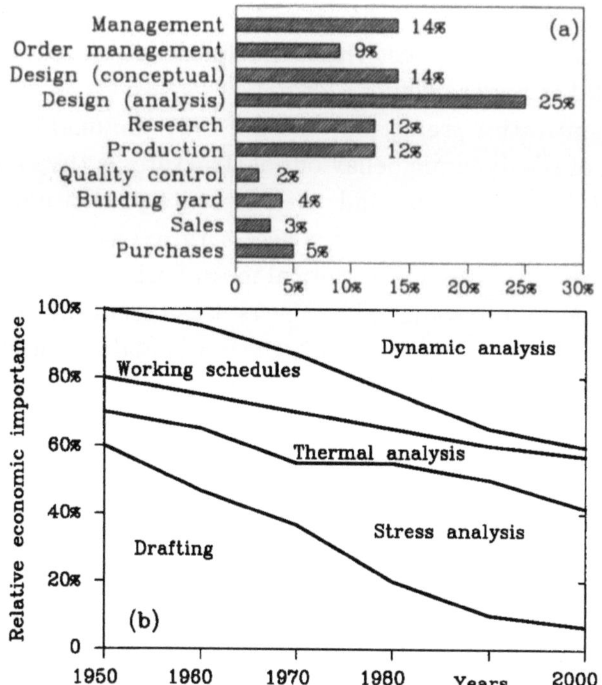

FIGURE 3. (a) Tasks in which engineers are employed in an Italian industrial group working in the field of energy systems; (b) relative economic weight of the various activities linked with structural design (P.G. Avanzini, *La formazione universitaria nel campo delle grandi costruzioni meccaniche*, Giornata di studio sull'insegnamento della costruzione delle macchine, Pisa, 31/3/1989).

ysis, mainly dynamic analysis, is increasing, while that of activities generally indicated as drafting is greatly reduced (Figure 3b).

Economic reasons advocate the use of predictive methods for the study of the dynamic behaviour of machines from the earliest stages of design, without having to wait until prototypes are built and experimental data are available. The cost of design changes increases rapidly during the progress of the development of a machine, from the very low cost of changes introduced very early in the design stage to the dreadful costs (also in terms of loss of image) that occur when a product already on the market has to be recalled to the factory to be modified. On the other hand, the effectiveness of the changes decreases while new constraints due to the progress of the design process are stated. This situation is summarized in the graph of Figure 4. Because many design changes can be necessary as a result of dynamic structural analysis, it must be started as early as possible in the design process, at least in the form of first-approximation studies. The analysis must then be refined and detailed when the machine takes a more definite form.

The quantitative prediction, and not only the qualitative understanding, of the dynamic behaviour of structures is then increasingly important. To understand and, even more, to predict quantitatively the behaviour of any system, it is necessary to resort to models that can be analyzed using mathematical tools. Such analysis work is unavoidable, even if in some of its aspects it can seem that the physical nature of the problem is lost within the mathematical intricacy of the analytical work. After the analysis has been performed it is necessary to extract results and interpret them to obtain a synthetic picture of the relevant phenomena. The analytical work is necessary to ensure a correct interpretation of the relevant phenomena, but if it is not followed by a synthesis, it remains only a sterile mathematical exercise. The tasks designers are facing in modern technology force them to understand increasingly complex analytical techniques. They must, however, retain the physical insight and engineering common sense without which no sound synthesis can be performed.

If technological advancement forces the designer to perform increasingly complex tasks, it also provides the instruments for the fulfilment of the new duties with powerful means of theoretical and experimental analysis.

The availability of more powerful computers has deeply changed the methods, the mathematical means, and even the language of structural analysis, while extending the ability of mathematical study of problems that previously could be tackled only through experiments. However, the basic concepts and theories of structural dynamics have not changed: Its roots are very deep and strong and can doubtless sustain the new rapid growth. Moreover, only the recent increase of computational power enabled a deeper utilization of the body of knowledge that accumulated in the last two centuries and often remained unexploited owing to the impossibility of handling the exceedingly complex computations. The numerical solution of problems that, until a few years ago, required an experimental approach can only be attempted by applying the aforementioned methods of theoretical mechanics.

At the same time, together with computational instruments, there was a striking progress in test machines and techniques. Designers

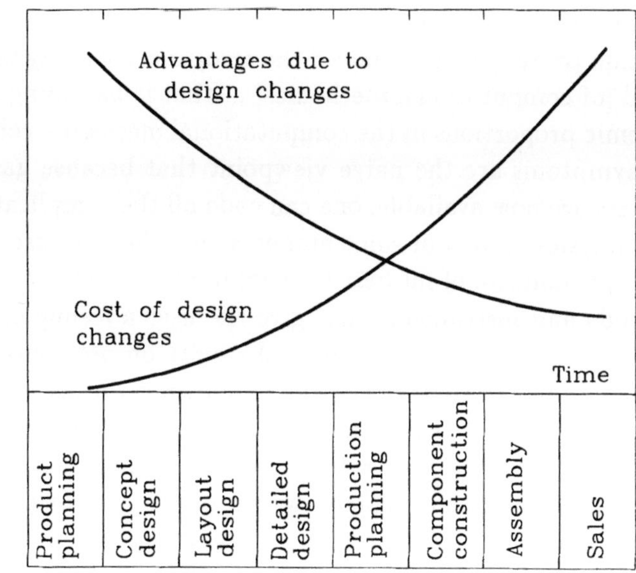

FIGURE 4. Cost and effectiveness of design changes as a function of the stage at which the changes are introduced.

can now base their choices on large quantities of experimental data obtained on machines similar to those being studied, which are often not only more plentiful, but also more detailed and less linked with the ability and experience of the experimenter than those that were available in the past. Tests on prototypes or on physical models of the machine (even if numerical experimentation is increasingly replacing physical experimentation) not only yield a large amount of information on the actual behaviour of machines, but also allow validation of theoretical and computational techniques.

Modern instrumentation is increasingly used to allow a more or less continuous monitoring of machines in operating conditions. This allows designers to collect a great deal of data on how machines work in their actual service conditions and to reduce safety margins without endangering, but actually with increasing, safety.

As already said, designers can now rely on very powerful computational instruments that are widely used in structural analysis. Their use is not, however, free of dangers. Often a sort of disease, called *number crunching syndrome*, has been identified as affecting those who deal with computational mechanics. Oden and Bathe[1] defined it as "blatant overconfidence, indeed the arrogance, of many working in the field [of computational mechanics] ... that is becoming a disease of epidemic proportions in the computational mechanics community. Acute symptoms are the naive viewpoint that because gargantuan computers are now available, one can code all the complicated equations of physics, grind out some numbers, and thereby describe every physical phenomena of interest to mankind."

Methods and instruments that give the user a feeling of omnipotence, because they supply numerical results on problems that can be of astounding complexity without allowing the user to control the various stages of the computation, are clearly potentially dangerous. They give the user a feeling of confidence and objectivity, because the computer cannot be wrong or have its own subjective bias. The finite element method, perhaps the most powerful computational method used for many tasks, among which the solution of problems of structural dynamics is one of the most important, is, without doubt, the most dangerous from this viewpoint.

In the beginning, computers entered the field of structural analysis in a quiet and reserved way. From the beginning of the 1950s computers were used to automatically perform those computational procedures that required long and tedious work, for which electromechanical calculators were widely used. Because the computations required for the solution of many problems (like the evaluation of the critical speeds of complex rotors or the torsional vibration analysis of crankshafts) were very long, the use of automatic computing machines was an obvious improvement. At the end of the 1950s computations that nobody could even think to perform without using computers became routine work. Programs of increasing complexity were often prepared by specialists, and analysts started to concentrate their attention on the preparation of data and the interpretation of results more than on how the computation was performed. In the 1960s the situation evolved further, and the first commercial finite element codes appeared on the market. Soon they had some sort of preprocessors and postprocessors to help the user handle the large amounts of data and results.

In the 1970s general-purpose codes that can tackle a wide variety

---

[1]Oden T.J., Bathe K.J., *A Commentary on Computational Mechanics*, Applied Mechanics Review, 31, p.1053, 1978.

of different problems were commonly used. These codes, which are often prepared by specialists who have little knowledge of the specific problems for which the code can be used, are generally considered by the users to be tools to use without bothering to find out how they work and the assumptions on which the work is based. More often the designer who must use these commercial codes tends to accept noncritically any result that comes out of the computer. Moreover, these codes allow a specialist in a single field to design a complex system without seeking the cooperation of other specialists in the relevant matters in the belief that the code can act as a most reliable and unbiased consultant. The user must, on the contrary, know very well what the code can do and the assumptions at its foundation. The user must have a good physical perception of the meaning of the data being introduced and the results obtained in order to be able to give a critical evaluation.

There are two main possible sources of errors in the results obtained from a code. First there can be errors (*bugs*, in the jargon of computer users) in the code itself. This can even happen in well-known commercial codes, particularly if the problem being studied requires the use of parts of the code that are seldom used or insufficiently tested. The user can try to solve problems the programmer never imagined the code could be asked to tackle and may thus, follow (without having the least suspicion of doing so) paths that were never imagined and thus never tested.

More often, it is the modeling of the physical problem that is to blame for poor results. The user must always be aware that even the most sophisticated code always deals with a simplified model of the real world, and it is a part of the user's task to ascertain that the model retains the relevant features of the actual problem.

Generally speaking, a model is acceptable only if it yields predictions close to the actual behaviour of the physical system. Other than this, only its internal consistency can be unquestionable, but internal consistency alone has little interest for the applications of a model.

The availability of programs that automatically prepare data (preprocessors) can make things worse. Together with the advantages of reducing the work required from the user and avoiding the errors linked with the manual preparation and introduction of a large mass of data, there is the drawback of giving a false confidence. The mathemratical model prepared by the machine is neither better nor more objective than a handmade one, and it is always the operator who must use engineering knowledge and common sense to reach a satisfactory model. The use of general-purpose codes requires the designer to have a knowledge of the physical features of the actual systems and of the modeling methods not much less than that required to prepare the code. The designer must also be familiar with the older simplified methods through which approximate results or at least an order of magnitude can be quickly obtained, allowing the designer to keep a close control over a process in which he has little influence.

The use of sophisticated computational methods must not decrease the skill of building very simple models that retain the basic feature of the actual system with a minimum of complexity. Some very ingenious analysts can create models, often with only one, or very few, degrees of freedom, which can simulate the actual behaviour of a complicated physical system. The need for this skill is actually increasing, and such models often constitute a base for a physical insight that cannot be reached using complex numerical procedures. The latter are then mandatory for the collection of quantitative information, whose interpretation is made easier by the insight already gained.

Concern about vibration and dynamic analysis is not restricted to designers. No matter how good the dynamic design of a machine is, if it is not properly maintained, the level of vibration it produces can increase to a point at which it becomes dangerous or causes discomfort. The balance conditions of a rotor, for example, can change in time, and periodic rebalancing can be required. Maintenance engineers must be aware of vibration-related problems to the same extent as design engineers. The analysis of the vibrations produced by a machine can be a very powerful tool for the engineer who has to maintain a machine in working condition. It has the same importance that the study of the symptoms of disease has for medical doctors.

In the past, the experimental study of the vibration characteristics of a machine was a matter of experience and was more an art than a science: Some maintenance engineers could immediately recognize problems developed by machines and sometimes even foretell future problems just by pressing an ear against the back of a screwdriver whose blade is in contact with carefully chosen parts of the outside of the machine. The study of the motion of water in a transparent bag put on the machine or of a white powder distributed on a dark vibrating panel could give other important indications. Modern instrumentation, particularly electronic computer-controlled instruments, gives a scientific basis to this aspect of the mechanics of machines.

The ultimate goal of "preventive maintenance" is that of continuously obtaining a complete picture of the working conditions of a machine in such a way as to plan the required maintenance operations in advance, without having to wait for malfunctions to actually take place. In some more advanced fields of technology, such as aerospace or nuclear engineering, this approach is already entering everyday practice. In other fields, these are more indications for future developments than current reality. Unfortunately, the subject of vibration analysis is complex and the use of modern instrumentation requires a theoretical background beyond the knowledge of many maintenance or practical engineers.

However, the revolution in all fields of technology, and increasingly in everyday life, due to the introduction of computers did not only change the way machines and structures are designed, built, and monitored but it also had an increasingly important impact on how they work and will deeply change the very idea of machines. A typical exaple is the expression "intelligent machines," which until a few years ago would have been considered an intrinsically contradictory statement, but now is commonly accepted.

The recent developments in the fields of electronics, information, and control systems have made it possible to tackle dynamic problems of structures in a new and often more effective way. While the traditional approach for reducing dynamic stressing has always been that of changing (usually increasing) the stiffness of the structure or adding damping, now control systems that can either adapt the behaviour of the structure to the changing dynamic requirements or fight vibrations directly by applying adequate dynamic forces to the structure are increasingly common. This trend is widespread in all fields of structural mechanics, with civil, mechanical, and aeronautical engineering applications. For example, structural control has been successfully attempted in tall buildings and bridges, machine tools, aeronautics, bearing systems for rotating machinery, robots, space structures, and ground vehicles. In the last case, the term *active suspensions* has even become popular among the general public.

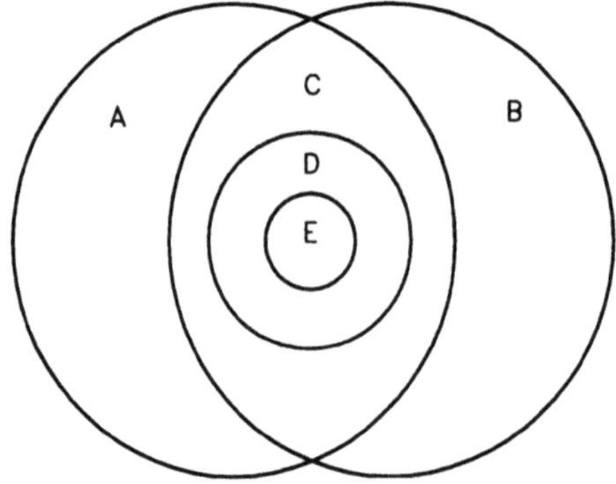

FIGURE 5. Tentative classification of adaptive and sensory structures. A: adaptive structures; B: sensory structures; C: controlled structures; D: active structures; and E: intelligent structures.

The advantages of this approach over the conventional one are clear and can be easily evidenced by the example of a large lightweight structure designed to be deployed in space in a microgravity environment. The absence or the low entity of static forces allows the design of very light structures, and lightness is a fundamental prerequisite for any structure that has to be brought into orbit. This leads to very low natural frequencies and corresponding vibration modes that can easily be excited and are very lightly damped. Any attempt to maintain the dynamic stresses and displacements within reasonable limits with conventional techniques, i.e., by stiffening the structure and adding damping, would lead to large increases in the mass and, hence, the cost of the structure. The application of suitable control devices can achieve the same goals in a far lighter and cheaper way.

A structure provided with actuators that can adapt its geometric shape or modify its mechanical characteristics to stabilize a number of working parameters (e.g., displacements, stresses, and temperatures) is said to be an *adaptive structure*. An adaptive structure can be better defined as a structure with actuators allowing controlled alterations of the system states and characteristics.

If there are sensors, the structure can be defined a *sensory structure*. The two things need not go together, as in the case of a structure provided with embedded optical fibers that supply information about the structural integrity of selected components or in the case of a machine with a built-in diagnostic system. If, however, the structure is both adaptive and sensory, it is a *controlled structure*.

Active structures are a subset of controlled structures in which there is an external source of power, aimed at supplying the control energy and modulated by the control system using the information supplied by the sensors. Another typical characteristic is that the integration between the structure and the control system is so strong that the distinction between structural functionality and control functionality is blurred and no separate optimization of the parts is possible.

Intelligent structures can be tentatively differentiated from active structures by the presence of a highly distributed control system that takes care of most of the functions. Most biological structures fall in this category; a good example of the operation of an intelligent structure is the way in which the wing of a bird regulates the aerodynamic forces needed to fly. Not only is the shape constantly adapted, but the dynamic behaviour of the structure is also controlled. Although a central control system coordinates all this, most of the control action is committed to peripheral subsystems, distributed on the whole structure.

A tentative classification of adaptive and sensory structures is shown in Figure 5.

In all types of controlled structures the control system may need to perform different tasks, with widely different requirements. For example, it can be used to change some critical parameter in order to adapt the characteristics of the system to the working conditions, as a device that varies the stiffness of the supports of a rotor with the aim of changing its critical speed during start-up to allow a shift from subcritical to supercritical conditions without having to actually pass a critical speed. In this case, there is no need to have a very complicated control system, and even a manual control can be used, if a slow start-up is predicted. Other examples requiring a slow control system are the suspension systems for ground vehicles that are able to maintain the vehicle body in a prescribed attitude even when variations of static or quasi-static forces (e.g., centrifugal forces in road bends) occur.

In the case where the control system has to supply forces to control vibrations, its response has to be faster. If only a few modes of a large and possibly very soft structure are to be controlled, as in the case of tall buildings, bridges, and some space structures, the requirements for the control system can be not very severe, but they become tougher when the characteristic time of the phenomena to be kept under control gets shorter as the relevant frequencies are high.

In other cases, when the structural elements are movable and a control system is already present to control the rigid-body motions, the control of the dynamic behaviour of the system can be achieved by suitably modulating the inputs to the devices that operate the machine. This is the case of robot arms or deployable space structures in which the dynamic behaviour is strongly affected by the way in which the actuators perform their task of driving the structural elements to the required positions.

It is easy to predict that the application of structural control, particularly using active control systems, will become more popular in the future. The advances in performance and cost reduction of control systems is going to make it very cost-effective, but a key factor for its success will be the incorporation into machines of complex microprocessor-based control systems that, although basically introduced for reasons different from structural control, can also take care of the latter in an effective and economical way. The recent advances in the field of neural networks also open very promising prospectives in the field of structural control.

However, if the control system must perform the vibration control of a structure, a malfunctioning of the first can cause a structural failure. The reliability required is that typical of control systems that perform vital functions, as in, for example, fly-by-wire systems, and this requirement can have heavy effects on costs, on both the component and the system level, and can slow down the application of structural control in low-cost, mass-production applications.

As already stated, the trend is toward an increasing integration between the structural and control functions, and this leads to the need for a unified approach at the design and analysis stages. The control subsystem can no longer be seen as something added to an already existing structural subsystem or to one that has been designed independently. There is a trend toward a unified approach to many aspects of structural dynamics and control, both from the theoretical viewpoint and it their practical applications. A further interdisciplinary effort must also include those aspects that are more strictly linked with the electrical and electronic components that are increasingly found in all kinds of machinery.

This interdisciplinary approach to the design of complex machines is increasingly referred to as *mechatronics*.

Although there are many definitions of what mechatronics is, it

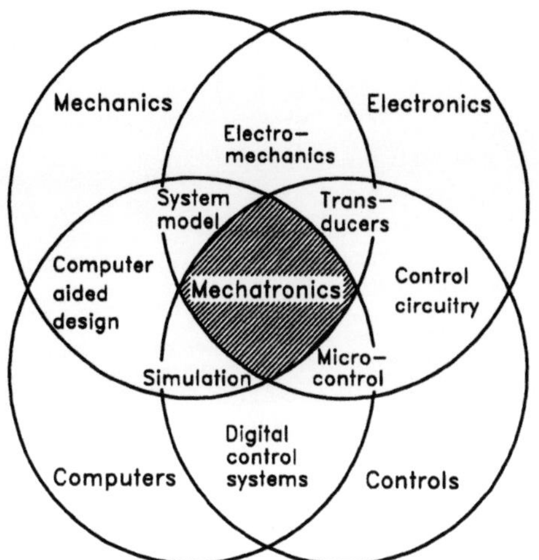

FIGURE 6. Tentative definition of mechatronics.

can be safely stated that it deals with the integration of mechanics,
electronics, and control science to design products that reach their
specifications maily through a deep integration of their structural
and control subsystems. A tentative graphical definition is shown in
Figure 6, which must be regarded to as an approximation[2]. Firstly,
the sets defining the various component technologies are not crisply
defined; they are fuzzy sets. Secondly, it is questionable whether
computer technology is to be so much stressed, as in this way analogic
devices seem to be ruled out.

But what is actually lacking in Figure 6 are the economic aspects,
which must eneter such an interdisciplinary approach from the onset of any practical application. The very need for an integrated
approach that allows a true simultaneous engineering of the various components of any machine to be reached comes from economic
consideration, even before even thinking of the performance or other
technical aspects of a machine.

It is the integration of a sound mechanical design, which includes
static and dynamic analyses and simulation, with electronic and control design which allows machines that offer better perfomances with
increased safety levels at potentially lower costs.

---

[2]S. Ashley, "Getting a hold on mechatronics", *Mechanical Engineering*, 119
(5), May 1997.

1

Discrete Linear Systems

1.1 Systems with a single degree of freedom

The simplest of the systems studied by structural dynamics is the linear mechanical oscillator with a single degree of freedom consisting of a point mass suspended by a linear spring and, possibly, a viscous damper (Figure 1.1a). Historically, however, the mathematical pendulum (Figure 1.1c) represented for centuries the most common paradigm of an oscillator, which could be assumed to be linear, at least within adequate limitations.

A linear spring is an element that reacts with a force equal to $-k(l - l_0)$ when stretched of the quantity $l - l_0$, where l_0 is the length at rest of the spring. k is a constant, usually referred to as stiffness of the spring, expressing the ratio between the force and the elongation. In S.I. units, it is expressed in N/m. To have true restoring force that opposes the displacement of point P, constant k must be positive. The system is then statically stable, in the sense that, when displaced from its equilibrium position, it tends to return to it[1].

A linear viscous damper is a device that introduces into the system

---

[1]For a more detailed definition of stability see Chapter 3. Only stable systems will be dealt with in this chapter.

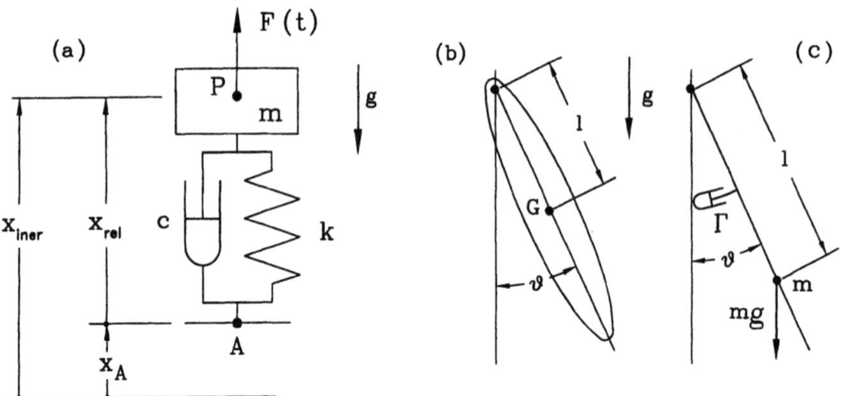

FIGURE 1.1. Linear oscillators with one degree of freedom: (a) Spring-mass-damper system; the coordinate x for the study of the motion of point P can expressed in an inertial reference system or be a relative displacement; (b) physical pendulum; (c) mathematical pendulum.

a force whose amplitude is proportional to the relative velocity of its end points and whose direction is opposite to that of the relative velocity through the damping coefficient c: $F_d = -c\dot{l}$. If the damping coefficient (in S.I. units expressed in Ns/m) is positive, the damper is a device that dissipates energy, and the amplitude of the free oscillations of the system decays in time. If the system is statically stable, it is also dynamically stable as it actually returns to the equilibrium position, at least asymptotically.

On the point mass m can act a force function of time $F(t)$ and the supporting point A can move in x direction with time history $x_A(t)$.

The importance of the study of the simple linear oscillators of Figure 1.1 is great for more than just historical reasons. In the first instance it is customary to start the study of mechanics of vibration with a model that is very simple but demonstrates, at least qualitatively, the behaviour of more complex systems. The systems of Figure 1.1 also have a great practical importance: They constitute models that can often be used to study, with good approximation, the behaviour of systems of greater complexity. Moreover, systems with many degrees of freedom, and even continuous systems, can be reduced, under fairly wide simplifying assumptions, to a set of independent systems with a single degree of freedom.

Owing to the linearity of the system of Figure 1.1a, the length at rest of the spring l_0, and all constant forces, such as those due to the gravitational acceleration g, affect the static equilibrium position but not the dynamic behaviour. The dynamic problem can then be separated from the static problem and all constant forces neglected when dealing with the first. The dynamic equilibrium equation, stating that the inertial force must be, at any time, in equilibrium with the elastic reaction of the spring and the force due to the damper added to the external forces, written with reference to the inertial x-coordinate, is simply

$$m\ddot{x} + c\dot{x} + kx = c\dot{x}_A(t) + kx_A(t) + F(t)\,, \qquad (1.1)$$

Equation (1.1) expresses the motion of point P in terms of its displacement from the position of static equilibrium $x = 0$ characterized by $F=0$ and $x_A=0$. The excitation provided by the motion of the supporting point and that provided by an external force can be dealt with in exactly the same way.

Often, when the excitation to the system is provided by the motion of the supporting point, it can be useful to express the position of point P with reference to point A. The absolute acceleration of point P is now expressed as $\ddot{x}_{iner} = \ddot{x}_{rel} + \ddot{x}_A$, and, neglecting the terms that are constant with only the effect of displacing the static position of equilibrium, the equation of motion,

$$m\ddot{x} + c\dot{x} + kx = -m\ddot{x}_A + F(t)\,, \qquad (1.2)$$

is very similar to equation (1.1), the only difference being the way the displacement of the supporting point is taken into account.

The solution of equation (1.1) (or of equation (1.2)) can be obtained by adding the complementary function, i.e., the general solution of the homogeneous equation, to a particular integral of the complete equation. The first describes the behaviour of the system when no external excitation acts on it (free behaviour) and, consequently, it is influenced only by its internal parameters. It can be obtained from the homogeneous equation associated with the equation of motion, which is a second-order autonomous differential equation because the independent variable, time, does not appear explicitly. Note that for the study of the free behaviour it is immaterial whether equation (1.1) or (1.2) is used, because their homogeneous parts are identical. The particular integral describes the motion under the effect of an external excitation and then is influenced by both the characteristics of the system and the time history of the excitation.

	Displacement x [m]	Mass m [kg]	Stiffness k [N/m]	Damping c [Ns/m]	Force F [N]
	Rotation θ [rad]	Moment of inertia J [kgm^2]	Torsional stiffness χ [Nm/rad]	Torsional damping Γ [Nms/rad]	Moment M [Nm]

TABLE 1.1. Formal equivalence between mechanical oscillators with translational and rotational motion. Quantities entering the equation of motion, common symbols, and S.I. units.

Because the equation of motion is a second-order differential equation, two conditions on the initial values must be stated in order to obtain a unique solution.

Instead of the translational oscillator of Figure 1.1a, a torsional oscillator can be devised. It consists of a rigid body free to rotate about an axis that passes through its center of gravity, constrained by a torsional spring and perhaps damped by a torsional damper. Equation (1.1) also holds for the present case, provided that the parameters involved are changed according to Table 1.1.

Example 1-1
Consider the pendulum shown in Figure 1.1b. It can be considered as a rotational oscillator, with moment of inertia J, torsional damping coefficient Γ, and restoring generalized force, due to the gravitational field, equal to $mgl\sin(\theta)$. In the case of the mathematical pendulum of Figure 1.1c, the rigid body reduces to a point mass suspended to a massless rod, and the moment of inertia reduces to $J = ml^2$. The equation of motion for the free oscillations can be easily computed from equation (1.1):

$$J\ddot{\theta} + \Gamma\dot{\theta} + mgl\sin(\theta) = 0 \ .$$

If the amplitude of the oscillations is small enough, the equation of motion can be linearized by substituting $\sin(\theta)$ with θ.

1.2 Systems with many degrees of freedom

Consider a discrete system consisting of a number of masses connected to each other and to a supporting frame by linear springs and viscous dampers. Using matrix notation, the dynamic equilibrium equations can be written in the compact form

$$\mathbf{M}\ddot{\mathbf{x}} + \mathbf{C}\dot{\mathbf{x}} + \mathbf{K}\mathbf{x} = \mathbf{f}(t) \ . \qquad (1.3)$$

$\mathbf{M}$ is the mass matrix of the system. It is diagonal if all coordinates x_i are related to translational degrees of freedom and measured with reference to an inertial frame.

$\mathbf{C}$ is the viscous damping matrix of the system.

$\mathbf{K}$ is the stiffness matrix. Generally it is not a diagonal matrix, although it usually has a band structure. In some cases, it is possible to resort to a set of generalized coordinates for which the stiffness matrix is diagonal (e.g., using as coordinates the length of the various springs), but such choice results in a nondiagonal mass matrix. The only exception is that of the modal coordinates, which allow the use of mass and stiffness matrices, which are both diagonal.

The vectors included in equation (1.3) are $\mathbf{x}$, a vector in which the generalized coordinates are listed, and $\mathbf{f}$, a time-dependent vector containing the forcing functions due to external forces or to the motion of the supporting points.

If a simple discrete system made of point masses connected to each other and to the ground by springs and dampers is considered, the generalized coordinates x_i can be just the components along the directions of the reference axes of the displacements, in exactly the same way as for the system with a single degree of freedom. Matrices $\mathbf{M}$, $\mathbf{C}$, and $\mathbf{K}$ are symmetrical matrices[2] of order n, where n is the number of degrees of freedom of the system.

Generally they are positive semidefinite, but in many cases $\mathbf{M}$ and $\mathbf{K}$ can be positive definite. The mass matrix is such when a non- vanishing mass is associated to all degrees of freedom, which is always possible, and more accurately reflects the actual situation. The stiffness matrix is positive defined when no rigid body motion is allowed. Sometimes a system in which the constraints prevent all rigid body motions is said to be a *structure*, and the term *mechanism* is used for the opposite case. However this distinction has faded because many devices, such as spacecraft, aircraft and simple drivelines, are actually unconstrained, and if modeled as a whole, they give rise to

---

[2]Note that this symmetry can be destroyed if some equations are substituted by linear combinations of the equations or are just multiplied by a constant. The equations of motion can then be written in forms in which the relevant matrices are not symmetrical.

singular stiffness matrices.

Sometimes the difficulties linked with the presence of a singular matrix can be circumvented by adding very soft constraints, which cause low-frequency rigid body oscillatory motions, but this can be done only when the vibrational behaviour of the structure is studied as uncoupled with the attitude dynamics of the system. If their coupling is accounted for, there is no way of removing the singularity of the stiffness matrix.

Equation (1.3) is, however, more general and holds for linear systems of any type. If a number of constraints are located between the point masses, their displacement vectors $\vec{r}_i$ can be expressed as functions of a number n of parameters x_i

$$\vec{r}_i = f(x_1, x_2, \ldots, x_n) \, . \tag{1.4}$$

Because the number of parameters needed to state the configuration of the system is n, it has a number n of degrees of freedom. The vector $\mathbf{x}$ is thus the vector of the generalized coordinates, and the corresponding elements of vector $\mathbf{f}$ are the generalized forces. Some of the x_i can be true displacements or rotations, but they can also have a less direct meaning, as in the case where they are coefficients of a series expansion. Correspondingly, the generalized forces are true forces, moments, or just mathematical expressions linked to the forces and moments acting on the system in a less direct way.

The choice of the generalized coordinates is in a way arbitrary, and different sets of generalized coordinates can be devised for a given system. However, the choice is not immaterial; the complexity of the mathematical model can strongly depend on it.

In this way, what has been seen for a system made of point masses can be extended to systems made by any number of rigid bodies, provided that a finite number of generalized coordinates can express their configuration. However, in general, matrices $\mathbf{C}$ and $\mathbf{K}$ are not symmetrical; by separating their symmetrical and skew symmetrical components, the equation of motion can be written in the form

$$\mathbf{M}\ddot{\mathbf{x}} + (\mathbf{C} + \mathbf{G})\dot{\mathbf{x}} + (\mathbf{K} + \mathbf{H})\mathbf{x} = \mathbf{f}(t) \, , \tag{1.5}$$

where $\mathbf{C}$ and $\mathbf{K}$ are the viscous damping and the stiffness matrices which are always symmetrical, $\mathbf{G}$ is the skew-symmetrical gyroscopic matrix and $\mathbf{H}$ is the skew-symmetric circulatory matrix.

In this chapter, only sytems with vanishing gyroscopic and circulatory matrices will be dealt with and the form (1.3) of the equations of motion will always be used.

1.3 Lagrange equations

The equations of motion can be obtained directly by writing the dynamic equilibrium equations for each of the masses m_i, i.e., by imposing that the sum of all forces internal and external to the system due to the springs, the dampers, and any other source and inertia forces acting on each mass is equal to zero. Although this approach is straightforward if the system is simple enough, if the number of degrees of freedom is high or if some of the generalized coordinates are not easily linked with the displacements and rotations of masses m_i, it is convenient to resort to the methods of analytical mechanics as the principle of virtual works, Hamilton's principle, or Lagrange equations in order to write the equations of motion. In this book, Lagrange equations

$$\frac{d}{dt}\left(\frac{\partial T}{\partial \dot{x}_i}\right) - \frac{\partial T}{\partial x_i} + \frac{\partial U}{\partial x_i} + \frac{\partial F}{\partial \dot{x}_i} = \frac{\partial(\delta L)}{\partial(\delta x_i)} \qquad (1.6)$$

will be used extensively, although the choice of one of these techniques is often just a matter of personal preference.

To understand the equivalence of two approaches (Lagrange equations and dynamic equilibrium equations), it is sufficient to observe that the first two terms of equation (1.6) are the expression of inertia forces as functions of the kinetic energy T, the third term expresses conservative forces obtainable from the potential energy U, and the fourth deals with dissipative forces obtainable from the so-called Rayleigh dissipation function F, and that on the right-hand side is a generic expression of forces that are not included in the previous categories. They are obtained from the virtual work δL performed when the virtual displacement $\delta \mathbf{x}$ is given to the system.

The potential energy does not depend on the generalized velocities; thus its derivatives with respect to the generalized velocities $\dot{x}_i$ vanish. By using the symbol Q_i to express all nonconservative forces, equation (1.6) is then often written by resorting to the Lagrangian function or Lagrangian $(T - U)$

$$\frac{d}{dt}\left[\frac{\partial(\mathcal{T}-\mathcal{U})}{\partial \dot{x}_i}\right] - \frac{\partial(\mathcal{T}-\mathcal{U})}{\partial x_i} = Q_i \,. \tag{1.7}$$

The kinetic energy is usually assumed to be a quadratic function of the generalized velocities

$$\mathcal{T} = \mathcal{T}_0 + \mathcal{T}_1 + \mathcal{T}_2 \,, \tag{1.8}$$

where $\mathcal{T}_0$ does not depend on the generalized velocities, $\mathcal{T}_1$ is linear, and $\mathcal{T}_2$ is a homogeneous quadratic expression.

In the case of linear systems, the kinetic energy, as well as the potential energy and the Rayleigh function, must contain terms in which no power greater than two of the displacements and velocities is present. As a consequence, $\mathcal{T}_2$ cannot contain the displacements, i.e.,

$$\mathcal{T}_2 = \frac{1}{2}\sum_{i=1}^{n}\sum_{j=1}^{n} m_{ij}x_i x_j = \frac{1}{2}\dot{\mathbf{x}}^T \mathbf{M}\dot{\mathbf{x}} \,, \qquad . \tag{1.9}$$

$\mathcal{T}_0$ does not contain the generalized velocities and then has a structure similar to that of the potential energy. The term $\mathcal{U} - \mathcal{T}_0$ is usually referred to as *dynamic potential*.

A system in which $\mathcal{T}_0$ and $\mathcal{T}_1$ vanish is said to be a *natural system*; such is the case for example of linear nonrotating structures like the ones studied in this chapter, in which the kinetic energy is expressed by equation (1.9) and the potential energy and the viscous dissipation function can be expressed as

$$\mathcal{U} = \frac{1}{2}\mathbf{x}^T \mathbf{K}\mathbf{x} \,, \qquad \mathcal{F} = \frac{1}{2}\dot{\mathbf{x}}^T \mathbf{C}\dot{\mathbf{x}} \,. \tag{1.10}$$

The gyroscopic matrix vanishes in natural (or nongyroscopic) systems and the circulatory matrix is not present if damping is of the viscous type, i.e. the Rayleigh function is expressed by equation (1.10).

The equivalence between equation (1.3) and equation (1.6) is easily demonstrated.

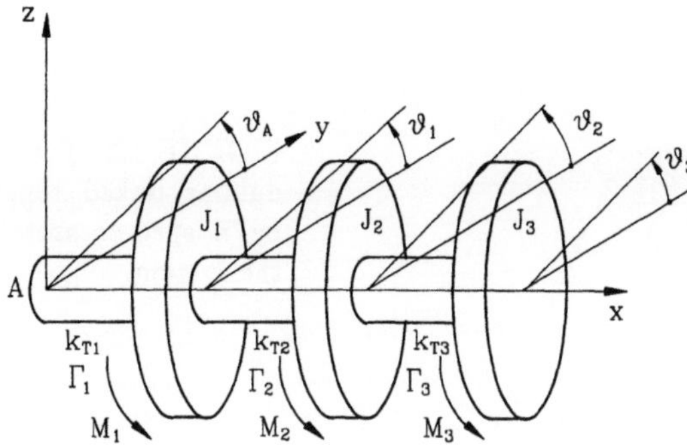

FIGURE 1.2. System with three degrees of freedom. J_1 = 1 kg m^2; J_2 = 4 kg m^2; J_3 = 0.5 kg m^2; k_{T1} = 10 Nm/rad; k_{T2} = 10 Nm/rad; k_{T3} = 4 Nm/rad; $\Gamma_1 = 0$; $\Gamma_2 = 1$ Ns/m; $\Gamma_3 = 0.4$ Ns/m.

Example 1-2

Write the equation of motion of the system sketched in Figure 1.2. It consists of three discs linked with each other by shafts that are torsionally deformable and have damping properties; the first shaft is clamped in point A to a fixed frame.

The system has been modeled as a lumped-parameter system, with three rigid inertias and three massless springs and dampers. Note that the numerical values reported in the figure are unrealistic for a system of that type and were chosen only in order to work with simple numbers. Consider the rotations θ_1, θ_2, and θ_3 as generalized coordinates.

By remembering the equivalences in Table 1.1, the equation of motion of the third disc is

$$J_3\ddot{\theta}_3 + \Gamma_3(\dot{\theta}_3 - \dot{\theta}_2) + k_{T3}(\theta_3 - \theta_2) = M_3.$$

The equations for the other two discs can be written in a similar way, obtaining a set of three second-order differential equations, which, after introducing the numerical values of the parameters, is

$$\begin{bmatrix} 1 & 0 & 0 \\ 0 & 4 & 0 \\ 0 & 0 & 0.5 \end{bmatrix} \begin{Bmatrix} \ddot{\theta}_1 \\ \ddot{\theta}_2 \\ \ddot{\theta}_3 \end{Bmatrix} + \begin{bmatrix} 1 & -1 & 0 \\ -1 & 1.4 & -0.4 \\ 0 & -0.4 & 0.4 \end{bmatrix} \begin{Bmatrix} \dot{\theta}_1 \\ \dot{\theta}_2 \\ \dot{\theta}_3 \end{Bmatrix} +$$

$$+ \begin{bmatrix} 20 & -10 & 0 \\ -10 & 14 & -4 \\ 0 & -4 & 4 \end{bmatrix} \begin{Bmatrix} \theta_1 \\ \theta_2 \\ \theta_3 \end{Bmatrix} = \begin{Bmatrix} M_1 \\ M_2 \\ M_3 \end{Bmatrix}.$$

Alternatively, the equations of motion can be obtained from Lagrange equations. The kinetic and potential energies, the Rayleigh dissipation function, and the virtual work of the external moments due to a virtual displacement $[\delta\theta_1, \delta\theta_2, \delta\theta_3]^T$ are

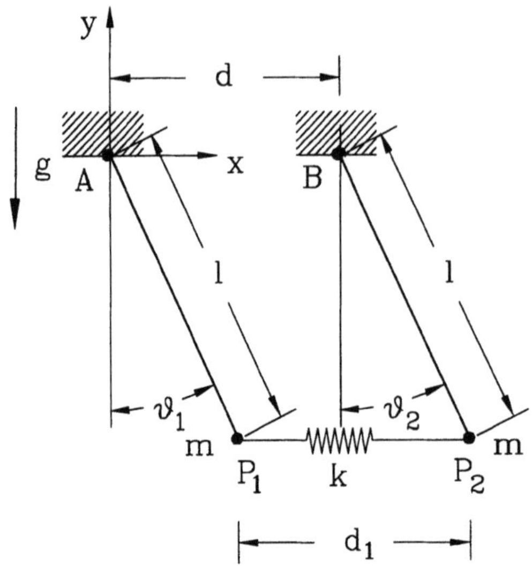

FIGURE 1.3. Two pendulums linked together by a spring: sketch of the system.

$$2T = J_1\dot{\theta}_1^{\ 2} + J_2\dot{\theta}_2^{\ 2} + J_3\dot{\theta}_3^{\ 2},$$
$$2\mathcal{U} = k_{T_1}\theta_1^2 + k_{T_2}(\theta_2 - \theta_1)^2 + k_{T_3}(\theta_3 - \theta_2)^2,$$
$$2\mathcal{F} = \Gamma_1\dot{\theta}_1^{\ 2} + \Gamma_2(\dot{\theta}_2 - \dot{\theta}_1)^2 + \Gamma_3(\dot{\theta}_3 - \dot{\theta}_2)^2,$$
$$\delta\mathcal{L} = M_1\delta\theta_1 + M_2\delta\theta_2 + M_3\delta\theta_3.$$

By performing the relevant derivatives, the same equation seen above is obtained.

Example 1-3

Consider the two identical pendulums connected by a spring shown in Figure 1.3.

Write the kinetic and potential energy of the system and obtain the equation of motion through Lagrange equations. Linearize the equation of motion in order to study the small oscillations about the static equilibrium position.

There are four dynamic equilibrium equations; they state that all forces, including inertia forces, that act on the two point masses in x- and y-directions balance each other. Two constraint equations, stating that the distances $(\overline{P_1 - A})$ and $(\overline{P_2 - B})$ are equal to the lengths l of the two pendulms, must be added to the dynamic equilibrium equations.

The system then has two degrees of freedom, and angles θ_1 and θ_2 can be chosen as generalized coordinates. Equations (1.4) linking the positions of the point masses with the generalized coordinates are

$$\vec{r}_1 = (\overline{P_1 - A}) = \left\{ \begin{array}{c} l\sin(\theta_1) \\ -l\cos(\theta_1) \end{array} \right\} , \quad \vec{r}_2 = (\overline{P_2 - A}) = \left\{ \begin{array}{c} d + l\sin(\theta_2) \\ -l\cos(\theta_2) \end{array} \right\} ,$$

Note that the relationship between the positions of the point masses and the generalized coordinates is nonlinear. The kinetic energy is then

$$\mathcal{T} = \frac{1}{2}m\left(\dot{x}_1^2 + \dot{y}_1^2 + \dot{x}_2^2 + \dot{y}_2^2\right) = \frac{1}{2}ml^2\left(\dot{\theta}_1^{\,2} + \dot{\theta}_2^{\,2}\right).$$

The gravitational potential energy can be defined with reference to any zero level, for example, that of point A. The potential energy is

$$\mathcal{U} = -mgl[\cos(\theta_1) + \cos(\theta_2)] + \frac{1}{2}k(d_1 - d)^2,$$

where the distance d_1 between points P_1 and P_2 can be easily shown to be

$$d_1 = \sqrt{d^2 + 2l^2[1 - \cos(\theta_1 - \theta_2)] - 2dl[\sin(\theta_2) - \sin(\theta_1)]}.$$

By performing all the relevant derivatives of the Lagrangian function, $\mathcal{T} - \mathcal{U}$, the equations of motion of the system are then obtained:

$$\begin{cases} ml^2\ddot{\theta}_1 + mgl\sin(\theta_1) + k(d_1 - d)\frac{\partial d_1}{\partial \theta_1} = 0 \\ ml^2\ddot{\theta}_2 + mgl\sin(\theta_2) + k(d_1 - d)\frac{\partial d_1}{\partial \theta_2} = 0 \end{cases}$$

The derivatives of d_1 with respect to θ_1 and θ_2 can be easily computed and the equations can be written in explicit form. They are clearly nonlinear, but, as angles θ_1 and θ_2 are assumed to be small, they can be linearized. As

$$d_1 \approx \sqrt{d^2 - 2dl(\theta_2 - \theta_1)} \approx d - l(\theta_2 - \theta_1)\,,\;\partial d_1/\partial\theta_1 \approx l\,,\;\partial d_1/\partial\theta_2 \approx -l\,,$$

the linearized equation of motion can be written in the form

$$ml\begin{bmatrix} 1 & 0 \\ 0 & 1 \end{bmatrix}\begin{Bmatrix} \ddot{\theta}_1 \\ \ddot{\theta}_2 \end{Bmatrix} + \begin{bmatrix} mg + kl & -kl \\ -kl & mg + kl \end{bmatrix}\begin{Bmatrix} \theta_1 \\ \theta_2 \end{Bmatrix} = \begin{Bmatrix} 0 \\ 0 \end{Bmatrix}.$$

1.4   State space

The configuration of the system is known once vector $\mathbf{x}$ is stated. As the latter defines a point in an n-dimensional space, each point of this space, referred to as the *configuration space*, represents a possible configuration of the system. During motion, the point representing the system configuration moves in the configuration space and its trajectory is referred to as the *dynamical path*. The dynamical paths corresponding to different time histories of the system can intersect each other, and a given configuration can be istantaneously taken during different motions. Knowledge of the system configuration at a given time and of the time history of the forcing function does not allow one to predict its future evolution or to know its past time history.

If, on the contrary, the generalized velocities are known, the state of motion of the system is completely known at any time. Positions and velocities, taken together, are thus the state variables of the system, even if the choice is not unique and other pairs of variables correlated with them can be used (e.g., position and momentum).

A state vector

$$\mathbf{z} = \left\{ \begin{array}{c} \dot{\mathbf{x}} \\ \mathbf{x} \end{array} \right\}$$

containing the displacements and velocities can thus be defined[3]. It has $2n$ components and defines a point in a space with $2n$ dimensions, the *state space*, i.e., a space defined by a reference frame whose coordinates are the state variables of the system. In the case of systems with a single degree of freedom, the state space has only two dimensions and is called the *state plane*. The configuration space is a subspace of the state space.

With reference to the state space, the equation of motion of a linear system can be transformed into a set of $2n$ first-order linear differential equations, the state equations of the system

$$\dot{\mathbf{z}}(t) = \boldsymbol{A}\mathbf{z}(t) + \boldsymbol{B}\mathbf{u}(t) \,, \tag{1.11}$$

where

$$\boldsymbol{A} = \left[\begin{array}{cc} -\mathbf{M}^{-1}\mathbf{C} & -\mathbf{M}^{-1}\mathbf{K} \\ \mathbf{I} & \mathbf{0} \end{array} \right]$$

is the dynamic matrix of the system. It is neither symmetrical nor positive defined.

Vector $\mathbf{u}(t)$, whose size need not be equal to the number of degrees of freedom of the system, is the vector in which the inputs affecting the behaviour of the system are listed. $\boldsymbol{B}$ is the input gain matrix; if the number of inputs is r, it has $2n$ rows and r columns. If the

---

[3]The state vector can be alternatively defined as

$$\mathbf{z} = \left\{ \begin{array}{c} \mathbf{x} \\ \dot{\mathbf{x}} \end{array} \right\} \,.$$

There is no difficulty in modifying all relevant matrices to cope with this definition.

inputs $\mathbf{u}(t)$ are linked with the generalized forces $\mathbf{f}(t)$ acting on the various degrees of freedom by the relationship

$$\mathbf{f}(t) = \mathbf{T}\mathbf{u}(t) , \qquad (1.12)$$

then the expression of the input gain matrix is

$$\mathcal{B} = \begin{bmatrix} \mathbf{M}^{-1}\mathbf{T} \\ \mathbf{0} \end{bmatrix} . \qquad (1.13)$$

If the output of the system consists of a linear combination of the state variables, to which a linear combination of the inputs can be added, a second equation can be added to equation (1.11)

$$\mathbf{y}(t) = \mathcal{C}\mathbf{z}(t) + \mathcal{D}\mathbf{u}(t) , \qquad (1.14)$$

where $\mathbf{y}$ is the vector in which the m outputs of the system are listed and $\mathcal{C}$ is a matrix with m rows and n columns, often referred to as the *output gains matrix*[4]. Matrix $\mathcal{D}$ has m rows and r columns and expresses the direct influence of the inputs on the outputs; it is therefore referred to as the *direct link matrix*. If all generalized displacements are taken as outputs of the system, matrix $\mathcal{C}$ is simply $\mathcal{C} = [\mathbf{0},\mathbf{I}]$.

If $r = 1$, i.e., there is a single input $u(t)$, and $m = 1$, i.e. there is a single output $y(t)$, the system is referred to as a *single input, single output* (SISO) *system*. Otherwise, if there are several inputs and outputs, the system is a *multiple input, multiple output* (MIMO) one. This distinction has nothing to do with the number of degrees of freedom or number of state variables, a single-degree-of-freedom system can be a MIMO one, where the input-output relationship is concerned.

Note that the state equation is a differential equation, but the output equation is simply algebraic. The set of matrices $\mathcal{A}, \mathcal{B}, \mathcal{C}$, and $\mathcal{D}$ constitutes what is usually called the *quadruple* of the system.

The points that represent the state of the system in subsequent instants describe a trajectory in the state space. The trajectory defines the motion. The various trajectories obtained with different initial conditions constitute the state portrait of the system. In the case

---

[4] The output gain matrix is usually referred to as $\mathcal{C}$. This symbol is used here even though it is similar to that used for the damping matrix C because the author thinks no confusion between them is possible.

of autonomous systems, it is possible to demonstrate that, with the exception of possible singular points, only one trajectory can pass through any given point of the state space.

Equation (1.11) is nonautonomous, as time appears explicitly in the input vector. In this case, one more dimension, namely, time, is added to the state space in order to prevent the trajectories from crossing each other, as would happen in $(\mathbf{x},\dot{\mathbf{x}})$ space. The state space for a nonautonomous system with a single degree of freedom is consequently a tridimensional space $(x,\dot{x},t)$. Even in this case, in order to reduce the complexity of the state portrait, only the state projection in the $(x,\dot{x})$ plane is often represented. Another technique is that of representing only some selected points of the trajectories, chosen at fixed time intervals, usually the period of the forcing function when the latter is periodic, as if a strobe were used. This strobed map is usually referred to as a *Poincaré section* or *Poincaré map*.

A point in the state space such that $\boldsymbol{A}\mathbf{z} + \boldsymbol{B}\mathbf{u} = 0$ for any value of time is an equilibrium point. Because it is a static solution, it can be defined only if the input vector $\mathbf{u}$ is constant in time. All generalized velocities are identically equal to zero and then the equilibrium point lies in the configuration space. Although a nonlinear system can have a number of equilibrium points, in the case of linear systems a single equlibrium point exists. If $\mathbf{u}$ is equal to zero, the equilibrium point is the solution of the homogeneous algebraic equation $\boldsymbol{A}\mathbf{z} = 0$, i.e., the trivial solution $\mathbf{z} = 0$, except in the case where the dynamic matrix $\boldsymbol{A}$ is singular.

If the inputs acting on the system are due to n generalized forces, the static inflected shape can be immediately computed as

$$\mathbf{x} = \mathbf{K}^{-1}\mathbf{F} = \mathbf{BF} \,, \tag{1.15}$$

where the inverse of the stiffness matrix $\mathbf{B} = \mathbf{K}^{-1}$ is the compliance matrix[5] or matrix of the coefficients of influence. The generic element β_{ij} of matrix $\mathbf{B}$ has an obvious physical meaning: It is the ith generalized displacement due to a unit jth generalized force, i.e., it is what is commonly called an *influence coefficient*.

---

[5]In the literature, the compliance matrix is often referred to with the symbol $\mathbf{C}$. Here, an alternative symbol ($\mathbf{B}$, i.e., capital β) had to be used to avoid confusion with the viscous damping matrix $\mathbf{C}$ and the output gain matrix $\mathcal{C}$. Unfortunately, this symbol is similar to $\boldsymbol{B}$, i.e., capital b, used for the input gain matrix.

Matrix **B** exists only if the stiffness matrix is not singular.

In the case of nonlinear systems, the equations of motion can usually be linearized about any given equilibrium points. The motion of the linearized system about an equilibrium point is usually referred to as *motion in the small*.

Note that the equation of motion in the state space can be written in many different forms, but the current formulation is standard for the study of dynamic systems in general. When the generalized momenta are used instead of the generalized velocities, usually the term *phase* is used instead of *state*.

Example 1-4

Write the equation of motion in the state space of the system of Example 1-2, assuming that the only input $u(t)$ is the moment M_3 acting on the third moment of inertia. Write the output equation, assuming that only one output is considered, the rotation of the third moment of inertia. Because there is only one input ($r=1$), matrix **T** has three rows and one column: $\mathbf{T} = [0,0,1]^T$. The equation of motion is then

$$\begin{Bmatrix} \ddot{\theta}_1 \\ \ddot{\theta}_2 \\ \ddot{\theta}_3 \\ \dot{\theta}_1 \\ \dot{\theta}_2 \\ \dot{\theta}_3 \end{Bmatrix} = \begin{bmatrix} -1 & 1 & 0 & -20 & 10 & 0 \\ 0.25 & -0.35 & 0.1 & 2.5 & -3.5 & 1 \\ 0 & 0.8 & -0.8 & 0 & 8 & -8 \\ 1 & 0 & 0 & 0 & 0 & 0 \\ 0 & 1 & 0 & 0 & 0 & 0 \\ 0 & 0 & 1 & 0 & 0 & 0 \end{bmatrix} \begin{Bmatrix} \dot{\theta}_1 \\ \dot{\theta}_2 \\ \dot{\theta}_3 \\ \theta_1 \\ \theta_2 \\ \theta_3 \end{Bmatrix} + \begin{Bmatrix} 0 \\ 0 \\ 2 \\ 0 \\ 0 \\ 0 \end{Bmatrix} M_3 .$$

Because the output is only θ_3, matrix $\mathcal{D}$ vanishes while the output gain matrix has one row and six columns

$$\mathcal{C} = \begin{bmatrix} 0 & 0 & 0 & 0 & 0 & 1 \end{bmatrix} .$$

1.5   Free behaviour

1.5.1   Systems with a single degree of freedom

The solution of the homogeneous equation associated to the equation of motion of a system with a single degree of freedom (1.1) is of the type

$$x = x_0 e^{st} . \tag{1.16}$$

By introducing solution (1.16) into the equation of motion, the following algebraic equation is obtained

$$x = x_0(ms^2 + cs + k) = 0 . \qquad (1.17)$$

The condition for the existence of a solution other than the trivial solution $x_0 = 0$ leads to the following characteristic equation

$$ms^2 + cs + k = 0 , \qquad (1.18)$$

whose two solutions, s_1 and s_2, are

$$s = \frac{-c \pm \sqrt{c^2 - 4mk}}{2m} . \qquad (1.19)$$

Generally speaking, s is expressed by a complex number; the solution (1.16) of the equation of motion is then

$$x = x_0 e^{\Re(s)t} + e^{i\Im(s)t} . \qquad (1.20)$$

The real part of s is then the decay rate σ changed in sign[6] , i.e. the velocity at which the amplitude of the oscillations grows in time. For stability σ must be positive, i.e. $\Im(s)$ must be negative.

The imaginary part of s is the frequancy λ of the damped oscillations of the system.

If condition

$$c > 2\sqrt{km} \qquad (1.21)$$

is satisfied, the solutions of the characteristic equation (1.18) are real. The motion of the system is not oscillatory, but simply the combination of two terms that decrease monotonically in time, because both roots are negative. The value of the damping expressed by equation (1.21) is often referred to as *critical damping*, the highest value of c that allows the system to show an oscillatory free behaviour. When condition (1.21) is satisfied, the system is said to be *overdamped*.

Introducing the damping ratio or relative damping ζ, i.e., the ratio between the value of the damping c and its critical value, c_{cr}

$$\zeta = \frac{c}{c_{cr}} = \frac{c}{2\sqrt{km}} , \qquad (1.22)$$

---

[6]In some cases the decay rate is defined as $\Re(s)$. The definition ($\sigma = -\Re(s)$) has been preferred so σ is positive when the motion actually decays in time.

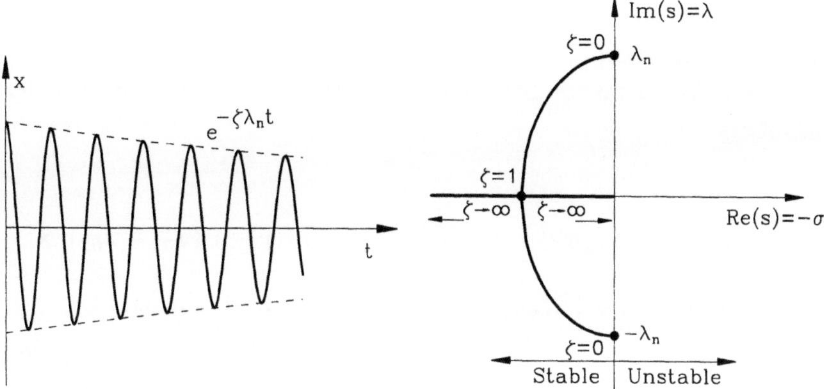

FIGURE 1.4. Damped oscillations of a lightly damped system.

FIGURE 1.5. Roots locus for a damped system with a single degree of freedom.

the two values of the decay rate σ, i.e., the speed at which the amplitude of the free oscillations fades out, of an overdamped system ($\zeta > 1$) are equal to the values of s changed of sign

$$\sigma = -\Re(s) = \sqrt{\frac{k}{m}}\left(\zeta \pm \sqrt{\zeta^2 - 1}\right) . \tag{1.23}$$

If the damping of the system is lower than the critical damping ($\zeta < 1$), the system is said to be *underdamped*. Equation (1.19) still holds but leads to a pair of complex conjugate solutions for s. The system performs damped harmonic oscillations (Figure 1.4) whose circular frequency λ is equal to the absolute value of the imaginary part of s and the decay rate σ is equal to the real part of s changed sign

$$\lambda = |\Im(s)| = \sqrt{\frac{k}{m}}\sqrt{1 - \zeta^2} , \qquad \sigma = -\Re(s) = \zeta\sqrt{\frac{k}{m}} . \tag{1.24}$$

If the system is undamped, i.e., $c = 0$, the two solutions of the characteristic equation are imaginary, yielding an undamped harmonic oscillation whose circular frequency is the natural frequency λ_n of the system

$$\lambda_n = \sqrt{\frac{k}{m}} . \tag{1.25}$$

The complementary function is the sum of two terms of the type shown in equation (1.16) with two constants, x_{01} and x_{02}, that depend on the initial conditions and are usually expressed by complex numbers

$$x = x_{01}e^{s_1 t} + x_{02}e^{s_2 t} = e^{-\sigma t}\left\{\left[\Re(x_{01} + x_{02}) + i\Im(x_{01} + x_{02})\right] \times\right.$$
$$\left.\times \cos(\lambda t) + \left[\Im(x_{02} - x_{01}) + i\Re(x_{01} - x_{02})\right]\sin(\lambda t)\right\}.$$
$$(1.26)$$

Displacement x is a real quantity and then the two constants, x_{01} and x_{02}, must be complex conjugate. If at time $t = 0$ the position $x(0)$ and the velocity $\dot{x}(0)$ are known, the values of the real and imaginary parts of constant x_{01} are

$$\Re(x_{01}) = \frac{x(0)}{2}, \qquad \Im(x_{01}) = \frac{-1}{2\lambda_n\sqrt{1 - \zeta^2}}\left[\dot{x}(0) + \zeta\lambda_n x(0)\right].$$
$$(1.27)$$

The equation describing the damped oscillations of the system is then

$$x = e^{-\zeta\lambda_n t}\left\{x(0)\cos\left(\lambda_n\sqrt{1 - \zeta^2}\,t\right) + \right.$$
$$\left. + \frac{1}{\sqrt{1 - \zeta^2}}\left[\frac{\dot{x}(0)}{\lambda_n} + \zeta x(0)\right]\sin\left(\lambda_n\sqrt{1 - \zeta^2}\,t\right)\right\}.$$
$$(1.28)$$

The roots of the characteristic equation can be reported on the Argand plane, i.e., in a plane in which the x-axis is taken as the real axis and the y-axis is the imaginary axis. The points that represent the solutions of the characteristic equation in the Argand plane are usually referred to as *poles* of the system. When the behaviour of the system depends on a parameter, as in the current case it depends on the damping ratio ζ, the plot of the roots with varying values of the parameter is said to be the *root locus*. The locus of the roots of a damped system with a single degree of freedom is shown in Figure 1.5.

In the case of underdamped systems, the two solutions have the same real part (decay rate) and imaginary parts equal in sign but opposite in modulus. For overdamped systems, on the contrary, the two solutions, which are real (lying on the real axis), are not coincident. One of them tends to infinity and the other tends to zero when damping tends to infinity. In the case of critically damped systems ($\zeta = 1$), the two solutions are coincident: The two branches of the locus for underdamped systems meet on the real axis to separate again for overdamped systems. The point where the two branches meet to remain on the real axis is said to be a *break-in point*. When the opposite occurs and two branches lying on the real axis meet to depart from the axis, the point is said to be a *breakaway point*.

Solutions on the left of the Argand plane, as in the case of systems with positive damping, are stable, while solutions lying on the right part of the plane denote unstable behaviour. Solutions located on the imaginary axis are neither stable nor unstable, as the oscillations are not damped, and the system, once set in motion, cannot reach an equilibrium condition.

The systems studied in structural dynamics are usually lightly damped; consequently, factor $e^{-\zeta\lambda_n t}$ decreases monotonically very slowly with time. The frequency of the oscillations is only slightly smaller than the natural frequency of the undamped system λ_n, owing to the presence of factor $\sqrt{1 - \zeta^2}$. In the case of lightly damped systems, this factor is almost equal to unity, and the frequency of the damped free oscillations is almost equal to that of the free oscillations of the undamped system.

The ratio between the amplitudes of two subsequent peaks is constant in time. Its natural logarithm, computed assuming that the peaks occur when the harmonic part of the response reaches its maxima, is

$$\delta = \ln\left(\frac{x_i}{x_{i+1}}\right) = 2\pi\frac{\zeta}{\sqrt{1 - \zeta^2}} \approx 2\pi\zeta \ . \tag{1.29}$$

It is usually defined as logarithmic decrement and gives a measure of the damping of the system that is not too difficult to evaluate from the recording of the amplitude versus time (Figure 1.4). If the system is lightly damped, the mentioned assumption leads to accurate results.

The trajectories of the free oscillations of a damped linear system in the state plane are logarithmic spirals. They wind up toward the origin, which is a singular point.

Instead of using equation (1.16), the solution of the equation of motion could also be written in the form $x = x_0 e^{i\nu t}$, where $i\nu = s$ obviously. The real and imaginary parts of the complex frequency ν are the actual frequency of the motion and the decay rate, respectively:

$$\Re(\nu) = \lambda \,, \qquad \Im(\nu) = \sigma \,. \tag{1.30}$$

Example 1-5

An instrument whose mass is 20 kg must be mounted on a space vehicle through a cantilever arm of annular cross section made of light alloy (Young's modulus $E = 72 \times 10^9$ N/m^2; density $\rho = 2{,}800$ kg/m^3), 600 mm long. Choose the dimensions of the cross section in such a way that the first natural frequency is higher than 50 Hz.

If the mass of the beam is neglected and a model with a single degree of freedom is used, in order to obtain a natural frequency higher than 50 Hz (314 rad/s) the stiffness of the beam must be $k \geq m\lambda_n^2 = 1.97 \times 10^6$ N/m.

The arm can be modeled as a cantilever beam clamped at one end and loaded by the inertia force of the instrument at the other end. The well-known formula giving its stiffness is $k = 3EI/l^3$, where l and I are the length of the beam and the area moment of inertia of the cross section, respectively.

The minimum value of the moment of inertia I can be easily computed: $I = kl^3/3E = 1.97 \times 10^{-6}$ m^4. By using a beam with annular cross section with inner and outer diameters of 100 mm and 110 mm respectively, the value of the moment of inertia is $I = 2.28 \times 10^{-6}$ m^4, yielding the following values of the stiffness and natural frequency: $k = 2.28 \times 10^6$ N/m; $\lambda_n = 337$ rad/s $= 53.7$ Hz. The mass of the beam is 2.77 kg, which is only slightly larger than 1/10 of the concentrated mass at the free end. The single-degree-of-freedom model in which the mass of the beam has been neglected can then be used with confidence.

1.5.2 *Systems with many degrees of freedom*

The solution of the homogeneous equation associated to the equation of motion in the state space (1.11) is of the type

$$\mathbf{z} = \mathbf{z}_0 e^{st} \,. \tag{1.31}$$

By introducing the solution (1.31) into equation (1.11), the following set of homogeneous linear algebraic equations is obtained

$$(\boldsymbol{A} - s\mathbf{I})\mathbf{z}_0 = \mathbf{0} \,. \tag{1.32}$$

To obtain solutions other than the trivial solution $\mathbf{z}_0 = \mathbf{0}$, the determinant of the matrix of the coefficients must vanish. An eigenproblem of order $2n$ is so obtained: The eigenvalues s yield the frequencies of oscillation and the decay rates, and the eigenvectors yield the complex mode shapes $\mathbf{z}_0$. All considerations of stability of the system seen in the previous section still hold, with the only difference being that now there are n pairs of complex conjugate solutions. If some of them are real, the corresponding modes are nonoscillatory; if they are imaginary, undamped oscillations occur. The fact that the eigenvectors are complex[7] causes the time histories related to the various degrees of freedom to be out of phase.

If the system has no damping, the solution can be performed directly with reference to the configuration space instead of resorting to the state space. The homogeneous equation that describes the free motion is

$$\mathbf{M}\ddot{\mathbf{x}} + \mathbf{K}\mathbf{x} = \mathbf{0} \ . \tag{1.33}$$

Equation (1.33) is a set of linear homogeneous second order differential equations. Such equations are coupled, because at least one of the matrices $\mathbf{M}$ or $\mathbf{K}$ is usually not diagonal. If matrix $\mathbf{M}$ is not diagonal, the system is said to have an *inertial coupling*; if $\mathbf{K}$ is not diagonal, the coupling is said to be *elastic*. Again, a solution similar to equation (1.31) can be assumed, and an eigenproblem of the same type as equation (1.32) can be obtained. However, because the system is undamped, all solutions s are imaginary and the use of a solution with the form $\mathbf{x} = \mathbf{x}_0 e^{i\lambda t}$, in which the frequency of oscillation λ is explicitly included, is expedient.

The characteristic equation of the relevant eigenproblem, then, is

$$\det\left(\mathbf{K} - \lambda^2\mathbf{M}\right) = 0 , \tag{1.34}$$

which can be reduced in standard form in one of two following ways

$$\det\left(\mathbf{K}^{-1}\mathbf{M} - \frac{1}{\lambda^2}\mathbf{I}\right) = 0 , \qquad \det\left(\mathbf{M}^{-1}\mathbf{K} - \lambda^2\mathbf{I}\right) = 0 . \tag{1.35}$$

Both matrices $\mathbf{K}^{-1}\mathbf{M}$ and $\mathbf{M}^{-1}\mathbf{K}$ are often referred to as dynamic matrices and symbol $\mathbf{D}$ is used for them. They should not be confused

---

[7] An interesting discussion on the meaning of complex modes can be found in G.F. Lang, "Demistifying complex modes", *Sound and Vibration*, January, 1989.

with either the dynamic matrix $\boldsymbol{A}$ or the direct link matrix $\boldsymbol{\mathcal{D}}$ defined with reference to the state space. Note that the dynamic matrices $\mathbf{D}$ are not symmetrical, even if both $\mathbf{M}$ and $\mathbf{K}$ are.

Equations (1.35) are algebraic equations of degree n in λ^2 (or in $1/\lambda^2$) that yield the n values of the natural frequencies of the system. The eigenvectors give the mode shapes, i.e., the amplitudes of oscillation of the various masses at the corresponding natural frequency. All eigenvalues are real and positive; the natural frequencies then are real, and undamped oscillations of the system are obtained. Also, the eigenvectors $\mathbf{q}_i$ are real, which means that all masses move in phase, or with a phase lag of $180°$. Because there are n eigenvectors, a square matrix, the matrix of the eigenvectors $\boldsymbol{\Phi} = [\mathbf{q}_1, \mathbf{q}_2, \ldots, \mathbf{q}_n]$, can be written. Each one of its columns is one of the eigenvectors.

By transforming the exponentials with imaginary argument into trigonometric functions, the complete solution of the equation of motion is

$$\mathbf{x} = \sum_{i=1}^{n} \left[\Re(K_i^*)\mathbf{q}_i \cos(\lambda_i t) - \Im(K_i^*)\mathbf{q}_i \sin(\lambda_i t) \right] , \qquad (1.36)$$

where the n complex constants K_i^* can be determined from the $2n$ initial conditions. If at time $t = 0$ the positions $\mathbf{x}_0$ and the velocities $\dot{\mathbf{x}}_0$ are known, it follows

$$\Re\{K_i^*\} = \boldsymbol{\Phi}^{-1}\mathbf{x}_0 , \qquad \Im\{\lambda_i K_i^*\} = -\boldsymbol{\Phi}^{-1}\dot{\mathbf{x}}_0 . \qquad (1.37)$$

The free motion of a damped system can be studied in the state space through the equation

$$\mathbf{z} = e^{\boldsymbol{A}t}\mathbf{z}_0 , \qquad (1.38)$$

where $\mathbf{z}_0$ is the state vector at time t_0 and $e^{\boldsymbol{A}t}$ is the so-called transition matrix at time t. It can be expressed by the series

$$e^{\boldsymbol{A}t} = \mathbf{I} + t\boldsymbol{A} + \frac{t^2}{2!}\boldsymbol{A}^2 + \frac{t^3}{3!}\boldsymbol{A}^3 + \ldots , \qquad (1.39)$$

which converges for any value of t. The computation of the transition matrix can become impractical for large-order systems, and the number of terms in the series (1.39), which must be considered, grows rapidly with increasing time t.

Example 1-6

Compute the time history of the free motion of a single-degree-of-freedom undamped system through the series (1.39) for the transition matrix.

For an undamped single-degree-of-freedom system, the state vector and the dynamic matrix are

$$\mathbf{z} = \left\{ \begin{array}{c} v \\ x \end{array} \right\}, \qquad \mathcal{A} = \begin{bmatrix} 0 & -\frac{k}{m} \\ 1 & 0 \end{bmatrix} = \begin{bmatrix} 0 & -\lambda_n^2 \\ 1 & 0 \end{bmatrix}.$$

Equation (1.39) yields the transition matrix

$$e^{\begin{bmatrix} 0 & -\lambda_n^2 \\ 1 & 0 \end{bmatrix} t} = \begin{bmatrix} 1 & 0 \\ 0 & 1 \end{bmatrix} + t \begin{bmatrix} 0 & -\lambda_n^2 \\ 1 & 0 \end{bmatrix} + \frac{t^2}{2} \begin{bmatrix} -\lambda_n^2 & 0 \\ 0 & -\lambda_n^2 \end{bmatrix} +$$

$$+ \frac{t^3}{3!} \begin{bmatrix} 0 & \lambda_n^4 \\ -\lambda_n^2 & 0 \end{bmatrix} + \frac{t^4}{4!} \begin{bmatrix} -\lambda_n^4 & 0 \\ 0 & -\lambda_n^4 \end{bmatrix} + \ldots =$$

$$= \begin{bmatrix} 1 + \frac{t^2}{2}\lambda_n^2 + \frac{t^4}{4!}\lambda_n^4 + \ldots & -\lambda_n^2\left(t\lambda_n - \frac{t^3}{3!}\lambda_n^3 + \ldots\right) \\ \frac{1}{\lambda_n}\left(t\lambda_n - \frac{t^3}{3!}\lambda_n^3 + \ldots\right) & 1 + \frac{t^2}{2}\lambda_n^2 + \frac{t^4}{4!}\lambda_n^4 + \ldots \end{bmatrix} =$$

$$= \begin{bmatrix} \cos(\lambda_n t) & -\lambda_n \sin(\lambda_n t) \\ \frac{1}{\lambda_n}\sin(\lambda_n t) & \cos(\lambda_n t) \end{bmatrix}.$$

After obtaining the product $e^{\mathcal{A}t}\mathbf{z}_0$, the following time histories for the displacement and the velocity are obtained:

$$\left\{ \begin{array}{l} v = v_0 \cos(\lambda_n t) - \lambda_n x_0 \sin(\lambda_n t) \\ x = \frac{v_0}{\lambda_n} \sin(\lambda_n t) + x_0 \cos(\lambda_n t). \end{array} \right.$$

1.6 Uncoupling of the equations of motion: Space of the configurations

Consider an undamped system and refer to the space of the configurations. The eigenvectors are orthogonal with respect to both the stiffness and mass matrices. This propriety can be demonstrated simply by writing the dynamic equilibrium equation in harmonic oscillations for the ith mode

$$\mathbf{K}\mathbf{q}_i = \lambda_i^2 \mathbf{M}\mathbf{q}_i. \tag{1.40}$$

Equation (1.40) can be premultiplied by the transpose of the jth eigenvector

$$\mathbf{q}_j^T \mathbf{K} \mathbf{q}_i = \lambda_i^2 \mathbf{q}_j^T \mathbf{M} \mathbf{q}_i \ . \tag{1.41}$$

The same can be done for the equation written for the jth mode and premultiplied by the transpose of the ith eigenvector

$$\mathbf{q}_i^T \mathbf{K} \mathbf{q}_j = \lambda_j^2 \mathbf{q}_i^T \mathbf{M} \mathbf{q}_j \ . \tag{1.42}$$

By subtracting equation (1.42) from equation (1.41) it follows that

$$\mathbf{q}_j^T \mathbf{K} \mathbf{q}_i - \mathbf{q}_i^T \mathbf{K} \mathbf{q}_j = \lambda_i^2 \mathbf{q}_j^T \mathbf{M} \mathbf{q}_i - \lambda_j^2 \ \mathbf{q}_i^T \mathbf{M} \mathbf{q}_j \ . \tag{1.43}$$

Remembering that $\mathbf{q}_j^T \mathbf{K} \mathbf{q}_i = \mathbf{q}_i^T \mathbf{K} \mathbf{q}_j$ and $\mathbf{q}_j^T \mathbf{M} \mathbf{q}_i = \mathbf{q}_i^T \mathbf{M} \mathbf{q}_j$, owing to the symmetry of matrices $\mathbf{K}$ and $\mathbf{M}$, it follows

$$\left(\lambda_i^2 - \lambda_j^2 \right) \mathbf{q}_j^T \mathbf{M} \mathbf{q}_i = 0 \ . \tag{1.44}$$

In the same way, it can be shown that

$$\left(\frac{1}{\lambda_i^2} - \frac{1}{\lambda_j^2} \right) \mathbf{q}_j^T \mathbf{K} \mathbf{q}_i = 0 \ . \tag{1.45}$$

From equations (1.45) and (1.44) it follows that, if $i \neq j$,

$$\mathbf{q}_i^T \mathbf{M} \mathbf{q}_j = 0 \ , \qquad \mathbf{q}_i^T \mathbf{K} \mathbf{q}_j = 0 \ , \tag{1.46}$$

which are the relationships defining the orthogonality properties of the eigenvectors with respect to the mass and stiffness matrices, respectively.

If $i = j$, the results of the same products are not zero:

$$\mathbf{q}_i^T \mathbf{M} \mathbf{q}_i = \overline{M_i} \ , \qquad \mathbf{q}_i^T \mathbf{K} \mathbf{q}_j = \overline{K_i} \ . \tag{1.47}$$

Constants $\overline{M_i}$ and $\overline{K_i}$ are the modal mass and modal stiffness of the ith mode, respectively. They are linked to the natural frequencies by the relationship

$$\lambda_i = \sqrt{\frac{\overline{K_i}}{\overline{M_i}}} \ , \tag{1.48}$$

stating that the ith natural frequency coincides with the natural frequency of a system with a single degree of freedom whose mass is the ith modal mass and whose stiffness is the ith modal stiffness.

The modal mass matrix and the modal stiffness matrix can be obtained from the following relationships based on the matrix of the eigenvectors $\mathbf{\Phi}$:

$$\begin{cases} \mathbf{\Phi}^T\mathbf{M}\mathbf{\Phi} = \mathrm{diag}[\overline{M_i}] = \overline{\mathbf{M}}, \\ \mathbf{\Phi}^T\mathbf{K}\mathbf{\Phi} = \mathrm{diag}[\overline{K_i}] = \overline{\mathbf{K}}. \end{cases} \tag{1.49}$$

The matrix of the eigenvectors can be used to perform a coordinate transformation that is particularly useful

$$\mathbf{x} = \mathbf{\Phi}\boldsymbol{\eta}, \qquad \boldsymbol{\eta} = \mathbf{\Phi}^{-1}\mathbf{x}. \tag{1.50}$$

This amounts to expressing the generic n-dimensional vector $\mathbf{x}$, which states the configuration of the system, as a linear combination of the eigenvectors using n coefficients of proportionality η_i. This is possible because the eigenvectors are linearly independent and form a possible system of reference in the space of the configurations of the system. It must be explicitly stated that the eigenvectors are orthogonal with respect to the mass and stiffness matrices (they are said to be m-orthogonal and k-orthogonal), but they are not orthogonal to each other. The product $\mathbf{\Phi}^T\mathbf{\Phi}$ does not yield a diagonal matrix, and the inverse $\mathbf{\Phi}^{-1}$ of matrix $\mathbf{\Phi}$ does not coincide with the transpose $\mathbf{\Phi}^T$. The eigenvectors are, however, orthogonal if the dynamic matrix is symmetrical, as happens for the system of Figure 1.2 if the moments of inertia are equal.

In the space of the configurations, the eigenvectors are n vectors that can be taken as a system of reference. However, as already stated, they are not orthogonal. Equations (1.50) are, consequently, nothing other than a coordinate transformation in the space of the configurations, as the n values η_i are the n coordinates of the point representing the configuration of the system, with reference to the system of the eigenvectors. They are said to be principal, modal, or normal coordinates.

Although the eigenvectors are not orthogonal to each other, it is possible to perform a coordinate transformation that yields a system whose dynamic matrix is symmetrical and has orthogonal eigenvectors. By performing a Cholesky decomposition of the mass matrix,

$$\mathbf{M} = \mathbf{L}\mathbf{L}^T, \tag{1.51}$$

where $\mathbf{L}$ is a lower triangular nonsingular matrix, a new set of generalized coordinates $\mathbf{x}^*$ can be defined by the relationship

$$\mathbf{x}^* = \mathbf{L}^T\mathbf{x}. \tag{1.52}$$

By introducing the generalized coordinates (1.52) into the equation of motion and premultiplying it by $\mathbf{L}^{-1}$, the mass matrix transforms into an identity matrix, while the stiffness and damping matrices and the force vector reduce to

$$\mathbf{K}^* = \mathbf{L}^{-1}\mathbf{K}(\mathbf{L}^{-1})^T \ , \qquad \mathbf{C}^* = \mathbf{L}^{-1}\mathbf{C}(\mathbf{L}^{-1})^T \ , \qquad \mathbf{f}^* = \mathbf{L}^{-1}\mathbf{f} \ . \tag{1.53}$$

The system so obtained has a unit mass matrix, and its eigenvectors are orthogonal. The eigenvalues are not changed by the transformation, and the eigenvectors are obtained from those of the original system using a simple linear combination. With simple computations, the modal mass matrix can be shown to be an identity matrix, and the modal stiffness matrix is a diagonal matrix with all elements equal to the squares of the natural frequencies ($\overline{\mathbf{M}} = [\lambda^2]$).

If the modal coordinates are introduced into the equation of motion (1.3) and the equation so obtained is premultiplied by $\mathbf{\Phi}^T$, it follows that

$$\overline{\mathbf{M}}\ddot{\boldsymbol{\eta}} + \overline{\mathbf{C}}\dot{\boldsymbol{\eta}} + \overline{\mathbf{K}}\boldsymbol{\eta} = \overline{\mathbf{f}} \ , \tag{1.54}$$

where $\overline{\mathbf{M}}$ and $\overline{\mathbf{K}}$ are the modal mass and modal stiffness matrices, defined by equation (1.47). They are diagonal. $\overline{\mathbf{C}}$ is the modal damping matrix

$$\overline{\mathbf{C}} = \mathbf{\Phi}^T\mathbf{C}\mathbf{\Phi} \ . \tag{1.55}$$

It is generally not diagonal, because the eigenvectors of the undamped system, although orthogonal with respect to the stiffness and mass matrices, are not orthogonal with respect to the damping matrix. The modal-damping matrix is, however, symmetrical, at least if the original damping matrix is. It has been demonstrated that a condition that is both necessary and sufficient to obtain a diagonal modal damping matrix is that matrix $\mathbf{M}^{-1}\mathbf{C}$ commutes with matrix $\mathbf{M}^{-1}\mathbf{K}$, or

$$\mathbf{C}\mathbf{M}^{-1}\mathbf{K} = \mathbf{K}\mathbf{M}^{-1}\mathbf{C} \ . \tag{1.56}$$

In this case, it is possible to define a modal damping $\overline{C_i} = \overline{C_{ii}}$ for each mode. A particular case that satisfies condition (1.56) is the so-called proportional damping, i.e., the case in which the damping matrix can be expressed as a linear combination of the mass and stiffness matrices:

$$\mathbf{C} = \alpha\mathbf{M} + \beta\mathbf{K} \,. \tag{1.57}$$

Because condition (1.56) is more general than condition (1.57), a system whose damping satisfies the first will be said to possess generalized proportional damping.

$\overline{\mathbf{f}}(t)$ is the modal force vector,

$$\overline{f_i}(t) = \mathbf{q}_i^T \mathbf{f}(t) \,. \tag{1.58}$$

In cases of undamped systems or systems with generalized proportional damping, all matrices are diagonal, and equation (1.54) is a set of n uncoupled second-order differential equations. Each of them is

$$\overline{M_i}\ddot{\eta}_i + \overline{C_i}\dot{\eta}_i + \overline{K_i}\eta_i = \overline{f_i} \,,$$

and the system with n degrees of freedom is broken down into a set of n uncoupled systems, each with a single degree of freedom.

The eigenvectors are the solutions of a linear set of homogeneous equations and, thus, are not unique: For each mode, an infinity of eigenvectors exists, all proportional to each other. Because the eigenvectors can be seen as a set of n vectors in the n-dimensional space providing a system of reference, the length of such vectors is not determined, but their directions are known. In other words, the scales of the axes are arbitrary. There are many ways to normalize the eigenvectors. The simplest is by stating that the value of one particular element or of the largest one is set to unity. Each eigenvector can be divided by its Euclidean norm, obtaining unit vectors in the space of the configurations. Another way is to normalize the eigenvectors in such a way that the modal masses are equal to unity. This can be done, simply by dividing each eigenvector by the square root of the corresponding modal mass. In the latter case, each modal stiffness coincides with the corresponding eigenvalue, i.e., with the square of the natural frequency. In the case of generalized proportional damping, equation (1.54) reduces to

$$\ddot{\boldsymbol{\eta}} + 2[\zeta\lambda]\dot{\boldsymbol{\eta}} + [\lambda^2]\boldsymbol{\eta} = \overline{\mathbf{f}'} \,, \tag{1.59}$$

where $[\lambda^2] = \mathrm{diag}\{\lambda_i^2\}$ is the matrix of the eigenvalues; matrix $[\zeta\lambda] = \mathrm{diag}\{\zeta_i\lambda_i\}$ contains the damping ratios for the various modes; and the modal forces $\overline{\mathbf{f}'}(t)$ are

$$\overline{f_i'} = \frac{\overline{f_i}}{\overline{M_i}} = \frac{\mathbf{q}_i^T \mathbf{f}}{\mathbf{q}_i \mathbf{M} \mathbf{q}_i} \ . \tag{1.60}$$

Often the modal damping matrix is not obtained by building matrix $\mathbf{C}$ and then performing the modal transformation, but directly by assuming reasonable values for the modal damping ratios ζ_i.

The time history of the response to an arbitrary excitation can be computed by:

1. Computing the eigenvalues and eigenvectors of the system and normalizing the eigenvectors.

2. Computing the modal forces as functions of time.

3. Solving the n equations (1.59), in order to obtain the time histories of the response in terms of modal coordinates.

4. Combining the responses computed through equation (1.50), which yields the time history of the system in terms of the coordinates $\mathbf{x}$.

This procedure has the notable advantage of dealing with n uncoupled equations, while a direct solution would require the integration of a set of n coupled differential equations. There is, however, another advantage: Not all modes are equally important in determining the response of the system. If there are many degrees of freedom, a limited number of modes (usually those characterized by the lower natural frequencies) is sufficient for obtaining the response with good accuracy. If only the first m modes are considered, the savings in terms of computation time, and hence cost, are usually noticeable, because only m eigenvalues and eigenvectors need to be computed and m systems with one degree of freedom have to be studied. Usually the modes that are more difficult to deal with are those characterized by the highest natural frequencies, particularly if the equations of motion are integrated numerically. The advantage of discarding the higher-order modes is, in this case, great.

When some modes are neglected, the reduced matrix of the eigenvectors, which will be referred to as $\mathbf{\Phi}^*$, is not square because it has n rows and m columns. The first coordinate transformation (1.50) still holds, and the m modal masses, stiffnesses, damping coefficients, and forces can be computed as usual. However, the inverse transformation (second equation 1.50) is not possible, because the inversion of matrix $\boldsymbol{\Phi}^*$ cannot be performed.

The modal coordinates $\boldsymbol{\eta}$ can be computed from the physical coordinates $\mathbf{x}$ by premultiplying both sides of the first equation (1.50) by $\boldsymbol{\Phi}^{*T}\mathbf{M}$ and then by the inverse of the matrix of the modal masses, obtaining

$$\boldsymbol{\eta} = \overline{\mathbf{M}}^{-1}\boldsymbol{\Phi}^{*T}\mathbf{M}\mathbf{x}\,. \tag{1.61}$$

Equation (1.61) is the required inverse modal transformation.

Although in undamped systems modal uncoupling does not introduce approximations, the equations of motion of damped systems can be uncoupled exactly only in the case of generalized proportional damping. This statement obviously does not exclude the possibility of uncoupling the equations of motion by introducing adequate approximations. The simplest way is by computing the modal damping matrix using equation (1.57) and then neglecting all its terms except those on the diagonal. Another possibility is to compute the eigenvalues related to the damped system solving the eigenproblem (1.32) and then using equation (1.24) to compute the modal damping for each mode. These procedures, which usually produce very similar results, introduce errors that are very small if the system is lightly damped.

There are, however, cases in which neglecting the out-of-diagonal elements of the modal damping matrix leads to unacceptable results, mainly when the system is highly damped and the damping distribution is far from proportional. In this case, however, it is still possible to distinguish between a proportional part $\overline{\mathbf{C}}_p$ of the damping matrix, which is a diagonal matrix containing the elements of $\overline{\mathbf{C}}$ on the main diagonal, and a nonproportional part $\overline{\mathbf{C}}_{np}$, containing all other elements of $\overline{\mathbf{C}}$. The latter is symmetrical with all elements on the main diagonal equal to zero.

By applying the inverse modal transformation, it is possible to show also that the original damping matrix $\mathbf{C}$ can be split into a proportional and a nonproportional part, $\mathbf{C}_p = (\boldsymbol{\Phi}^{-1})^T\overline{\mathbf{C}}_p\boldsymbol{\Phi}^{-1}$ and $\mathbf{C}_{np} = (\boldsymbol{\Phi}^{-1})^T\overline{\mathbf{C}}_{np}\boldsymbol{\Phi}^{-1}$, respectively. Note that $\mathbf{C}_p$ is not strictly proportional but only proportional in a generalized way.

Once the natural frequencies and mode shapes of the undamped system are known, the equation of motion of the system can be rewritten in modal coordinates in the form

$$\overline{\mathbf{M}}\ddot{\boldsymbol{\eta}} + \overline{\mathbf{C}}_p\dot{\boldsymbol{\eta}} + \overline{\mathbf{K}}\boldsymbol{\eta} = -\overline{\mathbf{C}}_{np}\dot{\boldsymbol{\eta}} + \overline{\mathbf{f}}(t)\,. \tag{1.62}$$

Note that all matrices on the left-hand side are diagonal and coupling terms are present only on the right-hand side. It is thus possible to devise an iterative procedure allowing the computation of the eigenvalues of the damped system without solving the eigenproblem (1.32), whose size is $2n$. Because a solution of the type of equation (1.26) can be assumed for the time history of the response in terms of modal coordinates, the differential homogeneous equation associated to equation (1.62) can be transformed into the following algebraic equation:

$$\left(s^2\overline{\mathbf{M}} + s\overline{\mathbf{C}}_p + \overline{\mathbf{K}}\right)\boldsymbol{\eta}_0 = -s\overline{\mathbf{C}}_{np}\boldsymbol{\eta}_0\,. \tag{1.63}$$

To compute the ith eigenvalue s_i, assume a set of n complex modal coordinates that are all zero, except for the real part of the ith coordinate, which is assumed with unit value. This amounts to assuming a complete uncoupling between the modes, at least where the ith mode is concerned. From the ith equation a first approximation of the complex eigenvalue can be obtained. The remaining $i-1$ complex equations can be used to obtain new values for the $n-1$ complex elements of the eigenvector; the ith element is assumed to have unit real part and zero imaginary part. Once the eigenvector has been computed, a new estimate for the ith eigenvalue can be obtained from the ith equation while the other equations yield a new estimate for the eigenvector. This procedure can be repeated until convergence is obtained.

This iterative procedure does not rely on any small damping assumption. The starting values of the eigenfrequencies are not those of the undamped system, but rather are those of a system that has a generalized proportional distribution of damping. Consequently, it can also be applied to systems with very high damping in which some modes show a nonoscillatory free response. Although no theoretical proof of the convergence of the technique has been attempted, a number of tests did show that convergence is very quick. The reduction of computation time with respect to the conventional state-space approach is particularly remarkable when the root loci are to be obtained and the eigenproblem has to be solved several times.

Also, equation (1.62), yielding the forced response of the system, can be solved using an iterative approach. The equation of motion obtained by neglecting matrix $\mathbf{C}_{np}$ is first solved. A solution $\boldsymbol{\eta}(t)^{(0)}$, corresponding to a system with generalized proportional damping, is thus obtained. This solution is introduced on the right-hand side of equation (1.62), and a second-approximation solution $\boldsymbol{\eta}(t)^{(1)}$ is obtained. The iterative procedure can continue until the difference between two subsequent solutions is smaller than any given small quantity. Either a Jacobi or a Gauss Siedel iterative scheme can be used; the second is generally faster (see Appendix 1). The convergence of the iterative scheme is very fast, even if the distribution of damping is far from proportional. The simplification of the computations obtainable in this way is very noticeable, particularly in the case of systems with many degrees of freedom.

It must, however, be stressed that the presence of damping couples the equations of motion and makes the possibility of considering a limited number of modes much more remote than it is possible for undamped or proportionally damped systems. The eigenvectors of the undamped system can be used as a system of reference in the space of the configurations; consequently, the motion of the damped system can be expressed in terms of modal coordinates, whether or not the damping is proportional. All eigenvectors can, however, be present in the response of the system at any frequency; it is not possible to understand a priori how much each of them affects the global response.

Example 1-7

Perform the modal analysis of the system in Example 1-2.
Because the mass matrix is diagonal, the formulation of the dynamic matrix involving the inversion of the mass matrix is used:

$$\mathbf{D} = \mathbf{M}^{-1}\mathbf{K} = \begin{bmatrix} 20 & -10 & 0 \\ -2.5 & 3.5 & -1 \\ 0 & -8 & 8 \end{bmatrix}.$$

The characteristic equation yielding the natural frequencies is easily obtained and solved:

$$\lambda^6 - 31.5\lambda^4 + 225\lambda^2 - 200 = 0 \, ;$$

$$\lambda_1 = 1.0166 \, ; \qquad \lambda_2 = 3.0042 \, ; \qquad \lambda_3 = 4.6305 \, .$$

The eigenvalues of the dynamic matrix $\mathbf{D}$ are 1.03353, 9.02522 and 21.44125. The corresponding eigenvectors, normalized by setting to unity the largest element, are

$$\left\{ \begin{array}{c} 0.45913 \\ 0.87081 \\ 1 \end{array} \right\} \, ; \qquad \left\{ \begin{array}{c} 0.11677 \\ 0.12815 \\ -1 \end{array} \right\} \, ; \qquad \left\{ \begin{array}{c} 1 \\ -0.14412 \\ 0.03578 \end{array} \right\} \, .$$

The products $\mathbf{q}_i^T \mathbf{M} \mathbf{q}_j$ are easily computed. Those with $i = j$ yield the three modal masses $\overline{M_1} = 3.74404$, $\overline{M_2} = 0.57933$, and $\overline{M_3} = 1.08616$, and those with different subscripts yield the values -1.02×10^{-5}, 1.54×10^{-5}, and 4.09×10^{-6}. The values obtained are very small, compared to the modal masses; they represent the deviations from orthogonality (with respect to the mass matrix) due to computational approximations.
The eigenvectors can be normalized by dividing each one of them by the square root of the corresponding modal mass. The matrix of the eigenvectors is

$$\mathbf{\Phi} = \left[\begin{array}{ccc} 0.23728 & 0.15342 & 0.95925 \\ 0.45004 & 0.16837 & -0.13825 \\ 0.51681 & -1.31383 & 0.08228 \end{array} \right] \, .$$

The modal mass matrix, then, is the identity matrix, while the modal stiffness matrix is a diagonal containing the squares of the natural frequencies computed earlier.

$$\overline{\mathbf{M}} = \left[\begin{array}{ccc} 1 & 0 & 0 \\ 0 & 1 & 0 \\ 0 & 0 & 1 \end{array} \right] \, , \qquad \overline{\mathbf{K}} = \left[\begin{array}{ccc} 1.03353 & 0 & 0 \\ 0 & 9.02522 & 0 \\ 0 & 0 & 21.44125 \end{array} \right] \, .$$

Note that in this case the Cholesky transformation of the mass matrix is very simple, because the mass matrix is diagonal. It yields

$$\mathbf{L} = \left[\begin{array}{ccc} 1 & 0 & 0 \\ 0 & 2 & 0 \\ 0 & 0 & 0.7071 \end{array} \right] \, .$$

By using the transformation (1.53) of the stiffness matrix, it follows that

$$\mathbf{K}^* = \left[\begin{array}{ccc} 20 & -5 & 0 \\ -5 & 3.5 & -2.8284 \\ 0 & -2.8284 & 8 \end{array} \right] \, .$$

In this case, because the mass matrix is an identity matrix, the dynamic matrix $\mathbf{D} = \mathbf{M}^{-1}\mathbf{K}^*$ coincides with the stiffness matrix $\mathbf{K}^*$ and is symmetrical.

The eigenvectors

$$\mathbf{L}^T\boldsymbol{\Phi} = \begin{bmatrix} 0.23728 & -0.15342 & 0.95925 \\ 0.90008 & -0.33674 & -0.27650 \\ 0.36544 & 0.92901 & 0.05818 \end{bmatrix}$$

are orthogonal, as is easily verified.

The modal damping matrix is not diagonal because the damping of the system cannot be considered proportional damping:

$$\overline{\mathbf{C}} = \boldsymbol{\Phi}^T\mathbf{C}\boldsymbol{\Phi} = \begin{bmatrix} 0.0471 & -0.0364 & -0.2276 \\ -0.0364 & 0.8790 & -0.1472 \\ -0.2276 & -0.1472 & 1.2240 \end{bmatrix}.$$

Even if the damping matrix is not diagonal, approximate modal uncoupling can be performed by neglecting all elements of the modal damping matrix lying outside the main diagonal. The fact that the neglected elements of matrix $\overline{\mathbf{C}}$ are of the same order of magnitude as the terms considered should not give the impression of a rough approximation. Actually, the behaviour of the system is similar to that of the undamped system, owing to the low value of damping, except when a mode is excited near its resonant frequency. Damping is important only in near-resonant conditions, and even then only in one of the equations: the one related to the resonant mode. In the equation in which damping is important, the element on the diagonal of the modal damping matrix is multiplied by a generalized coordinate that is far greater than the other ones, i.e., by the modal coordinate of the resonant mode. From the three uncoupled modal systems, the damped frequencies and decay rates of the free motions can be obtained easily:

$$s_1 = -0.0236 \pm i1.016\,, \; s_2 = -0.4395 \pm i2.971\,, \; s_3 = -0.6120 \pm i4.581,$$

which are very close to the values

$$s_1 = -0.0235 \pm i1.018\,, \; s_2 = -0.4399 \pm i2.974\,, \; s_3 = -0.6115 \pm i4.580,$$

obtained directly as eigenvalues of the dynamic matrix in the state space. The imaginary parts of the eigenvalues can be compared with the natural frequencies of the undamped system: It is clear that the presence of damping does not greatly affect the frequency of the free oscillations of the system.

1.7 Uncoupling of the equations of motion: State space

The eigenvectors of equation (1.32) do not uncouple the equations of motion in the state space, because matrix $\boldsymbol{A}$ is not symmetrical.

It is, however, possible to uncouple the state equations by resorting to the eigenvectors of the adjoint eigenproblem, i.e., the eigenvectors of matrix $\boldsymbol{A}^T$. Let $\mathbf{q}_{Ri}$ be the ith right eigenvector (i.e., the ith eigenvector of matrix $\boldsymbol{A}$) and let $\mathbf{q}_{Li}$ be the ith left eigenvector (i.e., the ith eigenvector of matrix $\boldsymbol{A}^T$). They are biorthogonal, i.e., all products $\mathbf{q}_{Lj}^T\mathbf{q}_{Ri}$ are equal to zero, if $i \neq j$.

The eigenvectors can be normalized in such a way that

$$\mathbf{q}_{Li}^T\mathbf{q}_{Ri} = 1\,, \tag{1.64}$$

in which case they are said to be *biorthonormal*.

By introducing into the state equation the modal states $\overline{\mathbf{z}}$ defined by the relationship

$$\mathbf{z} = \boldsymbol{\Phi}_R\overline{\mathbf{z}}\,, \tag{1.65}$$

where $\boldsymbol{\Phi}_R$ is the matrix of the right eigenvectors, and premultiplying by the matrix of the left eigenvectors $\boldsymbol{\Phi}_L$ transposed, the following modal uncoupled set of equations is obtained

$$\dot{\overline{\mathbf{z}}} = \overline{\boldsymbol{A}}\overline{\mathbf{z}} + \overline{\boldsymbol{B}}\mathbf{u}\,, \tag{1.66}$$

where $\overline{\boldsymbol{A}} = \boldsymbol{\Phi}_L^T\boldsymbol{A}\boldsymbol{\Phi}_R$ is a diagonal matrix listing the eigenvalues and $\overline{\boldsymbol{B}} = \boldsymbol{\Phi}_L^T\boldsymbol{B}$. Note that this is possible because $\overline{\boldsymbol{\Phi}}_L^T\boldsymbol{\Phi}_R = \mathbf{I}$.

The transition matrix $e^{\overline{\boldsymbol{A}}t}$ in this case is a diagonal matrix, which can be easily computed because each of its elements is simply e^{s_it}.

1.8 Excitation due to the motion of the constraints

Consider a discrete system where all degrees of freedom are translational, constrained to a rigid frame whose motion can be expressed as a translation x_A, y_A, z_A along the directions of the three axes of the inertial system of reference xyz, and express the generalized coordinates x_i with reference to the moving frame. A two-dimensional

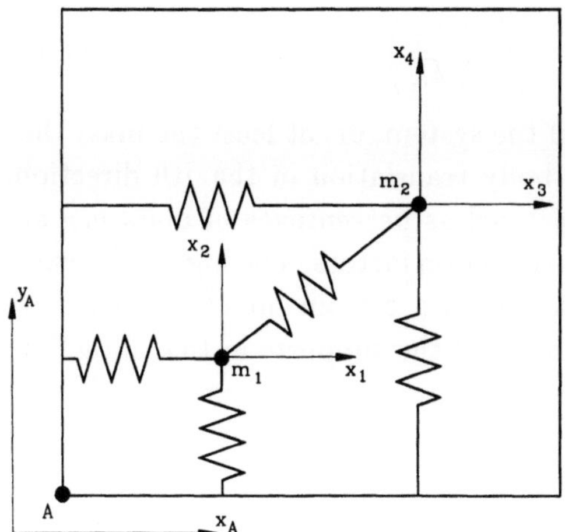

FIGURE 1.6. Example of a two-dimensional system excited by the motion of the constraints. In this case, $\boldsymbol{\delta}_x = [1, 0, 1, 0]^T$; $\boldsymbol{\delta}_y = [0, 1, 0, 1]^T$.

example is shown in Figure 1.6. The vector containing the absolute accelerations can be obtained from that containing the second derivatives of the coordinates x_i by the relationship

$$\ddot{\mathbf{x}}_{iner} = \ddot{\mathbf{x}} + \boldsymbol{\delta}_x \ddot{x}_A + \boldsymbol{\delta}_y \ddot{y}_A + \boldsymbol{\delta}_z \ddot{z}_A \,, \tag{1.67}$$

where the terms δ_{x_i}, δ_{y_i}, and δ_{z_i} are simply the direction cosines of the displacement x_i in the system of reference xyz. The equation of motion of the system written with reference to the relative coordinates is

$$\mathbf{M}\ddot{\mathbf{x}} + \mathbf{C}\dot{\mathbf{x}} + \mathbf{K}\mathbf{x} = -\mathbf{M}\boldsymbol{\delta}_x \ddot{x}_A - \mathbf{M}\boldsymbol{\delta}_y \ddot{y}_A - \mathbf{M}\boldsymbol{\delta}_z \ddot{z}_A + \mathbf{f}(t) \,. \tag{1.68}$$

Equation (1.68) is the generalization of equation (1.2) to multi-degree-of-freedom systems. Performing the modal transformation reduces it to

$$\overline{\mathbf{M}}\ddot{\boldsymbol{\eta}} + \overline{\mathbf{C}}\dot{\boldsymbol{\eta}} + \overline{\mathbf{K}}\boldsymbol{\eta} = -\mathbf{r}_x \ddot{x}_A - \mathbf{r}_y \ddot{y}_A - \mathbf{r}_z \ddot{z}_A + \overline{\mathbf{f}}(t) \,, \tag{1.69}$$

where $\mathbf{r}_j = \boldsymbol{\Phi}^T \mathbf{M}\boldsymbol{\delta}_j$ is a vector containing the so-called modal participation factors in the jth direction ($j{=}x, y, z$). Each term of this vector gives a measure of how much of the mass of the system participates to the ith mode when the system is excited by a motion of the supporting frame in the relevant direction.

It is easy to verify that

$$m_T = \sum_{i=1}^{n} \left(\frac{r_{j_i}^2}{\overline{M_i}} \right). \tag{1.70}$$

where m_T is the total mass of the system, or, at least the mass that can be associated to a rigid body translation in the jth direction. Often, the ratios $r_{j_i}^2 / \overline{M_i}$, expressed as percentages of mass m_T are used instead of the modal participation factors. The higher the value of the modal participation factor in a certain direction, the more that mode is excited by a motion of the supports in that direction. In computing the response, it is often sufficient to consider the few modes that are characterized by a high value of the corresponding modal participation factor.

If only a limited number m of modes is considered, the order of vector $\mathbf{r}_j$ is obviously m. All the considerations discussed in this section still hold, but equation (1.70) is only approximated. The precision that can be attained considering only a limited number of modes when computing the response to an excitation due to the motion of the constraints is measured by the approximation with which the sum in equation (1.70) approximates the total mass of the system.

When using the state-space approach to compute the response of the system to the motion of the constraints in the jth direction, the input vector $\mathbf{u}(t)$ contains a single element, the acceleration in the relevant direction, and matrix $\mathbf{T}$ to be introduced into equation (1.13) reduces to a vector of order n: $\mathbf{T} = -\mathbf{M}\boldsymbol{\delta}_j$.

1.9   Forced oscillations with harmonic excitation

The motion of a system with a single degree of freedom under the effect of an external excitation can be obtained by adding the solution of the homogeneous equation describing the free motion to a particular integral of equation (1.1) or (1.2). Among the simplest types of laws $F(t)$, one is of particular interest: Harmonic excitation. By resorting to the complex notation, it can be expressed in the form $F = f_0 e^{i\lambda t}$ if the excitation is provided by an external force, or $x_A = x_{A_0} e^{i\lambda t}$ in the case of motion of the constraint.

Note that the force $F(t)$ is a real quantity and should be expressed as $F = \Re(f_0 e^{i\lambda t})$; similarly, the expression of the displacements should mention explicitly the real part, as the complex notation is used to express quantities that have a harmonic time history as projections on the real axis of vectors that rotate in the Argand plane (Figure 1.7); the symbol $\Re$ is, however, usually omitted. The amplitudes f_0 and x_0 are also complex quantities, allowing consideration of arbitrary phasing of the excitation and of the response. If x_0 is real, for example, the amplitude reaches its maximum at $t = 0$; otherwise, it leads the function $\cos(\lambda t)$ of a phase angle equal to $\arctan[\Im(x_0)/\Re(x_0)]$.

It is easy to verify that the particular integral of the equation of motion is of the same type of equation (1.16), where $i\lambda$ is substituted for s. The response to a harmonic excitation is harmonic, with the same frequency of the forcing function. By introducing a harmonic time history for both excitation and response, the differential equation of motion can be transformed into an algebraic equation yielding the complex amplitude of the response

$$(- m\lambda^2 + i\lambda c + k)x_0 = \begin{cases} f_0, \\ (i\lambda c + k)x_{A_0}, \\ -m\lambda^2 x_{A_0}, \end{cases} \tag{1.71}$$

for excitation provided by a force, by the motion of the supporting point A using an inertial coordinate, and by the motion of the supporting point A using a relative coordinate, respectively. The coefficient of the unknown x_0 in equation (1.71) is the dynamic stiffness of the system

$$k_{dyn} = (- m\lambda^2 + ic\lambda + k) = k\left[1 - \left(\frac{\lambda}{\lambda_n}\right)^2 + 2i\zeta\left(\frac{\lambda}{\lambda_n}\right)\right]. \tag{1.72}$$

The dynamic stiffness is a function of the forcing frequency, and, in the case of damped systems, it is a complex quantity. Its reciprocal is usually referred to as *dynamic compliance* or *receptance* and expresses the ratio between the amplitude of the displacement x_0 and the amplitude of the exciting force f_0. When the forcing frequency tends to zero, the dynamic stiffness and compliance tend to their static counterparts, the stiffness k and the compliance $1/k$.

The ratio between the dynamic and the static compliance of the system is usually referred to as the *frequency response $H(\lambda)$* of the system, which, in cases of damped systems, is complex. The real part of the frequency response gives the component of the response that is in phase with the excitation. The imaginary part gives the component in quadrature, which lags the excitation by a phase angle of $90°$. The expressions for the real and imaginary parts of $H(\lambda)$, its amplitude, and phase are:

$$\Re(H) = k\frac{k - m\lambda^2}{(k - m\lambda^2)^2 + c^2\lambda^2} = \frac{1 - \left(\frac{\lambda}{\lambda_n}\right)^2}{\left[1 - \left(\frac{\lambda}{\lambda_n}\right)^2\right]^2 + \left(2\zeta\frac{\lambda}{\lambda_n}\right)^2},$$

$$\Im(H) = k\frac{-c\lambda}{(k - m\lambda^2)^2 + c^2\lambda^2} = \frac{-2\zeta\frac{\lambda}{\lambda_n}}{\left[1 - \left(\frac{\lambda}{\lambda_n}\right)^2\right]^2 + \left(2\zeta\frac{\lambda}{\lambda_n}\right)^2},$$

$$|H| = \frac{k}{\sqrt{(k - m\lambda^2)^2 + c^2\lambda^2}} = \frac{1}{\sqrt{\left[1 - \left(\frac{\lambda}{\lambda_n}\right)^2\right]^2 + \left(2\zeta\frac{\lambda}{\lambda_n}\right)^2}},$$

$$\Phi = \arctan\left(\frac{-c\lambda}{k - m\lambda^2}\right) = \arctan\left[\frac{-2\zeta\left(\frac{\lambda}{\lambda_n}\right)}{1 - \left(\frac{\lambda}{\lambda_n}\right)^2}\right].$$

$$(1.73)$$

The situation in the Argand plane at time t is described in Figure 1.7.

The absolute value of the frequency response is often called the *magnification factor*. It is plotted together with the phase angle Φ as a function of the forcing frequency in Figures 1.8a and b for different values of the damping factor. Logarithmic axes are often used, and the scale of the ordinates is expressed in decibels (Figure 1.8c). This plot is referred to as the *Bode diagram*. The value in decibels of the magnification factor $|H(\lambda)|$ is defined as

$$H_{dB} = 20log_{10}(|H|). \qquad (1.74)$$

In the logarithmic plot, the frequency response tends, for very low frequency, to the straight line $H=1$, and for very high frequency to a straight line sloping down with a slope equal to -2. This last situation is often referred to as an attenuation of 12 dB/oct (decibel per octave), even if the actual value is 12.041 dB/oct, or 40 dB/dec (decibel per decade). Where the response follows the first straight line, the system is said to be controlled by the stiffness of the spring because

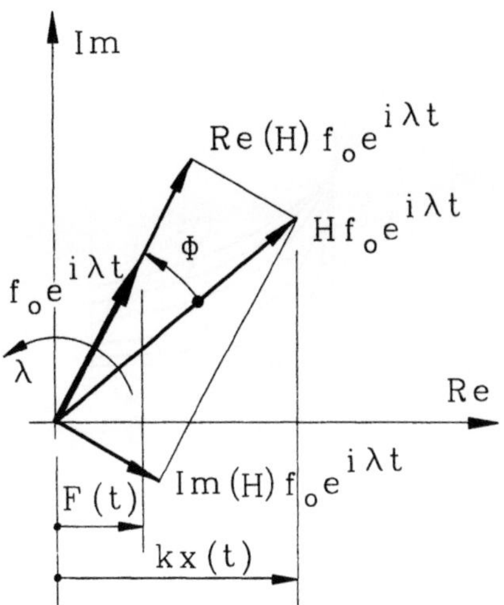

FIGURE 1.7. Response of a system with viscous damping as seen in the Argand plane. Situation at time t.

the inertia forces are negligible, and, in cases of lightly damped systems, the external force $F(t)$ is in equilibrium with the elastic force; in the case of excitation due to the motion of point A, mass m follows the displacement of the support. The phase angle is very small, tending to 0 for undamped systems.

When the system follows the sloping straight line, its behaviour is said to be controlled by the inertia of the mass m, because in cases of lightly damped systems, it is the inertia force that balances the external force $F(t)$. The phase angle is very close to $-180°$, tending to this value for undamped systems.

When the excitation frequency λ is close to the natural frequency of the undamped system λ_n, a resonance occurs and amplitudes can be large. In this zone the damping of the system, however small it can be, becomes the governing factor because, at resonance, the inertia force exactly balances the elastic force and, consequently, only the damping force can balance the excitation $F(t)$. In this zone, which is said to be controlled by damping, the presence of damping cannot be neglected; in the other frequency ranges, the behaviour of the system can often be very well approximated using an undamped model. At the natural frequency of the undamped system, the phase lag is exactly $90°$, regardless of the value of the damping.

If $\zeta < 1/2$, the frequency at which the peak amplitude occurs is $\lambda_p = \lambda_n\sqrt{1 - 2\zeta^2}$. It shifts toward the lower values of λ with increasing damping and does not coincide with the frequency of the free oscillations of the system. For greater values of ζ, the curve $H(\lambda)$

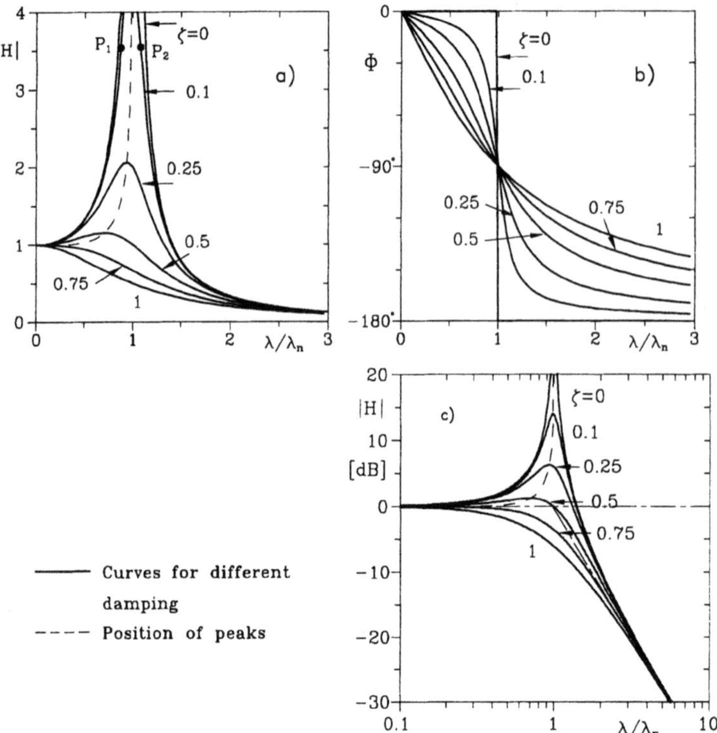

FIGURE 1.8. Bode diagram, i.e., magnification factor and phase as functions of the forcing frequency: (a) and (b) linear scales, (c) logarithmic scale for frequency and dB scale for amplitudes.

does not show a peak and the maximum value occurs at $\lambda=0$. If the system is lightly damped and ζ^2 is negligible compared with unity, then the maximum values of the amplitude and magnification factor are, respectively,

$$|x_0|_{\max} \approx \frac{f_0}{c\lambda_n}, \qquad |H|_{\max} \approx \frac{1}{2\zeta}. \qquad (1.75)$$

The term $1/2\zeta$ is often called the *quality factor*; Q is used to represent it.

On the curve obtained for $\zeta = 0.1$ in Figure 1.8a, points P_1 and P_2, at which the amplitude is equal to the peak amplitude divided by $\sqrt{2}$, are reported. They are often defined as half-power points and correspond to an attenuation of about 3 dB with respect to the maximum amplitude. The frequency interval between points P1 and P2 is often called the *half-power bandwidth* and is used as a measure of the sharpness of the resonance peak. If damping is small enough to

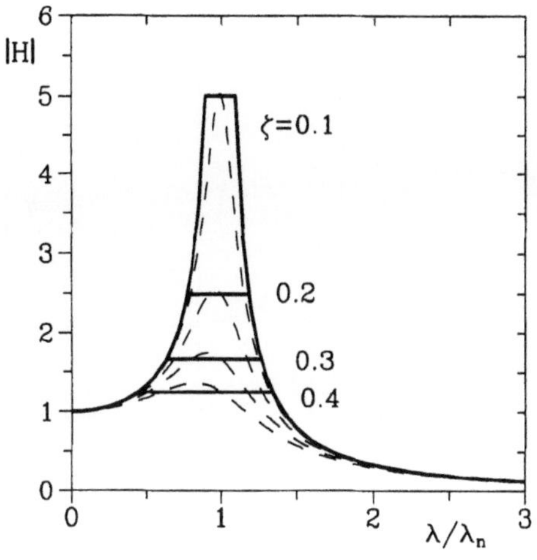

FIGURE 1.9. Frequency response of lightly damped systems approximated by using the response of the undamped system and shaving the peak at the value expressed by equation (1.40). Comparison with the exact solution.

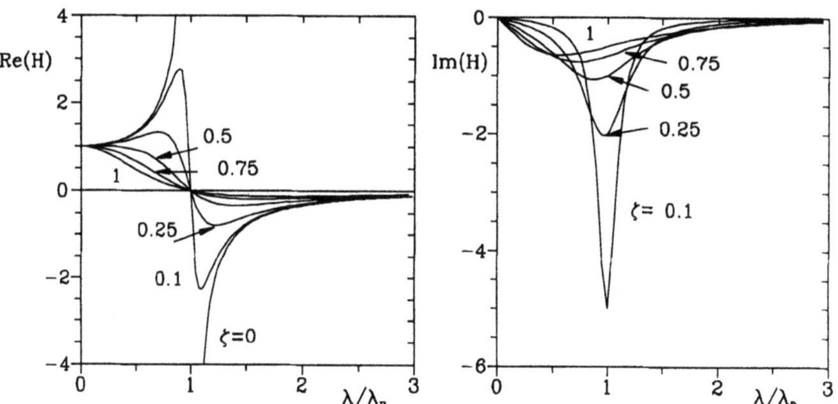

FIGURE 1.10. Real and imaginary parts of the frequency response as functions of the forcing frequency.

allow the usual simplifications (i.e., ζ^2 is negligible compared with unity), the frequencies corresponding to such points and the half-power bandwidth are

$$\lambda_{P_1} \approx \lambda_n \sqrt{1 - 2\zeta}, \qquad \lambda_{P_2} \approx \lambda_n \sqrt{1 + 2\zeta}, \qquad \Delta\lambda \approx 2\zeta\lambda_n. \quad (1.76)$$

The frequency response of a lightly damped system can be approximated by the frequency response of the corresponding undamped system except for the frequency range spanning from point P_1 to point P_2, where the amplitude can be considered constant, its value being expressed by equation (1.75). As shown in Figure 1.9, this approximation still holds for values of damping as high as

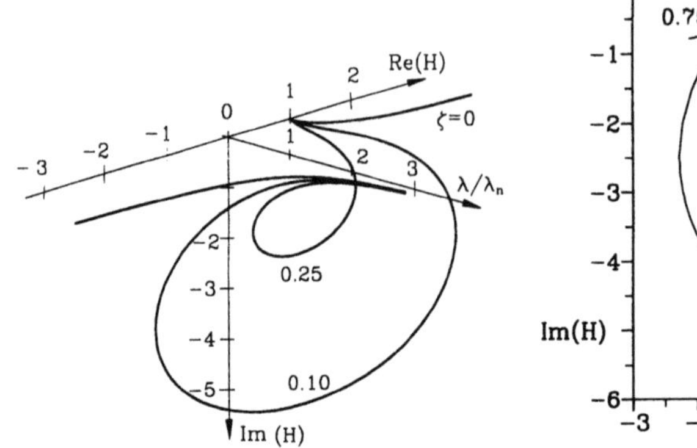

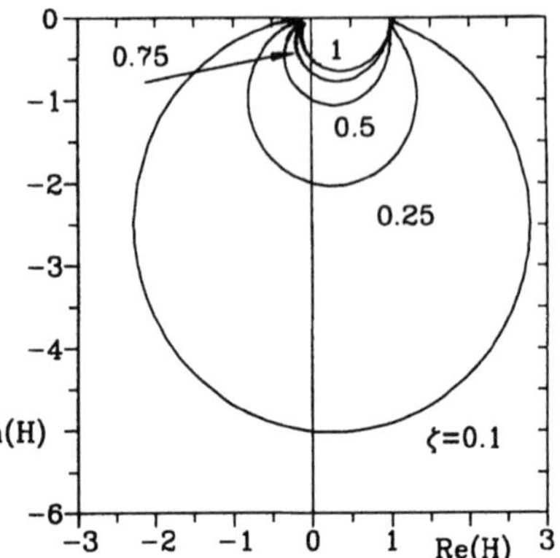

FIGURE 1.11. Same as Figure 1.8, but as a tridimensional plot.

FIGURE 1.12. Nyquist diagram for the system of Figure 1.1.

$\zeta = 0.25 - 0.30$.

Instead of plotting the amplitude and the phase of the frequency response, it is possible to separately plot its real and imaginary parts (Figure 1.10). The two plots of Figure 1.10 can be combined in the three-dimensional plot of Figure 1.11. The projection of the latter on the Argand plane is the so-called Nyquist diagram (Figure 1.12).

The complete solution of the equation of motion is obtained by adding the solutions found for the free and forced oscillations (i.e., adding a particular integral to the complementary function)

$$x = K^* e^{-\zeta \lambda_n t} e^{i\lambda_n \sqrt{1-\zeta^2} t} + H(\lambda)\frac{f_0}{k} e^{i\lambda t}. \qquad (1.77)$$

The complex constant K^* can be determined from the initial conditions. The first term of equation (1.77) tends to zero, often quite quickly, and the second one has a constant amplitude. When studying the response of a damped system to harmonic excitation, usually only the latter is considered. There are, however, cases in which the initial transitory cannot be neglected, as in the case in which the forcing function is applied to a system that is at rest: Oscillations with growing amplitude usually result, until the steady-state conditions are reached. To state that when resonance occurs the amplitude of the response becomes infinitely large is an oversimplification, even in the idealized case of undamped systems. In that case the amplitude grows linearly and an infinite time is required to reach an infinite amplitude. In practice, very large values of the amplitude can be reached in a very short time, but there are cases, particularly when

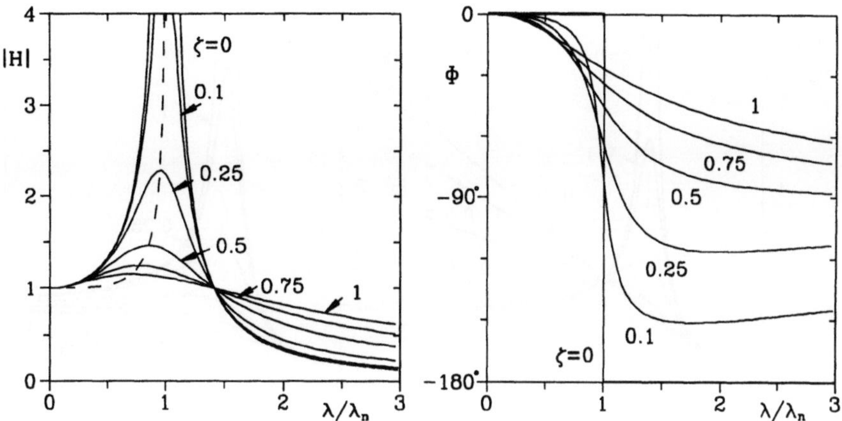

FIGURE 1.13. Same as Figure 1.8, but with the excitation provided by the harmonic motion of the supporting point. Full line indicates curves for different values of damping; dashed line indicates line connecting the peaks.

very lightly damped systems are excited by weak forcing functions, in which the amplitude buildup is very slow.

If the excitation is provided by the harmonic motion of the supporting point A, a frequency response $H(\lambda) = x_0/x_{A_0}$ can be defined. By separating the real part from the imaginary part, the following values of the magnification factor and phase lag are readily obtained:

$$\Re(H) = \frac{k(k - m\lambda^2) + c^2\lambda^2}{(k - m\lambda^2)^2 + c^2\lambda^2},$$

$$\Im(H) = \frac{-cm\lambda^3}{(k - m\lambda^2)^2 + c^2\lambda^2},$$

$$|H| = \frac{\sqrt{k^2 + c^2\lambda^2}}{\sqrt{(k - m\lambda^2)^2 + c^2\lambda^2}},$$

$$\Phi = \arctan\left(\frac{-cm\lambda^3}{k(k - m\lambda^2) + c^2\lambda^2}\right). \tag{1.78}$$

The amplitude and the phase of the frequency response are plotted as functions of the forcing frequency in Figure 1.13. A very common example of a system excited by the motion of the supports is that of a rigid body supported by compliant mountings whose aim is that of insulating it from vibrations that can be transmitted from the surrounding environment. In this case, the magnification factor $|H|$ is referred to as *transmissibility* of the suspension system.

The transmissibility is the ratio between the amplitude of the absolute displacement of the suspended object and the amplitude of

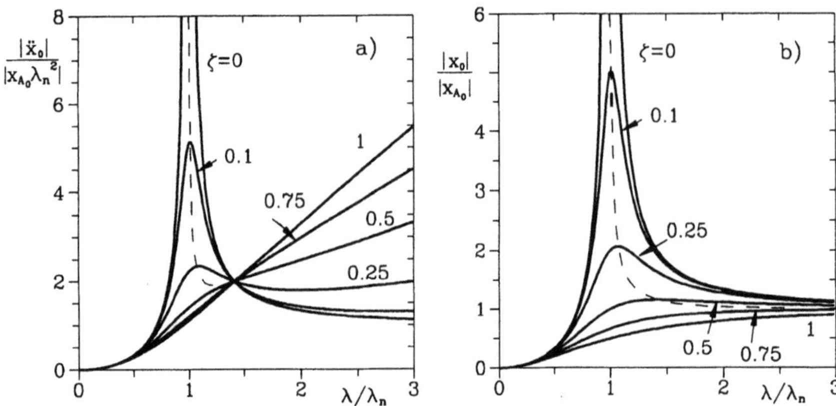

FIGURE 1.14. Nondimensional response of a system excited by the motion of the supporting point: (a) amplitude of the absolute acceleration of point P as a function of the driving frequency, (b) displacement expressed in relative coordinates. The same response holds for the case of a system excited by a forcing function whose amplitude is proportional to the square of the frequency. Full line indicates curves for different values of damping; dashed line indicates line connecting the peaks.

the displacement of the supporting points. In many cases, the amplitude of the acceleration is more important than the amplitude of the displacement (Figure 1.14a).

Another problem related to the insulation of mechanical vibrations is that of reducing the amount of excitation exerted on the supporting structure by a rigid body on which a force variable in time is acting. The ratio between the amplitude of the force exerted by the spring-damper system on the supporting point $kx + c\dot{x}$ and the amplitude of the excitation $F(t)$ is also referred to as transmissibility of the suspension. With simple computations, it is possible to show that the value of the transmissibility so defined is the same as obtained in equation (1.78). This explains why the two ratios are referred to by the same name.

From Figure 1.14a, it is clear that any increase of the damping causes a decrease of the transmissibility if the exciting frequency is lower than the natural frequency of the suspension, but it causes an increase of the vibration amplitude at higher frequencies. When the system is excited by the motion of the supporting point, it can be expedient to resort to relative coordinates. If the amplitude of the displacement of the supporting point is independent from the forcing frequency, the acceleration is proportional to the square of λ. The nondimensional frequency response for this case is shown in Figure

Frequency response	Definition	S.I. units	Lim$(\lambda \to 0)$
Dynamic compliance	x_0/f_0	m/N	$1/k$
Dynamic stiffness	f_0/x_0	N/m	k
Mobility	$\dot{x}_0/f_0 = \lambda x_0/f_0$	m/sN	0
Mechanical impedance	$f_0/\dot{x}_0 = f_0/\lambda x_0$	Ns/m	∞
Inertance	$\ddot{x}_0/f_0 = \lambda^2 x_0/f_0$	m/s^{2}N	0
Dynamic mass	$f_0/\ddot{x}_0 = f_0/\lambda^2 x_0$	s^{2}N/m	∞

TABLE 1.2. Frequency responses.

1.14b. The same figure can be used for the more general case of the response to a forcing function whose amplitude is proportional to the square of the frequency. Note that the peak is located at a frequency higher than the natural frequency of the undamped system.

Apart from the dynamic compliance and the dynamic stiffness, other frequency responses can be defined. The ratio between the amplitude of the velocity and that of the force F is said to be mobility; its reciprocal is the mechanical impedance. The ratio between the amplitude of the acceleration and that of the force F is said to be inertance, and its reciprocal is the dynamic mass. The aforementioned frequency responses are summarized in Table 1.2. The most widely used are the dynamic compliance and the inertance.

In case of multi-degree-of-freedom systems, when the damping is proportional, the equations of motion for forced vibrations can be uncoupled and the study is reduced to the computation of the response of n uncoupled linear damped systems. Also, such a procedure can often be applied in cases where the equations of motion could not be uncoupled theoretically. In particular, if the system is lightly damped, as is common in the dynamics of structures, the response of each mode to a harmonic excitation will be near to that of the corresponding undamped system except in a narrow frequency range centered on the resonance of the mode itself, i.e., except in that frequency range in which the response is governed by damping. The response of each mode can consequently be approximated as shown in Figure 1.9. However, while the amplitudes are approximated very well in this way, the error in the computation of the phases can be large.

It must be noted that even if all the harmonic exciting forces are in phase, i.e., the excitation is said to be monophase or coherently phased, the response of the system is harmonic but not coherently phased. Not even if the damping is proportional do the various parts of the system oscillate in phase when subjected to forces that are in phase with each other, except in the case of near-resonant conditions.

If the forcing frequency is near one of the natural frequencies, the shape of the response of a system with proportional damping is usually very close to the relevant mode shape. Because the mode shapes used for the modal transformation are the real mode shapes of the undamped system and not the complex modes of the damped system, the modal response of the resonant mode is coherently phased even if the forces acting on the system are not. However, even if damping is proportional, the response is not a pure mode shape even exactly in resonance, because the amplitude of the resonant mode is not infinitely larger than the amplitude of the other modes, as would happen in undamped systems.

At resonance, the phase lag between the modal force and the modal response of the resonant mode is exactly $90°$. By writing the generic harmonic excitation in the form $\mathbf{f}(t) = \mathbf{f}_0 e^{i\lambda t}$ and the response in the form $\mathbf{x}(t) = \mathbf{x}_0 e^{i\lambda t}$, the aforementioned observation can be traduced into the statement that $\mathbf{x}_0$ is generally not real even if $\mathbf{f}_0$ is real.

To fully understand the meaning of a complex vector $\mathbf{x}_0$, consider the case of a massless beam on which a number of masses are located (Figure 1.15a). Consider a representation in which the plane of oscillation of the system (the vertical plane in Figures 1.15b and c) is the real plane and the plane perpendicular to it is the imaginary plane. Any plane perpendicular to them can be considered an Argand plane, in which the real and imaginary axes are defined by the intersections with the real and imaginary planes. At the location of each mass, there is an Argand plane in which the vector representing the displacement of the mass rotates.

If, as in the case of undamped systems with coherently phased excitation, vector $\mathbf{x}_0$ is real, the situation in a space in which the coordinate planes are the real and imaginary planes is that shown in Figure 1.15b. The actual deformed shape is then the projection on the real plane of a plane line that rotates at the angular speed λ. The shapes it takes at various instants are consequently all similar; only their amplitudes vary in time.

If vector $\mathbf{x}_0$ is complex (Figure 1.15c), the deformed shape is the projection on the real plane of a rotating skew line. Consequently, its shape varies in time and no stationary point of minimum deformation (node) or maximum deformation (loop or antinode) exists. As

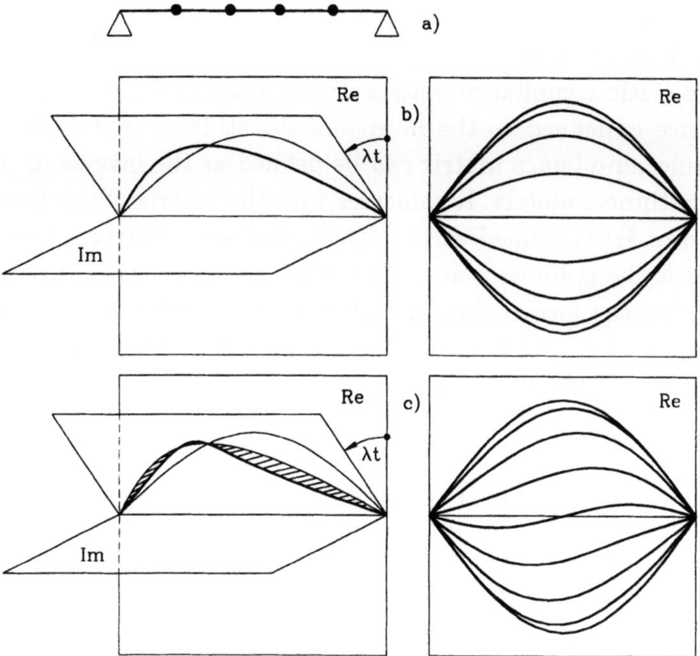

FIGURE 1.15. Meaning of the complex displacement vector $\mathbf{x}_0$: (a) Sketch of a beam modeled as a massless beam with concentrated masses, (b) undamped case: the deflected configuration is the projection on the real plane of a line lying on a plane that rotates at angular speed λ, (c) damped case: The line that generates the deflected shape is skew.

already seen, this last situation also characterizes the case of proportional damping and all the cases that can be reduced to it, at least in an approximate way.

The amplitude of the response and of the excitation are linked by the dynamic stiffness matrix of the system,

$$\mathbf{K}_{dyn}\mathbf{x}_0 = \mathbf{f}_0 \, ,$$

whose expression is of the same type as equation (1.72)

$$\mathbf{K}_{dyn} = -\lambda^2\mathbf{M} + \mathbf{K} + i\lambda\mathbf{C} \, . \tag{1.79}$$

The dynamic stiffness matrix is real only in the case of undamped systems; usually it is symmetrical but can be nonpositive defined.

A system with n degrees of freedom can be excited using n harmonic generalized forces corresponding to the n generalized coordinates, and, for each exciting force, n responses can be obtained. The frequency responses $H_{ij}(\lambda) = x_{0i}(\lambda)/f_{0j}$, where f_{0j} is the amplitude of the jth generalized force and x_{0i} is the response at the ith degree of freedom, are thus n^2.

The static compliance matrix or the matrix of the coefficients of influence is defined as the inverse of the stiffness matrix $\mathbf{K}$, and a dynamic compliance matrix can be defined as the inverse of the dynamic stiffness matrix. It coincides with the matrix of the frequency responses $\mathbf{H}(\lambda)$ defined earlier. The compliance matrix is symmetrical, as is the stiffness matrix, but while the latter often has a band structure, the former does not show useful regularities. In cases of damped systems, matrix $\mathbf{H}(\lambda)$ is complex; each frequency response can thus be defined in amplitude and phase or as real and imaginary parts. A number n^2 of Bode and Nyquist diagrams can be plotted in the same way seen for systems with a single degree of freedom.

The amplitude plot of the Bode diagrams has n peaks with infinite height in the case of undamped systems, which correspond to the natural frequencies. Some of the curves $|H(\lambda)|$ can cross the frequency axis, i.e., the amplitude can get vanishingly small at certain frequencies. This condition is usually referred to as antiresonance. It must be noted that while the resonances are the same for all the degrees of freedom, the antiresonances are different and can be absent in some of the responses. The number of antiresonances in $n - 1$ for the transfer functions on the main diagonal and decrease by one on each diagonal above or below it. No antiresonance is so found in $H_{1,n}$ and $H_{n,1}$.

If there is damping, the resonance peaks get smaller with increasing damping and the amplitude at the antiresonances grows. If the system is highly damped, some of the peaks can disappear completely.

The Nyquist diagrams usually have as many loops as degrees of freedom, if the system is lightly damped. With increasing damping, some loops can disappear.

Example 1-8
Compute the elements H_{11}, H_{13}, and H_{33} of the frequency response of the system in Example 1-2, neglecting damping. Repeat the computation for H_{33} taking also damping into account.

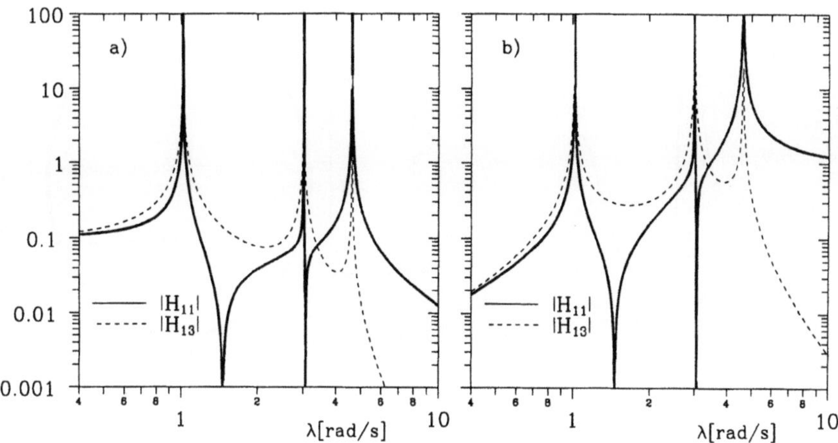

FIGURE 1.16. Two elements of the (a) dynamic compliance matrix and (b) the inertance matrix of a system with three degrees of freedom. The units are rad/Nm for the dynamic compliance and rad^3/Nms^2 for the inertance.

The dynamic compliance is easily computed by inverting the dynamic stiffness matrix. Two of the required frequency responses are plotted in Figure 1.16a using logarithmic scales. By multiplying the dynamic compliance by λ^2 the inertance is easily obtained (Figure 1.16b). The response H_{33} is reported in Figure 1.17 (dashed line).

Because the system is lightly damped, the use of modal damping can be expected to yield results that are not too far from the correct ones, even if damping is not proportional.

The frequency response H_{33} is reported in Figure 1.17, comparing the results directly obtained with those computed using the values of the modal damping, which were obtained earlier, to obtain the modal responses and then transforming the results to physical coordinates.

The two curves are almost everywhere exactly superimposed, showing the very good approximation obtainable when using modal damping.

The first mode is less damped than the other two, and the third one is so much damped that the resonance peak disappears completely. It must be noted that the third peak is, at any rate, very narrow in the response of the undamped system. In the Nyquist plot (Figure 1.17b) the first peak generates a loop that is far larger than the one related to the second resonance.

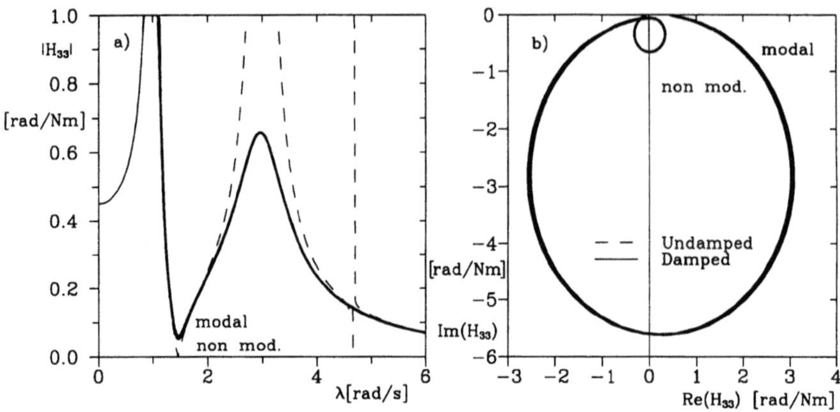

FIGURE 1.17. Frequency response H_{33}: (a) amplitude as a function of frequency; (b) Nyquist diagram.

Example 1-9: Two identical pendulums connected by a spring
Consider the two identical pendulums connected by a spring in Example 1-3. Compute the natural frequencies of the linearized system and the time history of the free oscillations when the system is released from a standstill with the first pendulum displaced of θ_0 and the second in the vertical position. The main data are $m=1$ kg, $l=600$ mm, $k=2$ N/m, and $g=9.81$ m/s^2.

By introducing the data, the equation of motion becomes

$$\begin{bmatrix} 0.6 & 0 \\ 0 & 0.6 \end{bmatrix} \begin{Bmatrix} \ddot{\theta}_1 \\ \ddot{\theta}_2 \end{Bmatrix} + \begin{bmatrix} 11.01 & -1.2 \\ -1.2 & 11.01 \end{bmatrix} \begin{Bmatrix} \theta_1 \\ \theta_2 \end{Bmatrix} = \begin{Bmatrix} 0 \\ 0 \end{Bmatrix}.$$

The eigenfrequencies and the corresponding eigenvectors can be easily computed:

$$\lambda_1^2 = \frac{g}{l} = 16{,}35 \qquad \lambda_1 = 4.04 \, \text{rad/s}$$
$$\lambda_2^2 = \frac{mg + 2kl}{ml} = 20{,}35 \qquad \lambda_2 = 4.51 \text{rad/s}$$

$$\mathbf{q}_1 = \begin{Bmatrix} 1 \\ 1 \end{Bmatrix}, \qquad \mathbf{q}_2 = \begin{Bmatrix} 1 \\ -1 \end{Bmatrix}.$$

In the first mode the two pendulums move together, without stretching the spring, with the same frequency they would have if they were not connected.

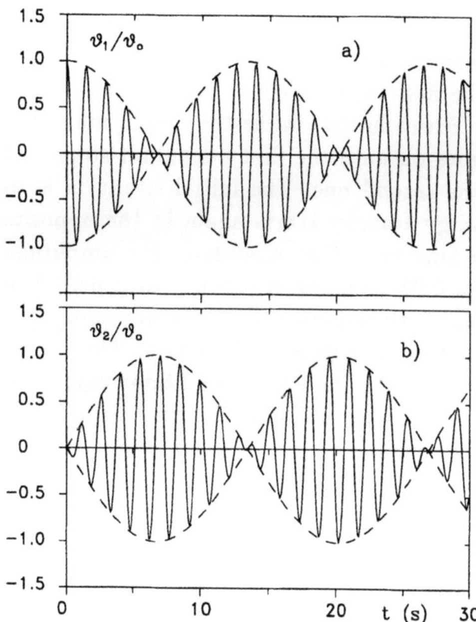

FIGURE 1.18. Two pendulums linked together by a spring: time history of the response.

This motion is not affected by the characteristics of the spring. In the second mode the pendulums oscillate in opposition with a frequency affected by the characteristics of both the spring and the pendulums. If the spring is very soft (k/m much smaller than g/l) the two natural frequencies are very close to each other.

The initial conditions are $\theta_1 = \theta_0$ and $\theta_2 = \dot{\theta}_1 = \dot{\theta}_2 = 0$. The time histories of the free oscillations are then easily obtained:

$$\theta_1(t) = \frac{\theta_0}{2}\left[\cos(\lambda_1 t) + \cos(\lambda_2 t)\right], \qquad \theta_2(t) = \frac{\theta_0}{2}\left[\cos(\lambda_1 t) - \cos(\lambda_2 t)\right],$$

or, remembering some trigonometric identities,

$$\begin{cases} \theta_1(t) = \theta_0\left[\cos\left(\frac{\lambda_2 - \lambda_1}{2}t\right)\cos\left(\frac{\lambda_2 + \lambda_1}{2}t\right)\right], \\ \theta_2(t) = \theta_0\left[\sin\left(\frac{\lambda_2 - \lambda_1}{2}t\right)\sin\left(\frac{\lambda_2 + \lambda_1}{2}t\right)\right]. \end{cases}$$

The motion can then be considered an oscillation with a frequency equal to the average of the natural frequencies of the system $(\lambda_2 + \lambda_1)/2$ with an amplitude that is modulated with a frequency equal to $(\lambda_2 - \lambda_1)/2$, as clearly shown in Figures 1.18a and b.

The system does not include any damping: The energy of the two pendulums is therefore conserved. The initial conditions are such that at time $t = 0$ all energy is concentrated in the first pendulum. The spring slowly transfers energy from the first to the second, in such a way that the amplitude of the former decreases in time while the amplitude of the latter increases. This process goes on until the first pendulum stops and all energy is concentrated in the second one. The initial situation is so reversed and the process of energy transfer starts again in the opposite direction. The frequency of the sine wave that modulates the amplitude is $(\lambda_2 - \lambda_1)/2 = 0.235$ rad/s $= 0.037$ Hz, corresponding to a period of 26.72 s. The frequency of the beat is then double the frequency computed earlier, i.e., 0.072 Hz, corresponding to a period of 13.37 s. The occurrence of the beat is, however, linked with the initial conditions that must be able to excite both modes. If this does not happen, the oscillation is monoharmonic, and no beat takes place.

1.10   Systems with structural damping

The damping of many systems can be described in terms of viscous damping only as a very rough approximation. Many materials, when subjected to cyclic loading, show a behaviour that can be described as structural or hysteretic damping. Assuming that the time history of the stress cycles is harmonic and that also the time hystory of the deformation follows a similar pattern, although slightly out of phase owing to the internal damaping of the material, an elliptical hysteresis cycle results in the $(\sigma\epsilon)$ plane (Figure 1.19). The strain lags the stress of a phase angle Φ, which is assumed to be independent from the frequency.

In these condition the time histories of the stress and of the strain can be written as

$$\begin{cases} \sigma = \sigma_0 \cos(\lambda t) = \sigma_0 e^{i\lambda t}\,, \\ \epsilon = \epsilon_0 \cos(\lambda t - \Phi) = \epsilon_0 e^{-i\Phi} e^{i\lambda t}\,, \end{cases} \tag{1.80}$$

where λ is the frequency at which the hysteresis cycle is gone through.

The ratio σ/ϵ, which is static conditions is the Young's modulus of the material, can thus be expressed by a complex quantity, the complex modulus

$$\frac{\sigma}{\epsilon} = \frac{\sigma_0}{\epsilon_0}[\cos(\Phi) + i\sin(\Phi)] = E' + iE''\,. \tag{1.81}$$

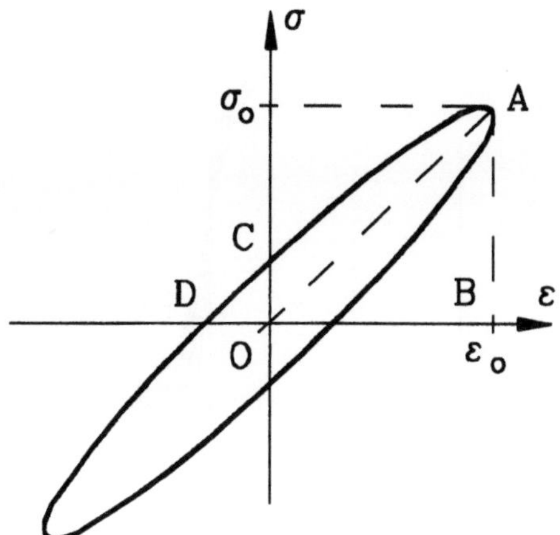

FIGURE 1.19. Structural damping: hysteresis cycle in $(\sigma\epsilon)$ plane.

In the same way also the stiffness of a structural member can be expressed by a complex number $k' + ik''$. The complex stiffness k^* then has a real part, the in-phase stiffness k', which gives the measure of the elastic stiffness of the material and is often referred to as storage stiffness and an imaginary part, the in-quadrature stiffness k'', which is linked with damping and is said to be loss stiffness. Their ratio is the loss factor or loss ratio η

$$\eta = \frac{k''}{k'} = \arctan(\Phi) \,. \tag{1.82}$$

Another parameter that is sometimes used to quantify the internal damping of materials is the specific damping capacity ψ. It is defined as the ratio between the energy dissipated in a cycle (area of the ellipse in Figure 1.19) and the elastic energy stored in the system in the condition of maximum amplitude (area of the OAB triangle in the same figure).

The damping of most engineering materials (except some elastomers) is quite small, and the trigonometric functions of the phase angle Φ can be linearized. The expressions of the quantities defined earlier can, consequently, be simplified:

$$\begin{aligned}
k' &\approx k, & \eta &\approx \Phi, \\
k^* &\approx k(1 + i\eta), & \Psi &\approx 2\pi\Phi \approx 2\pi\eta \,.
\end{aligned} \tag{1.83}$$

The loss factor of a structural member can be equal to that of the material (as in the case of a homogeneous monolithic spring) or greater, if some damping mechanisms other than material hysteresis

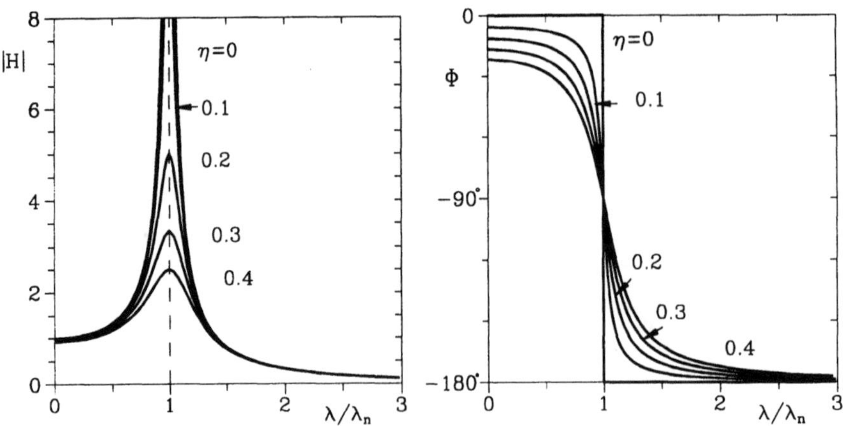

FIGURE 1.20. Same as Figure 1.8, but for a system with structural damping. Full line indicates curves for different values of damping; dashed line indicates line connecting the peaks.

are present (as in built-up members, with rivets or threaded joints, elements in viscous fluids, and so on).

The complex stiffness and the complex modulus have been introduced in connection with harmonic loading of a structural element. It is, however, very easy to demonstrate that the definition can be extended to the more general case of periodic loading, because any periodic time history can be expressed as the sum of harmonic terms. The extension to nonperiodic time histories through Fourier transform is more complex, because it can lead to noncausal results.

The dynamic stiffness of a system with structural damping is

$$k_{dyn} = -m\lambda^2 + k(1 + i\eta).\tag{1.84}$$

The complex frequency of the free oscillations of the system can be obtained by equating the dynamic stiffness to zero. Remembering that systems with structural damping are usually very lightly damped, the characteristic equation so obtained yields the following values of the frequency and decay rate

$$\Re(\lambda) = \lambda_n\sqrt{\frac{1 + \sqrt{1 + \eta^2}}{2}} \approx \lambda_n$$

$$\sigma = \Im(\lambda) = \lambda_n\sqrt{\frac{-1 + \sqrt{1 + \eta^2}}{2}} \approx \lambda_n\frac{\eta}{2}.\tag{1.85}$$

The frequency of the free oscillations is higher than the natural frequency of the undamped system. This result is different from that found for viscous damping, but, also in this case, the frequency shift due to the presence of damping is negligible for lightly damped systems. The logarithmic decrement δ takes the value

$$\delta = \frac{\pi\eta}{\sqrt{1 + (\eta^2/4)}} \approx \pi\eta. \tag{1.86}$$

The steady-state solution for harmonic excitation is readily obtained from the expression (1.84) of the dynamic stiffness: In the case of excitation provided by a force $F(t)$, the following expressions for the real and imaginary parts of the frequency response $H(\lambda)$, its amplitude, and the phase angle can be obtained:

$$\Re(H) = \frac{k(k - m\lambda^2)}{(k - m\lambda^2)^2 + k^2\eta^2}, \quad |H| = \frac{k}{\sqrt{(k - m\lambda^2)^2 + k^2\eta^2}},$$
$$\Im(H) = \frac{-k^2\eta}{(k - m\lambda^2)^2 + k^2\eta}, \quad \Phi = \arctan\left(\frac{-k\eta}{k - m\lambda^2}\right). \tag{1.87}$$

The magnification factor is plotted together with the phase angle as functions of the forcing frequency in Figure 1.20 for different values of the loss factor η.

The quality factor of a system with a single degree of freedom with structural damping is simply given by

$$Q = |H|_{max} = \frac{1}{\eta}. \tag{1.88}$$

Structural damping is a form of linear damping that does not differ much from viscous damping. It is possible to define an equivalent viscous damping coefficient

$$c_{eq} = \frac{\eta k}{\lambda}, \tag{1.89}$$

through which structural damping can be assimilated to viscous damping with a coefficient inversely proportional to the frequency at which the hysteresis cycle is gone through. If damping is small, as is usually the case, the shift of the resonance peak between viscous and structural damping is small and the behaviour of systems with the two different types of damping is, at least near the peak, similar. As in lightly damped systems, i.e., when η^2 can be neglected compared with unity, the effect of damping is important only near the resonance, it is possible to define a constant equivalent damping as

$$c_{eq} = \frac{\eta k}{\lambda_n}, \qquad \zeta_{eq} = \frac{\eta}{2}, \tag{1.90}$$

This why the expression 2ζ is sometimes called "loss factor" in systems with viscous damping.

In the case of systems with many degrees of freedom, the stiffness matrix $\mathbf{K}^*$ is complex, with a real part $\mathbf{K}'$ defining the conservative properties of the system and an imaginary part $\mathbf{K}''$ defining the dissipative properties. The dynamic stiffness matrix for a system with structural damping is

$$\mathbf{K}_{dyn} = -\lambda^2 \mathbf{M} + \mathbf{K}' + i\mathbf{K}''. \tag{1.91}$$

If the loss factor is constant throughout the system, matrices $\mathbf{K}''$ and $\mathbf{K}'$ are proportional and the complex stiffness matrix reduces to $\mathbf{K}^* = (1 + i\eta)\mathbf{K}$. This is a form similar to that of proportional damping, and the equations of motion can be uncoupled exactly, yielding n equations of motions of the type seen for systems with a single degree of freedom with structural damping.

In the case of systems with many degrees of freedom, it is also possible to define an equivalent viscous damping matrix equal to $\mathbf{K}''$ divided by λ. Because systems with structural damping are very lightly damped, the effects of damping are restricted only in the fields of frequency that are near the natural frequencies, and the modal uncoupling holds with good approximation. The behaviour of the system can thus be studied by uncoupling the equations of motion using the eigenvectors of the undamped system and introducing a constant equivalent damping that does not depend on the frequency $\overline{C_{ieq}} = \eta_i \overline{K_i}/\lambda_i$. The modal damping can easily be measured during a dynamic test by measuring the amplitude at resonance or the half-power bandwidth, or evaluated from data that can be found in the literature.

Structural damping is just a linear model that, while allowing the modeling of many actual systems better than viscous damping, in many cases gives only a rough approximation of the behaviour of structural members. The Young's modulus E and the loss factor η of most engineering materials are independent of the frequency only in an approximated way. Most metals stick to this rule with a fair or even good approximation, while elastomers often show very strong dependence of mechanical characteristics with changing frequency.

The loss factor of all materials is a function of many parameters and is particularly influenced by the amplitude of the stress cycle. During the life of a structural member, strong variations of the damping characteristics with the progress of fatigue phenomena are expected. Actually, damping can be used to obtain information on the extent of fatigue damage.

The behaviour of materials is only approximately linear, but while the nonlinearities of the stress-strain curve are usually only found at high stresses and a wide linearity field exists, the nonlinearities in the damping characteristics are found at all values of the load. While the dependence of the characteristics of the material on the frequency can be taken into account without great complications, the last consideration would lead to nonlinear equations and, consequently, is usually neglected. With all the aforementioned limitations, the structural damping model remains a powerful tool for structural analysis and finds a very wide application.

1.11   Systems with frequency-dependent parameters

In the same way the complex stiffness was defined, a complex viscous damping

$$c^* = c' + ic'' \tag{1.92}$$

can be introduced. The real part c' is coincident with the damping coefficient and the imaginary part c'' is actually a stiffness in the sense that it does not involve any dissipation of energy. As an equivalent damping was defined for the imaginary part of the complex stiffness, an equivalent stiffness can be defined for the imaginary part of the complex damping: $k_{eq} = -\lambda c''$. Consequently, the complex damping model allows the introduction of a stiffness that grows linearily with increasing frequency into the linear equation of motion (if c'' is expressed by a negative number).

More generally, it is possible to introduce into the model coefficients c and k, which are general functions of the forcing frequency. The equation of motion for a damped linear system is sometimes written in the form

$$m\ddot{x} + c(\lambda)\dot{x} + k(\lambda)x = F(t)\,. \tag{1.93}$$

This equation is written partly in the time domain, with the time histories $x(t)$ (and its time derivatives) and $F(t)$ and partly in the frequency domain, as the laws $c(\lambda)$ and $k(\lambda)$ enter explicitly into the equation. The frequency λ is, however, defined only when the time history $x(t)$ is harmonic. Consequently, while the frequency domain relationship corresponding to equation (1.93)

$$\left[-m\lambda^2 + ic(\lambda)\lambda + k(\lambda) \right]x_0 = f_0 \tag{1.94}$$

is useful for the study of harmonic motion, including all time histories that can be reduced to it through Fourier transform, even the response to stationary random excitation, equation (1.93) should not be used. For a more detailed discussion of this issue, see, for example, the well-known paper by S.H. Crandall[8].

A case in which the dependence of the elastic and damping characteristic of the system on the frequency is very important is that of structural members made of elastomeric materials. The in-phase and in-quadrature stiffness k' and k'' of an elastomeric element are reported as functions of the frequency in Figure 1.21. The loss factor $\eta = \tan(\Phi)$ has also been plotted in the figure. Three regions are usually defined: At low frequency (rubbery region), the material shows a very low stiffness; at high frequency (glassy region), its stiffness is substantially higher. Between the two regions there is a zone (transition region) in which the loss factor has a maximum.

The modulus of the material and the stiffness of a structural member in dynamic conditions are often defined as dynamic modulus and dynamic stiffness, as opposed to the static quantities measured under constant load. This use of the term *dynamic stiffness* must not be confused with that in Section 1.7: Here, the effect of frequency is due to the changes in the elasticity of the material, while there it was due to the inertia forces. However, the two definitions coincide when the system (only a spring or the spring-mass-damper system) is seen as a whole.

The stiffness and damping of elastomeric materials is also strongly influenced by temperature, and the effects of an increase of frequency

---

[8]S.H.Crandall, "The role of damping in vibration theory", *J. of Sound and Vibration*, 11(1), (1970), 3-18.

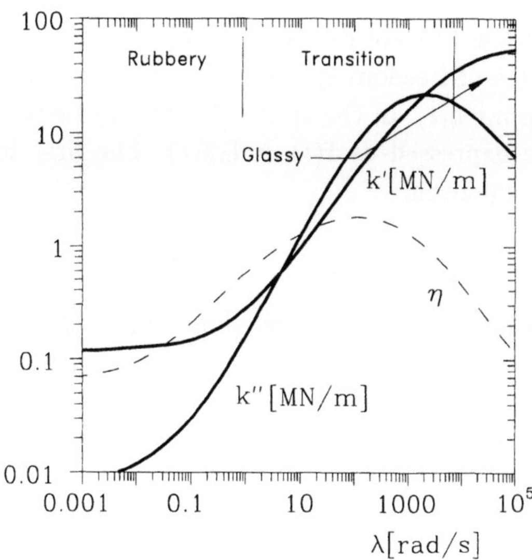

FIGURE 1.21. In-phase and in-quadrature stiffness and loss factor of an elastomeric spring as a function of the frequency.

or a decrease of temperature are so similar that it is possible to obtain the curves related to changes in temperature at constant frequency from those related to frequency at constant temperature and vice versa. Note that in the case of Figure 1.21, the value of the loss factor is quite high, at least in the transition region, and no small damping assumption can be made.

1.12 Coordinate transformation based on Ritz vectors

The dynamic behaviour of systems with many degrees of freedom was studied in the preceding section using either physical coordinates or modal coordinates. It is, however, obvious that any other coordinate transformation can be applied and that any nonsingular matrix of order n could be used to perform a coordinate transformation. This statement simply means that any set of linearly independent vectors can be assumed as a reference frame in the space of configurations. The modal transformation based on the eigenvectors has the drawbacks of requiring the solution of an eigenproblem, which sometimes is quite complex, and often requiring a large number of modes to compute the response to a generic forcing function. When the response to an excitation due to the motion of the supports has to be computed, this means that many modes have a modal participation factor high enough to prevent neglecting them. In this case, the use of Ritz vectors constitutes a different choice worthy of consideration.

Consider a multi-degree-of-freedom system excited in such a way that there is only one input $u(t)$. In the space of configurations, the forcing function can be expressed as $\mathbf{f}(t) = \mathbf{f}_0 f(t)$. The first Ritz vector is defined by the equation

$$\mathbf{r}_1 = \mathbf{K}^{-1}\mathbf{f}_0 , \qquad (1.95)$$

and then it coincides with the static deflected shape under the effect of the constant force distribution $\mathbf{f}_0$. Ritz vectors, like eigenvectors, are normalized. The simplest way to normalize Ritz vectors is by making the products $\mathbf{r}^T\mathbf{Mr}$ equal to unity. This can easily be performed by dividing each Ritz vector $\mathbf{r}$ by the square root of $\mathbf{r}^T\mathbf{Mr}$.

The following vectors can be computed using the following recursive equation

$$\mathbf{r}_i = \mathbf{K}^{-1}\mathbf{Mr}_{i-1} . \qquad (1.96)$$

Once a set of m Ritz vectors has been computed, a matrix with n rows and m columns $\mathbf{R} = [\mathbf{r}_1, \mathbf{r}_2, \ldots, \mathbf{r}_m]$ can be written. It can be used instead of the matrix of the eigenvectors to perform the coordinate transformation $\mathbf{x} = \mathbf{R}\bar{\mathbf{x}}$. The coordinates $\bar{\bar{x}}_i$ are m, as in the case of the use of a reduced number of modes. The transformed mass matrix is diagonal, due to the way the Ritz vectors are derived. All other transformed matrices are, however, nondiagonal and, consequently, the undamped equations of motions are coupled (elastic coupling).

The physical interpretation of Ritz vectors is straightforward. The first vector represents the static deformation under the effect of force distribution $\mathbf{f}_0$. No allowance is taken for inertia forces. Inertia forces due to harmonic motion with frequency λ and deformed shape $\mathbf{r}_1$ are $\lambda^2\mathbf{Mr}_1$. The second Ritz vector is then proportional to the deformed shape due to inertia forces consequent to a harmonic oscillation with a deformed shape corresponding to the first vector. In the same way, all other vectors are computed.

The advantage of Ritz vectors with respect to the true eigenvectors of the undamped system in the computation of the time history of the response is then clear: Fewer coordinates are generally required, and the amount of computational work needed to compute them is much smaller. The equation of motion obtained using Ritz vectors can then be subjected to modal analysis or reduced in a way that will be seen in Chapter 2.

There are, however, also disadvantages. First, it is common to perform an eigenanalysis before computing the time history of the response to obtain the natural frequencies and the mode shapes. In this case, the modal transformation involves very little additional computational work. In the undamped case, the equations of motion obtained through modal transformation are uncoupled, while, through Ritz vectors, a set of equations with elastic coupling is obtained. When damping is taken into consideration, and even more when nonlinearities are included in the model, the number of Ritz vectors needed can increase, and it is very difficult to assess how many must be considered, as happens with the true eigenvectors.

Although they are sometimes used in the computation of the response of structures to seismic excitation, Ritz vectors are not widely used in structural dynamics.

1.13   Structural modification

Many techniques aimed at computing the natural frequencies of a system after some of its characteristics have been modified without solving a new eigenproblem are listed under the general name of *structural modification*. Sometimes, the inverse problem is also considered: To compute the modifications needed to obtain required values of some natural frequencies.

Consider an undamped discrete system and introduce some small modifications in such a way that the mass and stiffness matrices can be written in the form $\mathbf{M} + \Delta\mathbf{M}$ and $\mathbf{K} + \Delta\mathbf{K}$. If the modifications introduced are small enough, the eigenvectors of the new system can be approximated by the eigenvectors of the old one and the ith modal mass, stiffness, and natural frequency of the new system can be approximated as

$$\begin{aligned}
\overline{M}_{i_{mod}} &= \mathbf{q}_i^T \left(\mathbf{M} + \Delta\mathbf{M} \right) \mathbf{q}_i = 1 + \mathbf{q}_i^T \Delta\mathbf{M}\mathbf{q}_i \,, \\
\overline{K}_{i_{mod}} &= \mathbf{q}_i^T \left(\mathbf{K} + \Delta\mathbf{K} \right) \mathbf{q}_i = \lambda_i^2 + \mathbf{q}_i^T \Delta\mathbf{K}\mathbf{q}_i \,, \\
\lambda_{i_{mod}}^2 &= \frac{\overline{K}_{i_{mod}}}{\overline{M}_{i_{mod}}} = \frac{\lambda_i^2 + \mathbf{q}_i^T \Delta\mathbf{K}\mathbf{q}_i}{1 + \mathbf{q}_i^T \Delta\mathbf{M}\mathbf{q}_i} \,,
\end{aligned} \tag{1.97}$$

where the eigenvectors have been normalized in such a way that the modal masses of the original system have unit values. The modifications are assumed to be very small. In this case, the series that expresses the square root of λ_i^2 can be truncated after the first term, yielding

$$\lambda_{i_{mod}} \approx \lambda_i \left(1 + \frac{\mathbf{q}_i^T \Delta \mathbf{K} \mathbf{q}_i}{2\lambda_i^2} - \frac{\mathbf{q}_i^T \Delta \mathbf{M} \mathbf{q}_i}{2} \right) . \qquad (1.98)$$

Equation (1.98) can be used to compute the new value of the ith eigenvalue, knowing the modifications that have been introduced into the system. However, the inverse problem can also be solved. If matrices $\Delta \mathbf{M}$ and $\Delta \mathbf{K}$ are functions of a few unknown parameters, a suitable set of equations (1.98) can be used to find the values of the unknowns, which allow the solution for some stated values of the natural frequencies. The procedure described here is approximated and can be used only for small modifications.

Another procedure that can be used to compute the effect of stiffness modifications that do not need to be small is the following. The eigenproblem allowing the computation of the natural frequencies of the modified system expressed in terms of modal coordinates of the original system is

$$\ddot{\boldsymbol{\eta}} + [\lambda^2]\boldsymbol{\eta} + \boldsymbol{\Phi}^T \Delta \mathbf{K} \boldsymbol{\Phi} \boldsymbol{\eta} = \mathbf{0} , \qquad (1.99)$$

where matrix $\boldsymbol{\Phi}^T \Delta \mathbf{K} \boldsymbol{\Phi}$ is, in general, not diagonal, because the eigenvectors of the original system do not uncouple the equations of motion of the modified system.

If only one modification is introduced, matrix $\Delta \mathbf{K}$ can be expressed in the form $\Delta \mathbf{K} = \alpha \mathbf{u} \mathbf{u}^T$ where, if the modification consists of the addition of a spring linking degrees of freedom i and j, constant α is nothing other than the stiffness of the spring and all elements of vector $\mathbf{u}$ are zero except elements i and j, which are equal to 1 and -1, respectively. This is actually not a limitation: Because the procedure is not approximated, several modifications can be performed in sequence without losing precision.

The modal matrix linked with the modification can be expressed as

$$\boldsymbol{\Phi}^T \Delta \mathbf{K} \boldsymbol{\Phi} = \alpha \boldsymbol{\Phi}^T \mathbf{u} \mathbf{u}^T \boldsymbol{\Phi} = \alpha \bar{\mathbf{u}} \bar{\mathbf{u}}^T , \qquad (1.100)$$

where, obviously, $\bar{\mathbf{u}} = \boldsymbol{\Phi}^T \mathbf{u}$.

The eigenproblem $(-\lambda^2 \mathbf{I} + [\lambda^2] + \alpha \bar{\mathbf{u}} \bar{\mathbf{u}}^T)\boldsymbol{\eta}_0 = \mathbf{0}$ linked with equation (1.99) yields a set of n equations of the type

$$\left(-\lambda^2 + \lambda_i^2\right)\frac{\eta_{0_i}}{\bar{u}_i} = -\alpha \sum_{k=1}^{n} \bar{u}_k \eta_{0_k}, \qquad (i = 1,2,\ldots,n). \qquad (1.101)$$

Note that λ^2 equals the eigenvalues of the modified system, while λ_i^2 equals those of the original one. The term on the left-hand side of (1.101) is the same in all equations. It then follows that

$$\left(-\lambda^2 + \lambda_1^2\right)\frac{\eta_{0_1}}{\bar{u}_1} = \left(-\lambda^2 + \lambda_2^2\right)\frac{\eta_{0_2}}{\bar{u}_2} = \ldots = \left(-\lambda^2 + \lambda_n^2\right)\frac{\eta_{0_n}}{\bar{u}_n}. \quad (1.102)$$

The eigenvector $\boldsymbol{\eta}_0$ can thus be easily computed. By stating that the ith element is equal to unity, the remaining elements can be computed from equation (1.102):

$$\eta_{0_k} = \frac{\left(-\lambda^2 + \lambda_i^2\right)\bar{u}_k}{\left(-\lambda^2 + \lambda_k^2\right)\bar{u}_i} \qquad (k \neq i). \qquad (1.103)$$

By introducing the eigenvector expressed by equation (1.103) into equation (1.101), it follows that

$$\frac{1}{\alpha} = \sum_{k=1}^{n} \frac{\bar{u}_k^2}{\left(-\lambda^2 + \lambda_k^2\right)} = 0. \qquad (1.104)$$

Equation (1.104) can be regarded as a nonlinear equation in λ, yielding the eigenfrequencies of the modified system. The same equation can, however, be used to compute the value of α once a value for the natural frequency of the modified system has been stated. Note that although it is possible to obtain any given value of the eigenfrequency, it is, however, impossible to be sure that a given eigenfrequency is modified as needed.

1.14   Parameter identification

In the preceding sections, attention was paid to the computation of the dynamic response of a system whose characteristics are known. Very often, however, the opposite problem must be solved: The behaviour of the system has been investigated experimentally and a mathematical model has to be obtained from the experimental results. First consider the case of a system with a single degree of freedom and assume that the response $x(t)$ and the excitation $F(t)$ are known in a number m of different instants and that the corresponding velocities and accelerations are also known. By writing the equation of motion of the system m times, the following equation can be obtained

$$\begin{bmatrix} \ddot{x}_1 & \dot{x}_1 & x_1 \\ \ddot{x}_2 & \dot{x}_2 & x_2 \\ \cdots & \cdots & \cdots \\ \ddot{x}_m & \dot{x}_m & x_m \end{bmatrix} \begin{Bmatrix} m \\ c \\ k \end{Bmatrix} = \begin{Bmatrix} f_1 \\ f_2 \\ \cdots \\ f_m \end{Bmatrix}. \tag{1.105}$$

Equation (1.105) is a set of m linear equations with three unknowns, the parameters of the system to be determined. A subset of three equations is then required to solve the problem. Actually, the situation is more complex. All measurements are affected by some errors, and the results obtainable from a set of three measurements are very unreliable. To obtain more reliable results, it is better to retain all rows of the matrix of the coefficients of equation (1.105) and to resort to its pseudo-inverse

$$\begin{Bmatrix} m \\ c \\ k \end{Bmatrix} = \begin{bmatrix} \ddot{x}_1 & \dot{x}_1 & x_1 \\ \ddot{x}_2 & \dot{x}_2 & x_2 \\ \cdots & \cdots & \cdots \\ \ddot{x}_m & \dot{x}_m & x_m \end{bmatrix}^\dagger \begin{Bmatrix} f_1 \\ f_2 \\ \cdots \\ f_m \end{Bmatrix}. \tag{1.106}$$

The pseudo-inverse $\mathbf{A}^\dagger$ of matrix $\mathbf{A}$ can be computed as $(\mathbf{A}^T\mathbf{A})^{-1}\mathbf{A}^T$, but this simple approach based on matrix inversion is increasingly less efficient for large matrices. Algorithms based on singular value decomposition or QR factorization are both more accurate and computationally efficient.

To avoid introducing into equation (1.106) the velocities and accelerations together with the displacements, it is possible to work in the frequency domain. In this case, the complex amplitudes of the response $x_0(\lambda)$ and the corresponding complex amplitudes of the excitation $f_0(\lambda)$ at m values of the frequency are measured, and equation (1.105) can be transformed into a set of m complex equations or $2m$ real equations

$$\begin{bmatrix} -\lambda^2 x_{0_1} & i\lambda x_{0_1} & x_{0_1} \\ -\lambda^2 x_{0_2} & i\lambda x_{0_2} & x_{0_2} \\ \cdots & \cdots & \cdots \\ -\lambda^2 x_{0_m} & i\lambda x_{0_m} & x_{0_m} \end{bmatrix} \begin{Bmatrix} m \\ c \\ k \end{Bmatrix} = \begin{Bmatrix} f_{0_1} \\ f_{0_2} \\ \cdots \\ f_{0_m} \end{Bmatrix} . \tag{1.107}$$

In the case of systems with many degrees of freedom, everything gets more complex as the number of parameters to be estimated becomes greater, but the computations can follow the same lines shown earlier for systems with a single degree of freedom. Also, in this case both time-domain and frequency-domain methods are possible, and many procedures have been proposed and implemented. The identification of the modal parameters of large mechanical systems is the main object of experimental modal analysis, which is, in itself, a specialized branch of mechanics of vibrations, which has been the subject of many books and papers in recent years. The algorithms used are often influenced by the hardware that is available for the acquisition of relevant data and subsequent computations. The recent advances in the field of computers and electronic instrumentation are causing continual advancements to take place in this field.

1.15 Laplace transforms, block diagrams, and transfer functions

Consider a function of time $f(t)$ defined for $t \geq 0$. Its Laplace transform $\mathcal{L}[f(t)] = F(s)$ is defined as

$$\mathcal{L}[f(t)] = F(s) = \int_0^\infty f(t)e^{-st}dt , \tag{1.108}$$

where s is a complex variable. For the mathematical details on Laplace transforms and the conditions on function $f(t)$, which make the transform possible, see one of the many textbooks on the subject[9].

Laplace transform is a linear transform, i.e., the transform of a linear combination of functions is equal to the linear combination of the transforms of the various functions. The main property that

---

[9]For example, W.T. Thompson, *Laplace trasformation*, Prentice Hall, Englewood Cliffs, 1960.

makes the Laplace transform useful in structural dynamics is that regarding the transform of the derivatives of function $f(t)$:

$$\mathcal{L}[\dot{f}(t)] = s\mathcal{L}[f(t)] - f(0), \qquad \mathcal{L}[\ddot{f}(t)] = s^2\mathcal{L}[f(t)] - sf(0) - \dot{f}(0).$$
$$(1.109)$$

The transform thus enables the changing of a differential equation into an algebraic equation without actually assuming the time history of the excitation, as was seen for harmonic excitation. Given the equation of motion of a linear system with a single degree of freedom,

$$m\ddot{x} + c\dot{x} + kx = f(t),$$

by transforming both functions $f(t)$ and $x(t)$ into $F(s)$ and $X(s)$, the following equation in the Laplace domain is obtained:

$$(s^2 m + sc + k)X(s) - msx(0) - m\dot{x}(0) - cx(0) = F(s). \quad (1.110)$$

The Laplace transform of the solution can be easily computed once the Laplace transform of the excitation is known

$$X(s) = \frac{F(s)}{s^2 m + sc + k} - \frac{m\dot{x}(0) + (ms + c)x(0)}{s^2 m + sc + k}. \qquad (1.111)$$

Since the Laplace transforms of the most common functions $f(t)$ are tabulated, equation (1.111) can be used to compute the Laplace transform of the response of the system. The time history $x(t)$ can then be obtained through the inverse transformation or, more simply, by using the same Laplace transform tables. The main limitation of the Laplace transform approach is that of being restricted to the solution of linear differential equations with constant coefficients.

By introducing the ordinary differential operator $D = d/dt$, the equation of motion of a damped system with a single degree of freedom excited by the forcing function $f(t)$ can be written in the form

$$mD^2 x + cDx + kx = f \qquad (1.112)$$

or

$$(mD^2 + cD + k)x = f.$$

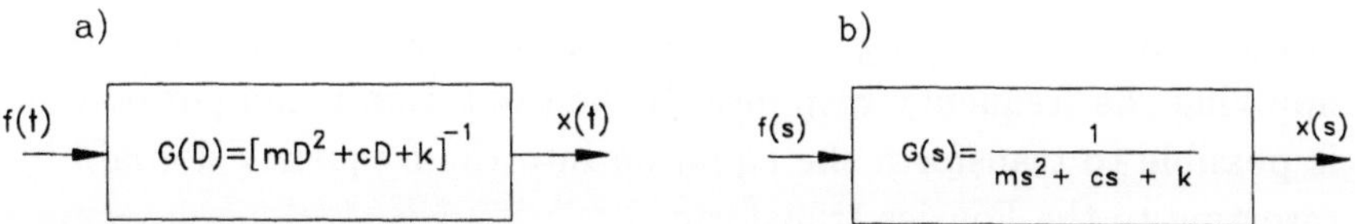

FIGURE 1.22. Block diagram of the system of Figure 1.1 in terms of the (a) transfer operator and (b) with reference to the Laplace domain.

If the operator $mD^2 + cD + k$ can be inverted, equation (1.112) can be solved as

$$x = [mD^2 + cD + k]^{-1} f \,, \qquad (1.113)$$

or

$$x = G(D) f \,,$$

where the transfer operator $G(D)$ is clearly the inverse of the operator $mD^2 + cD + k$.

A simple way to represent the system of Figure 1.1, i.e., a general damped linear system with a single degree of freedom, is the block diagram of Figure 1.22a, in which the output of the system, i.e., the time history $x(t)$, is related to its input, the time history of the excitation $f(t)$, by the transfer operator.

Equation (1.113) is, however, only a formal statement because the transfer operator cannot be written in explicit form. As a consequence, it cannot be used for the actual computation of the response of the system. There are two cases in which, as has already been stated, the differential operator can be transformed into an algebraic operator and the response can be obtained in explicit form, namely, when the time history of the excitation and that of the response are assumed to be harmonic and when the time histories of the excitation and of the response are transformed through the Laplace transform into the Laplace domain.

In the first case, the derivative of function $x(t)$ is obtained by simply multiplying it by $i\lambda$, where λ is the frequency. The transfer operator is simply the frequency response $H(\lambda)$

$$G(D) = H(\lambda) = \frac{1}{-m\lambda^2 + ic\lambda + k} \,. \qquad (1.114)$$

The response to a generic periodic forcing function can thus be studied by decomposing the excitation in Fourier series and then applying the frequency response. If the excitation is nonperiodic, it is possible to transform the equation into the frequency domain by resorting to the Fourier transform

$$\mathcal{F}[f(t)] = F(\lambda) = \int_{-\infty}^{\infty} f(t)e^{-i\lambda t}\,dt$$

of both functions $f(t)$ and $x(t)$.

In the second case, the derivative of function $x(t)$ is again obtained by simply multiplying it by the Laplace variable s, and the differential transfer operator reduces to an algebraic function, the transfer function $G(s)$

$$G(D) = G(s) = \frac{1}{ms^2 + cs + k}. \tag{1.115}$$

The block diagram of the system can be drawn with reference to the Laplace domain using the transfer function as shown in Figure 1.22b. The transfer function and the frequency response are strictly related to each other: The second can be obtained from the first by substituting the frequency multiplied by the imaginary unit $i\lambda$ for the Laplace variable s.

Also, in the case of systems with many degrees of freedom, the equations of motion can be written with reference to the Laplace domain instead of to the frequency domain and a transfer function can be defined as

$$\mathbf{x}(s) = \mathbf{G}(s)\mathbf{f}(s), \tag{1.116}$$

where

$$\mathbf{G}(s) = (s^2\mathbf{M} + s\mathbf{C} + \mathbf{K})^{-1}.$$

A number n^2 of transfer functions is included in matrix $\mathbf{G}(s)$, so the latter is often referred to as a *transfer matrix*. To avoid confusion with the transfer matrix $\mathbf{T}$ defined in Chapter 2, here it will be referred to as a matrix of the transfer functions.

By resorting to the state-space approach, equations (1.11) and (1.14) can be transformed into the Laplace domain. Consider a linear

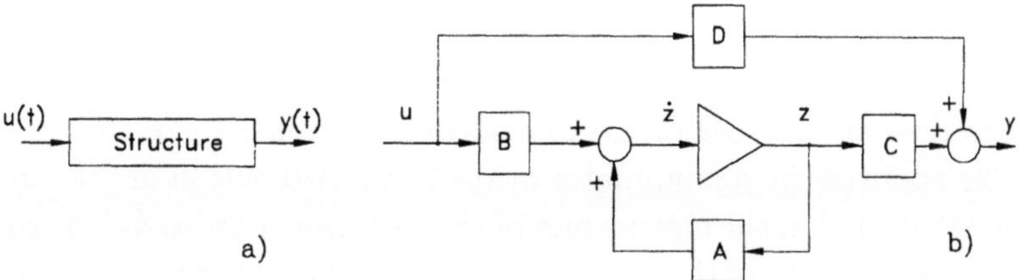

FIGURE 1.23. Block diagram of a multi-degree-of-freedom structure: (a) simplified input-output diagram; (b) detailed diagram corresponding to equations (1.11) and (1.14).

mechanical system with many degrees of freedom (Figure 1.23)[10], subject to a number r of independent inputs $\mathbf{u}(t)$ and providing a number m of outputs $\mathbf{y}(t)$ that can be different from the n state variables, but, in the case of a linear system, can be expressed in general as linear combinations of the inputs and the state variables.

The state equation (1.11) can be transformed into an algebraic equation through Laplace transform. Assuming that at time $t = 0$ the value of all state variables is zero, the state and output equations of the system in the Laplace domain are

$$\begin{cases} s\mathbf{z} = \boldsymbol{A}\mathbf{z}(s) + \boldsymbol{B}\mathbf{u}(s) \\ \mathbf{y}(s) = \boldsymbol{C}\mathbf{z}(s) + \boldsymbol{D}\mathbf{u}(s). \end{cases} \tag{1.117}$$

The equation that links the output of the system with the input then is

$$\mathbf{y}(s) = [\boldsymbol{C}\,(s\mathbf{I} - \boldsymbol{A})^{-1}\,\boldsymbol{B} + \boldsymbol{D}]\mathbf{u}(s)\,, \tag{1.118}$$

and the transfer function is

$$\mathbf{G}(s) = \boldsymbol{C}\,(s\mathbf{I} - \boldsymbol{A})^{-1}\,\boldsymbol{B} + \boldsymbol{D}\,. \tag{1.119}$$

The generic transfer function $\mathbf{G}_{ij}(s)$, which links the ith output with the jth input, can be written as the ratio of two polynomials,

$$\mathbf{G}_{ij}(s) = \frac{\beta_m s^m + \beta_{m-1} s^{m-1} + \ldots + \beta_1 s + \beta_0}{s^n + \alpha_{n-1} s^{n-1} + \ldots + \alpha_1 s + \alpha_0}\,, \tag{1.120}$$

---

[10]In system dynamics and control terminology the system is usually referred to as a *plant*. Here the more specific term *structure* will be used, as no attempt to deal with system dynamics in general is intended.

where n is the order of the system and m (with $m \leq n$) is the order of the numerator of the transfer function. The difference $n - m$ is referred to as the *pole excess* or *relative order* of the system.

The roots of the denominator of the transfer function are the poles of the system, i.e. the eigenvalues of the dynamic matrix $\boldsymbol{A}$. The roots of the numerator are the zeros. Note that the poles are characteristics of the system, and the zeros are typical of each transfer function.

The transfer functions can be written in the form

$$\mathbf{G}_{ij}(s) = k\frac{(s + z_1)(s + z_2)\dots}{(s + p_1)(s + p_2)\dots}, \tag{1.121}$$

where z_i and p_i are the zeros and poles, respectively. Note that the poles and zeros are either real or complex conjugate pairs, at least if the quadruple is real. In the case of single-degree-of-freedom systems, whose dynamic matrix is expressed by equation (1.11), if the displacement is taken as output and the force as input, it is easy to verify that the transfer function is

$$G(D) = G(s) = \frac{1}{ms^2 + cs + k},$$

which coincides with the frequency response of the system $H(\lambda)$, once s has been substituted for $i\lambda$.

1.16   Response to nonharmonic excitation

When an external force with nonharmonic time history acts on the systems, the response can be computed by resorting to Laplace or Fourier transform. There are, however, two cases in which a simple closed-form solution exists: Those of the impulse and step excitations.

When a force acts on the system for a very short time with a very large intensity, as in the case of shock loads, the impulsive model consisting of a force acting with an intensity that tends to infinity for a time tending to zero can be used. The unit-impulse function $\delta(t)$ (or Dirac's δ) is defined by the relationships

$$\begin{cases} \delta = 0 & \text{for } t \neq 0 \\ \delta \to \infty & \text{for } t = 0 \end{cases} \qquad \int_{-\infty}^{\infty} \delta(t)dt = 1. \tag{1.122}$$

The impulse excitation can be expressed as $F = f_0\delta(t)$. δ has the dimension of the reciprocal of a time $[\mathrm{s}^{-1}]$ and f_0 has the dimensions of an impulse $[\mathrm{Ns}]$. As the impulse of the function $\delta(t)$ has a unit value, the value of f_0 is that of the total impulse of force $F(t)$. The response to an impulse excitation is easily computed: It is sufficient to observe that in the infinitely short period of time in which the impulsive force acts, all other forces to which the system is subject are negligible, compared to it. The momentum theorem can be applied to compute the conditions of the system just after the impulsive force has been applied from those related to the instant that precedes its application. After the impulse, the time history can be computed from the equations governing the free behaviour of the system, obtaining

$$x(t) = \frac{f_0}{m\lambda_n} h(t) , \qquad (1.123)$$

where

$$\begin{cases} h(t) = \dfrac{1}{\sqrt{1-\zeta^2}} e^{-\zeta\lambda_n t} \sin\left(\lambda_n\sqrt{1-\zeta^2}\,t\right) , \\[2mm] h(t) = \lambda_n t e^{-\lambda_n t} , \\[2mm] h(t) = \dfrac{1}{2\sqrt{1-\zeta^2}} \left\{ -e^{-\left[\zeta+\sqrt{1-\zeta^2}\right]\lambda_n t} + e^{-\left[\zeta-\sqrt{1-\zeta^2}\right]\lambda_n t} \right\} . \end{cases}$$

The three expressions of the impulse response $h(t)$ hold for under-damped, critically damped, and overdamped systems, respectively. The impulse responses with different values of the damping factor are reported in nondimensional form in Figure 1.24.

Another case for which a closed-form solution is available is that of the response to a step excitation. The unit step function $u(t)$ can be defined by the expression

$$\begin{cases} u = 0 & \text{for } t < 0, \\ u = 1 & \text{for } t > 0, \end{cases} \qquad (1.124)$$

and is just the integral of the impulse function $\delta(t)$.

The response of the system to the excitation $F = f_0 u(t)$ can be computed by adding the solution obtained for free oscillations to the steady-state response to the constant force f_0. Also, in this case a simple expression is commonly used

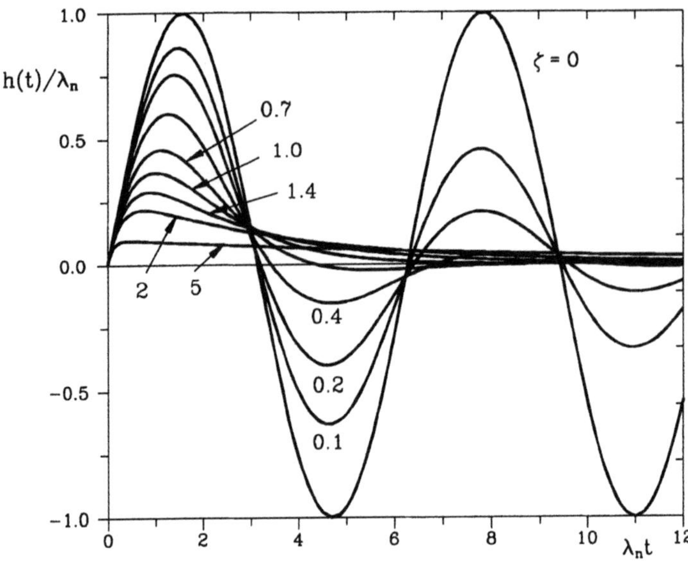

FIGURE 1.24. Response to an impulse excitation $h(t)$ for different values of the damping.

$$x(t) = \frac{f_0}{k} g(t) \,, \tag{1.125}$$

where

$$\begin{cases} g(t) = 1 - e^{-\zeta \lambda_n t} \left[\cos\left(\lambda_n \sqrt{1-\zeta^2}\,t\right) + \frac{\zeta}{\sqrt{1-\zeta^2}} \sin\left(\lambda_n \sqrt{1-\zeta^2}\,t\right) \right] , \\ g(t) = 1 - (1 - \lambda_n t) e^{-\lambda_n t} \,, \\ g(t) = 1 - \frac{1}{2} \left\{ -e^{-\left[\zeta + \sqrt{1-\zeta^2}\right]\lambda_n t} + e^{-\left[\zeta - \sqrt{1-\zeta^2}\right]\lambda_n t} \right\} . \end{cases}$$

The three expressions for the response to unit step $g(t)$ hold for underdamped, critically damped, and overdamped systems, respectively. They are plotted in nondimensional form in Figure 1.25a.

From the response to a step forcing function, some characteristics of the system that can be used to formulate performance criteria can be stated. With reference to Figure 1.25b, they are the peak time T_p (time required for the response to reach its peak value), the rise time T_r (time required for the response to rise from 10% to 90% of the steady-state value, sometimes from 5% to 95% or from 0 to 100%), the delay time T_d (time required for the response to reach

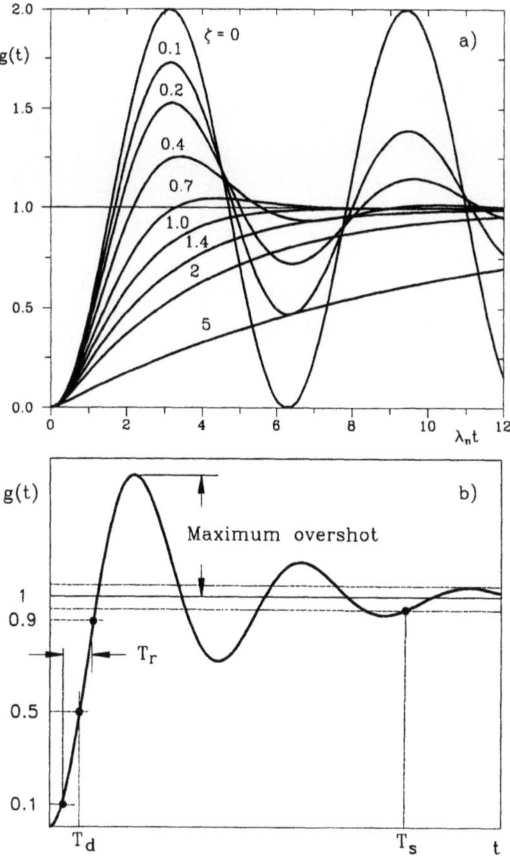

FIGURE 1.25. (a) Response to a step forcing function $g(t)$; (b) dynamic characteristics of a system with a single degree of freedom obtained from the response to a step forcing function.

50% of the steady-state value), the setting time T_s (time required for the response to settle within a certain range, usually 5% but sometimes 2%, of the steady-state value), and the maximum overshot (maximum deviation of the response with respect to the steady-state value). The last item is usually expressed as a percentage of the steady-state value.

If the excitation $F(t)$ is periodic with period T, the response can easily be computed by decomposing the forcing function in a Fourier series

$$F(t) = a_0 + \sum_{i=1}^{\infty} a_i \cos\left(\frac{2\pi i}{T}\right) + \sum_{i=1}^{\infty} b_i \sin\left(\frac{2\pi i}{T}\right). \qquad (1.126)$$

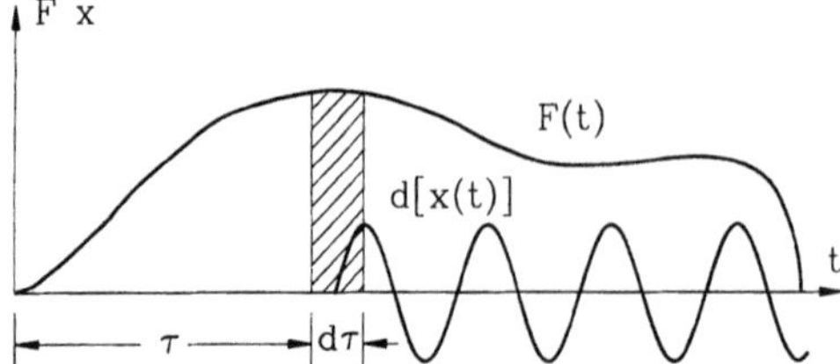

FIGURE 1.26. Response to the impulse exerted by force $F(t)$ between time τ and time $\tau + d\tau$.

Because the system is linear, the response to the polyharmonic excitation (1.126) can be obtained by adding the responses to all terms of the forcing functions. By separating the real and imaginary parts of the frequency response, the following expression for the particular integral of the equation of motion is easily obtained:

$$
x(t) = \frac{1}{k}\left\{ a_0 + \sum_{i=1}^{\infty}[a_i\Re(H(\lambda_i)) + b_i\Im(H(\lambda_i))]\cos(\lambda_i t) + \right.
$$
$$
\left. + \sum_{i=1}^{\infty}[b_i\Re(H(\lambda_i)) - a_i\Im(H(\lambda_i))]\sin(\lambda_i t)\right\}, \tag{1.127}
$$

where the frequency of the ith harmonic of the forcing function is $\lambda_i = 2\pi i/T$. In the actual computations, only a finite number of terms can be considered, and then the series must be truncated.

A different approach, applicable to both periodic and nonperiodic forcing functions, is the use of Duhamel's integral. The impulse of force $F(t)$ acting on the system, computed between time τ and time $\tau + d\tau$ (Figure 1.26) is simply $F(\tau)d\tau$. The response of the system to such an impulse can be easily expressed in the form

$$
d[x(t)] = \frac{F(\tau)d\tau}{m\lambda_n}h(t - \tau), \tag{1.128}
$$

where function $h(t)$ is the response to a unit impulse defined earlier.

The response to the forcing function $F(t)$ can be computed by adding (or better, integrating, as there is an infinity of vanishingly small terms) the responses to all the impulses taking place at all times up to time t

$$
x(t) = \int_0^t d[x(t)]d\tau = \frac{1}{m\lambda_n}\int_0^t F(\tau)h(t - \tau)d\tau. \tag{1.129}
$$

The integral of equation (1.129), usually referred to as Duhamel's integral, allows the computation of the response of any linear system to a force $F(t)$ with a time history of any type. Only in a few selected cases can the integration be obtained in closed form; however, numerical integration of equation (1.129) is simpler than the direct numerical integration of the equation of motion.

By introducing the impulse response of an underdamped system, the particular integral of the equation of motion can be expressed in the more compact form

$$x(t) = A(t)\sin\left(\sqrt{1-\zeta^2}\lambda_n t\right) - B(t)\cos\left(\sqrt{1-\zeta^2}\lambda_n t\right) , \quad (1.130)$$

where functions $A(t)$ and $B(t)$ are expressed by the following integrals:

$$A(t) = \frac{1}{m\lambda_n\sqrt{1-\zeta^2}}e^{\zeta\lambda_n t}\int_0^t F(\tau)e^{\zeta\lambda_n\tau}\cos\left(\sqrt{1-\zeta^2}\lambda_n\tau\right)d\tau,$$

$$B(t) = \frac{1}{m\lambda_n\sqrt{1-\zeta^2}}e^{\zeta\lambda_n t}\int_0^t F(\tau)e^{\zeta\lambda_n\tau}\sin\left(\sqrt{1-\zeta^2}\lambda_n\tau\right)d\tau .$$

$$(1.131)$$

An increasingly popular approach to the computation of the time history of the response from the time history of the excitation is the numerical integration of the equation of motion. It must be clearly stated that while all other approaches seen (Laplace and Fourier transforms, Duhamel's integral) can be applied only to linear systems, the numerical integration of the equation of motion can also be performed for nonlinear systems (see Chapter 3). However, any solution obtained through the numerical approach must be considered the result of a numerical experiment and usually gives little general insight to the relevant phenomena. The numerical approach does not substitute other analytical methods, but rather provides a very powerful tool to deal with cases that cannot be studied in other ways.

There are many different methods that can be used to perform the integration of the equation of motion. All of them operate following the same guidelines: The state of the system at time $t+\Delta t$ is computed from the known conditions that characterize the state of the system at time t. The finite time interval Δt must be small enough to allow the use of simplified expressions of the equation of motion without incurring errors that are too large. The mathematical simulation

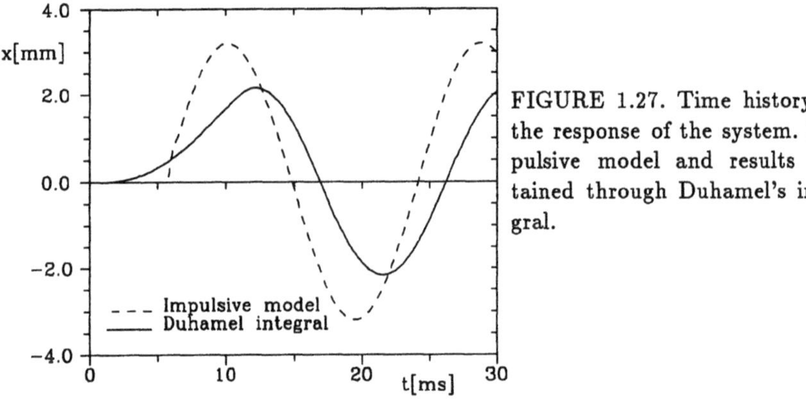

FIGURE 1.27. Time history of the response of the system. Impulsive model and results obtained through Duhamel's integral.

of the motion of the system is performed step by step, increasing the independent variable t with subsequent finite increments Δt.

Example 1-10

Check the ability of the system studied in Example 1-5 to withstand a shock corresponding to the prescriptions of MIL-STD 810 C, basic design, i.e., an acceleration of the supporting structure that increases linearly in time for 11 ms up to a value of 20 g. The stresses due to the shock must not exceed the ultimate strength of the material, 328 MN/m^2.

As the period of the free oscillations of the system is T=18.6 ms, the duration of the shock is not much shorter than the period of the free oscillations, and good accuracy cannot be expected if the shock is studied as an impulse loading. The computation will, therefore, be performed using both an impulsive model and the Duhamel integral.

The force acting on the beam is given by the mass multiplied by the acceleration. It increases in time from 0 to the value ma_{max} =3,924 N, which is reached after 11 ms. The total impulse is equal to 21.6 Ns. If damping is neglected, the impulsive model (equation (1.123) with f_0=21.6 Ns) yields an amplitude of the harmonic motion that follows the impulse $x_0 = 3.2 \times 10^{-3}$ m. The maximum value of the stress, which takes place at the clamped end, can be computed by dividing the maximum value of the bending moment klx_0 by the section modulus of the beam, obtaining $\sigma_{max} = 105.6 \times 10^6$ N/m^2.

To compute the displacement through the Duhamel integral, function $F(t)$ must be explicitly computed: F=357,000 t for $0 \leq t \leq 0.011$; F=0 for $t >$ 0.011. By neglecting the presence of damping, which gives a conservative result, functions $A(t)$ and $B(t)$ are, for the first 11 ms,

$$
A(t) = \frac{357{,}000}{m\lambda_n} \int_0^t \tau \cos\left(\lambda_n \tau\right) d\tau = \frac{357{,}000}{m\lambda_n^2} \left[\lambda_n t \sin\left(\lambda_n t\right) + \cos\left(\lambda_n t\right) - 1\right],
$$

$$
B(t) = \frac{357{,}000}{m\lambda_n} \int_0^t \tau \sin\left(\lambda_n \tau\right) d\tau = \frac{357{,}000}{m\lambda_n^2} \left[-\lambda_n t \cos\left(\lambda_n t\right) + \sin\left(\lambda_n t\right)\right].
$$

When no more exciting force is present, i.e., after 11 ms, the values of
$A(t)$ and $B(t)$ remain constant. By introducing a value of time $t=11$ ms
in the expressions of $A(t)$ and $B(t)$, the following equation for the free
motion of the system is obtained

$$
x(t) = -0.0018 \sin(\lambda_n t) - 0.0012 \cos(\lambda_n t).
$$

The response of the system is plotted in Figure 1.27. The maximum value
of the displacement is equal to 2.2 mm, and the corresponding value of
the maximum stress is 72.6 MN/m^2.
As predicted, the impulsive model in this case does not allow a good
approximation of the results, as the duration of the shock is not much
shorter than the period of the free oscillations. The value of the stress is
far smaller than the allowable value and, consequently, it is not necessary
to repeat the computation taking into account the presence of damping.

In the case of multi-degree-of-freedom systems, the response to a
generic input $u(t)$ can be expressed using the transition matrix e^{At}
in the form

$$
\mathbf{z}(t) = e^{At}\mathbf{z}_0 + \int_0^t e^{A(t-\tau)}\boldsymbol{B}\mathbf{u}(\tau)d\tau, \qquad (1.132)
$$

which can be regarded as a generalization of Duhamel's integral.
Some difficulties can be encountered in computing the transition
matrix; they increase with increasing time t and with increasing absolute value of the highest eigenvalue of the dynamic matrix A A.
Clearly, the time interval t can be subdivided into subintervals and
equation (1.132) can be applied in sequence, one subinterval after
the other. If the input is constant at the value $\mathbf{u}_0$ in the subinterval
from time t_0 to time t_1, the state $\mathbf{z}_1$ at the end can be computed
from the one at the beginning $\mathbf{z}_0$ as

$$
\mathbf{z}_1 = e^{A(t_1-t_0)}\left\{\mathbf{z}_0 + \boldsymbol{A}^{-1}\left[\mathbf{I} - e^{-A(t_1-t_0)}\right]\boldsymbol{B}\mathbf{u}_0\right\}. \qquad (1.133)
$$

If the input varies linearly from $\mathbf{u}_0$ to $\mathbf{u}_1$, the integral can be
solved in closed form, yielding

$$\mathbf{z}_1 = e^{\boldsymbol{A}(t_1 - t_0)} \left(\mathbf{z}_0 + \mathbf{R}\mathbf{u}_0 + \mathbf{S}\mathbf{u}_1 \right), \qquad (1.134)$$

where

$$\mathbf{R} = \boldsymbol{A}^{-1} \left\{ \mathbf{I} - \frac{1}{t_1 - t_0} \boldsymbol{A}^{-1} \left[\mathbf{I} - e^{-\boldsymbol{A}(t_1 - t_0)} \right] \right\} \boldsymbol{B},$$

$$\mathbf{S} = \boldsymbol{A}^{-1} \left\{ \frac{1}{t_1 - t_0} \boldsymbol{A}^{-1} \left[\mathbf{I} - e^{-\boldsymbol{A}(t_1 - t_0)} \right] - e^{-\boldsymbol{A}(t_1 - t_0)} \right\} \boldsymbol{B}.$$

By resorting to the left and right eigenvectors, the equations of motion can be easily uncoupled. Each time history of the modal state variables is of the type

$$\overline{z}_i(t) = \overline{z}_{i_0} e^{s_i t} + \int_0^t e^{s_i(t-\tau)} \mathbf{q}_{Li}^T \boldsymbol{B}\mathbf{u}(\tau) d\tau. \qquad (1.135)$$

1.17   Short account of random vibrations

There are many cases in which the forcing function acting on the system is known only statistically; typical examples are seismic excitation on buildings, excitation on the structures of ground vehicles due to road irregularities, and excitation of the structure of ships due to sea waves. In all these cases and in the many others that occur in many fields of technology, it is possible to measure experimentally the time history of the excitation for a more or less prolonged time and to perform a statistical analysis. The study of the response of dynamic systems to an excitation of this kind is quite complex and requires a good background in statistics. There are many excellent books devoted to the subject of random vibrations where the interested reader can find a more complete analysis. In this section only a brief outline, mainly on the qualitative aspects of the relevant phenomena, will be given.

Consider a function $y(t)$ characterized by a random behaviour. Given a sample whose duration is T, the average value, the r.m.s. value and the variance, usually referred to by the symbol σ^2, of the function are defined as

$$\bar{y} = \frac{1}{T} \int_0^T y(t)dt,$$

$$y_{rms} = \sqrt{\bar{y}^2} = \sqrt{\frac{1}{T} \int_0^T y^2(t)dt}, \qquad (1.136)$$

$$\sigma^2 = \overline{(y - \bar{y})^2} = \frac{1}{T} \int_0^T \left[y(t) - \bar{y}\right]^2 dt.$$

The standard deviation σ is the square root of the variance. Usually both the forcing function and the response are assumed to have a vanishing average value. Also, in the case of random vibrations it is preferable to separate the static response of the system under the action of a constant force equal to the average value of the actual force from the dynamic problem. This approach is possible only in the case of linear systems. The variance then coincides with the square of the r.m.s. value and the standard deviation with the r.m.s. value.

The statistical parameters just defined depend on the sample used for the analysis. If the phenomenon studied is stationary, i.e., if its characteristics do not change when the study is performed starting at different times, and ergodic, i.e., any sample can be considered typical of the whole set of available samples, the average, the r.m.s. value, the variance, and all other statistical parameters can be considered independent of the particular sample used for their computation. Obviously, these assumptions are oversimplifications of a more complex phenomenon, but in most cases they allow for results that are in close accordance with experimental evidence. If the phenomenon is stationary and ergodic, the values of the average and of the variance computed in time T are coincident with the same values obtained for T tending to infinity.

Another very important statistical parameter is the autocorrelation function

$$\Psi(\tau) = \lim_{T \to \infty} \frac{1}{T} \int_0^T y(t)y(t + \tau)dt, \qquad (1.137)$$

which states how the value of function $y(t)$ at time t is linked with the value it takes at time $t + \tau$. In the case of an ideal random phenomenon, in which the value of function $y(t)$ in every instant is completely independent of the value it takes in any other instant, the autocorrelation is equal to zero for every value of τ except $\tau = 0$, where its value is equal to the square of the r.m.s. value of function $y(t)$.

Another characteristic of a random variable is the power spectral density $S(\lambda)$, defined as the Fourier transform of the autocorrelation function

$$S(\lambda) = \int_{-\infty}^{\infty} \Psi(\tau)e^{i\lambda\tau}d\tau. \tag{1.138}$$

The integral of function $S(\lambda)$ is the variance of function $y(t)$, i.e., if the average value is equal to zero, the square of its r.m.s. value

$$y_{rms} = \sqrt{\int_{-\infty}^{\infty} S(\lambda)d\lambda}. \tag{1.139}$$

If the random vibrations are excited by a force, the dimension of the power spectral density $S(\lambda)$ is that of the square of a force divided by a frequency. In S.I. units it is therefore measured in $N^2/(rad/s)$ $= N^2 s/rad$ or in N^2/Hz. If the system is excited by the motion of the supporting point, the forcing function is an acceleration and its power spectral density is measured in $(m/s^2)^2/(rad/s) = m^2/s^3 rad$ or in g^2/Hz.

The simplest type of random excitation is a random forcing function with constant power spectral density. This type of forcing function, which contains all possible frequencies in the same measure, is often referred to as *white noise*. It is just a mathematical model, as its r.m.s. value would be infinitely large. Its spectrum should extend for all the frequency field from 0 to infinity. Its autocorrelation function has a zero value for all values of τ and goes to infinity for $\tau = 0$. It is, therefore, a Dirac impulse function $\delta(\tau)$. Often the power spectral density is assumed to have the shape of a trapezium in a bilogarithmic plane. In the central frequency field, it is, consequently, constant, and the forcing function has the characteristics of white noise.

A random forcing function is usually defined as a narrow-band or wide-band excitation, depending on the width of the frequency field involved.

The quantities defined earlier are not yet sufficient to completely characterize a random forcing function. It is also necessary to define a function expressing the probability density function $p(y)$ related to the amplitude. Usually, such a function is assumed to be a normal or Gaussian probability distribution.

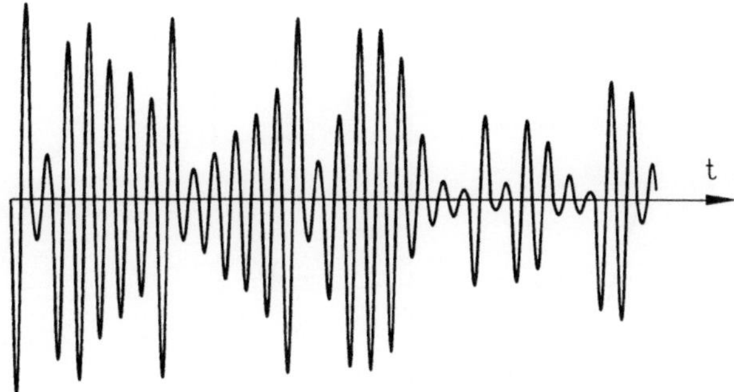

FIGURE 1.28. Pattern of the time history of the response of a lightly damped system to a random forcing function of the white-noise type.

Consider a linear system with a single degree of freedom on which a random forcing function is acting, which is stationary, ergodic, and characterized by a normal probability distribution. In such conditions the average (which is assumed to be equal to zero), the variance, and the power spectral density characterize completely the forcing function. The behaviour of the system is completely characterized by its frequency response $H(\lambda)$, which is complex if the system is damped.

The response, which can be measured in terms of displacement $x(t)$, velocity, or acceleration, has itself a random nature, with the same characteristics of stationarity and ergodicity and the same normal probability distribution as the forcing function. Also, the mean value of the response is equal to zero. The power spectral density and the r.m.s. value of the response can be computed directly from the power spectral density of the excitation and the frequency response of the system:

$$S_x(\lambda) = S_f(\lambda)|H(\lambda)|^2, \qquad x_{rms} = \sqrt{\int_{-\infty}^{\infty} S_x(\lambda)d\lambda}. \qquad (1.140)$$

A linear system with a single degree of freedom acts as a sort of filter, amplifying the input signal in a very narrow band about the resonant frequency and cutting off all other components. The lower is the damping of the system, the narrower is the passing band. The response is consequently a narrow-band random vibration, particularly if the system is lightly damped. The time history of the response follows the pattern sketched in Figure 1.28: an oscillation that is almost harmonic with strongly variable amplitude and slightly variable

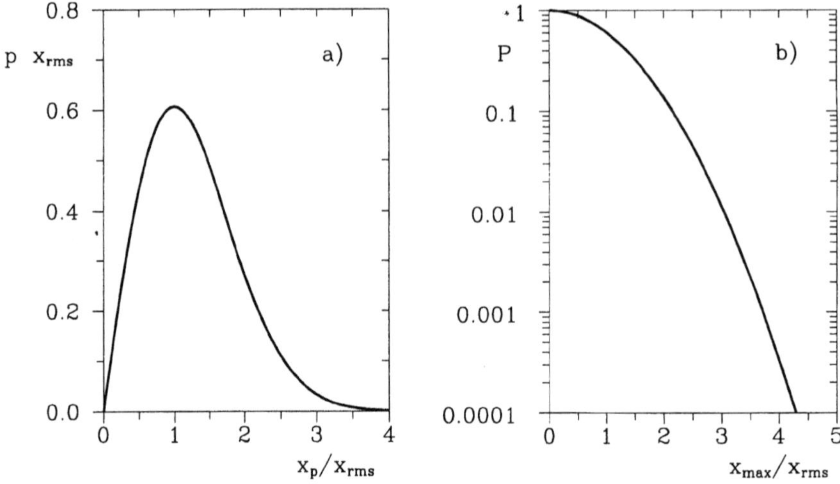

FIGURE 1.29. Probability density of the peak values of (a) a narrow-band random response and (b) probability that one of the peaks is higher than a given value x_{max}. Note the logarithmic scale in Figure 1.29b.

frequency. The frequency is very near the natural frequency of the system.

Under the aforementioned conditions, the probability density of the peaks of the response $x(t)$ and the probability that a peak is higher than the generic value x_{max} can be shown to be, respectively

$$p\left(\frac{x_p}{x_{rms}}\right) = \frac{x_p}{x_{rms}^2}e^{-x_p^2/2x_{rms}^2} \ , \qquad P\left(\frac{x_{max}}{x_{rms}}\right) = e^{-x_{max}^2/2x_{rms}^2} \ .$$

$$(1.141)$$

Equations (1.141) are plotted in Figure 1.29. From Figure 1.29a it is clear that very low and very high values of the peaks are unlikely and that the maximum probability is that of having peaks roughly as high as the r.m.s. value. From the second equation (1.141), it is possible to directly compute the probability that the maximum amplitude of the response reaches any given value in a given working time. Because the response is a narrow-band random signal, its frequency is very near the natural frequency of the system and the number of oscillations taking place in time t is $t\lambda_n/2\pi$. The probability that in one of these periods the peak value is greater than x_{max} is

$$P\left(\frac{x_{max}}{x_{rms}}\right) = \frac{t\lambda_n}{2\pi}e^{-x_{max}^2/2x_{rms}^2} \ . \qquad (1.142)$$

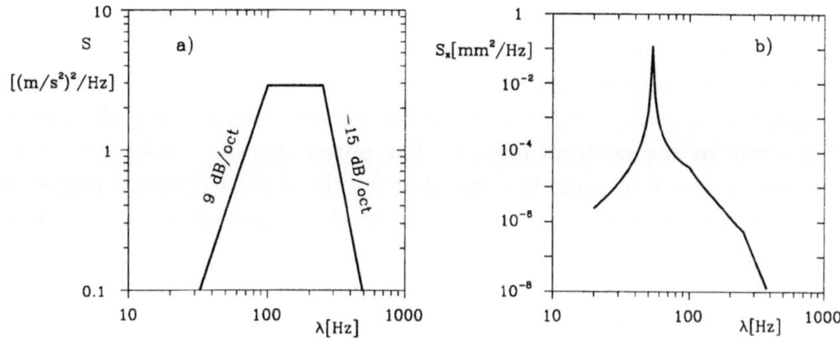

FIGURE 1.30. Power spectral density of (a) the excitation and (b) the response.

Example 1-11

Check the ability of the system studied in Example 1-5 to withstand for a time of 30 s the random excitation provided by the motion of the supporting point defined in Figure 1.30a. The power spectral density is constant in a frequency field between 100 and 250 Hz at a value of 0.03 g^2/Hz, increases at 9 dB/oct between 20 and 100 Hz, and decreases at -15 dB/oct between 250 and 2,000 Hz (Figure 1.30a).

The stresses must not exceed the ultimate strength of 328 MN/m^2 divided by a safety factor of 1.575, or the yield strength of 216 MN/m^2 divided by a safety factor of 1.155, or the allowable fatigue strength for the prescribed duration. The latter for 10^7 cycles is 115 MN/m^2. The relative displacement between the instrument at the end of the beam and the supporting structure must not exceed 4 mm.

The power spectral density of the excitation includes the natural frequency of the system. As a consequence, the computation of the response cannot be performed when neglecting the presence of damping, which will be assumed to be of the hysteretic type. For safety, a low value of the loss factor $\eta = 0.01$ will be assumed, so that conservative results will be obtained.

Consider the first frequency field. The relationship between power spectral density and frequency is linear in a bi-logarithmic plane; its expression is then of the type $S(\lambda) = a\lambda^n$. As the power spectral density increases of 9 dB/oct and at a frequency of 100 Hz its value is 0.03 g^2/Hz $= 2.88$ $(m/s^2)^2$/Hz, its expression is $S(\lambda) = 2.88 \times 10^{-6}\lambda^3$, where frequencies are measured in Hz. In the field between 100 and 250 Hz, the power spectral density is constant at a value 2.88 $(m/s^2)^2$/Hz, while between 250 Hz and 2,000 Hz, where the power spectral density decreases at -15 dB/oct, the expression $S(\lambda) = 2.81 \times 10^{12}\lambda^{-5}$ is readily obtained.

The r.m.s. value of the acceleration is obtained by integrating the power spectral density between 20 and 2,000 Hz: $a_{rms} = 26.15$ m/s^2.

Because the response has to be computed in terms of relative displacement, the frequency response (1.73) can be used, substituting the inertial force ma for the external force F. The power spectral density of the response can be immediately computed for the three frequency ranges by multiplying the power spectral density of the excitation by the square of the frequency response

$$S_x(\lambda) = 2.23 \times 10^{-16} \frac{\lambda^3}{\left[1 - \left(\frac{\lambda}{\lambda_n}\right)^2\right]^2 + \eta^2},$$

$$S_x(\lambda) = 2.23 \times 10^{-10} \frac{1}{\left[1 - \left(\frac{\lambda}{\lambda_n}\right)^2\right]^2 + \eta^2},$$

$$S_x(\lambda) = 2.17 \times 10^2 \frac{\lambda^{-5}}{\left[1 - \left(\frac{\lambda}{\lambda_n}\right)^2\right]^2 + \eta^2}.$$

where frequencies are expressed in Hz and power spectral densities in m^2/Hz. The power spectral density of the response is plotted in Figure 1.30b. The r.m.s. value of the response can be computed by integrating the power spectral density: $x_{rms} = 0.539$ mm.

By comparing the contributions of the three integrals related to the various frequency ranges, it is clear that the only frequency field that contributes significantly to the response is the first one, because it was predictable as the natural frequency of the system falls in it and the response is of the narrow-band type.

The r.m.s. value of the stress, computed in the same way as in Example 1-9, is $\sigma_{rms} = 17.8 \times 10^6$ N/m^2. The narrow-band response can be assimilated to a harmonic oscillation with random varying amplitude and frequency equal to the natural frequency 53.7 Hz. The total number of cycles occurring in the prescribed 30 s is 1,611.

As the r.m.s. value of the stress is 6.5 times smaller than the fatigue strength at 10^7 cycles, the third condition is surely satisfied. For the first condition, the allowable stress is 208 MN/m^2. The r.m.s. value is, therefore, 11.7 times smaller than the allowable value of the stress. The probability that the stress reaches the allowable value in the prescribed 30 s can be computed using equation (1.142), obtaining $P = 3.6 \times 10^{-27}$. The situation is slightly more critical for the second condition, regarding the yield strength: The allowable strength is 187 MN/m^2, the ratio between the allowable stress and the r.m.s. value is 10.5, and the probability of reaching the critical condition in the prescribed time is 1.7×10^{-21}.

The probability that the structure fails under the effects of the random excitation prescribed is extremely low. For the critical condition on the displacement, the probability is also very low, namely, 1.8×10^{-9}.

1.18   Concluding examples

Example 1-12: Dynamic vibration absorber
A dynamic vibration absorber is basically a spring-mass-damper system
that is added to any vibrating system with the aim of reducing the amplitude of the vibrations of the latter. If the damper or the spring is missing,
an undamped vibration absorber or a Lanchester damper (springless vibration absorber) is obtained.

Consider a system consisting of a mass m suspended on a spring with
stiffness k on which is acting a force varying harmonically in time with
frequency λ and maximum amplitude f_0. The vibration absorber, consisting of a second mass m_s, a spring of stiffness k_s, and a damper with
damping coefficient c, is connected to mass m (Figure 1.31a). The equation yielding the amplitude of the harmonic response of the system is

$$
\left[-\lambda^2 \begin{bmatrix} m_s & 0 \\ 0 & m \end{bmatrix} + \begin{bmatrix} k_s & -k_s \\ -k_s & k_s + k \end{bmatrix} + i\lambda \begin{bmatrix} c & -c \\ -c & c \end{bmatrix} \right] \left\{ \begin{array}{c} x_{s_0} \\ x_0 \end{array} \right\} = \left\{ \begin{array}{c} 0 \\ f_0 \end{array} \right\}.
$$

By introducing the mass ratio $\mu = m_s/m$, the stiffness ratio $\chi = k_s/k$, the
tuning ratio $\tau = \chi/\mu$, and the nondimensional frequency $\lambda^* = \lambda/\lambda_n = \lambda\sqrt{m/k}$, the frequency response $H_{22}(\lambda)$ can be easily computed:

$$
|H_{22}| = \frac{1}{k} \sqrt{ \frac{\left(\tau - \lambda^{*2}\right)^2 + c^{*2}\lambda^{*2}}{f^2(\lambda^*) + c^{*2}\lambda^{*2} g^2(\lambda^*)} },
$$

where $f(\lambda^*) = \lambda^{*4} - \lambda^{*2}(1+\chi+\tau)$, $g(\lambda^*) = 1 - \lambda^*(1+\mu)$ and $c^* = c/\mu\sqrt{km}$.
The tuning ratio is the square of the ratio between the natural frequency
of the original system λ_n and that of the vibration absorber. If the vibration absorber is undamped, the amplitude of the motion of mass m
is vanishingly small if the square of the nondimensional frequency of the
excitation coincides with the tuning ratio. The frequency response computed for two particular values of the mass and the tuning ratios is shown
in Figure 1.31b, for curve $c = 0$. The presence of an undamped vibration
absorber is successful in completely damping the vibration at a given frequency but produces two new resonance peaks at the frequencies at which
function $f(\lambda^*)$ vanishes (Figure 1.31c).
The working of the undamped vibration absorber can be easily understood by noting that at the frequency at which the vibration absorber is
tuned, the motion of mass m_s is large enough to produce a force on mass
m that balances force F. Consequently, the amplitude of the motion of
mass m_s increases when the mass ratio μ decreases and tends to infinity
when m_s tends to zero.

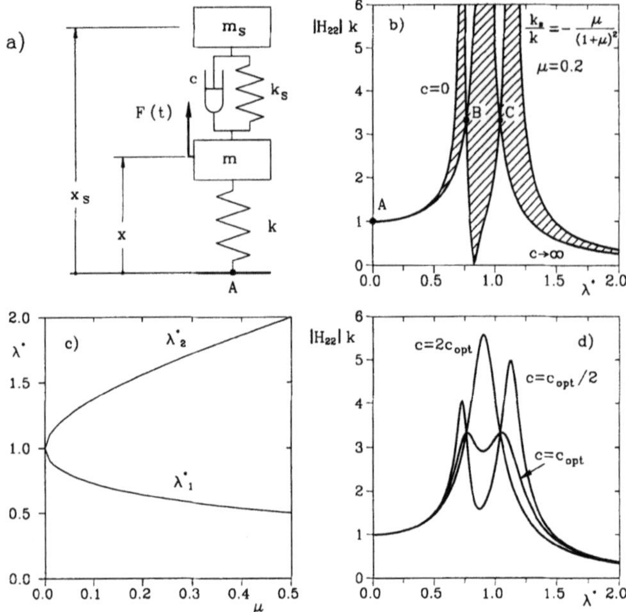

FIGURE 1.31. Vibration absorber: (a) Sketch of the system; (b) limiting cases (damping tending to zero and infinity) for systems with optimum tuning; (c) natural frequencies of the undamped system as functions of the mass ratio μ; (d) amplitude of the response of the system in (b), but with three different values of damping.

If the amplitude of motion of mass m is to be reduced also outside a narrow range near frequency λ_n, the use of a damper is mandatory.

All response curves, obtained with any value of c, pass through points A, B, and C and lie in the shaded zone in Figure 1.31b bounded by the two limiting cases of the undamped system and that with infinitely large damping, which coincides with a system with a single degree of freedom with mass $m + m_s$ and stiffness k. Such curves have a maximum in the zone included between points B and C in the case of high damping and two maxima outside the field BC in the case of small damping.

A reasonable way of optimizing the vibration absorber is to look for a value of the damping causing the maxima to coincide with points B and C and to tune the system (i.e., select the value of k_s) in a way so as to obtain the same value of the response in B and C.

The latter condition can be shown to be obtained (Den Hartog, *Mechanical vibrations*, McGraw-Hill, New York, 1956, pp. 93 and following) for the optimum value of the tuning ratio $\chi_{opt} = \mu/(1 + \mu)^2$. Strictly speaking, no value of the damping can cause the two peaks to be located simultaneously at points B and C. The value of the damping allowing one to meet this condition with good approximation, and the approximated value of the maximum amplitude are

$$c_{opt} = \sqrt{km}\sqrt{\frac{3\mu^3}{2(1+\mu^3)}}\,, \qquad |H|_{max} = \sqrt{1+\frac{2}{\mu}}.$$

The frequency response of a system with mass ratio $\mu = 0.1$, optimum tuning ratio, and three values of damping is plotted in Figure 1.31d.

Another different type of dynamic vibration absorber is the so-called Lanchester damper, which consists of a mass connected to the system through a damper. Originally it had a sort of dry-friction damper, but if the damper is of the viscous type, it is basically a damped vibration absorber without restoring spring. The frequency response can be easily computed, obtaining

$$|H_{22}| = \frac{1}{k}\sqrt{\frac{\lambda^{*2}c^{*2}}{\lambda^{*2}(\lambda^{*2}-1)^2 + \lambda^{*2}[1-(\mu+1)\lambda^{*2}]^2}}.$$

Following the procedure used for the preceding case, the optimum value of the damping and the corresponding maximum value of the frequency response can be computed:

$$c_{opt} = \sqrt{km}\sqrt{\frac{2\mu^2}{(2+\mu)(1+\mu)}}\,, \qquad |H|_{max} = 1+\frac{2}{\mu}.$$

Example 1-13: Quarter-car model

One of the simplest models used to study the dynamic behaviour of motor vehicle suspensions is the so-called quarter car with two degrees of freedom (Figure 1.32a). The upper mass simulates the part of the mass of the car body (the sprung mass) that can be considered supported by a given wheel, and the lower one simulates the wheel and all the parts that can be considered as rigidly connected with the unsprung mass. The two masses are connected by a spring-damper system simulating the spring of the suspension and the shock absorber. The unsprung mass is connected to the ground with a second spring simulating the stiffness of the tire. The point at which the tire contacts the ground is assumed to move in a vertical direction with a given law $h(t)$, in order to simulate the motion on uneven ground. Assume the following values of the parameters: Sprung mass $m_s = 250$ kg; unsprung mass $m_u = 25$ kg; spring stiffness $k = 25$ kN/m; tire stiffness $k_t = 100$ kN/m; damping coefficient of the shock absorber $c = 2{,}150$ Ns/m.

The following analyses will be performed: (1) modal analysis; (2) computation of the frequency response; and (3) computation of the response of the system due to the road excitation when traveling at a speed of 30 m/s on a road whose surface can be considered normal following ISO (International Standards Organization) standards.

Modal analysis. If coordinates x_s and x_u are defined with reference to an inertial frame, the equation yielding the response of the system to a harmonic excitation with frequency λ and amplitude h_0 is

$$\left[-\lambda^2 \begin{bmatrix} m_s & 0 \\ 0 & m_n \end{bmatrix} + \begin{bmatrix} k & -k \\ -k & k + k_t \end{bmatrix} + i\lambda \begin{bmatrix} c & -c \\ -c & c \end{bmatrix} \right] \begin{Bmatrix} x_{s_0} \\ x_{n_0} \end{Bmatrix} = \begin{Bmatrix} 0 \\ k_t h_0 \end{Bmatrix}$$

where

$$\mathbf{M} = 25 \begin{bmatrix} 10 & 0 \\ 0 & 1 \end{bmatrix}, \qquad \mathbf{K} = 25{,}000 \begin{bmatrix} 1 & -1 \\ -1 & 5 \end{bmatrix}, \qquad \mathbf{C} = 2{,}150 \begin{bmatrix} 1 & -1 \\ -1 & 1 \end{bmatrix},$$

$$\mathbf{f} = 100{,}000 h_0 \begin{Bmatrix} 0 \\ 1 \end{Bmatrix}, \qquad \mathbf{D} = \mathbf{M}^{-1}\mathbf{K} = \begin{bmatrix} 100 & -100 \\ -1{,}000 & 5{,}000 \end{bmatrix}.$$

The characteristic equation that allows the computation of the natural frequencies of the undamped system is $\lambda^4 - 5{,}100\lambda^2 + 400{,}000 = 0$. Its solutions and the values of the natural frequencies are $\lambda_1^2 = 79.676$ ($\lambda_1 = 8.926$ rad/s $= 1.421$ Hz), $\lambda_2^2 = 5{,}020.32$, and ($\lambda_2 = 70.85 = 11.28$ Hz), respectively.

The eigenvectors and the modal masses are then easily computed. The latter are 6,052.397 and 25.1033, respectively. These values can be used to normalize the eigenvectors, obtaining

$$\boldsymbol{\Phi} = \begin{bmatrix} 0.062346 & -0.0040564 \\ 0.012854 & 0.199588 \end{bmatrix}.$$

The modal masses then have unit values, while the modal stiffnesses are equal to the squares of the natural frequencies. The modal damping matrix is not diagonal and can be split into its proportional and nonproportional parts:

$$\overline{\mathbf{C}}_p = \begin{bmatrix} 5.4596 & 0 \\ 0 & 89.1627 \end{bmatrix}, \qquad \overline{\mathbf{C}}_{np} = \begin{bmatrix} 0 & -22.0634 \\ -22.0634 & 0 \end{bmatrix}.$$

The inverse modal transformation yields the following expressions of the two parts of the original damping matrix

$$\mathbf{C}_p = \begin{bmatrix} 1.445 & -424 \\ -424 & 2{,}221 \end{bmatrix}, \qquad \mathbf{C}_{np} = \begin{bmatrix} 705 & -1{,}726 \\ -1{,}726 & -71 \end{bmatrix}.$$

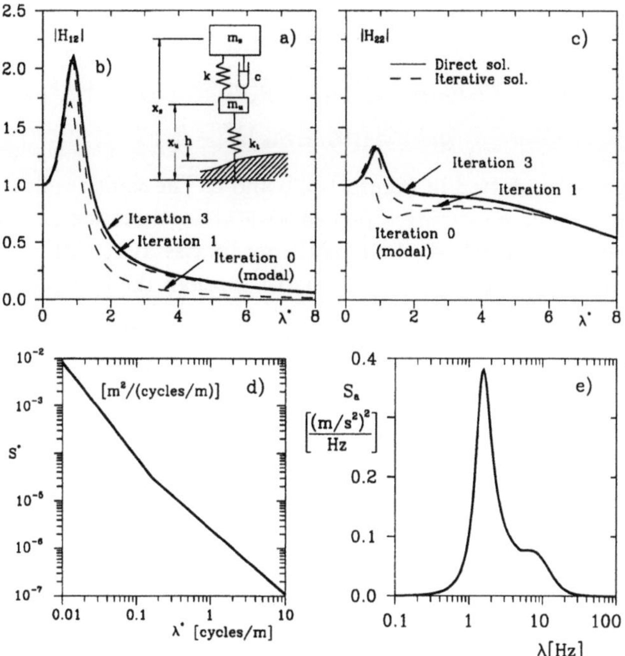

FIGURE 1.32. (a) Scheme of the quarter car model; (b), (c) frequency response; (d) power spectral density of a normal road following ISO standards; (e) power spectral density of the acceleration at a speed of 30 m/s on the road profile whose power spectral density is given in (d).

The distribution of the damping is far from proportional and the nonproportional damping matrix is nonpositive definite. Matrix $\mathbf{C}_p$ is, however, strictly proportional, with coefficients $\alpha = 4.084$ and $\beta = 0.01696$. The modal forces are then obtained:

$$\bar{\mathbf{f}} = \mathbf{\Phi}^T \mathbf{f} = h_0 \left\{ \begin{array}{c} 1{,}285 \\ 19.959 \end{array} \right\}.$$

Computation of the frequency response. The frequency response of the system can be computed directly from the equation of motion, obtaining

$$|H_{12}| = \frac{x_{s_0}}{h_0} = k_p \sqrt{\frac{k^2 + c^2\lambda^2}{f^2(\lambda)c^2\lambda^2 g^2(\lambda)}},$$

$$|H_{22}| = \frac{x_{n_0}}{h_0} = k_p \sqrt{\frac{(k - m\lambda^2)^2 + c^2\lambda^2}{f^2(\lambda)c^2\lambda^2 g^2(\lambda)}},$$

where

$$f(\lambda) = m_s m_n \lambda^4 - [k_p m_s + k(m_s + m_n)]\lambda^2 + kk_p \,,$$

$$g(\lambda) = (m_s + m_n)\lambda^2 - k_p \,.$$

The response is plotted in Figure 1.32b and c. The same result can also be obtained through an iterative procedure starting from modal uncoupling. Using the Jacobi method, the modal coordinates at the ith iteration can be obtained from those at the $(i-1)$-th through the formulae

$$\eta_1^{(i)} = \frac{\bar{f}_1 - i\lambda\bar{C}_{12}\eta_2^{(i-1)}}{\bar{K}_1 - \lambda^2\bar{M}_1 + i\lambda\bar{C}_{11}} \,, \qquad \eta_2^{(i)} = \frac{\bar{f}_2 - i\lambda\bar{C}_{21}\eta_1^{(i-1)}}{\bar{K}_2 - \lambda^2\bar{M}_2 + i\lambda\bar{C}_{22}} \,.$$

The response, computed after the first and third iterations, is plotted in Figure 1.32b and c. Three iterations are enough to obtain a very good approximation even in the current case, where damping is neither small nor close to proportional.

Response to random excitation (road excitation). The power spectral density of a normal road following ISO standards is shown in Figure 1.32d. The power spectral density of a road profile S^* is usually measured in $\mathrm{m}^2/(\text{cycles/m})$ and is a function of the space frequency λ^*, in cycles/m. Once the speed V of the vehicle has been stated, it is possible to obtain the various quantities defined with reference to the time frequency from the space frequency through the formulas $\lambda = V\lambda^*$, $S = S^*/V$.

The curve of Figure 1.32d can be expressed by the equation $S^* = a\lambda^{*^n}$, where constants a and n are $a = 8.33 \times 10^{-7}$ m and $n = -2$ if $\lambda^* \leq 1/6$ cycles/m, i.e., for road undulations with wavelength greater than 6 m, and $a = 2.58 \times 10^{-6}$ $\mathrm{m}^{1,63}$, $n = -1.37$ if $\lambda^* > 1/6$ cycles/m, i.e., for short wavelength irregularities.

The power spectral density (with reference to the time frequency) is $S = a'\lambda^n$ where $a' = a/V^{n+1}$. At a speed of 30 m/s it follows $a' = 2.5 \times 10^{-5}$ m^3 s^{-2} for $\lambda \leq 5$ Hz, $a' = 9.07 \times 10^{-6}$ $\mathrm{m}^{2,37}$ for $\lambda > 5$ Hz. The power spectral density of the acceleration of the sprung mass can be easily computed by multiplying the power spectral density of the road profile by the square of the inertance of the suspension. Because the inertance is equal to the dynamic compliance H_{12} (computed in 2) multiplied by λ^2, or, better, by $4\pi^2\lambda^2$ if λ is measured in Hz, it follows that

$$S_a = S\,16\pi^4\lambda^4 H_{12}^2 = 16\pi^4 a'\lambda^{n+4} H_{12}^2 \,.$$

To obtain the power spectral density of the acceleration in $(\mathrm{m/s}^2)^2/\mathrm{Hz}$, the frequency must be expressed in Hz, while in the equation yielding the dynamic compliance H_{12}, it is expressed in rad/s. The result is shown in Figure 1.32e.

The r.m.s. value of the acceleration can be obtained by integrating the power spectral density and extracting the square root $a_{rms} = 1.21$ m/s^2 $= 0.123$ g.

1.19 Exercises

Exercise 1-1

A ballistic pendulum of mass m and length l is struck by a bullet, having a mass m_b and velocity v, when it is at rest. Assuming that the bullet remains in the pendulum and neglecting damping, compute the frequency and amplitude of the oscillations.

If a viscous damper with damping coefficient c is applied to the pendulum, compute the maximum amplitude, the logarithmic decrement and the time needed to reduce the amplitude to $1/100$ of the original value. Data: $m = 100$ kg; $m_b = 0.05$ kg, $v = 200$ m/s, $l = 3$ m, $c = 6$ Ns/m.

Exercise 1-2

A tubular cantilever beam has a length l and inner and outer diameters d_i and d_o. At the end of the beam a mass m has been attached. Compute the dynamic compliance of the system using the structural damping model and substituting the structural damping with a constant viscous damping. Compare the results obtained. Data: $l = 1$ m, $d_i = 60$ mm, $d_o = 80$ mm, $m = 30$ kg, $E = 2.1 \times 10^{11}$ N/m^2, $\eta = 0.05$.

Exercise 1-3

A machine whose mass is m contains a rotor with an eccentric mass m_e at a radius r_e running at speed ω. Compute the stiffness of the supporting elements in such a way that the natural frequency of the machine is equal to $1/8$ of the forcing frequency due to the rotor; compute the maximum amplitude of the force transmitted to the supports at all speeds up to the operating speed, if the damping factor of the suspension is $\zeta = 0.1$.

Add a second unbalanced rotor identical to the first running in the same direction at a slightly different speed ω'. Write the equation of motion and show that a beat takes place. Compute the frequency of the beat and plot the time history of the response (assuming that the free part of the response has already decayed away) starting from the time at which the two unbalances are in phase. Data: $m = 500$ kg, $m_e = 0.050$ kg, $r_e = 1$ m, $\omega = 3,000$ rpm, $\omega' = 3,010$ rpm.

Exercise 1-4

Consider the same machine of Exercise 1-3 (with a single rotor). Use an elastomeric supporting system that, when tested at the frequencies of 50 and 200 Hz is found to have a stiffness of 500 kN/m and 800 kN/m, respectively, and a damping of 980 Ns/m and 450 Ns/m, respectively. Using the models of complex stiffness and complex damping to model the supporting system, compute the force exerted by the machine on the supporting structure at all speeds up to the maximum operating speed.

Exercise 1-5

A spring-mass-damper system is excited by a force whose time history is given by the equation $F = |f_0 \sin(\lambda t)|$. At time $t = 0$ the system is at rest in the equilibrium position. Compute the time history of the response (a) by numerical integration and (b) by computing a Fourier series for the forcing function $F(t)$. Data: $m = 1$ kg, $k = 1{,}000$ N/m, $c = 6$ Ns/m, $f_0 = 2$ N, $\lambda = 10$ rad/s.

Exercise 1-6

A spring-mass system is excited by a shock load with a duration of 11 ms and a linearly increasing intensity with peak acceleration of 20 g. Compute the response through the impulse model and Duhamel integral and compare the results. Repeat the computations, assuming that the duration of the impulse is 1.1 ms and the peak acceleration is 200 g. Perform the numerical integration of the equation of motion, with time steps of 0.05, 0.2, and 1 ms and compare the results. Data: $m = 20$ kg, $k = 2 \times 10^6$ N/m.

Exercise 1-7

A spring-mass-damper system is excited by a force whose time history is random. The power spectral density of the excitation is constant at a value of 10,000 N2/Hz between the frequencies of 50 and 200 Hz. At lower frequencies it increases at 12 dB/oct, while a decay of 12 dB/oct takes place at frequencies higher than 200 Hz. Compute the power spectral density of the response and the r.m.s. values of both excitation and response. Data: $m = 10$ kg, $k = 5$ MN/m, $c = 3$ kNs/m.

Exercise 1-8

Consider a system with two degrees of freedom, made by masses m_1 and m_2, connected by a spring k_{12} between each other and springs k_1 and k_2 to point A. Compute the natural frequencies, mode shapes, the modal masses, and stiffnesses. Plot the eigenvectors in the space of the configurations and show that the mode shapes are m-orthogonal and k-orthogonal.

Compute the modal participation factor for an excitation due to the motion of point A. Data: $m_1 = 5$ kg, $m_2 = 10$ kg, $k_1 = k_2 = 2$ kN/m, $k_{12} = 4$ kN/m.

Exercise 1-9

Compute the time history of the system of Exercise 1-8 for an excitation due to the motion of the supports $x_A = x_{A_0}\sin(\lambda t)$. Compute the time history of the response using physical and modal coordinates (neglect the free motion part of the response). Data: $x_{A_0} = 5$ mm, $\lambda = 30$ rad/s.

Exercise 1-10

Add a hysteretic damping with loss factor $\eta = 0.05$ to all springs of the system of Exercise 1-8.

Plot the four values of the dynamic compliance matrix as a function of the driving frequency. The system is excited by a shock applied through the supporting point A. The acceleration of point A has the same time history described in Example 1-9. Compute through numerical integration in time without resorting to modal coordinates the time histories of the displacement of both masses. Repeat the computation using modal coordinates and Duhamel integral.

Exercise 1-11

Study the effect of an undamped vibration absorber with mass $m_a = 0.8$ kg applied to mass m_1 of Exercise 1-8. Tune it on the first natural frequency of the system. Plot the dynamic compliance H_{22} with and without the dynamic vibration absorber.

Add a viscous damper between mass m_1 and the vibration absorber. Compute the response of the system with various values of the damping coefficient, trying to minimize the amplitude of the displacement of mass m_2 in a range of frequency between 5 and 30 rad/s.

Exercise 1-12

Add a viscous damper $c_{12} = 40$ Ns/m between masses m_1 and m_2 of the system of Exercise 1-9. Compute the modal damping matrix and say whether damping is proportional or not. In the case where it is not proportional, compute matrices $\mathbf{C}_p$ and $\mathbf{C}_{np}$. Compute the complex frequencies and the complex modes of the system, both in the direct way and through an iterative modal procedure.

Exercise 1-13

Compute the forced response of the system of Exercise 1-12 when excited by a motion of the supporting point defined in Exercise 1-9. Use both a nonmodal and an iterative approach. Add a further damper, equal to the one already existing between mass m_1 and m_2, between mass m_2 and point A, and repeat the analysis.

Exercise 1-14

Consider the system of Example 1-10. The points at which the two pendulums are connected move together in the x-direction with a harmonic

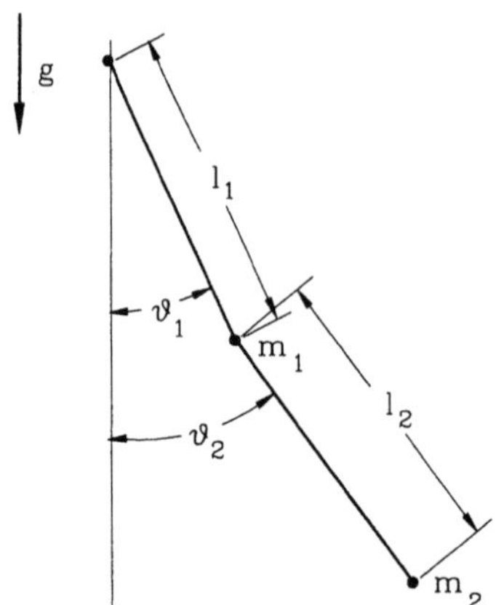

FIGURE 1.33. Double pendulum.

time history $x_A = x_B = x_{A_0} \sin(\lambda t)$, with $x_{A_0} = 30$ mm, $\lambda = 1$ Hz.

Compute the response of the system, assuming that at time $t = 0$ the whole system is at a standstill. Compute the modal participation factors and assess whether an excitation of the type here assumed, with the initial condition described, can give way to a beat.

Exercise 1-15

Consider the double pendulum of Figure 1.33. Write the kinetic and potential energy of the system and obtain the equation of motion through Lagrange equations. Linearize the equation of motion in order to study the small oscillations about the static equilibrium position. Perform the modal analysis of the linearized system. Data: $m_1 = 2$ kg, $m_2 = 4$ kg, $l_1 = 600$ mm, $l_2 = 400$ mm, $g = 9.81$ m/s^2.

2
Continuous Linear Systems

2.1 General considerations

The main feature of the discrete systems studied in the preceding chapter is that a finite number of degrees of freedom is sufficient to describe their configuration. The differential equations of motion can be easily substituted by a set of linear algebraic equations: The natural mathematical tool for the study of linear discrete systems is matrix algebra.

The situation is different when the behaviour of a deformable elastic body is studied as a continuous system. Usually a deformable body is modeled as an elastic continuum or, in the case in which its behaviour can be assumed to be linear, as a linear elastic continuum. It is clear that the elastic continuum is only a model, since no actual body is such at an atomic scale, but for most objects studied by structural dynamics the continuum model is more than adequate. Beside that, it is to exclude (not only from a practical point of view, but also for theoretical reasons) the possibility of using the discontinuous structure of matter to build a model with a very large, but finite, number of degrees of freedom.

An elastic body can be thought of as consisting of an infinity of points. To describe the undeformed (or initial) configuration of the body, a coordinate frame system is set in space. Many problems

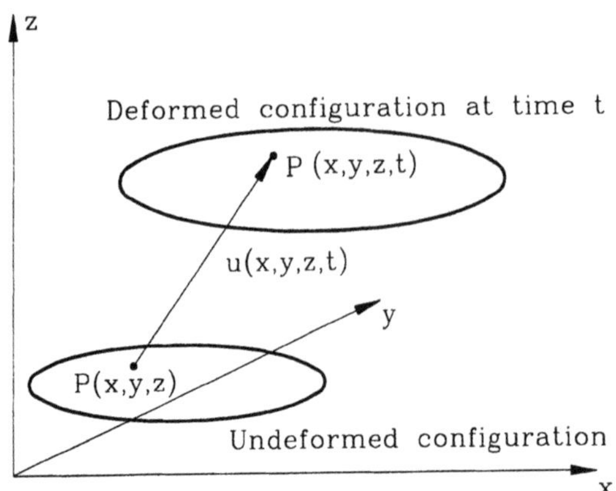

FIGURE 2.1. Deformation of an elastic continuum; system of reference and displacement vector.

can be studied with reference to a two-dimensional frame or even a single coordinate, but there are cases in which a full tridimensional approach is required. The characteristics of the material are defined by functions of the position in all the portions of space (or plane or line) that are occupied by the continuum. These functions need not, in general, be continuous.

The configuration at any time t can be obtained from the initial configuration once a vector function expressing the displacements of all points is known (Figure 2.1). The displacement of a point is a vector, with a number of components equal to the number of dimensions of the reference frame. Usually the components of this vector are taken as the degrees of freedom of each point, even if in some cases a different choice is considered; the number of degrees of freedom of an elastic (or, more generally, deformable) body is then infinite. The corresponding generalized coordinates can be manipulated as continuous functions of space coordinates and time, and the theory of continuous functions is the natural tool for dealing with deformable continua. The function $\vec{u}(x,y,z,t)$ describing the displacement of the points of the body is differentiable with respect to time at least twice; the first derivative gives the displacement velocity and the second the acceleration. Usually, however, higher-order derivatives also exist.

If displacements and rotations can be considered small, in order not to introduce geometrical nonlinearities, the usual definition of stresses and strains used in elementary theory of elasticity can be used. When the dynamics of an elastic body can be dealt with as a linear problem, as when the behaviour of the material is linear and no geometrical nonlinearity is considered, an infinity of natural frequencies exist as a consequence of the infinity of degrees of freedom of the model.

Assuming that the forces acting on the body are expressed by the function $\mathbf{f}(x,y,z,t)$, the equation of motion can generally be written as

$$D[\mathbf{u}(x,y,z,t)] = \mathbf{f}(x,y,z,t)\,, \qquad U[\mathbf{u}(x,y,z,t)]_B = 0\,. \qquad (2.1)$$

where the differential operator D completely describes the behaviour of the body and operator U, defined on the boundary, defines the boundary conditions [only homogeneous boundary conditions are described by equation (2.1)]. The actual form of the differential operator can be obtained by resorting directly to the dynamic equilibrium equations or by writing the kinetic and potential energies and using Lagrange equations, and the boundary conditions usually follow from geometrical considerations. The solution of equation (2.1) exists if an inverse operator D^{-1} can be defined

$$\mathbf{u}(x,y,z,t) = D^{-1}[\mathbf{f}(x,y,z,t)]\,. \qquad (2.2)$$

Equation (2.2) is just a formal statement; in most cases the relevant operator cannot be written in explicit form, particularly when the boundary conditions are not the simplest ones. Because no general approach to the dynamics of an elastic body is feasible, many different models for the study of particular classes of structural elements (beams, plates, shells, etc.) have been developed. In this chapter, only beams will be studied in detail, while a short outline will be given on other structural elements. This choice is only in part due to the fact that many machine elements are studied as beams (shafts, blades, springs, etc.); it comes from the need of showing some general properties of continuous systems in the simplest case in order to gain a good insight on the properties of deformable bodies.

The computational difficulties arising both from the differential equations themselves and, even more, from the boundary conditions, can, however, be overcome only in a few simple cases. The solution of most problems encountered in engineering practice requires dealing with complex structures and the use of continuous models is, consequently, ruled out. For complex shapes the only feasible approach is the discretization of the continuum and then the application of the methods seen for discrete systems. The substitution of a a continuous system, characterized by an infinite number of degrees of freedom, with a discrete system, sometimes with a very large but finite number of degrees of freedom, is usually referred to as *discretization*. This step is of primary importance in the solution of practical problems, because the accuracy of the results depends largely on the adequacy of the discrete model to represent the actual system.

Over the last two centuries, many discretization techniques have been developed with the aim of substituting the equation of motion consisting of a partial derivative differential equation (with derivatives with respect to time and space coordinates) with a set of linear ordinary differential equations containing only derivatives with respect to time. The resulting set of equations, generally of the second order, is of the same type seen for discrete systems (hence, the term *discretization*).

Another problem, which is now far less severe due to the growing power of computers, is linked with the size of the eigenproblem obtained through discretization techniques. Before the widespread use of electronic computers, the solution of an eigenproblem containing matrices whose order was greater than a few units was very difficult, if not insoluble. Many techniques were aimed at reducing to a minimum the size of the eigenproblem or transforming it into a form that could be solved using particular algorithms. The possibility of solving very large eigenproblems without long or costly computations did actually change the basic approach to dynamic problems, making many popular methods obsolete.

It is difficult to make a satisfactory classification of discretization techniques, and perhaps impossible to make a complete one. It is, however, possible to attempt to group them in three classes.

The first class is that of the methods in which the deformed shape of the system is assumed to be a linear combination of n known functions of the space coordinates, defined in the whole space occupied by the body. These methods could be labeled *assumed-modes methods*, owing to the similarity of these functions, which are arbitrarily assumed, with the mode shapes of the system.

The second class is that of the so-called *lumped-parameters methods*. The mass of the body is lumped in a certain number of rigid bodies (sometimes simply point masses) located at given stations in the deformable body. These lumped masses are then connected by massless fields that possess elastic, and sometimes damping, properties. Usually the properties of the fields are assumed to be uniform in space. Because the degrees of freedom of the lumped masses are used to describe the motion of the system, the model leads intuitively to a discrete system. While the mass matrix of such systems is easily obtained, it is often quite difficult to write the stiffness matrix, or, alternatively, the compliance matrix. To avoid such difficulty, together with that linked to the solution of large eigenproblems, an alternative approach can be followed. Instead of dealing with the system as a whole, the study can start at a certain station and proceed station by station using the so-called *transfer matrices*. Methods based on transfer matrices were very common in the recent past, because they could be worked out with tabular manual computations or implemented on very small computers. Their limitations are now making them yield to the finite element method. This change between the two approaches is not yet complete, and many computer codes based on transfer matrices are still in use.

It is also possible to model a continuous system by lumping the elastic properties at given stations leaving the inertia properties as distributed along the fields. Such methods were proposed but seldom used.

A separate class can be assigned to the finite element method (FEM). In the FEM, the body is divided into a number of regions, called *finite elements*, as opposed to the vanishingly small regions used in writing the differential equations for continuous systems. The deformed shape of each finite element is assumed to be a linear combination of a set of functions of space coordinates through a certain number of parameters, considered the degrees of freedom of the element. Usually such functions of space coordinates (called *shape functions*) are quite simple and the degrees of freedom have a direct physical meaning; generalized displacements at selected points of the element, usually referred to as *nodes*. The analysis then proceeds to writing a set of differential equations of the same type as those obtained for discrete systems.

Actually, the use of the FEM can be limited to the writing of the stiffness matrix to be introduced into a lumped-parameters approach. Alternatively, the mass matrix of the system can be obtained using the FEM approach. In this case, the mass matrix is said to be *consistent*, because it is obtained in such a way that is consistent with that used for the computation of the stiffness matrix.

Although often considered a separate approach, the dynamic stiffness method will be regarded here as a particular form of the FEM in which the shape functions are obtained from the actual deflected shape in free vibration. The facts that it can be used only for free vibration or harmonic excitation and that the dynamic stiffness matrix it yields is a function of frequency has prevented it from becoming very popular.

The FEM is gaining popularity for the unquestionable advantage of being implemented in the form of general-purpose codes, which can be used for static and dynamic analyses, and which can be interfaced with CAD and CAM codes.

The component mode synthesis method is here regarded as a method aimed at reducing the number of degrees of freedom and will be studied with the other procedures having the same goal.

The finite difference method will not be dealt with, because it is considered more a method for the solution of the differential equations obtained from continuous models than as a discretization technique yielding a model with many degrees of freedom from a continuous one. Actually, finite difference techniques are at present seldom used in structural dynamics.

2.2 Beams and bars

2.2.1 General considerations

The simplest type of continuous system is the prismatic beam. The study of the elastic behaviour of beams dates back to Galileo, with important contributions by Daniel Bernoulli, Euler, De Saint Venant, and many others. A *beam* is essentially an elastic solid in which one dimension is prevalent over the others. Often the beam is prismatic (i.e., the cross sections are all equal), homogeneous (i.e., with constant material characteristics), straight (i.e., its axis is a part of a straight line), and untwisted (i.e., the principal axes of elasticity of all sections are equally directed in space). The unidimensional nature of beams allows simplification of the study: Each cross section is considered a rigid body whose thickness in the axial direction is vanishingly small; it has six degrees of freedom, three translational and three rotational. The problem is then reduced to a unidimensional problem, in the sense that a single coordinate, namely, the axial coordinate, is required.

Setting the z-axis of the reference frame along the axis of the beam, the six generalized coordinates of each cross section are the axial displacement u_z, the lateral displacements u_x and u_y, the torsional rotation ϕ_z about the z-axis and the flexural rotations ϕ_x and ϕ_y about axes x and y. Displacements and rotations are assumed to be small, so rotations can be regarded as vector quantities, which simplifies all rotation matrices by linearizing trigonometrical functions. The three rotations will then be considered components of a vector in the same way as the three displacements are components of vector $\mathbf{u}$. The generalized forces acting on each cross section and corresponding to the six degrees of freedom defined earlier are the axial force F_z, shear forces F_x and F_y, the torsional moment M_z about the z-axis, and the bending moments M_x and M_y about the x- and y-axes.

From the aforementioned assumptions it follows that all normal stresses in directions other than z (σ_x and σ_y) are assumed to be small enough to be neglected. When geometric and material parameters are not constant along the axis they must change at a sufficiently slow rate in order not to induce stresses σ_x and σ_y, which could not be considered in this. If the axis of the beam is assumed to be straight, the axial translation is uncoupled for the other degrees of freedom, at least in first approximation. A beam that is loaded only in the axial direction and whose axial behaviour is the only one studied is usually referred to as a *bar*. The torsional-rotation degree of freedom is uncoupled from the others only if the area center of all cross sections coincides with their shear centre, which happens if all cross sections have two perpendicular planes of symmetry. If the planes of symmetry of all sections are equally oriented (the beam is not twisted) and the x- and y-axes are perpendicular to such planes, the flexural behaviour in the xz-plane is uncoupled from that in the yz-plane. The coupling of the degrees of freedom in straight, untwisted beams with cross sections having two planes of symmetry is summarized in Table 2.1.

2.2.2 *Axial behaviour of straight bars*

Consider a bar as defined in the preceding section, made of an isotropic linear elastic material. On each point of the bar acts an

Type of behaviour	Degrees of freedom	Generalized forces
Axial	Displacement u_z	Axial force F_z
Torsional	Rotation ϕ_z	Torsional moment M_z
Flexural (xz-plane)	Displacement u_x / Rotation ϕ_y	Shearing force F_x / Bending moment M_y
Flexural (yz-plane)	Displacement u_y / Rotation ϕ_x	Shearing force F_y / Bending moment M_x

TABLE 2.1. Degrees of freedom and generalized forces in beams.

axial force per unit length $f_z(z,t)$. The assumption that the cross section behaves as a rigid body is clearly a simplification of the actual behaviour, since a longitudinal stretching or compression is always accompanied by deformations in the plane of the cross section, but the assumption of small deformations makes these side effects negligible. The position of each point in the original configuration is defined by its coordinate z, while its displacement has only one component u_z.

Consider a length dz of a beam centered on the point with coordinate z (Figure 2.2). The inertia force due to motion in z direction is

$$\ddot{u}_z dm = \rho A \ddot{u}_z dz \,. \tag{2.3}$$

The resultant of the axial force F_z exerted by the other parts of the bar is

$$F_z + \frac{1}{2}\frac{\partial F_z}{\partial z}dz - F_z + \frac{1}{2}\frac{\partial F_z}{\partial z}dz \,. \tag{2.4}$$

The dynamic equilibrium equation can then be written in the form

$$\rho A \ddot{u}_z = \frac{\partial F_z}{\partial z} + f_z(z,t) \,. \tag{2.5}$$

The axial force F_z is easily linked with the displacement by the usual formula from the theory of elasticity

$$F_z = A\sigma_z = EA\epsilon_z = EA\partial u_z/\partial z \,. \tag{2.6}$$

By introducing equation (2.6) into equation (2.5), the dynamic equilibrium equation is then

$$m(z)\frac{d^2 u_z}{dt^2} - \frac{\partial}{\partial z}\left[k(z)\frac{\partial u_z}{\partial z}\right] = f_z(z,t) \,, \tag{2.7}$$

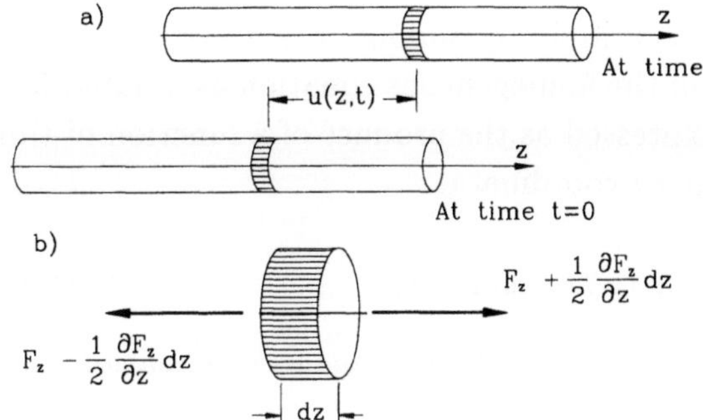

FIGURE 2.2. Axial behaviour of a straight bar: (a) System of reference and displacement; (b) forces acting on the length dz of the bar.

where the mass per unit length of the bar and the axial stiffness are, respectively,

$$m(z) = \rho(z)A(z), \qquad k(z) = E(z)A(z). \tag{2.8}$$

Equation (2.7) defines the differential operator D introduced into equation (2.1). If the bar is prismatic and homogeneous, i.e., all characteristics are constant, equation (2.7) reduces to

$$\frac{d^2 u_z}{dt^2} - \frac{E}{\rho}\frac{\partial^2 u_z}{\partial z^2} = f_z(z,t). \tag{2.9}$$

The homogeneous equation associated with equation (2.9) that describes the free behaviour of the bar is a wave propagation equation in a one-dimensional medium and constant E/ρ is the square of the speed of propagation of the waves, which in this case is the speed of sound v_s in the material. It can be used to study the propagation of elastic waves in the bar, such as those involved in shocks and other phenomena of interest in the field of structural dynamics, but stationary solutions can also be obtained. The two approaches lead to the same results, because it is possible to describe the propagation of elastic waves in terms of superimposition of stationary motions and vice versa. It is not, however, immaterial which way is followed in the actual study of any particular phenomena, because the complexity of the analysis can be quite different. As an example, in the study of the propagation of the elastic waves caused by a shock, the number of modes, which must be considered when the relevant phenomena are studied as the sum of stationary solutions, can be exceedingly high, and then it is advisable to consider the equation of motion a wave equation.

In most cases, however, the dynamic behaviour of continuous systems can be described accurately by adding only a limited number of modes. The solution of the homogeneous equation associated with equation (2.7) can be expressed as the product of a function of time and a function of the space coordinate

$$u_z(z,t) = q(z)\eta(t) \,. \tag{2.10}$$

Introducing equation (2.10) into the homogeneous equation associated with equation (2.7) and separating the variables, it follows that

$$\frac{1}{\eta(t)} \frac{d^2\eta(t)}{dt^2} = \frac{1}{m(z)q(z)} \frac{d}{dz} \left[k(z) \frac{dq(z)}{dz} \right] \,. \tag{2.11}$$

The function on the left-hand side depends on time but not on the space coordinate z. Conversely, the function on the right-hand side is a function of z but not of t. The only possibility of satisfying equation (2.11) for all values of time and of coordinate z is to state that both sides are constant and that the two constants are equal. This constant can be indicated as $-\lambda^2$. The condition on the function of time on the left-hand side is

$$\frac{1}{\eta(t)} \frac{d^2\eta(t)}{dt^2} = \text{constant} = -\lambda^2 \,, \tag{2.12}$$

which, neglecting a proportionality constant that will be introduced into function $q(z)$ later, yields a harmonic oscillation with frequency λ

$$\eta(t) = \sin(\lambda t + \phi) \,. \tag{2.13}$$

The solution of the equation of motion for the free axial oscillations of the bar is

$$u(z,t) = q(z) \sin(\lambda t + \phi) \,. \tag{2.14}$$

Function $q(z)$ is said to be the *principal function*. Each point of the bar performs a harmonic motion with frequency λ, while the amplitude is given by function $q(z)$. The resultant motion is then a standing wave, with all points of the bar vibrating in phase. By introducing equation (2.14) into equation (2.11), it follows that

$$-\lambda^2 m(z)q(z) = \frac{d}{dz}\left[k(z)\frac{dq(z)}{dz}\right] ,$$

$$-\lambda^2 q(z) = v_s^2 \frac{d^2 q(z)}{dz^2} .$$

$$(2.15)$$

The second equation here holds in the case of constant parameters. Equations (2.15) are eigenproblems. The second, for example, states that the second derivative of function $q(z)$ (with respect to the space coordinate z) is proportional to the function itself, the constant of proportionality being $-\lambda^2/v_s^2 = -\lambda^2\rho/E$. The values of such a constant that allows the equation to be satisfied by a solution other than the trivial solution $q(z)=0$ are the eigenvalues, and the corresponding functions $q(z)$ are the eigenfunctions. The first equation (2.15), although more complex, has a similar meaning. The eigenvalues are infinite in number, and the general solution of the equation of motion (2.15) can be obtained as the sum of an infinity of terms of the type of equation (2.14).

The eigenfunctions $q_i(z)$ are defined only as far as their shape is concerned, exactly as was the case for eigenvectors. The amplitude of the various modes can be computed only after the initial conditions have been stated. Although the number of eigenfunctions, and hence of modes, is infinite, a small number of principal functions is often sufficient to describe the behaviour of an elastic body with the required precision, in a way that is similar to what has already been said for eigenvectors. Eigenfunctions have some of the properties seen for eigenvectors, particularly that of orthogonality with respect to the mass $m(z)$ and to the stiffness $k(z)$.

In the case of a bar, the property of orthogonality with respect to mass and stiffness means that if $q_i(z)$ and $q_j(z)$ are two distinct eigenfunctions, it follows that, if $(i \neq j)$,

$$\int_0^l m(z)q_i(z)q_j(z)dz = 0 , \qquad \int_0^l k(z)\frac{dq_i(z)}{dz}\frac{dq_j(z)}{dz}dz = 0 . \quad (2.16)$$

As was the case for eigenvectors, eigenfunctions are not directly orthogonal, except for when the mass $m(z)$ is constant along the beam. If $i = j$, the integrals of equation (2.16) do not vanish:

$$\int_0^l m(z)[q_i(z)]^2 dz = \overline{M}_i \neq 0 \,, \qquad \int_0^l k(z) \left[\frac{dq_i(z)}{dz}\right]^2 dz = \overline{K}_i \neq 0 \,.$$

$$(2.17)$$

Constants $\overline{M}_i$ and $\overline{K}_i$ are the modal mass and the modal stiffness, respectively, for the ith mode. The meaning of the modal mass and stiffness in the case of continuous systems is exactly the same as for discrete systems; the only differences are that in the current case the number of modes, and then of modal masses and stiffnesses, is infinite.

Any deformed configuration of the system $u(z,t)$ can be expressed as a linear combination of the eigenfunctions. The coefficients of this linear combination, which are functions of time, are the modal coordinates $\eta_i(t)$:

$$u(z,t) = \sum_{i=0}^{\infty} \eta_i(t)q_i(z) \,. \qquad (2.18)$$

Equation (2.18) expresses the modal transformation for continuous systems and is equivalent to the first equation (1.50). The inverse transformation, needed to compute the modal coordinates $\eta_i(t_k)$ corresponding to any given deformed configuration $u(z,t_k)$ occurring at time t_k, can be obtained through a procedure closely following that used to obtain equation (1.61). Multiplying equation (2.18) by the jth eigenfunction and by the mass distribution $m(z)$ and integrating on the whole length of the bar, it follows that

$$\int_0^l [m(z)q_j(z)u(z,t_0)]dz = \sum_{i=0}^{\infty} \eta_i(t_0) \int_0^l [m(z)q_j(z)q_i(z)]dz \,. \quad (2.19)$$

Of the infinity of terms on the right-hand side, only the term with $i = j$ does not vanish as the integral yields the jth modal mass. Remembering the definition of the modal masses, it then follows that

$$\eta_i(t_0) = \frac{1}{\overline{M}_j} \int_0^l [m(z)q_j(z)u(z,t_0)]dz \,, \qquad (2.20)$$

which can be used to perform the inverse modal transformation, i.e., to compute the modal coordinates corresponding to any deformed shape of the system.

Eigenfunctions can be normalized in several ways, one being that leading to unit values of the modal masses. This is achieved simply by dividing each eigenfunction by the square root of the corresponding modal mass.

The vibration of the bar under the effect of the forcing function $f_z(z,t)$ can be obtained by solving the complete equation (2.7), whose general solution can be expressed as the sum of the complementary function obtained earlier, and a particular integral of the complete equation. Owing to the orthogonality properties of the normal modes $q_i(z)$, the latter can be expressed as a linear combination of the eigenfunctions. Equation (2.18) then also holds in the case of forced motion of the system.

By introducing the modal transformation (2.18) into the equation of motion (2.7), the latter can be transformed into a set of an infinite number of equations in the modal coordinates η_i

$$\overline{M}_i \ddot{\eta}_i(t) + \overline{K}_i \eta_i(t) = \overline{f}_i(t) , \qquad (2.21)$$

where the ith modal force is defined by the following formulas

$$\overline{f}_i(t) = \int_0^l q_i(z) f(z,t) dz , \qquad \overline{f}_i(t) = \sum_{k=1}^{m} q_i(z_k) f_k(t) , \qquad (2.22)$$

holding in the cases of a continous force distribution and m concentrated axial forces $f_k(t)$ acting on points of coordinates z_k, respectively. Note that equations (2.22) correspond exactly to the definition of the modal forces for discrete systems (1.58).

Equations (2.21) can be used to study the forced response of a continuous system to an external excitation of any type by reducing it to a number of systems with a single degree of freedom. In the case of continuous systems, their number is infinite, but usually a small number of them is enough to obtain the results with the required precision.

If the excitation is provided by the motion of the structure to which the bar is connected, it is expedient to resort to a coordinate system that moves with the supporting points. In the case of the axial behaviour of a bar, only the motion of the supporting structure in the axial direction is coupled with its dynamic behaviour. If the origin of the coordinates is displaced by the quantity z_A, the absolute displacement in z-direction $u_{z_{iner}}(z,t)$ is linked to the relative displacement

$u_z(z,t)$ by the obvious relationship $u_{z_{iner}}(z,t) = u_z(z,t) + z_A(t)$. By writing equation (2.7) using the relative displacement, it follows

$$m(z)\frac{d^2 u_z}{dt^2} - \frac{\partial}{\partial z}\left[k(z)\frac{\partial u_z}{\partial z}\right] = -m(z)\ddot{z}_A . \qquad (2.23)$$

This result is similar to that obtained for discrete systems: The excitation due to the motion of the constraints can be dealt with simply by using relative coordinates and applying an external force distribution equal to $-m(z)\ddot{z}_A$. The modal forces can be readily computed through equation (2.22)

$$\overline{f}_i(t) = -r_i\ddot{z}_A , \qquad (2.24)$$

where

$$r_i = \int_0^l q_i(z)m(z)dz$$

are the modal participation factors related to the axial motion of the bar. The physical meaning of the modal participation factors is the same as that already seen for discrete systems.

If the bar is prismatic and homogeneous, i.e., all characteristics are constant along the axial coordinate, the general solution of the second equation (2.15) is

$$q(z) = C_1 \sin\left(\frac{\lambda}{v_s}z\right) + C_2 \cos\left(\frac{\lambda}{v_s}z\right) . \qquad (2.25)$$

Equation (2.25) expresses the mode shapes of longitudinal vibration of the bar, after constants C_1 and C_2 have been computed from the boundary conditions. First consider the case in which both ends of the bar are free; the stress σ_z and, hence, the strain $\epsilon_z = \partial u_z/\partial z$ vanish at both ends

$$\left[\frac{dq(z)}{dz}\right]_{z=0} = \left[\frac{dq(z)}{dz}\right]_{z=l} = 0 , \qquad (2.26)$$

if the origin is set at the left end of the bar. With simple computations, equation (2.26) yields

$$C_1 = 0 , \qquad C_2 \sin\left(\frac{\lambda}{v_s}l\right) = 0 . \qquad (2.27)$$

The second equation (2.27) leads to a solution different from the trivial solution $C_2 = 0$ only if $\lambda_i l/v_s = i\pi$, i.e., if

$$\lambda_i = i\pi\frac{v_s}{l} = \frac{i\pi}{l}\sqrt{\frac{E}{\rho}}\,, \qquad i = 0,1,2,3,\dots\,. \tag{2.28}$$

Equation (2.28) yields the natural frequencies of the longitudinal vibrations of the bar. The first natural frequency is 0, which corresponds to a rigid-body longitudinal motion. This was to be expected, because no axial constraint was stated. Constant C_2 is not determined, which is obvious because it is the amplitude of the mode of free vibration. Some of the mode shapes

$$q_i(z) = q_{i_0}\cos(i\pi\zeta)\,, \tag{2.29}$$

where

$$\zeta = \frac{z}{l}$$

are plotted in Figure 2.3. The number of *nodes* (points in which the amplitude of motion vanishes) of each mode is equal to the order of the mode. The cases regarding other boundary conditions can be dealt with in a similar way. A general expression for the natural frequencies is

$$\lambda_i = \frac{\beta_{a_i}}{l}\sqrt{\frac{E}{\rho}}\,, \tag{2.30}$$

where constants β_{a_i} depend on the boundary conditions

$$\begin{cases} \beta_{a_i} = i\pi, & (i = 0,1,2,\dots), \text{ free-free;} \\ \beta_{a_i} = i\pi, & (i = 1,2,3,\dots), \text{ constrained-constrained;} \\ \beta_{a_i} = (i + 1/2)\pi, & (i = 0,1,2,\dots), \text{ constrained-free .} \end{cases} \tag{2.31}$$

The mode shapes for the last two cases are, respectively,

$$q_i(z) = q_{i_0}\sin(i\pi\zeta)\,, \qquad q_i(z) = q_{i_0}\cos\left[\pi\left(i + \frac{1}{2}\right)\zeta\right]\,. \tag{2.32}$$

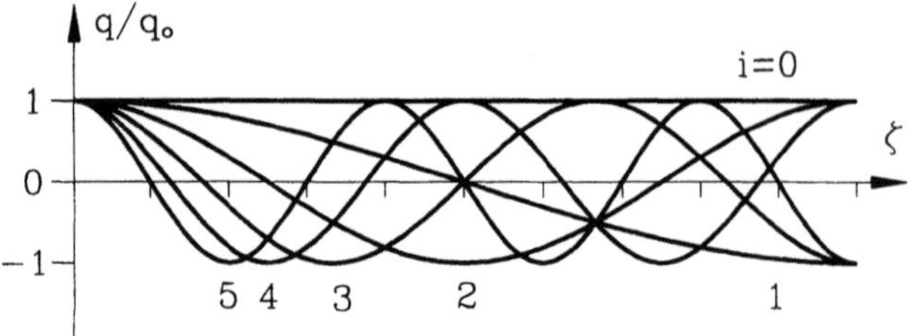

FIGURE 2.3. The first five normal modes of a prismatic homogeneous bar free at both ends. The plot must not confuse the reader; the displacements are in the axial direction.

Example 2-1

Consider a prismatic homogeneous bar free at both ends and at rest. At time $t = 0$ a concentrated force $F_0(t)$ is applied at the right end. Compute the equations of motion in modal form.

If the eigenfunctions (2.29) are normalized in such a way that their maximum amplitude is equal to one, all constants q_i have a unit value. The modal masses and stiffnesses are then

$$\overline{M}_i = \rho A \int_0^l \left[\cos(i\pi\zeta)\right]^2 = \begin{cases} \rho A l, & i = 0 \\ \dfrac{\rho A l}{2}, & i = 1,2,\ldots \end{cases}$$

$$\overline{K}_i = EA \int_0^l \left\{ \frac{d}{dz}\left[\cos(i\pi\zeta)\right] \right\}^2 = \begin{cases} 0, & i = 0 \\ i^2 \pi^2 \dfrac{EA}{2l}, & i = 1,2,\ldots \end{cases}$$

Note that the mass for the mode with $i = 0$, i.e., for the rigid-body motion in z-direction, is equal to the actual mass of the bar while the stiffness related to the same mode is zero, because the bar is unrestrained. The modal forces are

$$\overline{f}_i(t) = F_0(t)\cos(i\pi) = (-1)^i F_0(t).$$

All modes are then excited; the equations of motion are

$$\rho A l \ddot{\eta}_i = F_0(t)$$

for the first mode ($i = 0$) and

$$\frac{\rho A l}{2}\ddot{\eta}_i + i^2 \pi^2 \frac{EA}{2l} = (-1)^i F_0(t)$$

for all other modes.

Once force $F_0(t)$ has been stated, it is easy to obtain the modal coordinates as functions of time and superimpose the solutions to obtain the solution $u(z,t)$. The number of modes that need to be considered depends on function $F_0(t)$.

If instead of a concentrated force at one end of the bar, a force $f(z,t)$ is applied that is constant along the bar, the obvious result is that the bar moves along the z-axis without any axial vibration, because all points are subject to the same acceleration. In fact, if $f(z,t) = F_0/l =$ constant, the modal forces are

$$\overline{f}_i(t) = \int_0^l f(z,t)\cos(i\pi\zeta)dz = F_0$$

for the first mode ($i = 0$) and

$$\overline{f}_i(t) = 0$$

for all other modes. Only the first mode, i.e., the rigid-body mode, is excited.

2.2.3 Torsional vibrations of straight beams

Consider a beam of the same type studied in the preceding section. Under the assumptions seen in Section 2.2.1, the torsional behaviour of the beam is uncoupled from the axial and flexural behaviour and can be studied separately. Using the elementary theory of torsion, the problem is one-dimensional, because all parameters are functions only of the axial coordinate z. The dynamic equilibrium equation of a length dz of the beam is

$$\rho I_z \ddot{\phi}_z dz = \frac{\partial M_z}{\partial z}dz + m_z(z,t)\,. \tag{2.33}$$

The torsional moment M_z is linked to the rotation of the cross section by the usual formula

$$M_z = GI_p'\frac{\partial \phi_z}{\partial z}\,, \tag{2.34}$$

were the torsional moment of inertia I_p' coincides with the polar moment of inertia I_z if the cross section is circular or annular; in other cases the two quantities are different, and I_p' is tabulated for the cases of greater practical interest. The dynamic equilibrium equation is then

$$\rho I_z \frac{d^2\phi_z}{dt^2} = \frac{\partial}{\partial z}\left[GI_p'\frac{\partial\phi_z}{\partial z}\right] + m_z(z,t)\,, \qquad (2.35)$$

and is formally coincident with equation (2.7), related to axial behaviour, provided that u_z, f_z, ρA, and EA are substituted with ϕ_z, m_z, ρI_z, and GI_p', respectively.

The torsional behaviour of beams can thus be studied with the same equations seen for the axial behaviour, with the aforementioned substitutions. The only difference is that the equations reported for prismatic homogeneous bars can be used only if the beam has a circular or annular cross section, because the area (here the moments of inertia I_z and I_p') was canceled from both the mass and the stiffness.

2.2.4 Flexural vibrations of straight beams: The Euler-Bernoulli beam

With the assumptions in Section 2.2.1, the flexural behaviour in each lateral plane can be studied separately from the other degrees of freedom. The study of the flexural behaviour is more complicated than that of axial or torsional behaviour, because in bending, two of the degrees of freedom of each cross section are involved. If bending occurs in the xz-plane, the relevant degrees of freedom are displacement u_x and rotation ϕ_y. The simplest approach is that often defined as Euler-Bernoulli beam, based on the added assumptions that both shear deformation and rotational inertia of the cross sections are negligible compared with bending deformation and translational inertia, respectively. This assumption leads to a good approximation if the beam is very slender, i.e., if the thickness in the x-direction is much smaller than length l. Note that, at any rate, the thickness in the x-direction must be small enough to use beam theory.

The equilibrium equation for translations in x-direction of the length dz of the beam (Figure 2.4) is

$$\rho A \frac{d^2 u_x}{dt^2} = \frac{\partial F_x}{\partial z} + f_x(z,t)\,, \qquad (2.36)$$

If the rotational inertia of the length dz of the beam is neglected, and no distributed bending moment acts on the beam, the equilibrium equation for rotations about y-axis of the length dz of the beam is

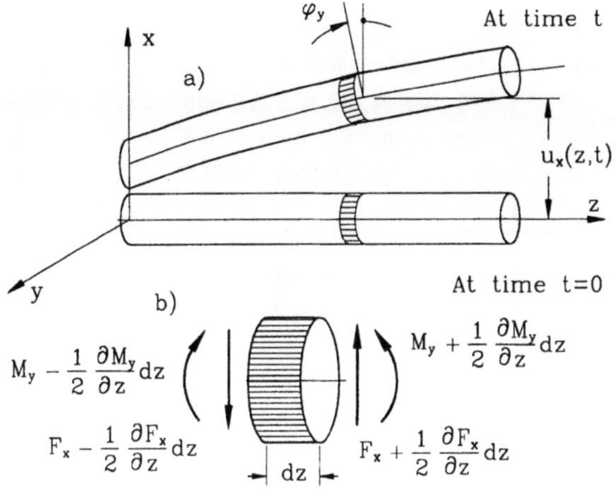

FIGURE 2.4. Flexural behaviour of a straight beam in the xz-plane; (a) sketch of the system; (b) forces and moments acting on the length dz of the beam

$$F_x dz + \frac{\partial M_y}{\partial z} dz = 0 \, . \tag{2.37}$$

By introducing equation (2.37) into equation (2.36), it follows

$$\rho A \frac{d^2 u_x}{dt^2} = -\frac{\partial^2 M_y}{\partial z^2} + f_x(z,t) \, . \tag{2.38}$$

The bending moment is proportional to the curvature of the inflected shape of the beam; neglecting shear deformation and using elementary beam theory the latter coincides with the second derivative of the displacement u_x

$$M_y = E I_y \frac{\partial^2 u_x}{\partial z^2} \, , \tag{2.39}$$

where I_y is the area moment of inertia of the cross section of the beam about its y axis. The following equilibrium equation can thus be obtained

$$m(z)\frac{d^2 u_x}{dt^2} + \frac{\partial^2}{\partial z^2}\left[k(z)\frac{\partial^2 u_z}{\partial z^2}\right] = f_x(z,t) \, , \tag{2.40}$$

where the mass per unit length and the bending stiffness are, respectively,

$$m(z) = \rho(z)A(z) \, , \qquad k(z) = E(z)I_y(z) \, . \tag{2.41}$$

In the case of a prismatic homogeneous beam, equation (2.40) reduces to

Boundary condition	$i=0$	$i=1$	$i=2$	$i=3$	$i=4$	$i>4$
Free-free	0	4.730	7.853	10.996	14.137	$\approx (i+1/2)\pi$
Supported-free	0	1.25π	2.25π	3.25π	4.25π	$(i+1/4)\pi$
Clamped-free	–	1.875	4.694	7.855	10.996	$\approx (i-1/2)\pi$
Supported-supported	–	π	2π	3π	4π	$i\pi$
Supported-clamped	–	3.926	7.069	10.210	13.352	$\approx (i+1/4)\pi$
Clamped-clamped	–	4.730	7.853	10.996	14.137	$\approx (i+1/2)\pi$

TABLE 2.2. Values of constants $\beta_{f_i} = a_i l$ for the various modes with different boundary conditions.

$$\rho A \frac{d^2 u_x}{dt^2} + E I_y \frac{\partial^4 u_x}{\partial z^4} = f_x(z,t) \,. \tag{2.42}$$

Operating in the same way seen for axial vibrations, the following general expression for the natural frequencies with any boundary condition are obtained

$$\lambda_i = \frac{\beta_{f_i}^2}{l^2} \sqrt{\frac{E I_y}{\rho A}} \,, \tag{2.43}$$

where the values of constants $\beta_{f_i} = a_i l$ depend on the boundary conditions (Table 2.2).

The eigenfunctions, expressed with reference to the nondimensional coordinate $\zeta = z/l$ and normalized in such a way that the maximum value of the displacement is equal to unity, are, for different boundary conditions, as follows:

1. *Free-free.* Rigid-body modes:

$$q_0^I(\zeta) = 1 \,, \qquad q_0^{II}(\zeta) = 1 - 2\zeta \,.$$

Other modes:

$$q_i(\zeta) = \frac{1}{2N} \left\{ \sin(\beta_i \zeta) + \sinh(\beta_i \zeta) + N \left[\cos(\beta_i \zeta) + \cosh(\beta_i \zeta) \right] \right\} \,,$$

where

$$N = \frac{\sin(\beta_i) - \sinh(\beta_i)}{-\cos(\beta_i) + \cosh(\beta_i)} \,.$$

2. *Supported-free.* Rigid-body mode:

$$q_0(\zeta) = \zeta \,.$$

Other modes:

$$q_i(\zeta) = \frac{1}{2\sin(\beta_i)}\left[\sin(\beta_i\zeta) + \frac{\sin(\beta_i)}{\sinh(\beta_i)}\sinh(\beta_i\zeta)\right].$$

3. *Clamped-free.*

$$q_i(\zeta) = \frac{1}{N_2}\left\{\sin(\beta_i\zeta) - \sinh(\beta_i\zeta) - N_1\left[\cos(\beta_i\zeta) - \cosh(\beta_i\zeta)\right]\right\},$$

where

$$N_1 = \frac{\sin(\beta_i) + \sinh(\beta_i)}{\cos(\beta_i) + \cosh(\beta_i)},$$

$$N_2 = \sin(\beta_i) - \sinh(\beta_i) - N_1\left[\cos(\beta_i) - \cosh(\beta_i)\right].$$

4. *Supported-supported.*

$$q_i(\zeta) = \sin(i\pi\zeta).$$

5. *Supported-clamped.*

$$q_i(\zeta) = \frac{1}{N}\left[\sin(\beta_i\zeta) - \frac{\sin(\beta_i)}{\sinh(\beta_i)}\sinh(\beta_i\zeta)\right],$$

where N is the maximum value of the expression within brackets and must be computed numerically.

6. *Clamped-clamped.*

$$q_i(\zeta) = \frac{1}{N_2}\left\{\sin(\beta_i\zeta) - \sinh(\beta_i\zeta) - N_1\left[\cos(\beta_i\zeta) - \cosh(\beta_i\zeta)\right]\right\},$$

where

$$N_1 = \frac{\sin(\beta_i) - \sinh(\beta_i)}{\cos(\beta_i) - \cosh(\beta_i)}$$

and N_2 is the maximum value of the expression between braces and must be computed numerically. The first four mode shapes (plus the rigid-body modes where they do exist) for each boundary condition are plotted in Figure 2.5.

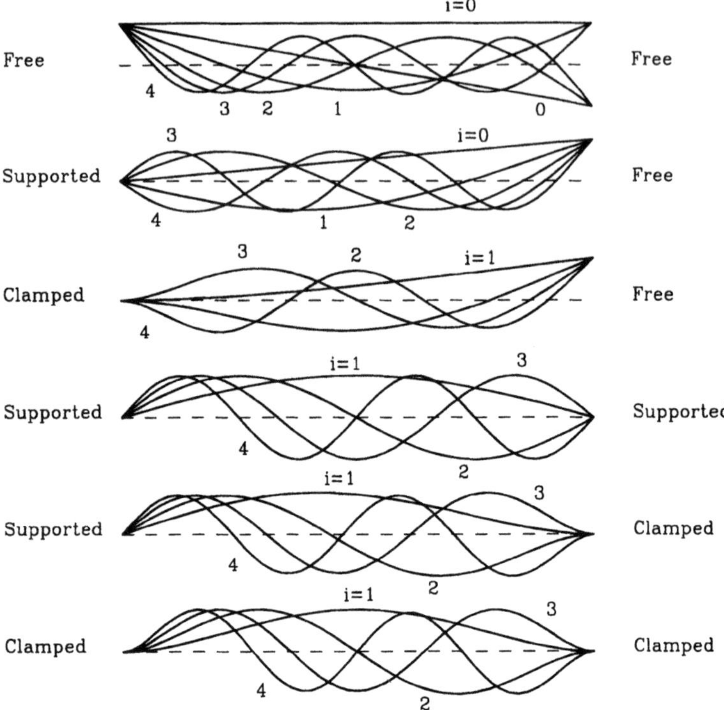

FIGURE 2.5. Normal modes of a straight beam with different end conditions. The first four modes plus the rigid-body modes, where they exist, are shown.

Eigenfunctions also have orthogonality properties in the case of bending. While the property of orthogonality with respect to the mass has the same expression as in the case of axial vibrations, the expression of orthogonality with respect to the stiffness is different

$$\int_0^l m(z)q_i(z)q_j(z)dz = 0\,, \qquad \int_0^l k(z)\frac{d^2q_i(z)}{dz^2}\frac{d^2q_j(z)}{dz^2}dz = 0\,, \tag{2.44}$$

for $i \neq j$.

As a consequence, the definitions of modal masses and stiffnesses are

$$\overline{M}_i = \int_0^l m(z)\left[q_i(z)\right]^2 dz\,, \qquad \overline{K}_i = \int_0^l k(z)\left[\frac{d^2q_i(z)}{dz^2}\right]^2 dz\,. \tag{2.45}$$

If function $m(z)$ is constant, the eigenfunctions are orthogonal. The expressions of the modal forces $\overline{f}_i(t)$ are similar to those seen for the axial behaviour

$$\overline{f}_i = \int_0^l q_i(z) f_x(z,t) dz \, . \tag{2.46}$$

In the case of excitation due to motion $x_A(t)$ of the supporting structure, the response of the beam can be computed by resorting to a reference frame fixed to the constraints and applying the modal forces

$$\overline{f}_i = -r_i \ddot{x}_A \, , \tag{2.47}$$

where

$$r_i = \int_0^l q_i(z) m(z) dz \, .$$

The situation in the yz-plane is clearly very similar to that already studied with reference to the xz-plane. The physical problem is clearly the same; however, there are some differences in the equations due to the fact that in the xz-plane the positive direction for rotations is from the z-axis to the x-axis while in the yz-plane a rotation is positive if it goes from the y-axis to the z-axis. Actually, the difficulties of notation arising from the differences existing in the two planes could be easily circumvented by assuming $-\phi_x$ instead of ϕ_x as the generalized coordinate for rotation. If the corresponding generalized force is taken as $-M_x$ instead of M_x, the situation in the two planes is exactly the same.

2.2.5   *The prismatic homogeneous Timoshenko beam*

In the preceding section the rotational inertia of the cross section and shear deformation were not taken into account. In this section this assumption will be dropped, with reference only to a prismatic homogeneous beam. A beam in which these effects are not neglected is usually referred to as a *Timoshenko beam*. Shear deformation can be taken into account as a deviation of the direction of the deflected shape of the beam not accompanied by a rotation of the cross section. The latter is then no more perpendicular to the deformed shape of the beam, and the rotation of the cross section can be expressed as

$$\phi_y = \frac{\partial u_x}{\partial z} - \gamma_x \, . \tag{2.48}$$

The shear strain γ_x is linked to the shear force by the relationship

$$\gamma_x = \frac{\chi F_x}{GA}, \tag{2.49}$$

where the shear factor χ depends on the shape of the cross section, even if there is not complete accord on its value. For a circular beam, a value of $10/9$ is usually assumed; for other shapes the expressions reported by Cowper[1] can be used.

Equation (2.48) can thus be written in the form

$$\phi_y = \frac{\partial u_x}{\partial z} - \frac{\chi F_x}{GA}. \tag{2.50}$$

The bending moment has no effect on the shear deformation: If the latter is accounted for, the relationship linking the bending moment to the inflected shape of the beam becomes

$$M_y = E I_y \frac{\partial \phi_y}{\partial z}. \tag{2.51}$$

The rotational inertia of the cross section is no more neglected, and the dynamic equilibrium equations for displacement and rotation of the length dz of the beam are

$$\begin{aligned}
\rho A \frac{d^2 u_x}{dt^2} &= \frac{\partial F_x}{\partial z} + f_x(z,t), \\
\rho I_y \frac{d^2 \phi_y}{dt^2} &= F_x + \frac{\partial M_y}{\partial z}.
\end{aligned} \tag{2.52}$$

By solving equation (2.50) in F_x and introducing it into equation (2.50), it follows:

$$\begin{aligned}
\rho A \frac{d^2 u_x}{dt^2} &= \frac{GA}{\chi} \left(\frac{\partial^2 u_x}{\partial z^2} - \frac{\partial \phi_y}{\partial z} \right), \\
\rho I_y \frac{d^2 \phi_y}{dt^2} &= \frac{GA}{\chi} \left(\frac{\partial u_x}{\partial z} - \phi_y \right) + E I_y \frac{\partial^2 \phi_y}{\partial z^2}.
\end{aligned} \tag{2.53}$$

By differentiating the second equation (2.53) with respect to z and eliminating ϕ_y, the following equation can be obtained

$$E I_y \frac{\partial^4 u_x}{\partial z^4} - \rho I_y \left(1 + \frac{E\chi}{G} \right) \frac{\partial^2}{\partial z^2} \left(\frac{\partial^2 u_x}{\partial t^2} \right) + \frac{\rho^2 I_y \chi}{G} \frac{d^4 u_x}{dt^4} + \rho A \frac{d^2 u_x}{dt^2} = 0 \tag{2.54}$$

---

[1] G.R. Cowper, "The shear coefficient in Timoshenko's beam theory", *J. Appl. Mech.*, June, 1966, 335-340.

and then, in the case of free oscillations with harmonic time history,

$$EI_y \frac{d^4 q(z)}{dz^4} + \rho \lambda^2 I_y \left(1 + \frac{E\chi}{G}\right) \frac{d^2 q(z)}{dz^2} - \rho \lambda^2 \left(A - \lambda^2 \frac{\rho I_y \chi}{G}\right) q(z) = 0 \, .$$

$$(2.55)$$

The same considerations regarding the form of the eigenfunctions (seen in the preceding section) also hold in this case. If the beam is simply supported at both ends, the same eigenfunctions seen in the case of the Euler-Bernoulli beam still hold, and equation (2.55) can be expressed in nondimensional form as

$$\left(\frac{\lambda}{\lambda^*}\right)^4 - \left(\frac{\lambda}{\lambda^*}\right)^2 \frac{\alpha^2}{i^2 \pi^2 \chi^*} \left(1 + \chi^* + \frac{\alpha^2}{i^2 \pi^2}\right) + \frac{\alpha^4}{i^4 \pi^4 \chi^*} = 0 \quad (2.56)$$

where the slenderness α of the beam and χ^* are $\alpha = l\sqrt{A/I_y} = l/r$; r is the radius of inertia of the cross section; $\chi^* = \chi E/G$; and λ_i^* is the ith natural frequency computed using the Euler-Bernoulli assumptions.

The results obtained from equation (2.56) for a beam with circular cross section and material with $\nu = 0.3$, are reported as functions of the slenderness in Figure 2.6. The effects of both shear deformation and rotational inertia tend to lower the value of the natural frequencies, the first being stronger than the second by a factor of about 3, as shown by Timoshenko[2]. It must be remembered, at any rate, that even the so-called Timoshenko beam model is an approximation, because it is based on the usual approximations of the beam theory, and the very model of a one-dimensional solid is no more satisfactory when the slenderness is very low.

2.2.6 Interaction between axial forces and flexural vibrations of straight beams

The uncoupling mentioned in the preceding sections is based on the assumption of geometrical linearity, among others. When studying a linear problem, the equilibrium equations are written applying the loads to the body in its undeformed configuration. Consider a

---

[2]S.P. Timoshenko et al., *Vibration problems in engineering*, Wiley, New York, 1974.

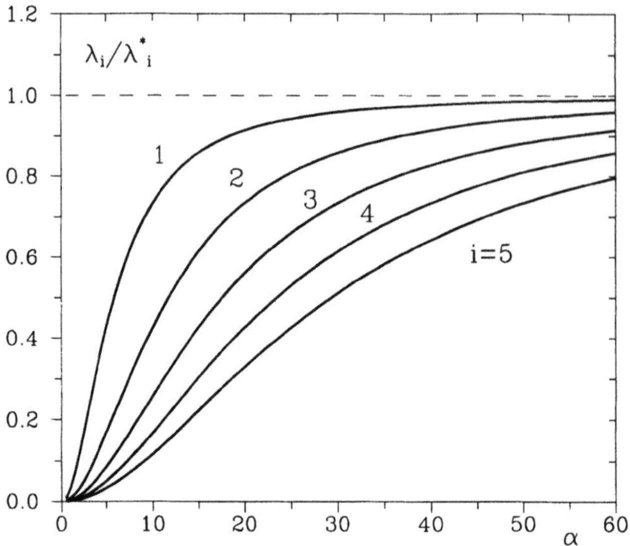

FIGURE 2.6. Effect of shear deformation on the first five natural frequencies of a simply supported beam. Ratio between the natural frequency computed taking into account rotational inertia and shear deformation and that computed using the Euler-Bernoulli model.

straight beam to which an axial force F_z, constant in time, is applied, and assume that the cross section of the beam is such that torsional and lateral behaviour and bending in the xz- and yz-planes are uncoupled. If the equilibrium equations in the xz-plane are written in the deflected position, the bending moment due to the axial force $F_z u_x(z,t)$ acting in the xz-plane must be taken into account. Similarly, in studying the behaviour in the yz-plane, the moment $-F_z u_y(z,t)$ must be considered. If shear deformation and rotational inertia of the cross sections are neglected, the dynamic equilibrium equation of a length dz of the beam for rotation about the y-axis (equation 2.37) becomes (Figure 2.7)

$$F_x dz + \frac{\partial M_y}{\partial z}dz - F_z\frac{\partial u_x}{\partial z}dz = 0 \,. \tag{2.57}$$

The equation of motion can thus be written in the form

$$\rho A\frac{d^2 u_x}{dt^2} + \frac{\partial^2}{\partial z^2}\left[EI_y\frac{\partial^2 u_z}{\partial z^2}\right] - F_z\frac{\partial^2 u_x}{\partial z^2} = f_x(z,t)\,. \tag{2.58}$$

In the case of a prismatic homogeneous beam simply supported at both ends, the eigenfunctions are still those seen for the Euler-

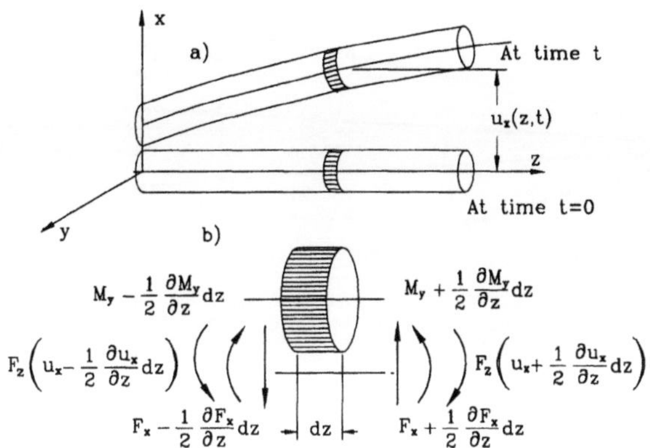

FIGURE 2.7. Effect of a constant axial force on the flexural behaviour of a straight beam in the xz-plane; (a) sketch of the system; (b) forces and moments acting on the length dz of the beam

Bernoulli beam, and the characteristic equation yielding the natural frequencies can be shown to be

$$\lambda_i^2 \rho A = E I_y \left(\frac{i\pi}{l}\right)^4 + F_z \left(\frac{i\pi}{l}\right)^2 . \tag{2.59}$$

By introducing the natural frequency λ_i^*, computed without taking into account the axial force F_z, the slenderness $\alpha = l/\sqrt{A/I_y}$ of the beam, and the axial stress $\sigma_z = F_z/A$, equation (2.58) yields

$$\frac{\lambda_i}{\lambda_i^*} = \sqrt{1 + \frac{\sigma_z}{E}\frac{\alpha^2}{i^2\pi^2}} . \tag{2.60}$$

A tensile axial force then causes an increase of the natural frequency, i.e., acts as if it increases the flexural stiffness of the beam. An opposite effect is due to axial compressive forces. The first natural frequency obtained from equation (2.60) is plotted as a function of the axial strain $\epsilon_z = \sigma_z/E$ for various values of the slenderness α in Figure 2.8. If the axial stress is compressive, equation (2.60) holds only if $|\sigma_z| < Ei^2\pi^2/\alpha^2$. When the axial stress reaches one of the mentioned values, the natural frequency reduces to zero; this means that no stable equilibrium condition is possible. Actually, when the first value is reached, buckling takes place and the expression obtained with $i = 1$ is the buckling stress of the beam.

An effect similar to that due to an axial force applied to the end of the beam is due to axial force fields, like the centrifugal field in propeller or turbine blades. Centrifugal field also causes a stiffening effect in discs and other parts of rotors.

The limiting case for an infinitely slender beam under tensile axial

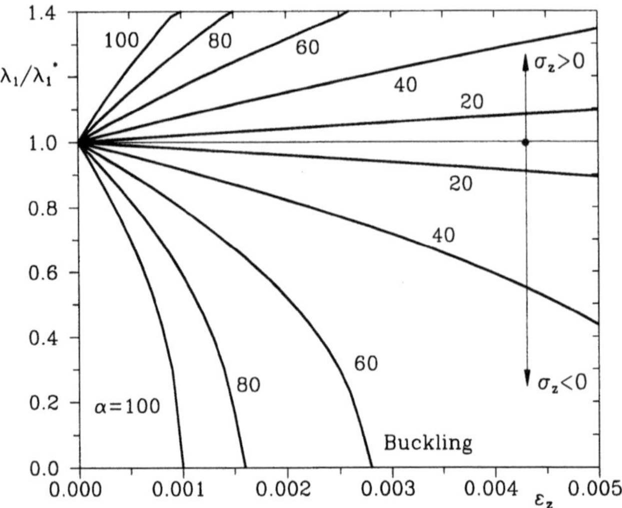

FIGURE 2.8. Effect of an axial force on the first natural frequency of a simply supported beam. Ratio between the natural frequency computed taking into account the axial force and that computed using equation (2.43) as a function of the axial strain for beams with different values of the slenderness α.

force, i.e., a beam whose bending stiffness is negligible when compared with the stiffening effect of the axial force, is that of a vibrating taut string. If both the area of the cross section of the string and the tensile force F_z are constant, the equation of motion (2.58) reduces to

$$\rho A \frac{d^2 u_x}{dt^2} = F_z \frac{\partial^2 u_x}{\partial z^2}. \tag{2.61}$$

The equation of motion is then identical to equation (2.11), which was obtained for the axial vibration of bars. The results obtained in Section 2.2.2 can be directly used by simply substituting the axial force F_z for the stiffness EA. Equation (2.61) has been obtained from the equation of motion of a beam subject to axial forces and, therefore, holds under the assumptions seen here. They are, however, not needed in the case of a vibrating taut string, and the same equation holds if the lateral displacements are not small and the tensile force along the string is not constant, provided the material has a linear stress-strain characteristic[3].

---

[3] R. Buckley, *Oscillations and waves*, Adam Hilger Ltd., Bristol, U.K., 1985.

Example 2-2

A string whose length is 500 mm has a cross section of 3 mm^2 and is made of a material whose density is $\rho = 8{,}000$ kg/m^3. Compute the tensile force in the string in such a way that the first natural frequency corresponds to the central A (nominal frequency 440 Hz) and the relative amplitudes of the harmonics of the vibration obtained by plucking the string at 1/3 of its length.

The first natural frequency is $\lambda_1 = \sqrt{\pi^2 F_z / \rho A l^2}$.

By stating that it is equal to 440 Hz $= 880\,\pi$ rad/s, it follows that $F_z = 4{,}646$ N. The other harmonics are simply multiples of the fundamental frequency.

If the string is plucked at 1/3 of its length, the configuration at time $t = 0$ is a triangle, with the vertex at 1/3 of the length. The equation of the deformed shape at time $t = 0$ is then

$$\begin{cases} u_x(\zeta,0) = 3\zeta & \text{for } 0 \le \zeta \le 1/3; \\ u_x(\zeta,0) = 3(1 - \zeta)/2 & \text{for } 1/3 < \zeta \le 1. \end{cases}$$

This shape can easily be expressed as a linear combination of the mode shapes. Because all modes oscillate in phase and at time $t = 0$, they are at their maximum amplitude, the shape of the string at any time can be expressed as

$$u_x(z,t) = \sum_{i=1}^{\infty} a_i \sin(i\pi\zeta) \cos(\lambda_i).$$

Constants a_i, expressing the amplitudes of the various harmonics, can be computed from equation (2.20)

$$a_i = \frac{m_t}{\overline{M}_i} \int_0^1 u_x(\zeta,0) \sin(i\pi\zeta) d\zeta =$$

$$= 6 \int_0^{1/3} \zeta \sin(i\pi\zeta) d\zeta + 3 \int_{1/3}^1 (1 - \zeta) \sin(i\pi\zeta) d\zeta.$$

The results for the first 10 natural frequencies are as follows:

Mode	λ_i [Hz]	a_i/a_1	Mode	λ_i [Hz]	a_i/a_1
1	440	1	6	2,640	0
2	880	0.250	7	3,080	0.0206
3	1,320	0	8	3,520	0.0158
4	1,760	−0.0628	9	3,960	0
5	2,200	−0.0402	10	4,400	−0.0102

Note that the third, sixth, ninth, ...harmonics are not excited. This could be exactly predicted because the string was plucked in a spot where these harmonics have a node.

2.3   Flexural vibration of rectangular plates

Consider a plate that is thin enough to be considered a two-dimensional system: It can be considered a two-dimensional equivalent of the beam studied in the preceding sections. In the case of plates, two basic formulations exist: In the simple one, the so-called *Kirchoff plate*, corresponding to the Euler-Bernoulli beam, shear deformation is neglected and any line perpendicular to the plate is assumed to remain perpendicular to the deflected midsurface during the deformation. The second formulation, in which shear deformation is not neglected, is referred to as the *Mindlin plate*, and is similar to the Timoshenko beam.

Even if the analysis is limited to the first formulation and the displacements are assumed to be small enough to allow a complete linearization of the problem, the analysis is far more complex than what has been seen for the beams. As a consequence, only a few results will be given here[4].

A system of reference with axes x and y contained in the midplane of the plate is stated. The force distribution acting on the plate is then $f_z(x,y,t)$, and the displacement is $u_z(x,y,t)$. Neglecting the rotational inertia of the cross section and shear deformation, the following dynamic equilibrium equation can be obtained

$$\frac{\rho h}{D}\frac{\partial^2 u_z}{\partial t^2} + \frac{\partial^4 u_z}{\partial x^4} + 2\frac{\partial^4 u_z}{\partial x^2 \partial y^2} + \frac{\partial^4 u_z}{\partial y^4} = \frac{f_z}{D}, \qquad (2.62)$$

where

$$D = \frac{Eh^3}{12(1-\nu^2)}$$

is the bending stiffness of the plate.

In the study of free vibration, the force $f_z(x,y,t)$ must be set to zero, and equation (2.62) reduces to a homogeneous equation. Its solution can be expressed as the product of a function of the space coordinates by a function of time

$$u_z(x,y,t) = q(x,y)\eta(t),$$

---

[4]For a complete description of the static behaviour of plates, see S. Timoshenko, S. Woinowsky-Krieger, *Theory of plates and shells*, McGraw-Hill, New York, 1959.

which is the two-dimensional equivalent of equation (2.10). The function of time $\eta(t)$ that solves equation (2.62) is harmonic, and the form of equation (2.13) can be assumed. Also, in this case function $q(x,y)$ can be regarded as a principal function and can be obtained from the equation

$$\lambda^2 \frac{\rho h}{D} q(x,y) + \frac{\partial^4 q(x,y)}{\partial x^4} + 2\frac{\partial^4 q(x,y)}{\partial x^2 \partial y^2} + \frac{\partial^4 q(x,y)}{\partial y^4} = 0 \,. \qquad (2.63)$$

An eigenproblem is thus obtained, although a far more complex one than those encountered in the study of one-dimensional systems. Remarkable difficulties can arise in the statement of boundary conditions, particularly when the geometrical shape of the boundary is not of the simplest type. In the case of a rectangular plate of length a (in the x-direction) and width b (in the y-direction) simply supported at all edges, it can be shown that the eigenvalues and the corresponding eigenfunctions are[5]

$$\lambda_{ij} = \pi^2 \left[\left(\frac{i}{a}\right)^2 + \left(\frac{j}{b}\right)^2 \right] \sqrt{\frac{D}{\rho h}} = \pi^2 \left[\left(\frac{i}{a}\right)^2 + \left(\frac{j}{b}\right)^2 \right] \sqrt{\frac{E h^2}{12\rho(1 - \nu^2)}}$$
$$(2.64)$$
$$q_{ij} = q_0 \sin\left(\frac{i\pi x}{a}\right)^2 \sin\left(\frac{j\pi y}{b}\right)^2 \,, \qquad i,j = 1,2,\ldots$$

Constants i and j are the number of half-waves of the deflected shape in the x- and y-directions, respectively. Also, in this case the eigenfunctions can be shown to possess the usual properties of orthogonality with respect to both stiffness and mass. Because the plate has uniform thickness and properties, the eigenfunctions are strictly orthogonal. Some eigenfunctions are shown in Figure 2.9. In the figure, the nodal lines, i.e., the lines along which the amplitude of the vibration vanishes, are clearly visible. They are straight lines; those parallel to the x-axis are $j - 1$, and those parallel to the y-axis are $i - 1$, excluding the boundaries of the plate, which are themselves nodal lines.

Note that if the plate is square ($a = b$), the eigenfrequency of order i,j is coincident with that of order j,i. In all cases in which

---

[5]L. Meirovitch, *Analytical methods in vibrations*, Macmillan, New York, 1967, p. 184.

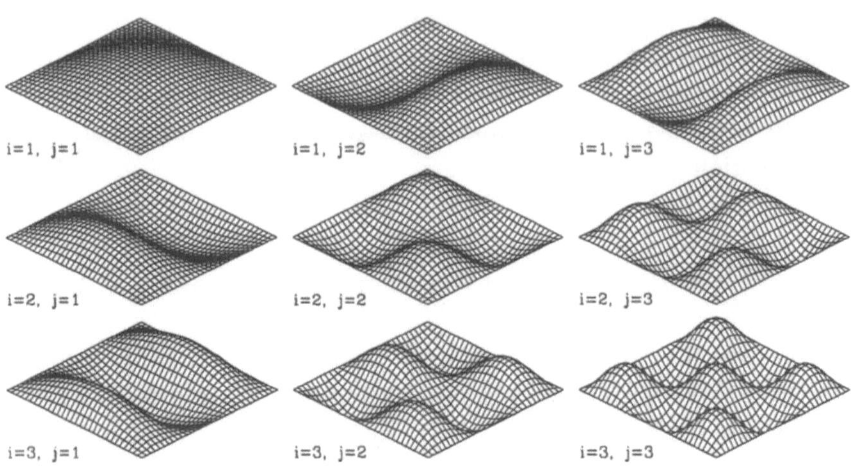

FIGURE 2.9. Nine eigenfunctions of a rectangular plate.

coincident eigenfrequencies are found, any linear combinations of the eigenfunctions is itself an eigenfunction, and it often happens that those obtained from the computations are not the more significant or obvious ones. No other case will be studied here because, even in the case of circular plates, the analysis is fairly complex and it is necessary to resort to Bessel's functions.

Similar results, particularly where the mode shapes are concerned, can be obtained in the case of vibrating membranes. A membrane is like a very thin plate, with a vanishingly small bending stiffness, in which the restoring forces towards the undeformed configuration is supplied by in-plane tension. As the plate is the two-dimensional equivalent of the beam, the membrane is the two-dimensional equivalent of the taut string. Note that in the case of a vibrating taut string the natural frequencies are in harmonic proportion, i.e., they are proportional to the sequence of the natural numbers, but the natural frequencies of plates and membranes are spaced in a more complicated and less regular way. This partially explains the peculiar characteristics of the sound of musical instruments based on a vibrating membrane, like the drum, as opposed to those based on vibrating strings.

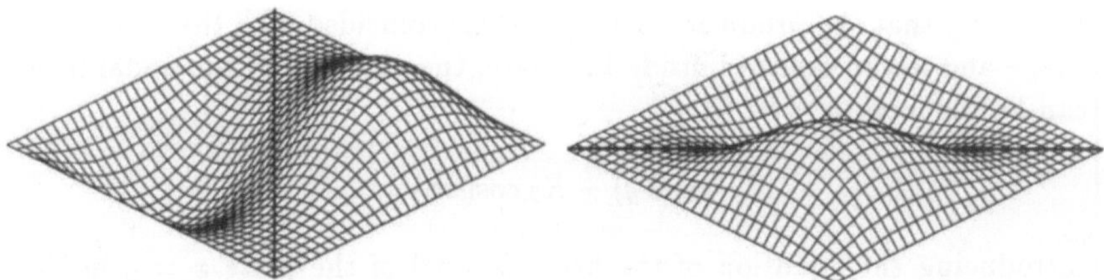

FIGURE 2.10. Mode shapes corresponding to the multiple eigenvalue chosen in such a way that the nodal lines coincide with the diagonals of the plate.

Example 2-3

Consider a square plate simply supported at all sides; compute the first frequencies corresponding to modes with up to three half-waves in both directions. Because the second and third modes correspond to identical eigenfrequencies (i.e., to a multiple eigenvalue), find the linear combinations of the eigenfunctions yielding nodal lines coinciding with the diagonals of the square plate. Data: $a = b = 1$ m; $h = 10$ mm; $E = 2.1 \times 10^{11}$ N/m^2; $\rho = 7{,}810$ kg/m^3; $\nu = 0.3$.

The natural frequencies can be computed using equation (2.64): $\lambda_{11} = 98.6$ rad/s $= 15.7$ Hz, $\lambda_{12} = \lambda_{21} = 246.4$ rad/s $= 39.2$ Hz, $\lambda_{13} = \lambda_{31} = 492.8$ rad/s $= 78.4$ Hz, $\lambda_{22} = 394.3$ rad/s $= 62.8$ Hz, $\lambda_{23} = \lambda_{32} = 640.7$ rad/s $= 102.0$ Hz, $\lambda_{33} = 887.1$ rad/s $= 141.2$ Hz.

The eigenfunctions related to the modes with $i = 1$, $j = 2$ and $i = 2$, $j = 1$ are

$$q_{12}(x,y) = q_0 \sin(\pi x) \sin(2\pi y)$$

and

$$q_{21}(x,y) = q_0 \sin(2\pi x) \sin(\pi y).$$

Any linear combination of these modes

$$K_1 \sin(\pi x) \sin(2\pi y) + K_2 \sin(2\pi x) \sin(\pi y)$$

is itself a mode. With simple trigonometric manipulations the expression of the generic linear combination becomes

$$2 \sin(\pi x) \sin(\pi y)[K_1 \cos(\pi y) + K_2 \cos(\pi x)].$$

The values of the coefficients yielding a mode shape whose nodal line coincides with the diagonal of the square expressed by the equation $x = y$ is readily obtained. The equation of the nodal line is simply obtained by equating to zero the expression of the mode shape given earlier.

By noting that the product outside brackets coincides with the first mode shape and never vanishes inside the plate, the equation of the nodal lines can be reduced to

$$K_1 \cos(\pi y) + K_2 \cos(\pi x) = 0.$$

Introducing the equation of the first diagonal of the plate $x = y$ in the equation so obtained, it follows that $K_1 + K_2 = 0$, i.e., $K_1 = -K_2$. The equation of the other diagonal is $x = 1 - y$. By introducing this expression into the equation of the nodal lines it follows that $K_1 = K_2$. The mode shapes so obtained are shown in Figure 2.10.

2.4   Propagation of elastic waves in taut strings and pipes

Consider the taut string studied in the last part of Section 2.2.6. The equation of motion of the string (2.61) can be considered a wave equation and constant $v = \sqrt{F_z/\rho A}$ is the speed of propagation of the waves along the string. This can be readily shown by taking an arbitrary function $f(\zeta)$, where $\zeta = z \pm vt$, and noting that its derivatives

$$\frac{d^2 f}{dt^2} = v^2 \frac{\partial^2 f}{\partial \zeta^2} \, ; \qquad \frac{\partial^2 f}{\partial z^2} = \frac{\partial^2 f}{\partial \zeta^2} \qquad (2.65)$$

satisfy equation (2.61). Any function of $z \pm vt$ is a solution of the equation of motion. At time $t = 0$ function $f(\zeta) = f(z)$ represents a deformed configuration of the string. As time goes on, the deformed configuration travels along the string toward the positive z-axis if the solution with sign $(-)$ is taken, or in the opposite direction if the solution with sign $(+)$ is considered. The motion of the deformed configuration, which does not involve any actual movement of matter in the z-direction, takes place with speed v. No change in shape takes place, apart from this displacement.

Note that the actual motion of the string takes place in the x-direction while the waves propagate in the z-direction. When this occurs, the wave is said to be a *transversal* wave. In the case of the longitudinal behaviour of bars, which is governed by equation (2.11), however, the direction of the motion and that of propagation coincide and the wave is said to be a *longitudinal* wave.

The general solution of equation (2.61) can be expressed as

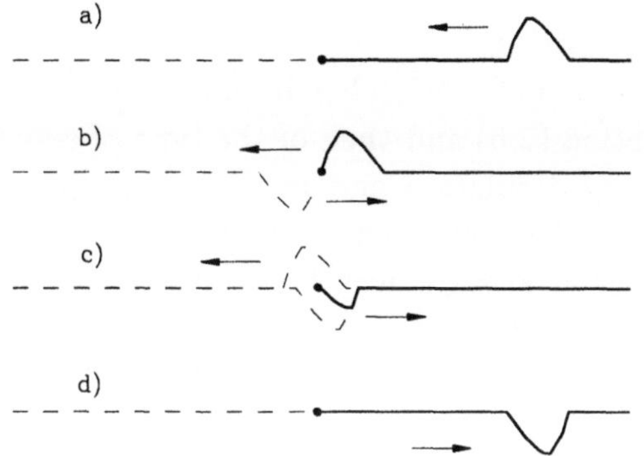

FIGURE 2.11. Reflection of a wave at a fixed end of a semi-indefinite string.

$$u_x(z,t) = f_1(z + vt) + f_2(z - vt), \qquad (2.66)$$

where f_1 and f_2 are two arbitrary functions that represent the shape of two waves traveling to the left and right, respectively. Owing to the complete linearity of the equation of motion, the solutions can simply be added to each other and, when the two waves meet along the string, the relevant amplitudes can simply be added for any value of time and z-coordinate.

In the analysis performed up to now, the string has been considered infinitely long and no boundary condition has been assumed. Consider an indefinite string on the positive half of the z-axis. Assume that the origin $(z = 0)$ is a fixed point and that a general disturbance $u_x = f_1(z + vt)$ travels toward the origin (Figure 2.11a). At a certain time the disturbance reaches the end of the string, which is fixed (Figure 2.11b).

The boundary condition can be introduced simply by adding a second function $f_2(z - vt)$, i.e., by adding a second wave traveling to the right and computing the unknown function f_2 in such a way that the displacement in the origin is always equal to zero: $u_x(0,t) = f_1(z + vt) + f_2(z - vt) = 0$, for $z = 0$, i.e., $f_2(\zeta) = -f_1(-\zeta)$.

The complete solution of the equation of motion is $u_x(z,t) = f_1(z + vt) - f_1(-z + vt)$. The first term is the original wave traveling toward the left and the second term is a reflected wave, reversed in space and with opposite sign, traveling toward the right. To show this reversal of shape, an unsymmetrical disturbance is shown in Figure 2.11. An easy way to visualize the reflection of a wave at a fixed point is to consider the string as if it were infinite by adding a ghost string to the left of the origin. The original wave, after reaching the origin, passes to the ghost string, while the reflected wave, coming from the latter, passes on the actual string.

The solution of the equation of motion has been assumed initially to be of the type of equation (2.6) and then of the type of equation (2.66), i.e., as a stationary oscillation and as a wave propagating along the system. Both have been claimed to be the general solution of the problem, so they must in some way be equivalent. Consider a string of length l extending from $z = 0$ to $z = l$ and give it a disturbance of the type $u_x(z,0) = q_0 \sin(i\pi z/l)$ at time $t = 0$. The solution can be expressed as a wave traveling to the left and reflecting in the origin, which is a fixed point,

$$u_x(z,t) = q_0 \sin\left[\frac{i\pi}{l}(z + vt)\right] - q_0 \sin\left[\frac{i\pi}{l}(-z + vt)\right] . \qquad (2.67)$$

Note that the boundary condition at the other end is satisfied because equation (2.67) yields $u_x(l,t) = 0$ for any value of t. By remembering some simple trigonometric identities, equation (2.67) can be transformed to

$$u_x(z,t) = 2q_0 \sin\left(\frac{i\pi z}{l}\right) \cos\left(\frac{i\pi v}{l}t\right) . \qquad (2.68)$$

The solution expressed by equation (2.68) is a standing wave because it does not move along the string and it coincides with the vibration at the ith natural frequency

$$\lambda_i = \frac{i\pi v}{l} = \frac{i\pi}{l}\sqrt{\frac{F_z}{\rho A}} . \qquad (2.69)$$

Its shape coincides with the ith mode shape of the string. A harmonic standing wave can thus be considered the sum of two identical harmonic waves traveling in opposite directions. The identity of the two ways of describing the free vibration of a continuous system that has been seen here for the case of a taut string is general and also applies to all other cases studied here, even if greater analytical difficulties can arise.

Although the detailed study of wave-propagation phenomena is well beyond the scope of this book, a simple model that allows the study of the one-dimensional propagation of sound waves will be summarized. Consider a pipe of cross section A, filled with a gas

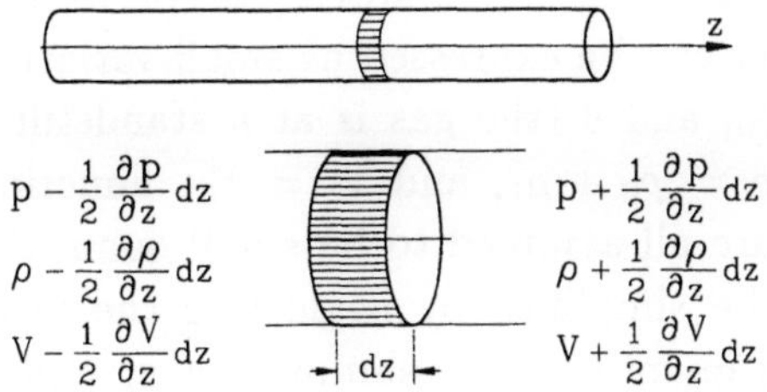

$$p - \frac{1}{2}\frac{\partial p}{\partial z}dz \qquad\qquad p + \frac{1}{2}\frac{\partial p}{\partial z}dz$$
$$\rho - \frac{1}{2}\frac{\partial \rho}{\partial z}dz \qquad\qquad \rho + \frac{1}{2}\frac{\partial \rho}{\partial z}dz$$
$$V - \frac{1}{2}\frac{\partial V}{\partial z}dz \qquad\qquad V + \frac{1}{2}\frac{\partial V}{\partial z}dz$$

FIGURE 2.12. Wave propagation in a pipe filled with gas.

with pressure p_0 and density ρ_0. Assume that all parameters related to the local properties of the gas (pressure p, density ρ, velocity in axial direction V) are functions of the axial z-coordinate alone (Figure 2.12).

This assumption is equivalent to those at the base of beam theory and allow a description of the phenomenon of wave propagation using only one space coordinate. A first equation can be stated by observing that the net flow into the control volume of length dz must be equal to the increase of the mass contained in the same volume. This equation, generally referred to as the conservation of mass equation, or simply, the continuity equation, can be written as

$$\frac{d\rho}{dt} = -\frac{\partial(\rho V)}{\partial z}. \tag{2.70}$$

The momentum of the gas contained in the control volume $\rho V\,A\,dz$ can change during time dt owing to the momentum of the gas entering the same volume and the total impulse of pressure forces. The equation expressing the conservation of momentum is

$$\frac{d(\rho V)}{dt} = -\frac{\partial(\rho V^2)}{\partial z} - \frac{\partial p}{\partial z}. \tag{2.71}$$

A third equation expressing the usual relationship between pressure and density during adiabatic changes

$$\frac{p}{p_0} = \left(\frac{\rho}{\rho_0}\right)^{\gamma}, \tag{2.72}$$

where γ is the ratio between the specific heats at constant pressure and volume, respectively, can be added. Without entering into details about the assumption of adiabatic wave propagation, equation (2.72) will be assumed to be an expression of the conservation of energy. Equations (2.70) to (2.72) are a set of three nonlinear equations describing sound-wave propagation in the pipe.

A linearization of these equations can, however, be performed. Pressure, density, and velocity can be expressed as small variations about the static values p_0, ρ_0, and 0 (the gas is at a standstill as an average): $p = p_0 + p_1$, $\rho = \rho_0 + \rho_1$, and $V = V_1$. Functions $p_1(z,t)$, $\rho_1(z,t)$, and $V_1(z,t)$ are all assumed to be small quantities. The relevant equations can be simplified by neglecting the terms containing products of small quantities. By eliminating p_1 and ρ_1 these equations yield the following wave equation

$$\frac{d^2 V_1}{dt^2} = \gamma \frac{p_0}{\rho_0} \frac{\partial^2 V_1}{\partial z^2}. \tag{2.73}$$

The speed of propagation of the waves, i.e., the speed of sound in the gas, is $v_s = \sqrt{\gamma p_0/\rho_0} = \sqrt{\gamma RT}$, where R is a constant depending on the nature of the gas and T is the absolute temperature. For air at standard temperature, pressure, and density, the value of γ is roughly 7/5, the pressure is about $p_0 = 1 \times 10^5$ Pa, and the density is $\rho_0 = 1.29$ kg/m^3; the speed of sound is then $v_s = 330$ m/s.

This model can be applied to the propagation of sound in an infinitely long pipe. If the length of the pipe is limited, appropriate boundary conditions must be stated, and standing waves can be found. Note that the speed of propagation of sound does not depend on the frequency, at least in the case of the linearized theory. This is, however, not a general rule for all wave propagation phenomena: An example of waves that are dispersive, i.e., their speed of propagation depends on the frequency, is that of water waves in deep waters. The waveform of any wave can be considered the superimposition of harmonic waves, each having its own frequency. If they propagate at different speeds, the waveform distorts continuously during propagation.

2.5   The assumed-modes methods

As a first class of discretization techniques, consider the assumed-modes methods. Assume that the displacement field $\vec{u}(x,y,z,t)$ of a general undamped continuous elastic body can be expressed as a linear combination of n arbitrarily assumed functions $\vec{q}_i(x,y,z)$, often referred to as *assumed modes*:

$$\vec{u}(x,y,z,t) = \sum_{i=1}^{n} a_i(t)\vec{q}_i(x,y,z) \,. \qquad (2.74)$$

Expression (2.74) yields exact results if the assumed functions $\vec{q}_i(x,y,z)$ coincide with the normal modes of the system and if their number is infinite. In this case, functions $a_i(t)$ are nothing other than the modal coordinates of the system.

If a finite number of coordinates is used, the results are clearly approximate. Within the approximation expressed by equation (2.74), functions $a_i(t)$ can be considered the generalized coordinates expressing the deformation of the system. From the expressions of the displacement (2.74), it is easy to obtain the expressions of the kinetic and potential energy of the system and of the virtual work due to a force distribution $\vec{f}(x,y,z,t)$ for the virtual displacement

$$\delta\vec{u}(x,y,z) = \sum_{i=1}^{n} \delta a_i \vec{q}_i(x,y,z)$$

corresponding to a virtual change of the generalized coordinates δa_i. In general, their expression is of the type

$$T = \frac{1}{2}\dot{\mathbf{a}}^T \mathbf{M}\dot{\mathbf{a}}\,, \qquad \mathcal{U} = \frac{1}{2}\mathbf{a}^T \mathbf{K}\mathbf{a}\,,$$

$$\delta\mathcal{L} = \int_V \sum_{i=1}^{n} \delta a_i \vec{f}(x,y,z,t) \times \vec{q}_i(x,y,z)dV\,,$$

where $\mathbf{a}$ is a vector containing the n generalized coordinates $a_i(t)$, and $\mathbf{M}$ and $\mathbf{K}$ are square matrices of order n that depend on the inertial and elastic properties of the system, respectively, and on the assumed functions q_i. The equations of motion of the system can then be computed through Lagrange equations

$$\mathbf{M}\ddot{\mathbf{a}} + \mathbf{K}\mathbf{a} = \mathbf{f}(t)\,, \qquad (2.75)$$

where the generalized forces f_i are

$$f_i = \int_V \vec{f}(x,y,z,t) \times \vec{q}_i(x,y,z)dV\,.$$

Equations (2.75) are formally identical to the equations of motion for discrete systems. If the assumed functions $q_i(z)$ are coincident with the mode shapes, the mass and stiffness matrices are diagonal and the nonzero elements are coincident with the modal masses and stiffnesses. Forces f_i coincide with the modal forces.

Many methods were developed following the lines summarized here, the best known are the Ritz, Rayleigh-Ritz, and Galerkin methods. The aim of the Rayleigh method, for example, is to compute an approximate value of the first natural frequency by transforming the original system, which can be a continuous system (if used as a true discretization method) or a system with many degrees of freedom (and then the method allows a drastic reduction in the number of degrees of freedom), into a system with a single degree of freedom.

Only one arbitrary mode shape is assumed. It must be chosen in such a way that it approximates the actual mode shape corresponding to the natural frequency that is looked for, usually the lowest natural frequency, and must be compatible with the constraints and the boundary conditions. To assume a deformed shape different from the actual shape the system takes in free vibration amounts to implicitly adding other constraints, which have no physical meaning, and then to make the system stiffer than it actually is. Increasing the stiffness causes the relevant natural frequency to grow, so the nearer the assumed mode is to the actual mode shape, the lower the value of the natural frequency computed through the Rayleigh method is. Among the infinite number of possible shapes that can be assumed, that corresponding to the actual mode shape of the first natural frequency (the first eigenfunction or eigenvector, if the system is discrete) yields the lowest value of the natural frequency.

The method is very easily applied: A deflected shape is assumed, and the corresponding kinetic and potential energies are computed. The mass and stiffness of the system with a single degree of freedom are then obtained and the natural frequency is computed. The Rayleigh method allows very quick computation of an approximate value of the first natural frequency, which is usually quite close to the correct one. Its ease of application allowed it to survive the introduction of numerical methods that allow detailed analysis of very complex systems but involve computation times and costs that are largely greater. It is possible to increase its precision by introducing iterative procedures aimed at obtaining assumed modes that are closer to the actual first mode, by loading the structure with a distribution of inertia forces corresponding to the deformation $q(x,y,z)$. The precision increases in this way, but the main advantage of the

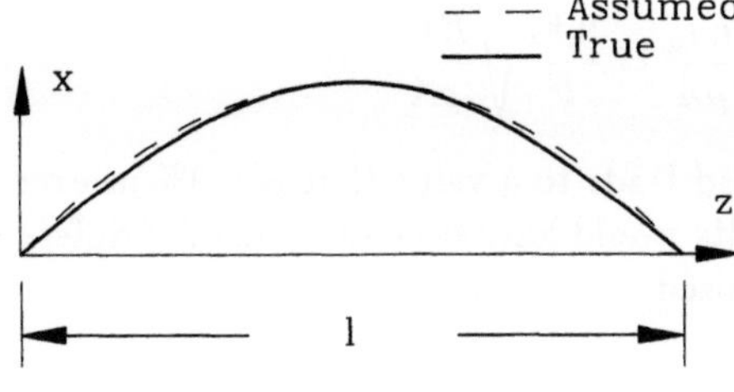

FIGURE 2.13. A simply supported beam; first mode shape and parabolic assumed shape. The two shapes have been normalized in order to have the same amplitude.

method, namely, its simplicity, is lost.

The greatest disadvantage of these approaches lies in their very principle: Because they are based on the approximation of the displacement field with functions defined in the whole field, they are not very well suited to model systems containing discontinuities or, at least, abrupt changes in characteristics, as often occurs in actual systems. They will not be dealt with in further detail, because their use in structural dynamics is now limited.

Example 2-4

Compute the first natural frequency of a prismatic, homogeneous simply supported beam using the Rayleigh method and compare the result with that obtained using the beam theory. Neglect the rotational inertia of the cross sections and shear deformation and assume a parabolic assumed-mode shape.

Using the system of reference shown in Figure 2.13, the parabolic assumed mode is expressed by the formula $q(\zeta) = \zeta(\zeta - 1)$, where $\zeta = z/l$. The kinetic and potential energies corresponding to the deflected shape $aq(\zeta)$ can be easily computed:

$$\mathcal{T} = \frac{1}{2} \int_0^l \rho A \left[\dot{a}q(z)\right]^2 dz\,, \qquad \mathcal{U} = \frac{1}{2} \int_0^l EI_y \left[a\frac{\partial^2 q(z)}{\partial dz^2}\right]^2 dz\,.$$

The mass and stiffness of the system with a single degree of freedom and its natural frequency are

$$m = \int_0^l \rho A \left[q(z)\right]^2 dz = \frac{\rho A l}{30}\,, \qquad k = \int_0^l EI_y \left[\frac{\partial^2 q(z)}{\partial dz^2}\right]^2 dz = \frac{4EI_y}{l^3}\,,$$

$$\lambda = \sqrt{\frac{k}{m}} = \frac{10{,}95}{l^2}\sqrt{\frac{EI_y}{\rho A}}\,.$$

The value computed through the beam theory is

$$\lambda = \frac{\pi^2}{l^2}\sqrt{\frac{EI_y}{\rho A}} = \frac{9,87}{l^2}\sqrt{\frac{EI_y}{\rho A}}.$$

so the use of the Rayleigh method leads to a value that is 9.9% in excess of the correct one. Different results would have been obtained if a different assumed-mode shape had been used.

2.6   Lumped-parameters methods

2.6.1   Discretization of the structure

Lumped-parameters methods are based on the discretization of continuous systems by discretizing their physical structure. The inertial properties are concentrated into a certain number of rigid bodies or point masses, located at chosen points, often called *stations*. They are connected by the fields to which the elastic properties of the structure are ascribed. There is no conceptual difficulty in also considering its damping properties. External force distributions are substituted by concentrated forces acting at the stations. The generalized displacements at the nodes are assumed to be generalized coordinates.

The determination of the mass matrix is a more or less arbitrary process, since it follows directly from the lumping of the system and is more a matter of common sense and engineering judgment than of structural analysis. The same can be said for the computation of the force vector. It is obvious that some systems are better suited than others for this type of modeling: In some cases the mass is already lumped in the actual system and then the modeling follows directly from the actual configuration. In other cases, however, the inertial properties of the actual system are distributed on a large part of it and many discretization schemes are possible. In this case it may be advisable to try different models to understand how the results are influenced by their complexity. The aim is clearly to reach the required precision with the simplest, and consequently the least costly, model.

The construction of the stiffness matrix is usually a difficult part of the computation. Traditionally, the compliance matrix $\mathbf{B}$ is obtained by computing the influence coefficients of the structure instead of the stiffness matrix $\mathbf{K}$. It must be noted that the compliance matrix can be obtained only if the stiffness matrix is nonsingular, i.e., if the system is constrained in such a way that no rigid-body motion is possible. If this does not occur, the compliance approach is not feasible, at least in a direct way. It is clear that if the system has n degrees of freedom, n static deflected shapes due to n different load conditions must be computed. Moreover, while the stiffness matrix of most structural systems has a band structure with a more or less narrow bandwidth, the compliance matrix does not have a similar structure. Its only regularity is usually that of being symmetrical. For the aforementioned reasons, it is customary to resort to the stiffness approach and to use the FEM to compute the stiffness matrix. Many FEM codes that can perform dynamic computations work, optionally or compulsorily, using the lumped-parameter approach instead of the consistent approach.

2.6.2   The transfer-matrices method

The lumped-parameters methods, as described in the preceding section, involve long computations, mainly in two phases: the construction of the stiffness matrix (or of the compliance matrix) and the solution of the eigenproblem. To avoid such difficulties, a different approach to the computation of the natural frequencies was evolved: The transfer-matrices method. Two applications of this approach are well known and were very widely used, particularly when no automatic computing machines were in use: Holtzer's method for torsional vibrations of shafts and Myklestadt's method for flexural vibrations of beamlike structures. Both were suited for tabular computations, even if they required much computational work, and, when computers became available, computer codes based on them were written and widely used. They are still used, at least in particular cases. Their main limitation lies in the very principle of the transfer-matrices method, which is suited only for in line systems, i.e., systems in which every station is linked by two fields to only two other stations, a leading station and a following station.

If the mass of the system is lumped in n nodes, the fields are $n - 1$. The computation starts at the first station (often called *station at the left*, as the structure is thought to go from left to right as a printed line) and ends at the last (right) station. It is clear that no branched systems or systems with multiple connections can be studied (unless by approximate approaches in which a branch is concentrated in the station at which it stems from the main structure)and this is a serious drawback. This leads to the difficulty of writing a general-purpose code that can deal with systems with different geometrical configurations.

The aforementioned characteristics explain why the transfer-matrices method is now yielding to methods based on the study of the system as a whole, without having to go through it step by step from one station to the next.

The method is based on the definition of state vectors and transfer matrices. A *state vector* is a vector that contains the generalized displacements and forces related to the degrees of freedom that characterize the ends of each field, considered insulated from the rest of the structure. Consequently, each field has two state vectors, one at the left end and one at the right end. The state vectors at the ends of a field are linked by the transfer matrix of the field

$$\mathbf{s}_{R_i} = \mathbf{T}_{f_i}\mathbf{s}_{Li}\,, \tag{2.76}$$

where subscript i refers to the ith field and R and L designate the right and left ends, respectively. $\mathbf{T}_{f_i}$ is the transfer matrix of the ith field. The left end of the ith field and the right end of the $(i-1)$-th field are located at the ith station, and between them there is the ith lumped mass. The corresponding state vectors are not coincident because the mass exerts generalized inertia forces on the node. They are linked by the transfer matrix of the ith station

$$\mathbf{s}_{Li} = \mathbf{T}_{ni}\mathbf{s}_{R_{i-1}}\,. \tag{2.77}$$

The station transfer matrix contains inertia forces due to the lumped mass that, in harmonic free vibrations, are functions of the square of the frequency of vibration. On the contrary, the field is massless and the field transfer matrix is independent of the frequency.

If there is a linear elastic constraint at the i-th node, its stiffness can be introduced into the expression of the station transfer matrix. In this way it is possible to use the transfer-matrix method for systems in which the constraints are applied to nodes other than the first and last ones. The case of rigid constraints can be dealt with by introducing elastic constraints with very high stiffness.

The state vector at the left of the first station $\mathbf{s}_0$ and that at the right of the last station $\mathbf{s}_n$ can be linked together by the equation

$$\mathbf{s}_n = \mathbf{T}_n\mathbf{T}_{n-1}\mathbf{T}_{n-2}\cdots\mathbf{T}_2\mathbf{T}_1\mathbf{s}_0 = \mathbf{T}_G\mathbf{s}_0\,, \tag{2.78}$$

where the transfer matrices $\mathbf{T}_i$ are related to all what is included between two subsequent nodes and usually are the product of the transfer matrix of a node and that of a field.

The overall transfer matrix

$$\mathbf{T}_G = \prod_{i=n}^{1} \mathbf{T}_i \tag{2.79}$$

is the product of all transfer matrices of all nodes and fields from the last to the first, in the correct order. Note that the product of matrices depends on the order in which it is performed, so the overall transfer matrix must be computed by strictly following the aforementioned rule. The overall transfer matrix is a function of the frequency λ of the oscillations of the system or, better, of λ^2.

Some elements of the state vectors $\mathbf{s}_0$ or $\mathbf{s}_n$ are known: If a degree of freedom is constrained, the corresponding generalized displacement is zero, while if it is free the corresponding generalized force vanishes. This leads to some simplifications, because some of the rows and columns of the overall transfer matrix can be canceled.

Equation (2.78) yields a solution different from the trivial one with all state vectors equal to zero only if the overall transfer matrix is singular. Because such a matrix is a function of λ^2, this condition leads to an equation in λ^2, which coincides with the characteristic equation of the eigenproblem yielding the natural frequencies of the system.

The transfer-matrix approach is used to avoid solving an intricate eigenproblem and an alternative approach is usually followed. A value of the frequency λ is assumed, and the transfer matrices are computed and multiplied to obtain the overall transfer matrix. After canceling the rows and columns following the constraint conditions, its determinant is computed: If it vanishes, the frequency assumed is one of the natural frequencies. If this does not occur, a new frequency is assumed and the computation is repeated. By plotting the value of the determinant as a function of the frequency, it is easy to obtain as many natural frequencies as needed. Operating along these lines, no matrix of an order greater than that of the state vector (usually not higher than 4) must be dealt with. When the appropriate rows and columns of the overall transfer matrix have been canceled, the size of the determinant to be computed is usually not greater than 2. This explains why the method could be used without resorting to computers, even if long computations are usually involved.

Note that cases in which the system is constrained in such a way that rigid-body motions are possible can be dealt with using the transfer-matrices method, obtaining a first natural frequency equal to zero.

2.6.3 Holtzer's method for torsional vibrations of shafts

Consider a lumped torsional system, i.e., a system consisting of rigid discs connected by straight shafts possessing all the properties needed for the uncoupling of the torsional modes from flexural ones (Figure 2.14a). The system can result from the lumping of a more or less uniform continuous system or can model a system that is actually made of concentrated rotors and lightweight shafts. Where the torsional behaviour is concerned, each station has only one degree of freedom, rotation ϕ_z. The order of state vectors is then 2: It contains rotation ϕ_z and moment M_z.

If the stations are located at the positions of the discs, the field transfer matrices are easily computed by considering that the moment at the left is equal to the moment at the right because no moment acts along the field, while the rotation at the right end is equal to the rotation at the left end increased by the twisting of the field. The station transfer matrix is obtained by considering that the rotation at the left of the ith station (i.e., the rotation at the right of the $(i-1)$-th field) is equal to the rotation at the right, while the moment at the right of the station is equal to that at the left of the station, increased by the concentrated moment acting at the station. If no external moment acts on the station, the latter is due only to the inertia reaction, which, in harmonic motion, is proportional to the square of the frequency and to the rotation $\phi_{z_{L_i}}$. If the station is constrained by a torsional spring of stiffness χ_c, the moment $\chi_c \phi_{z_{L_i}}$ also acts on the node.

The equations defining the field and station transfer matrices are, respectively,

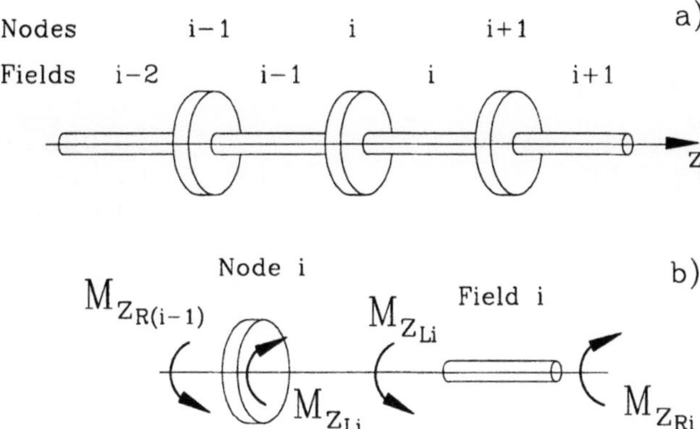

FIGURE 2.14. Holtzer method: (a) sketch of the system, (b) moments acting on the ith node and on the ith field.

$$\left\{ \begin{array}{c} \phi_z \\ M_z \end{array} \right\}_{R_i} = \left[\begin{array}{cc} 1 & \dfrac{l_i}{G_i I'_{P_i}} \\ 0 & 1 \end{array} \right] \left\{ \begin{array}{c} \phi_z \\ M_z \end{array} \right\}_{L_i},$$

$$\left\{ \begin{array}{c} \phi_z \\ M_z \end{array} \right\}_{L_i} = \left[\begin{array}{cc} 1 & 0 \\ -\lambda^2 J_{z_i} + \chi_c & 1 \end{array} \right] \left\{ \begin{array}{c} \phi_z \\ M_z \end{array} \right\}_{R_{i-1}}. \tag{2.80}$$

Once the transfer matrices have been obtained, it is easy to multiply them to obtain the overall transfer matrix. Equation (2.78) takes the simple form

$$\left\{ \begin{array}{c} \phi_z \\ M_z \end{array} \right\}_n = \left[\begin{array}{cc} T_{11} & T_{12} \\ T_{21} & T_{22} \end{array} \right] \left\{ \begin{array}{c} \phi_z \\ M_z \end{array} \right\}_0. \tag{2.81}$$

The boundary conditions are easily assessed. If the end at the left is free, moment M_{z_0} must vanish. The second column is of no interest, because it multiplies a vanishing moment. If the end at the right is also free, as often happens with torsional systems, the second equation (2.81) reduces to $M_{z_n} = T_{21}\phi_{z_0} = 0$, yielding a solution different from the trivial one $\phi_{z_0} = 0$ only if $T_{21} = 0$.

Working in the same way for the other end conditions, the following equations yielding the natural frequencies can be obtained:

Right end	Left end	Equation
Free	Free	$T_{21} = 0$
Free	Clamped	$T_{22} = 0$
Clamped	Free	$T_{11} = 0$
Clamped	Clamped	$T_{12} = 0$

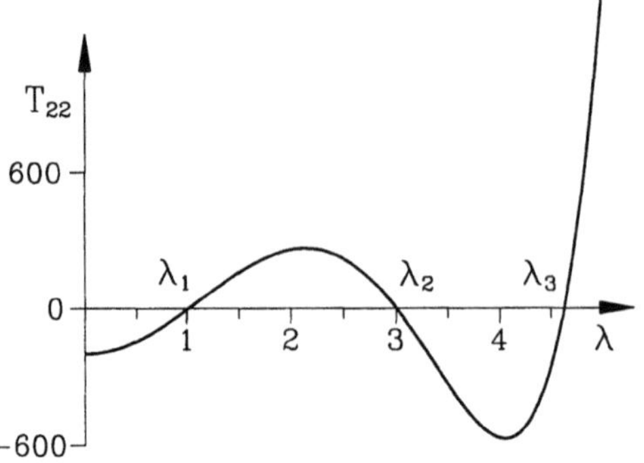

FIGURE 2.15. Plot used for the graphical computation of the natural frequencies of Example 1-2 using Holzer's method.

As already said, the solution is usually obtained numerically, plotting the appropriate element T_{ij} as a function of λ, and looking for the values of the frequency at which the curve crosses the frequency axis. The mode shapes can be obtained by setting an arbitrary value (usually a unit value) for M_{z_0}, if the left end is free, or for ϕ_{z_0} if it is clamped, and computing the state vectors after having introduced the value of the relevant natural frequency. Because only the shape of the modes is defined, the arbitrarity is not inconvenient, and it must be removed by normalizing the eigenvectors.

Example 2-5

Compute the natural frequencies of the system shown in Example 1-2 using Holtzer's method. Show that if the global transfer matrix is computed, it yields the same characteristic equation already obtained.

Starting the numeration of the nodes from the constrained end, the following transfer matrices are obtained:

$$\mathbf{T}_{f1} = \begin{bmatrix} 1 & 0.1 \\ 0 & 1 \end{bmatrix}, \quad \mathbf{T}_{n2} = \begin{bmatrix} 1 & 0 \\ -\lambda^2 & 1 \end{bmatrix},$$

$$\mathbf{T}_{f2} = \begin{bmatrix} 1 & 0.1 \\ 0 & 1 \end{bmatrix}, \quad \mathbf{T}_{n3} = \begin{bmatrix} 1 & 0 \\ -4\lambda^2 & 1 \end{bmatrix},$$

$$\mathbf{T}_{f3} = \begin{bmatrix} 1 & 0.25 \\ 0 & 1 \end{bmatrix}, \quad \mathbf{T}_{n4} = \begin{bmatrix} 1 & 0 \\ -0.5\lambda^2 & 1 \end{bmatrix}.$$

The global transfer matrix obtained by performing the relevant matrix multiplications is

$$\mathbf{T}_G = \left[\begin{array}{cc} 0.1\lambda^4 - 1.35\lambda^2 + 1 & 0.01\lambda^4 - 0.235\lambda^2 + 0.45 \\ -0.05\lambda^6 + 1.075\lambda^4 - 5.5\lambda^2 & -0.005\lambda^6 + 0.1575\lambda^4 - 1.125\lambda^2 + 1 \end{array} \right].$$

The right end is free while the left end is clamped and the condition allowing computation of the natural frequencies is $T_{22} = 0$, i.e.,

$$\lambda^6 - 31.5\lambda^4 + 225\lambda^2 - 200 = 0,$$

which coincides with that obtained in Example 1-5.

The plot of element T_{22} of the global transfer matrix as a function of λ is reported in Figure 2.15. The curve can be used to evaluate the natural frequencies without having to solve the characteristic equation.

2.6.4 Myklestadt's method for flexural vibrations of beams

Another well-known method based on the transfer-matrices approach is that introduced by Myklestadt for the flexural behaviour of beams, originally developed for the dynamic study of aircraft wings. It will be presented here in a form different from that of the original paper by Myklestadt with the aim of organizing it in the way generally seen for transfer-matrices methods.

Consider the bending behaviour of a beam, which is modeled as a lumped-parameters system, in the xz-plane. Because a moment of inertia J_y can be associated with each station, they are considered rigid bodies rather than point masses. Each end of a field is characterized by two degrees of freedom: displacement u_x and rotation ϕ_y; the order of state vectors is four, and they contain the shearing force F_x and bending moment M_y. The computation of the transfer matrices follows the same guidelines seen for Holtzer's method. Considering the ith field, it follows that the shear force at the right end is equal to the shear force at the left end, because no force acts on the field; the bending moment at the right end is equal to the bending moment at the left end, increased by the moment due to the shear force at the left end; the displacement at the right end is equal to the displacement at the left end, increased by the displacements due to the rigid rotation and to the deformation of the field; the rotation at the right end is equal to the rotation at the left end increased by the rotation due to bending deformation of the field.

The increments of displacement and rotation Δu_i and $\Delta \phi_i$ due to the deformation of the field can be computed through the beam theory as the ith field can be considered a prismatic beam clamped at the left end and loaded at the right end by a force $F_{x_{R_i}}$ and a moment $M_{y_{R_i}}$. The usual equations obtained for a clamped beam yield the following expression for the field transfer matrix

$$
\left\{ \begin{array}{c} u_x \\ \phi_y \\ F_x \\ M_y \end{array} \right\}_{R_i}
=
\left[\begin{array}{cccc}
1 & l_i & -\dfrac{l_i^3}{6EI_y} + \dfrac{l_i\chi}{GA} & \dfrac{l_i^2}{2EI_y} \\[2ex]
0 & 1 & -\dfrac{l_i^2}{2EI_y} & \dfrac{l_i}{EI_y} \\[2ex]
0 & 0 & 1 & 0 \\[1ex]
0 & 0 & -l_i & 1
\end{array} \right]
\left\{ \begin{array}{c} u_x \\ \phi_y \\ F_x \\ M_y \end{array} \right\}_{L_i}.
$$

$$(2.82)$$

Now consider the ith node. The displacement and the rotation at the left of the node are equal to the displacement at the right. The force at the right is equal to the force at the left increased by the inertia force due to mass m_i, which, in harmonic motion, is $-\lambda^2 m_i u_x$; the moment at the right is equal to the moment at the left increased by the inertia torque $-\lambda^2 J_{i_y}\phi_y$. If a constraint with stiffness k_c and angular stiffness χ_c is located at the node, the force at the left of the node must be increased by $k_c u_x$ and the moment must be increased by $\chi_c \phi_y$. Using matrix notation, the following expression for the transfer matrix of the ith node is obtained:

$$
\left\{ \begin{array}{c} u_x \\ \phi_y \\ F_x \\ M_y \end{array} \right\}_{L_i}
=
\left[\begin{array}{cccc}
1 & 0 & 0 & 0 \\
0 & 1 & 0 & 0 \\
-\lambda_i^2 m_i + k_c & 0 & 1 & 0 \\
0 & -\lambda^2 J_{i_y} + \chi_c & 0 & 1
\end{array} \right]
\left\{ \begin{array}{c} u_x \\ \phi_y \\ F_x \\ M_y \end{array} \right\}_{R_{i-1}}.
$$

$$(2.83)$$

Once the transfer matrices of all nodes and all fields have been obtained, there is no difficulty computing the global transfer matrix and introducing the boundary conditions in a way that is similar to that seen for the Holtzer method.

Consider, for example, that both ends are supported: The displacement and the bending moment must vanish in both s_n and s_0. Equation (2.78) then becomes

$$\left\{\begin{array}{c} 0 \\ \phi_y \\ F_x \\ 0 \end{array}\right\}_n = \left[\begin{array}{cccc} T_{11} & T_{12} & T_{13} & T_{14} \\ T_{21} & T_{22} & T_{23} & T_{24} \\ T_{31} & T_{32} & T_{33} & T_{34} \\ T_{41} & T_{42} & T_{43} & T_{44} \end{array}\right] \left\{\begin{array}{c} 0 \\ \phi_y \\ F_x \\ 0 \end{array}\right\}_0 . \qquad (2.84)$$

The first and fourth columns of the global transfer matrix can be canceled, because they multiply elements equal to zero in the state vector. The first and last equations (2.78) reduce to the homogenous equation

$$\left[\begin{array}{cc} T_{12} & T_{13} \\ T_{42} & T_{43} \end{array}\right] \left\{\begin{array}{c} \phi_y \\ F_x \end{array}\right\}_0 = \left\{\begin{array}{c} 0 \\ 0 \end{array}\right\}, \qquad (2.85)$$

which yields a solution different from the trivial one only if

$$\det \left[\begin{array}{cc} T_{12} & T_{13} \\ T_{42} & T_{43} \end{array}\right] = T_{12}T_{43} - T_{42}T_{13} = 0. \qquad (2.86)$$

In this case the possible boundary conditions are many; only a few of them are reported here:

Right end	Left end	Equation
Free	Free	$T_{31}T_{42} - T_{41}T_{32} = 0$
Free	Supported	$T_{32}T_{43} - T_{42}T_{33} = 0$
Free	Clamped	$T_{33}T_{44} - T_{43}T_{34} = 0$
Supported	Supported	$T_{12}T_{43} - T_{42}T_{13} = 0$
Clamped	Free	$T_{11}T_{22} - T_{21}T_{12} = 0$
Clamped	Clamped	$T_{13}T_{24} - T_{23}T_{14} = 0$

The value of the determinant of the matrix of the coefficients is plotted as a function of the frequency λ, and the values of the frequency for which it vanishes are obtained in a way similar to that seen in connection with Holtzer's method. It must be noted that the order in which the state variables are listed in the state vector can be different from the one shown here. Before using any formula from other books, it is necessary to verify the meaning of the subscripts of the elements of the transfer matrix.

a)

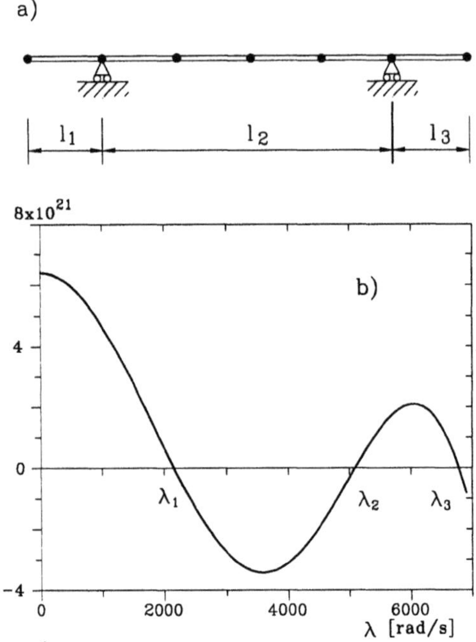

FIGURE 2.16. (a) Sketch of the beam of Example 2-6. (b)Plot used for the graphical computation of the natural frequencies using Myklestadt's method.

Example 2-6

Compute the natural frequencies of the beam shown in Figure 2.16a using Myklestadt's method. The beam has an annular cross section with inner and outer diameters of 80 mm and 100 mm, respectively, and the dimensions shown in the figure are $l_1 = l_3 = 200$ mm and $l_2 = 800$ mm. The main characteristics of the material are $E = 2.1 \times 10^{11}$ N/m^2; $\rho = 7{,}810$ kg/m^3; $\nu = 0.3$.

The mass of the beam is lumped in seven stations (4.417 kg in all nodes except the first and last ones where there is a mass of 2.209 kg). In the second and sixth station there are two supports. The rigid supports are modeled using elastic constraints with a stiffness of 1×10^{11} N/m. The computation of the global transfer matrix involves the multiplication of six field transfer matrices and seven node transfer matrices.

As both ends are free, the plot of the determinant $T_{31}T_{42} - T_{41}T_{32}$ of the reduced global transfer matrix as a function of λ is reported in Figure 2.16b, for a range up to the third natural frequency.

The natural frequencies, up to the fifth, computed by searching the points at which the plot of Figure 2.16b (extended to higher frequencies) crosses the frequency axis, are: $\lambda_1 = 2{,}158$ rad/s $= 343.4$ Hz; $\lambda_2 = 5{,}091$ rad/s $= 810.2$ Hz; $\lambda_3 = 6{,}780$ rad/s $= 1{,}079.1$ Hz; $\lambda_4 = 10{,}469$ rad/s $= 1{,}666.2$ Hz; $\lambda_5 = 15{,}858$ rad/s $= 2{,}523.9$ Hz;

2.7   The finite element method

2.7.1   Element characterization

The finite element method is a general discretization method for the
solution of partial derivative differential equations and, consequently,
it finds its application in many other fields beyond structural dynamics and structural analysis. The aim of this section is not to provide
a complete survey of the method, which can be dealt with only in a
specialized text, but simply to describe its main feature in order to
relate it to the other discretization techniques. The FEM is based on
the subdivision of the structure into finite elements, i.e., into parts
whose dimensions are not vanishingly small. Many different element
formulations have been developed, depending on their shape and
characteristics: beam elements, shell elements, plate elements, solid
elements, and many others. A structure can be built by assembling
elements of the same or different types, as dictated by the nature of
the problem and by the capabilities of the computer code used.

Each element is essentially the model of a small deformable solid.
The behaviour of the element is studied using an assumed-modes approach, as seen in Section 2.5. A limited number, usually quite small,
of degrees of freedom is then substituted to the infinity of degrees
of freedom of each element. Inside each element, the displacement
$\vec{u}(x,y,z)$ of the point of coordinates x,y,z is approximated by the
linear combination of a number n of functions, the shape functions,
which are assumed arbitrarily.

Because the FEM is usually developed using matrix notation, in
order to obtain formulas readily transferable to computer code, the
displacement is written as a vector of order 3 in the tridimensional
space (sometimes of higher order, if rotations are also considered),
and the equation expressing the displacement of the points inside
each element is

$$\mathbf{u}(x,y,z,t) = \mathbf{N}(x,y,z)\mathbf{q}(t)\,, \qquad (2.87)$$

where $\mathbf{q}$ is a vector in which the n generalized coordinates of the element are listed and $\mathbf{N}$ is the matrix containing the shape functions.
There are as many rows in $\mathbf{N}$ as in $\mathbf{u}$ and as many columns as the
number n of degrees of freedom.

Usually the degrees of freedom of the elements are the displacements at given points, which are referred to as *nodes*. In this case,

equation (2.87) is usually reduced to the simpler form,

$$\left\{ \begin{array}{c} u_x(x,y,z,t) \\ u_y(x,y,z,t) \\ u_z(x,y,z,t) \end{array} \right\} = \left[\begin{array}{ccc} \mathbf{N}(x,y,z) & 0 & 0 \\ 0 & \mathbf{N}(x,y,z) & 0 \\ 0 & 0 & \mathbf{N}(x,y,z) \end{array} \right] \left\{ \begin{array}{c} \mathbf{q}_x(t) \\ \mathbf{q}_y(t) \\ \mathbf{q}_z(t) \end{array} \right\},$$

$$(2.88)$$

where the displacements in each direction are functions of the nodal displacements in the same direction only. Matrix $\mathbf{N}$, in this case, has only one row and as many columns as the number of nodes of the element. Equation (2.88) has been written for a three-dimensional element; a similar formulation can also be easily obtained for one- or two-dimensional elements.

The shape functions are, as already stated, arbitrary. The freedom in the choice of such a function is, however, quite limited, because they must satisfy several conditions. A first requirement is a simple mathematical formulation, which is needed to lead to developments that are not too complex. Usually a set of polynomials in the space coordinates is assumed. To get results that are closer to the exact solution of the differential equations, which constitute the continuous model discretized by the FEM, while reducing the size of the elements, the shape functions must

- be continuous and derivable up to the required order, which depends on the type of element;
- be able to describe rigid-body motions of the element leading to vanishing elastic potential energy;
- lead to a constant strain field when the overall deformation of the element dictates so; and
- lead to a deflected shape of each element that matches the shape of the neighbouring elements.

This means that when the nodes of two neighbouring elements displace in a compatible way, all the interface between the elements must displace in a compatible way.

Another condition, which is not always satisfied, is that the shape functions must be isotropic, i.e., must not show particular geometrical properties that depend on the orientation of the reference frame. Sometimes not all the first conditions are completely met; in particular, there are elements that fail to completely satisfy the matching of the deflected shapes of neighbouring elements.

The nodes are usually located at the vertices or on the sides of the elements and are common to two or more of them, but points that are internal to an element can also be used.

The equation of motion of each element can be written following what was said about the assumed-modes methods. The strains can be expressed as functions of the derivatives of the displacements with respect to space coordinates. In general, it is possible to write a relationship of the type

$$\epsilon(x,y,z,t) = \mathbf{B}(x,y,z)\mathbf{q}(t)\,, \qquad (2.89)$$

where ϵ is a column matrix in which the various elements of the strain tensor are listed (it is commonly referred to as a *strain vector* but it is such only in the sense that it is a column matrix) and $\mathbf{B}$ is a matrix containing appropriate derivatives of the shape functions. $\mathbf{B}$ has as many rows as the number of components of the strain vector and as many columns as the number of degrees of freedom of the element.

If the element is free from initial stresses and strains and the behaviour of the material is linear, the stresses can be directly expressed from the strains

$$\boldsymbol{\sigma}(x,y,z,t) = \mathbf{E}\epsilon = \mathbf{E}(x,y,z)\mathbf{B}(x,y,z)\mathbf{q}(t)\,, \qquad (2.90)$$

where $\mathbf{E}$ is the stiffness matrix of the material. It is a symmetric square matrix whose elements can theoretically be functions of the space coordinates but are usually constant within the element. The potential energy of the element can be easily expressed as

$$\mathcal{U} = \frac{1}{2}\int_V \epsilon^T \boldsymbol{\sigma}\, dV = \frac{1}{2}\mathbf{q}^T\left(\int_V \mathbf{B}^T\mathbf{E}\mathbf{B}\, dV\right)\mathbf{q}\,. \qquad (2.91)$$

The integral in equation (2.91) is the stiffness matrix of the element

$$\mathbf{K} = \int_V \mathbf{B}^T\mathbf{E}\mathbf{B}\, dV\,. \qquad (2.92)$$

Because the shape functions do not depend on time, the generalized velocities can be expressed as $\dot{\mathbf{u}}(x,y,z,t) = \mathbf{N}(x,y,z)\dot{\mathbf{q}}(t)$. In the case where all generalized coordinates are related to displacements, the kinetic energy and the mass matrix of the element can be expressed as

$$T = \frac{1}{2}\dot{\mathbf{q}}^T \left(\int_V \rho \mathbf{N}^T \mathbf{N} dV \right) \dot{\mathbf{q}} ,$$
$$\mathbf{M} = \int_V \rho \mathbf{N}^T \mathbf{N} dV . \tag{2.93}$$

In the case that some generalized displacements are physically rotations, equation (2.93) must be changed in order to introduce moments of inertia, but its basic structure remains the same.

As already stated, the FEM is often used just to compute the stiffness matrix to be used in the context of the lumped-parameters approach. In this case, the consistent mass matrix (2.93) is not computed and a diagonal matrix obtained by lumping the mass at the nodes is used. The advantage is that of dealing with a diagonal mass matrix, whose inversion to compute the dynamic matrix is far simpler than that of the consistent mass matrix. The accuracy is, however, reduced or, better, a greater number of elements is needed to reach the same accuracy, and then the convenience between the two formulations must be assessed in each case. Generally speaking, the consistent approach leads to values of the natural frequencies that are in excess with respect to those computed using the elastic continuum model, while those obtained using the lumped-parameters approach are smaller.

If a force distribution $\mathbf{f}(x,y,z,t)$ acts on the body, the virtual work linked with the virtual displacement $\delta\mathbf{u} = \mathbf{N}\delta\mathbf{q}$ and the nodal force vector can be expressed in the form

$$\delta\mathcal{L} = \int_V \delta\mathbf{q}^T \mathbf{N}^T \mathbf{f}(x,y,z,t) dV ,$$
$$\mathbf{f}(t) = \int_V \mathbf{N}^T \mathbf{f}(x,y,z,t) dV . \tag{2.94}$$

In a similar way, it is possible to obtain the nodal force vectors corresponding to surface force distributions or to concentrated forces acting on any point of the element.

The equation of motion of the element is then the usual one for discrete undamped systems

$$\mathbf{M}\ddot{\mathbf{q}} + \mathbf{K}\mathbf{q} = \mathbf{f}(t) , \tag{2.95}$$

where vector $\mathbf{f}$ contains all forces acting on the element. The equations of motion and the relevant matrices have been obtained here by closely following the "assumed-mode" approach, basically using

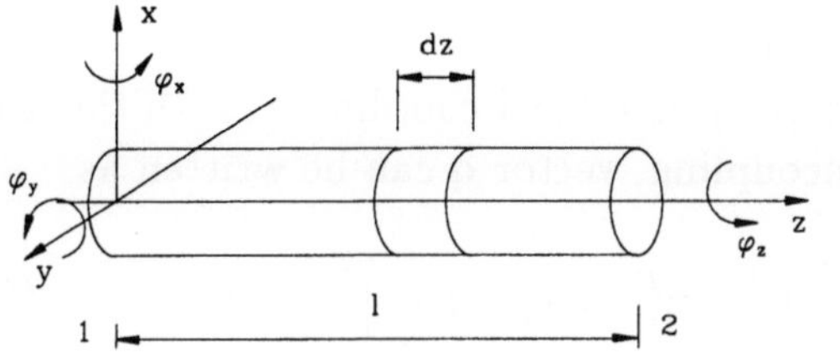

FIGURE 2.17. Beam element: geometrical definitions and reference frame.

Lagrange equations; this approach is neither the only one nor the most common.

2.7.2 Timoshenko beam element

The beam element is one of the most common elements and is generally available in all computer codes. Several beam formulations have been developed that differ owing to the theoretical formulation (some of them are Euler-Bernoulli element, i.e., do not take into account shear deformation, while others are Timoshenko elements) and the number of nodes and degrees of freedom per node. The element that will be studied here is often referred to as the *simple Timoshenko beam*. It has two nodes at the ends of the beam and six degrees of freedom per node, and it consists of a prismatic homogeneous beam to which all considerations seen in Section 2.2.1 apply. The relevant geometrical definition and the reference frame used for the study are shown in Figure 2.17.

Each cross section has six degrees of freedom, three displacements, and three rotations, and the total number of degrees of freedom of the element is 12. The vector of the nodal displacements, i.e., of the generalized coordinates of the element, is

$$\mathbf{q} = [u_{x_1}, u_{y_1}, u_{z_1}, \phi_{x_1}, \phi_{y_1}, \phi_{z_1}, u_{x_2}, u_{y_2}, u_{z_2}, \phi_{x_2}, \phi_{y_2}, \phi_{z_2}]^T . \qquad (2.96)$$

The beam has the properties needed to perform a complete uncoupling between axial, torsional, and flexural behaviour in each of the coordinate planes; it is then expedient to subdivide vector $\mathbf{q}$ into four smaller vectors: $\mathbf{q}_A = [u_{z_1}, u_{z_2}]^T$; $\mathbf{q}_T = [\phi_{z_1}, \phi_{z_2}]^T$; $\mathbf{q}_{F1} = [u_{x_1}, \phi_{y_1}, u_{x_2}, \phi_{y_2}]^T$; and $\mathbf{q}_{F2} = [u_{y_1}, \phi_{x_1}, u_{y_2}, \phi_{x_2}]^T$. Note that if the rotational degree of freedom $-\phi_x$ was used instead of ϕ_x the same equations could have been used to describe the flexural behaviour in both planes. This approach is, however, uncommon because it would make it more difficult to pass from the system of reference of the elements to that of the whole structure.

By reordering the various generalized coordinates with the aim of clearly showing such uncoupling, vector $\mathbf{q}$ can be written as

$$\mathbf{q} = \begin{bmatrix} \mathbf{q}_A^T & \mathbf{q}_T^T & \mathbf{q}_{F1}^T & \mathbf{q}_{F2}^T \end{bmatrix}^T . \tag{2.97}$$

The uncoupling between the various degrees of freedom makes it possible to split the matrix of the shape functions into a number of submatrices, most of which are equal to zero. The generalized displacement of an internal point of the element whose coordinate is z can be expressed in the form of equation (2.87) by the equation

$$\mathbf{u}(z,t) = \begin{Bmatrix} u_z \\ \phi_z \\ u_x \\ \phi_y \\ u_y \\ \phi_x \end{Bmatrix} = \begin{bmatrix} \mathbf{N}_A & 0 & 0 & 0 \\ 0 & \mathbf{N}_T & 0 & 0 \\ 0 & 0 & \mathbf{N}_{F1} & 0 \\ 0 & 0 & 0 & \mathbf{N}_{F2} \end{bmatrix} \begin{Bmatrix} \mathbf{q}_A \\ \mathbf{q}_T \\ \mathbf{q}_{F1} \\ \mathbf{q}_{F2} \end{Bmatrix} . \tag{2.98}$$

Axial behaviour. Where axial behaviour is concerned, the beam reduces to a bar.

Because each point of the element has a single degree of freedom, vector $\mathbf{u}$ has a single component u_z and matrix $\mathbf{N}_A$ has one row and two columns (the element has two degrees of freedom). u_z can be expressed as a polynomial in z, or, better, in the nondimensional axial coordinate $\zeta = z/l$:

$$u_z = a_0 + a_1\zeta + a_2\zeta^2 + a_3\zeta^3 + \dots . \tag{2.99}$$

The polynomial must yield the values of the displacements u_{z_1} and u_{z_2}, respectively at the left end (node 1, $\zeta=0$) and at the right end (node 2, $\zeta=1$). These two conditions allow computation of only two coefficients a_i and then the polynomial expression of the displacement must include only two terms, i.e., the constant and the linear terms. With simple computations, the matrix of the shape functions is obtained:

$$\mathbf{N}_A = [1 - \zeta, \, \zeta] . \tag{2.100}$$

The axial strain ϵ_z can be expressed as $\epsilon_z = du_z/dz$, or, using vector $\boldsymbol{\epsilon}$, which in this case has only one element

$$\epsilon_z = \left[\frac{d}{dz}(1-\zeta), \frac{d}{dz}\zeta\right] \left\{ \begin{array}{c} u_{z_1} \\ u_{z_2} \end{array} \right\}. \tag{2.101}$$

Matrix

$$\mathbf{B} = \left[\frac{d}{dz}(1-\zeta), \frac{d}{dz}\zeta\right] = \frac{1}{l}[-1,1] \tag{2.102}$$

has one row and two columns.

Also, vector $\boldsymbol{\sigma}$ and matrix $\mathbf{E}$ have only a single element: the axial stress σ_z and Young's modulus E, respectively. The stiffness and mass matrices can be obtained directly from equations (2.92) and (2.93). Remembering that $dV = Adz$, they reduce to

$$\mathbf{K}_A = \int_0^l A\mathbf{B}^T E\mathbf{B}dz = \frac{EA}{l}\int_0^1 \left[\begin{array}{cc} 1 & -1 \\ -1 & 1 \end{array} \right] d\zeta = \frac{EA}{l} \left[\begin{array}{cc} 1 & -1 \\ -1 & 1 \end{array} \right], \tag{2.103}$$

$$\mathbf{M}_A = \int_0^l \rho A\mathbf{N}^T\mathbf{N}dz = \rho Al\int_0^1 \left[\begin{array}{cc} (1-\zeta)^2 & \zeta(1-\zeta) \\ \zeta(1-\zeta) & \zeta^2 \end{array} \right] d\zeta =$$

$$= \frac{\rho Al}{6} \left[\begin{array}{cc} 2 & 1 \\ 1 & 2 \end{array} \right]. \tag{2.104}$$

If an axial force distribution $f_z(t)$ that is constant along the space coordinate z or a concentrated axial force $F_{z_k}(t)$ located in the point of coordinate z_k is acting on the bar, the nodal force vector is, respectively,

$$\mathbf{f}(t) = l\left[\int_0^l \left\{ \begin{array}{c} (1-\zeta) \\ \zeta \end{array} \right\} d\zeta\right] f_z(t) = \frac{1}{2}lf_z(t) \left\{ \begin{array}{c} 1 \\ 1 \end{array} \right\}, \tag{2.105}$$

or

$$\mathbf{f}(t) = F_{z_k}(t) \left\{ \begin{array}{c} 1 - \dfrac{z_k}{l} \\ \dfrac{z_k}{l} \end{array} \right\}. \tag{2.106}$$

In this case the distributed load has been reduced to two identical forces; each is equal to half of the total load acting on the bar. An identical result would have been obtained by simply lumping the load at the nodes. This is not, however, a general rule; in other cases the consistent approach leads to a load vector that is not the same, which could have been obtained from the lumped-parameters approach.

Torsional behaviour. As stated in Section 2.2.3, the equations of motion governing the torsional behaviour of beams are formally identical to those governing the axial behaviour. Using this identity, the characterization of the beam element in torsion can be obtained from what has been seen for the axial behaviour. Matrix $\mathbf{N}_T$ is identical to matrix $\mathbf{N}_A$: $\mathbf{N}_T = [1 - \zeta , \zeta]$, and the expressions of the relevant matrices and vectors are

$$\mathbf{M}_T = \frac{\rho I_p l}{6} \begin{bmatrix} 2 & 1 \\ 1 & 2 \end{bmatrix} , \qquad \mathbf{K}_T = \frac{G I'_p}{l} \begin{bmatrix} 1 & -1 \\ -1 & 1 \end{bmatrix} ,$$

$$\mathbf{f}(t)_T = \frac{1}{2} l m_z(t) \begin{Bmatrix} 1 \\ 1 \end{Bmatrix} . \tag{2.107}$$

Flexural behaviour in the xz-plane. The expressions of the shape function are in this case more complex; matrix $\mathbf{N}_{F1}$ has two rows and four columns because it must yield the displacement in the x-direction and the rotation about the y-axis of the generic cross section of the beam when it is multiplied by vector $\mathbf{q}_{F1}$, which contains four elements, namely, the displacements in the x-direction and the rotation about the y-axis of the two nodes. The simplest approach would be that of assuming polynomial expressions for the generalized displacements

$$\begin{cases} u_x = a_0 + a_1\zeta + a_2\zeta^2 + a_3\zeta^3 + \dots \\ \phi_y = b_0 + b_1\zeta + b_2\zeta^2 + b_3\zeta^3 + \dots . \end{cases} \tag{2.108}$$

These polynomial expressions must yield the values of the displacements u_{x1} and u_{x2} and of the rotations ϕ_{y1} and ϕ_{y2} at the left end (node 1, $\zeta=0$) and at the right end (node 2, $\zeta=1$), respectively. These four conditions allow computation of only four coefficients a_i and b_i to be introduced into the polynomial expressions. Each polynomial must then include only two terms, and both rotations and displacements must vary linearly along the z-coordinate. This element formulation, although sometimes used, leads to the very severe problem of locking, i.e., to the possibility of grossly overestimating the stiffness of the element.

Although locking will not be dealt with in detail here (the reader can find a detailed discussion on this matter in any good textbook on FEM), an intuitive explanation can be seen immediately: If the beam is slender the rotation of each cross section is very close to the derivative of the displacement, as stated by the Euler-Bernoulli approach for slender beams. The polynomial shape functions, when truncated at the second term, do not allow the rotation to be equal to the derivative of the displacement, and this can be shown to lead to a severe underestimate of the displacements, i.e., of the flexibility of the beam.

A simple cure for the problem is that of resorting to Euler-Bernoulli formulation, i.e., neglecting shear deformation and then assuming that the rotation is coincident with the derivative of the displacement. In this case, only the polynomial for u_x needs to be stated, and the aforementioned four conditions at the nodes can be used to compute the four coefficients of a cubic expression of the displacement. A Timoshenko beam element can be formulated using as shape functions the deformed shape computed using the continous model assuming that only end forces are applied to the beam. This Timoshenko beam element reduces to the Euler-Bernoulli element as the slenderness of the beam increases and no locking occurs. The relevant shape functions are

$$N_{11} = \frac{1 + \Phi(1 - \zeta) - 3\zeta^2 + 2\zeta^3}{1 + \Phi}, \quad N_{12} = l\zeta\frac{1 + \frac{1}{2}\Phi(1 - \zeta) - 2\zeta + \zeta^2}{1 + \Phi},$$

$$N_{13} = \zeta\frac{\Phi + 3\zeta - 2\zeta^2}{1 + \Phi}, \quad N_{14} = l\zeta\frac{-\frac{1}{2}\Phi(1 - \zeta) - \zeta + \zeta^2}{1 + \Phi},$$

$$N_{21} = 6\zeta\frac{\zeta - 1}{l(1 + \Phi)}, \quad N_{22} = \frac{1 + \Phi(1 - \zeta) - 4\zeta + 3\zeta^2}{1 + \Phi},$$

$$N_{23} = -6\zeta\frac{\zeta - 1}{l(1 + \Phi)}, \quad N_{24} = \frac{\Phi\zeta - 2\zeta + 3\zeta^2}{1 + \Phi},$$

$$(2.109)$$

where

$$\Phi = \frac{12EI_y\chi}{GAl^2}.$$

When the slenderness of the beam increases, the value of Φ decreases, tending to zero for a Euler-Bernoulli beam.

In this case, some of the generalized coordinates are related to rotations; as a consequence, equations (2.92) and (2.93) cannot be used directly to express the stiffness and mass matrices. The potential energy obtained can be computed by adding the contributions due to bending and shear deformations. By using the symbols $\mathbf{N}_1$ and $\mathbf{N}_2$ to express the first and second rows of matrix $\mathbf{N}_{F1}$, respectively, the two contributions to the potential energy of the length dz of the beam are

$$d\mathcal{U}_f = \frac{1}{2}EI_y\left(\frac{d\phi_y}{dz}\right)^2 dz = \frac{1}{2}EI_y\{q\}^T\left[\frac{d}{dz}\mathbf{N}_2\right]^T\left[\frac{d}{dz}\mathbf{N}_2\right]\{q\}dz,$$

$$d\mathcal{U}_t = \frac{1}{2}\frac{GA}{\chi}\left(\phi_y - \frac{du_x}{dz}\right)^2 dz = \tag{2.110}$$

$$= \frac{12EI_y}{2\Phi l^2}\{q\}^T\left[\mathbf{N}_2 - \frac{d}{dz}\mathbf{N}_1\right]^T\left[\mathbf{N}_2 - \frac{d}{dz}\mathbf{N}_1\right]\{q\}dz.$$

By introducing the expressions of the shape functions into the expression of the potential energy and integrating, the bending stiffness matrix is obtained

$$\mathbf{K}_{F_1} = \frac{EI_y}{l^3(1+\Phi)}\begin{bmatrix} 12 & 6l & -12 & 6l \\ & (4+\Phi)l^2 & -6l & (2-\Phi)l^2 \\ & & 12 & -6l \\ & \text{simm.} & & (4+\Phi)l^2 \end{bmatrix}. \tag{2.111}$$

The kinetic energy of the length dz of the beam is

$$d\mathcal{T} = \frac{1}{2}\rho A\dot{u}^2 dz + \frac{1}{2}\rho I_y\dot{\phi}_y^{\,2} dz =$$

$$= \frac{1}{2}\rho A\dot{\mathbf{q}}^T\mathbf{N}_1^T\mathbf{N}_1\dot{\mathbf{q}}dz + \frac{1}{2}\rho I_y\dot{\mathbf{q}}^T\mathbf{N}_2^T\mathbf{N}_2\dot{\mathbf{q}}dz. \tag{2.112}$$

The first term on the right-hand side is the translational kinetic energy, and the second expresses the rotational kinetic energy, which is often neglected in the case of slender beams. By introducing the expressions of the shape functions and integrating, the consistent mass matrix, which is made of two parts - one taking into account the translational inertia and the other the rotational inertia of the cross sections - is obtained

$$\mathbf{M}_{F1} = \frac{\rho A l}{420(1 + \varPhi)^2} \begin{bmatrix} m_1 & lm_2 & m_3 & -lm_4 \\ & l^2 m_5 & lm_4 & -l^2 m_6 \\ & & m_1 & -lm_2 \\ \text{simm.} & & & l^2 m_5 \end{bmatrix} +$$

$$+ \frac{\rho I_y}{30l(1 + \varPhi)^2} \begin{bmatrix} m_7 & lm_8 & -m_7 & lm_8 \\ & l^2 m_9 & -lm_8 & -l^2 m_{10} \\ & & m_7 & -lm_8 \\ \text{simm.} & & & l^2 m_9 \end{bmatrix} , \qquad (2.113)$$

where

$$
\begin{aligned}
m_1 &= 156 + 294\varPhi + 140\varPhi^2, & m_2 &= 22 + 38.5\varPhi + 17.5\varPhi^2, \\
m_3 &= 54 + 126\varPhi + 70\varPhi^2, & m_4 &= 13 + 31.5\varPhi + 17.5\varPhi^2, \\
m_5 &= 4 + 7\varPhi + 3.5\varPhi^2, & m_6 &= 3 + 7\varPhi + 3.5\varPhi^2, \\
m_7 &= 36, & m_8 &= 3 - 15\varPhi, \\
m_9 &= 4 + 5\varPhi + 10\varPhi^2, & m_{10} &= 1 + 5\varPhi - 5\varPhi^2.
\end{aligned}
$$

The consistent load vector due to a uniform distribution of shear force per unit length $f_x(t)$ or of bending moment $m_y(t)$ can be obtained directly from equation (2.94)

$$\mathbf{f}(t)_{F1} = l \left[\int_0^1 \mathbf{N}_{F1}^T d\zeta \right] \left\{ \begin{array}{c} f_x(t) \\ m_y(t) \end{array} \right\} = \frac{l f_x(t)}{12} \left\{ \begin{array}{c} 6 \\ l \\ 6 \\ -l \end{array} \right\} + \frac{m_y(t)}{1 + \varPhi} \left\{ \begin{array}{c} -l \\ \dfrac{\varPhi l}{2} \\ l \\ \dfrac{\varPhi l}{2} \end{array} \right\} .$$

$$(2.114)$$

Flexural behaviour in the yz-plane. As already pointed out, the flexural behaviour in the yz-plane must be studied using equations different from those used for the xz-plane, owing to the different signs of rotation in the two planes. Matrix $\mathbf{N}_{F2}$ can, however, be obtained from matrix $\mathbf{N}_{F1}$ simply by changing the signs of elements with subscripts 12, 14, 21, and 23, and the mass and stiffness matrices related to plane yz are the same as those computed for plane xz, except for the signs of elements with subscripts 12, 14, 23, and 34 and their symmetrical ones. In the force vectors related to external forces (distributed or concentrated) or external moments, the signs of elements 2 and 4 or 1 and 3, respectively, must be changed. If the beam is not axially symmetrical and the elastic and inertial properties in the two planes are not coincident, different values of the moments of inertia and the shear factors must be introduced.

Global behaviour of the beam. The complete expression of the mass and stiffness matrices and of the nodal force vector are, respectively

$$
\mathbf{M} = \begin{bmatrix} \mathbf{M}_A & 0 & 0 & 0 \\ 0 & \mathbf{M}_T & 0 & 0 \\ 0 & 0 & \mathbf{M}_{F1} & 0 \\ 0 & 0 & 0 & \mathbf{M}_{F2} \end{bmatrix} , \qquad (2.115)
$$

$$
\mathbf{K} = \begin{bmatrix} \mathbf{K}_A & 0 & 0 & 0 \\ 0 & \mathbf{K}_T & 0 & 0 \\ 0 & 0 & \mathbf{K}_{F1} & 0 \\ 0 & 0 & 0 & \mathbf{K}_{F2} \end{bmatrix} , \qquad \mathbf{f} = \begin{Bmatrix} \mathbf{f}_A \\ \mathbf{f}_T \\ \mathbf{f}_{F1} \\ \mathbf{f}_{F2} \end{Bmatrix} .
$$

Effect of axial forces. One of the simplest geometric nonlinearity effects is the interaction between a constant axial load and the flexural behaviour of beams. In Section 2.2.6 it was shown that if the axial load can be considered a constant, it is possible to study this interaction using a linearized model. The same also holds when FEM is used.

Consider the flexural vibration in the xz-plane of a Timoshenko beam element, that is subject to a constant and known axial force F_a. Such a force must not be confused with the axial force $F_z(t)$ acting on the element, which is usually one of the unknowns in the problem. If force F_a is not known, an iterative procedure must be followed: The axial force F_z due to the loads must be computed first, and then the same force can be introduced in a further dynamic computation. Taking into account the lateral inflection of the beam, the axial strain in correspondence of the neutral axis of the beam, which is usually expressed as $\epsilon_z = \partial u_z / \partial z$, becomes

$$
\epsilon_z = \frac{\partial u_z}{\partial z} + \frac{1}{2}\left(\frac{\partial u_y}{\partial z}\right)^2 . \qquad (2.116)
$$

If the small term linked with lateral deflection is also considered, the elastic potential energy due to the axial strain becomes

$$
\mathcal{U} = \frac{1}{2}\int_0^l EA\epsilon_z^2 dz = \frac{1}{2}\int_0^l EA\left[\frac{\partial u_z}{\partial z} + \frac{1}{2}\left(\frac{\partial u_y}{\partial z}\right)^2\right]^2 dz =
$$

$$= \frac{1}{2} \int_0^l EA \left(\frac{\partial u_z}{\partial z} \right)^2 dz + \frac{1}{2} \int_0^l EA \frac{\partial u_z}{\partial z} \left(\frac{\partial u_y}{\partial z} \right)^2 dz . \qquad (2.117)$$

The last expression has been obtained by neglecting the term in $(\partial u_y/\partial z)^4$. The first term of the potential energy has already been taken into account in the computation of the stiffness matrix of the element. The second one causes an increase of the potential energy, which can be considered by adding a suitable matrix, the so-called geometric stiffness matrix, or simply geometric matrix, to the stiffness matrix of the element. Assuming that the axial force F_a is constant, the increment of potential energy can be written in terms of matrix $\mathbf{K}_{g_1}$ for flexural behaviour in the xz-plane in the form

$$\Delta \mathcal{U} = \frac{1}{2} F_a \int_0^l EA \left(\frac{\partial u_y}{\partial z} \right)^2 dz = \frac{1}{2} F_a \mathbf{q}_{F1}^T \mathbf{K}_{g_1} \mathbf{q}_{F1} , \qquad (2.118)$$

where

$$\mathbf{K}_{g_1} = \frac{F_a}{30 l (1 + \Phi)^2} \begin{bmatrix} k_1 & l k_2 & -k_1 & l k_2 \\ & l^2 k_3 & -l k_2 & -l^2 k_4 \\ & & k_1 & -l k_2 \\ \text{simm.} & & & l^2 k_3 \end{bmatrix} ;$$

$k_1 = 36 + 60\Phi + 3\Phi^2$; $k_2 = 3$; $k_3 = 4 + 5\Phi + 2.5\Phi^2$; and $k_4 = 1 + 5\Phi + 2.5\Phi^2$.

As for all other matrices describing flexural behaviour, the matrix related to the yz-plane differs from that computed for the xz-plane for the signs of elements 12, 14, 23, and 34 and their symmetricals, apart from being computed using the appropriate value of Φ.

Apart from being used to take into account the effects of axial forces on the flexural behaviour, the geometric matrix allows the study of the elastic stability of a structure. Because an axial tensile force causes an increase of stiffness while a compressive force has the opposite effect, the value of the compressive axial force that causes the stiffness matrix to be singular is the buckling load of the element or of the structure. Geometric matrices can also be defined for other types of elements.

2.7.3 Mass and spring elements

Consider a concentrated mass, or better, a rigid body located at the ith node. Let $\mathbf{q}$ be the vector of the generalized displacements of the relevant node, which can also contain rotations, if the node is of the type of those seen in the case of beam elements: $\mathbf{q} = [u_x, u_y, u_z, \phi_x, \phi_y, \phi_z]$. In the latter case, if the mass also has moments of inertia, let the axes of the reference frame coincide with the principal axes of the rigid body. By writing the kinetic energy of the element, it can be shown that the mass matrix, which is diagonal because the axes of the reference frame coincide with the principal axes of the body, is

$$\mathbf{M} = \operatorname{diag}[m, m, m, J_x, J_y, J_z] . \tag{2.119}$$

A simpler expression is obtained when only the translational degrees of freedom of the node are considered. Note that in many computer programs, different values of the mass can be associated with the various degrees of freedom. This can account for particular physical layouts, such as the addition of a mass constrained to move with the structure in one direction and not in others.

Consider a spring element, i.e., an element that introduces a concentrated stiffness between two nodes, say node 1 and node 2, of the structure. When the nodes have a single degree of freedom, the generalized coordinates of the element are $q = [u_1, u_2]^T$, and the stiffness matrix is

$$\mathbf{K} = \begin{bmatrix} k & -k \\ -k & k \end{bmatrix} . \tag{2.120}$$

If the nodes, like those used with beam elements, have three translational and three rotational degrees of freedom, three values of translational stiffness k_x, k_y, and k_z and three values of rotational stiffness χ_x, χ_y, and χ_z can be stated and matrices of the type stated earlier can be written for each degree of freedom.

2.7.4 Assembling the structure

The equations of motion of the element are written with reference to a local or element reference frame that has an orientation determined by the features of the element. In a beam element, for example, the z-axis can coincide with the axis of the beam, while the x- and y-axes are principal axes of inertia of the cross section. The various local reference frames of the elements have, in general, a different orientation in space. To describe the behaviour of the structure as a whole, another system of reference, namely the global or structure reference frame, is defined. The orientation in space of any local frame can be expressed, with reference to the orientation of the global frame, by a suitable rotation matrix

$$\mathbf{R} = \begin{bmatrix} l_x & m_x & n_x \\ l_y & m_y & n_y \\ l_z & m_z & n_z \end{bmatrix}, \tag{2.121}$$

where l_i, m_i, and n_i are the direction cosines of the axes of the former in the global frame. The expressions $\mathbf{q}_{il}$ and $\mathbf{q}_{i_g}$ of the displacement vector $\mathbf{q}_i$ of the ith node in the local and global reference frames are linked by the usual coordinate transformation $\mathbf{q}_{il} = \mathbf{R}\mathbf{q}_{i_g}$. The generalized coordinates in the displacement vector of the element can be transformed from the local to the global reference frame using a similar relationship in which an expanded rotation matrix $\mathbf{R}'$ is used to deal with all the relevant generalized coordinates. It is essentially made by a number of matrices of the type of equation (2.121) suitably assembled together. Note that the assumption of small displacements and rotations allows consideration of the rotations about the axes as the components of a vector, which can be rotated in the same way as displacements.

The force vectors can also be rotated using the rotation matrix $\mathbf{R}'$, and the equation of motion of the element can be written with reference to the global frame and premultiplied by the inverse of matrix $\mathbf{R}'$, obtaining

$$\mathbf{R}'^{-1}\mathbf{M}\mathbf{R}'\ddot{\mathbf{q}}_g + \mathbf{R}'^{-1}\mathbf{K}\mathbf{R}'\mathbf{q}_g = \mathbf{f}_g. \tag{2.122}$$

Because the inverse of a rotation matrix is coincident with its transpose, the expressions of the mass and stiffness matrices of the element rotated from the local to the global frame are $\mathbf{M}_g = \mathbf{R}'^T\mathbf{M}_l\mathbf{R}'$ and $\mathbf{K}_g = \mathbf{R}'^T\mathbf{K}_l\mathbf{R}'$. Similarily, the nodal load vector can be rotated using the obvious relationship $\mathbf{f}_g = \mathbf{R}'^T\mathbf{f}_l$.

Once the mass and stiffness matrices, referring to the global frame, of the various elements have been computed, it is possible to easily obtain the matrices of the whole structure. The n generalized coordinates of the structure can be ordered in a single vector $\mathbf{q}_g$. The matrices of the various elements can be rewritten in the form of matrices of order $n \times n$, containing all elements equal to zero except those that are in the rows and columns corresponding to the generalized coordinates of the relevant element. Because the kinetic and potential energies of the structure can be obtained simply by adding the energies of the various elements, it follows that

$$
\begin{aligned}
\mathcal{T} &= \frac{1}{2} \sum_{\forall i} \dot{\mathbf{q}}_g^T \mathbf{M}_i \dot{\mathbf{q}}_g = \frac{1}{2} \dot{\mathbf{q}}_g^T \mathbf{M} \dot{\mathbf{q}}_g \, ; \\
\mathcal{U} &= \tfrac{1}{2} \sum_{\forall i} \mathbf{q}_g^T \mathbf{K}_i \mathbf{q}_g = \tfrac{1}{2} \mathbf{q}_g^T \mathbf{K} \mathbf{q}_g \, .
\end{aligned}
\tag{2.123}
$$

Matrices $\mathbf{M}$ and $\mathbf{K}$ are the mass and stiffness matrices of the whole structure and are obtained simply by adding all the mass and stiffness matrices of the elements. In practice, the various matrices of size $n \times n$ for the elements are never written: Each term of the matrices of all elements is just added into the global mass and stiffness matrices in the correct place. Actually, the matrices of the structure are very easily assembled, and this is one of the simplest steps. If the generalized coordinates are taken into a suitable order, the assembled matrices have a band structure; many general-purpose computer codes have a suitable routine that reorders the coordinates in such a way that the bandwidth is the smallest possible.

In a similar way the nodal force vector can be easily assembled: $\mathbf{f} = \sum_{\forall i} \mathbf{f}_i$.

The forces that are exchanged between the elements at the nodes cancel each other in this assembling procedure, and then the force vectors that must be inserted into the global equation of motion of the structure are only those related to the external forces applied to the structure.

2.7.5 Constraining the structure

One of the advantages of the FEM is the ease with which the constraints can be defined. If the ith degree of freedom is rigidly constrained, the corresponding generalized displacement vanishes and, as a consequence, the ith column of the stiffness and mass matrices can be neglected, because they multiply a displacement and an acceleration, respectively, that are equal to zero. Because one of the generalized displacements is known, one of the equations of motion can be neglected when solving for the deformed configuration of the system. The ith equation can thus be separated from the rest of the set of equations, which amounts to canceling the ith row of all matrices and of the force vector. Note that the ith equation could, in this case, be used after all displacements have been computed to obtain the value of the ith generalized nodal force, which, in this case, is the unknown reaction of the constraint.

To rigidly constrain a degree of freedom, it is sufficient to cancel the corresponding row and column in all matrices and vectors. This approach allows simplification of the formulation of the problem, which is particularly useful in dynamic problems, but this simplification is often marginal, as the number of constrained degrees of freedom is very small, compared with the total number of degrees of freedom. To avoid restructuring the whole model and rewriting all the matrices, rigid constraints can be transformed into very stiff elastic constraints.

If the ith degree of freedom is constrained through a linear spring with stiffness k_i, the potential energy of the structure is increased by the potential energy of the spring

$$\mathcal{U} = \frac{1}{2} k_i q_i^2 \ . \tag{2.124}$$

To take the presence of the constraint into account, it is sufficient to add the stiffness k_i to the element in the ith row and ith column of the global stiffness matrix. This procedure is very simple, which explains why a very stiff elastic constraint is often added instead of canceling a degree of freedom in the case of rigid constraints. An additional advantage is that the reaction of the constraint can be obtained simply by multiplying the very high generalized stiffness k_i by the corresponding very small generalized displacement q_i.

Example 2-7
Write the mass and stiffness matrices for the study of the flexural behaviour of the transmission shaft with a central joint sketched in Figure 2.18. The joint, whose mass is m_G, is supported by an elastic constraint with stiffness k_G. Use the FEM, neglecting shear deformation and rotational inertia of the cross sections.

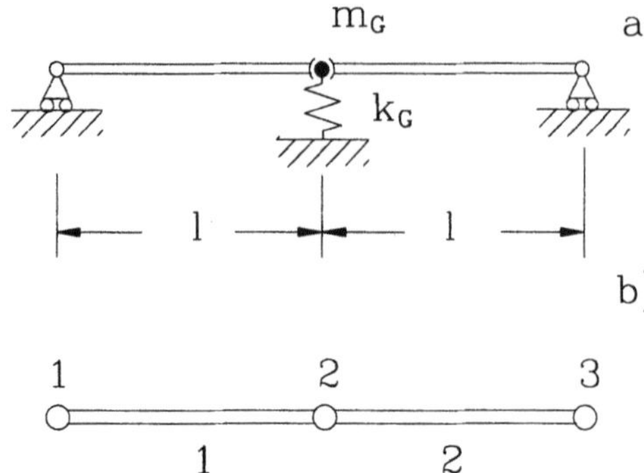

FIGURE 2.18. (a) Sketch of the system; (b) model based on two beam elements.

A two-beam element model is sketched in Figure 2.18b. There are four degrees of freedom of each element involved in flexural behaviour in the xz-plane. Using the Euler-Bernoulli approach, the stiffness matrices of the elements are

$$\mathbf{K} = \frac{EI_y}{l^3} \begin{bmatrix} 12 & 6l & -12 & 6l \\ & 4l^2 & -6l & 2l^2 \\ & & 12 & -6l \\ \text{simm.} & & & 4l^2 \end{bmatrix}.$$

The total number of degrees of freedom of the unconstrained system is seven, because the connection between the elements is performed through a joint that allows different rotations of the two elements at node 2. The two constraints lock translational degrees of freedom at nodes 1 and 3, reducing the number of degrees of freedom to five. Directly considering the constraints, the map of the degrees of freedom, i.e., the table stating the correspondence of the degrees of freedom of the elements to those of the structure, is as shown here:

Element 1	d.o.f.	1	2	3	4			
	type	transl.	rot.	transl.	rot.			
Element 2	d.o.f.		1		2	3	4
	type		transl.		rot.	transl.	rot.
Global	d.o.f.	constr.	1	2	3	4	constr.	5

Noting that the stiffness k_G and the mass m_G must be added to element 22, the stiffness and mass matrices are

$$\mathbf{K} = \frac{EI_y}{l^3} \begin{bmatrix} 4l^2 & -6l & 2l^2 & 0 & 0 \\ & 24 + k^* & -6l & 6l & 6l \\ & & 4l^2 & 0 & 0 \\ & & & 4l^2 & 2l^2 \\ \text{symm.} & & & & 4l^2 \end{bmatrix},$$

$$\mathbf{M} = \frac{\rho A}{420} \begin{bmatrix} 4l^2 & 13l & -3l^2 & 0 & 0 \\ & 312 + m^* & -22l & 22l & -13l \\ & & 4l^2 & 0 & 0 \\ & & & 4l^2 & -3l^2 \\ & \text{symm.} & & & 4l^2 \end{bmatrix},$$

where

$$k^* = \frac{k_G l^3}{EI_y}, \qquad m^* = \frac{420 m_G}{\rho A l}.$$

Example 2-8

Compute the natural frequencies of the beam of Example 2-6 using the FEM. Repeat the computations using both the lumped-parameters and the consistent approach and compare the results with those obtained using the Myklestadt method. Regard the supports as both rigid and very stiff elastic constraints.

The simplest approach is the lumped-parameters one. The model follows closely what was seen in Example 2-6: The stiffness matrices of the six Timoshenko beam elements are readily computed (they are all equal) and assembled. The assembly procedure is so simple that a map is not even needed. Because there are seven nodes and the element has two degrees of freedom per node (bending in one of the lateral planes), the order of the matrices is 14.

The stiffness of the constraints is assumed to be 1×10^{11} N/m. This number is then added to the element 3,3 and 11,11 of the stiffnness matrix. The mass matrix is diagonal, with the lumped masses in the positions with odd subscipts and zeros in those with even subscripts, because the moments of inertia are neglected. Because the the mass matrix is singular, the stiffness matrix is inverted to obtain the dynamic matrix $\mathbf{D}$.

The natural frequencies, up to the fifth, are: $\lambda_1 = 2{,}158$ rad/s $= 343.4$ Hz; $\lambda_2 = 5{,}091$ rad/s $= 810.2$ Hz; $\lambda_3 = 6{,}780$ rad/s $= 1{,}079.1$ Hz; $\lambda_4 = 10{,}469$ rad/s $= 1{,}666.2$ Hz; $\lambda_5 = 15{,}858$ rad/s $= 2{,}523.9$ Hz.

The computed values coincide up to the last digit reported with those computed using the transfer-matrix approach.

By canceling the third and eleventh row and column of the stiffness and mass matrices and again solving the eigenproblem, the following values of the first five natural frequencies can be found: $\lambda_1 = 2{,}158$ rad/s; $\lambda_2 = 5{,}093$ rad/s; $\lambda_3 = 6{,}790$ rad/s; $\lambda_4 = 10{,}488$ rad/s; $\lambda_5 = 15{,}869$ rad/s.

The results are close to the ones obtained earlier, showing that a stiffness of 1×10^{11} N/m for the constraints is high enough to model rigid supports. The consistent mass matrices of the elements were then built and the eigenanalysis was repeated.

The natural frequencies, computed assuming elastic constraints with $k_c = 1 \times 10^{11}$ N/m, are: $\lambda_1 = 2{,}239$ rad/s $= 356.3$ Hz; $\lambda_2 = 6{,}031$ rad/s $= 959.8$ Hz; $\lambda_3 = 8{,}241$ rad/s $= 1{,}311.6$ Hz; $\lambda_4 = 11{,}989$ rad/s $= 1{,}980.4$ Hz; $\lambda_5 = 20{,}552$ rad/s $= 3{,}271.0$ Hz.

The values of natural frequencies resulting from the consistent approach are higher than those obtained from the lumped model, as would be expected. The difference is not small, particularly for the higher-order modes, and this shows that the modelization with only six elements is too rough.

2.7.6 Dynamic stiffness matrix

As already stated, the finite element method yields approximate results even in the case of beam elements, because the shape functions cannot give way to the exact inflected shape. To overcome this limitation, which is more apparent than real, an exact method has been developed for beam elements, in which the expression of the deflected shape obtained through the continuous model is assumed.

In the case of a homogeneous, prismatic Euler-Bernoulli beam performing harmonic oscillations with frequency λ in the xz-plane, for example, the general solution of equation (2.42) yielding the deflected shape with any boundary condition is

$$q(z) = C_1 \sin(az) + C_2 \cos(az) + C_3 \sinh(az) + C_4 \cosh(az), \quad (2.125)$$

where $a = \sqrt{\lambda} = \sqrt[4]{\rho A / E I_y}$.

Constants C_i can be easily obtained from the boundary conditions, i.e., from the nodal generalized displacements. Note that in this case the shape functions that can be obtained from equation (2.125) depend on the frequency λ of the harmonic motion. They can be used to compute the mass and stiffness matrices of the element in the usual way. In this case, these matrices contain the frequency λ of the harmonic motion, and it is possible to directly write the dynamic stiffness matrix $\mathbf{K}_{dyn} = \mathbf{K} - \lambda^2 \mathbf{M}$, which for the Euler-Bernoulli beam, is[6]:

---

[6]See, for example, R.D. Henshell, G.B. Warburton, "Transmission of vibration in beam systems", *Int. J. for Numer. Meth. in Eng.*, Vol.1, (1969), 47-66. The dynamic matrices for both Euler-Bernoulli and Timoshenko beams are reported in detail.

$$\mathbf{K}_{dyn} = \frac{EI_y a}{\cos(al)\cosh(al) - 1} \begin{bmatrix} a^2 k_1 & ak^2 & a^2 k_3 & ak_4 \\ & -k_1 & ak_4 & k_5 \\ & & a^2 k_1 & ak_2 \\ \text{symm.} & & & -k_1 \end{bmatrix}.$$

$$(2.126)$$

where

$$k_1 = \sin(al)\cosh(al) - \cos(al)\sinh(al),$$

$$k_2 = \sin(al)\sinh(al), \qquad k_3 = \sin(al) + \sinh(al),$$

$$k_4 = \cos(al) - \cosh(al), \qquad k_5 = \sinh(al) - \sin(al).$$

A similar approach can be used, at least in principle, for any type of element, provided that an expression of the type of equation (2.125) can be obtained from a continuous model. The dynamic matrices of the various elements can be assembled in the same way seen for mass and stiffness matrices. When the response to harmonic excitation has to be computed, the frequency of the motion is known and the computation is straightforward. On the contrary, when the natural frequencies of the system have to be computed, the solution of the eigenproblem is quite complicated because it is impossible to express it in standard form and then apply the usual numerical methods. The solution follows the lines seen for methods based on transfer matrices: A value of the frequency is assumed and the determinant of the dynamic stiffness matrix is computed. The determinant is not equal to zero, unless a natural frequency has been chosen, and new values of the frequency are assumed. A plot of the value of the determinant as a function of the frequency is then drawn. From the plot, the values of the frequency that cause the determinant to vanish can be obtained. The problem here is that the determinant of a fairly large matrix has to be computed many times and this is, from a computational point of view, far more involving than solving an eigenproblem in standard form using standard techniques.

The dynamic stiffness matrix, which is now a function of the frequency, can be expressed as a power series in the frequency λ:

$$\mathbf{K}_{dyn} = \mathbf{K}_{d_0} + \lambda^2 \mathbf{K}_{d_2} + \lambda^4 \mathbf{K}_{d_4} + \ldots.$$

It is possible to show that only even powers of λ are present and that the first term $\mathbf{K}_{d_0}$ depends only on the elastic properties. In the case of beam elements it is coincident with the stiffness matrix. The second term $\mathbf{K}_{d_2}$ depends only on the inertial properties and is coincident, except for the sign, which is obviously changed, with the mass matrix. All other matrices depend on both inertial and stiffness properties and represent a correction of the potential and kinetic energy of the element due to a more precise formulation of the displacement field. The standard approach can thus be thought of as a truncation at the second term of the series for the exact dynamic stiffness matrix.

The drawbacks of the dynamic stiffness matrix approach, mainly those of leading to long and costly computations and of being applicable only to harmonic motion, are greater, in the opinion of the author, than its advantage of giving a better approximation, which can be obtained simply by using a finer mesh, with far less computational difficulty. Note that the term *exact* used in this context means simply that it leads to the same results of the beam theory, i.e., of the continuous model. All the approximations linked to that model itself are, however, present.

2.7.7 Damping matrices

It is possible to take into account the damping of the structure in a way that closely follows what has been said for the stiffness. If elements that can be modeled as viscous dampers are introduced into the structure between two nodes or between a node and the ground, a viscous damping matrix can be obtained using the same procedures used for the stiffness matrix of spring elements or elastic constraints. Actually, the relevant equations are equal, once the damping coefficient is substituted for the stiffness. If the damping of some of the elements can be modeled as hysteretic damping, within the limits of validity of the complex stiffness model, an imaginary part of the element stiffness matrix can be obtained by simply multiplying the real part by the loss factor. Note that if there is a geometric stiffness matrix, the imaginary part of the stiffness matrix must be computed before adding the geometric matrix to the stiffness matrix.

Viscous or structural damping matrices are then assembled following the same rules seen for mass and stiffness matrices. The real and imaginary parts of the stiffness matrices must be assembled separately, because, when the loss factor is not constant along the structure, they are not proportional to each other.

2.7.8 Finite elements in time

An alternative to the conventional finite element discretization as described in Section 2.7.1 is the space-time approach in which time is considered an added dimension of the problem and four-dimensional space-time is meshed in a way similar to the tridimensional meshing of the standard FEM.

Equation (2.87) defining the shape functions now becomes

$$\mathbf{u}(x,y,z,t) = \mathbf{N}(x,y,z,t)\mathbf{q}\,, \qquad (2.127)$$

where the elements of vector $\mathbf{q}$ are constants expressing the values of the generalized coordinates of the element at different times and the shape functions are functions of time and space coordinates.

The equation of motion of the element can be directly obtained from Hamilton's principle, stating that

$$\int_{t_1}^{t_2} \left[\delta(\mathcal{T} - \mathcal{U}) + \delta\mathcal{L}\right] dt = 0 \qquad (2.128)$$

where $\delta(\mathcal{T} - \mathcal{U})$ and $\delta\mathcal{L}$ are, respectively, the variation of the Lagrangian of the system and the virtual work of nonconservative forces corresponding to the virtual displacement δu. The generalized velocities, the strains, and the virtual displacements can be expressed as

$$\dot{\mathbf{u}}(x,y,z,t) = \left[\frac{\partial \mathbf{N}}{\partial t}\right] \mathbf{q}\,, \qquad \boldsymbol{\epsilon} = \mathbf{B}\mathbf{q}\,, \qquad \delta\mathbf{u} = \mathbf{N}\delta\mathbf{q}\,, \qquad (2.129)$$

where matrix $\mathbf{B}$ is of the same type as the matrix defined by equation (2.89), the only difference being that in this case it is also a function of time.

In the case where all generalized coordinates are related to displacements, the Lagrangian of the element can be expressed as

$$\mathcal{T} - \mathcal{U} = \frac{1}{2}\mathbf{q}^T \left(\int_V \rho \left[\frac{\partial \mathbf{N}}{\partial t}\right]^T \left[\frac{\partial \mathbf{N}}{\partial t}\right] dV\right) \mathbf{q} - \frac{1}{2}\mathbf{q}^T \left(\int_V \mathbf{B}^T \mathbf{E}\mathbf{B}dV\right) \mathbf{q}\,.$$
$$(2.130)$$

If a generalized force $\mathbf{f}(x,y,z,t)$ is acting on the element, Hamilton's principle can be written in the form

$$\int_{t_1}^{t_2} \int_V \left\{ \delta \mathbf{q}^{\mathrm{T}} \left[\rho \left[\frac{\partial \mathbf{N}}{\partial t} \right]^{\mathrm{T}} \left[\frac{\partial \mathbf{N}}{\partial t} \right] - \mathbf{B}^{\mathrm{T}} \mathbf{E} \mathbf{B} \right] \mathbf{q} + \delta \mathbf{q}^{\mathrm{T}} \mathbf{N}^{\mathrm{T}} \mathbf{f} \right\} \, d\mathrm{V} dt = 0 \,, \tag{2.131}$$

i.e.,

$$\mathbf{K}^* \mathbf{q} = \mathbf{f}^* \,, \tag{2.132}$$

where matrix $\mathbf{K}^*$ and vector $\mathbf{f}^*$ are

$$\mathbf{K}^* = \int_{V^*} \left\{ -\rho \left[\frac{\partial \mathbf{N}}{\partial t} \right]^{T} \left[\frac{\partial \mathbf{N}}{\partial t} \right] + \mathbf{B}^{T} \mathbf{E} \mathbf{B} \right\} dV^* \tag{2.133}$$

$$\mathbf{f}^* = \int_{V^*} \mathbf{N}^T \mathbf{f} dV^* \,.$$

and volume V^* is the four-dimensional volume of the element.

The element matrices $\mathbf{K}^*$ and force vectors $\mathbf{f}^*$ are then assembled in the usual way, allowing us to write an equation of the same type as equation (2.132) for the whole structure. Note that the number of degrees of freedom is far larger than that resulting from conventional FEM modeling.

An example of space-time modeling of a rectangular plate is shown in Figure 2.19. Because the structure is two-dimensional in space, the model has three dimensions. In the case shown in the figure, the elements are rectangular in time and are arranged in regular layers along the time coordinate. This is not necessarily the case, and the current method demonstrates its greatest advantages when the mesh is tailored in time to suit the local needs and triangular (hypertetrahedral) elements are used.

In the case of rectangular elements, the equations can be assembled layer by layer first. In this way, the generalized coordinates can be partitioned according to the time to which they refer, and the equation of the ith layer can be written as

$$\begin{bmatrix} \mathbf{A}_i & \mathbf{B}_i \\ \mathbf{C}_i & \mathbf{D}_i \end{bmatrix} \left\{ \begin{array}{c} \mathbf{q}_{i-1} \\ \mathbf{q}_i \end{array} \right\} = \left\{ \begin{array}{c} \mathbf{f}_{1_i} \\ \mathbf{f}_{2_i} \end{array} \right\} \,, \tag{2.134}$$

where matrices $\mathbf{A}_i$, $\mathbf{B}_i$, $\mathbf{C}_i$, and $\mathbf{D}_i$[7] are submatrices of $\mathbf{K}^*$ for the

---

[7]Matrices $\mathbf{A}$, $\mathbf{B}_i$, $\mathbf{C}_i$, and $\mathbf{D}_i$ obviously have nothing to do with the matrices with the same name referred to as the quadruple of the system.

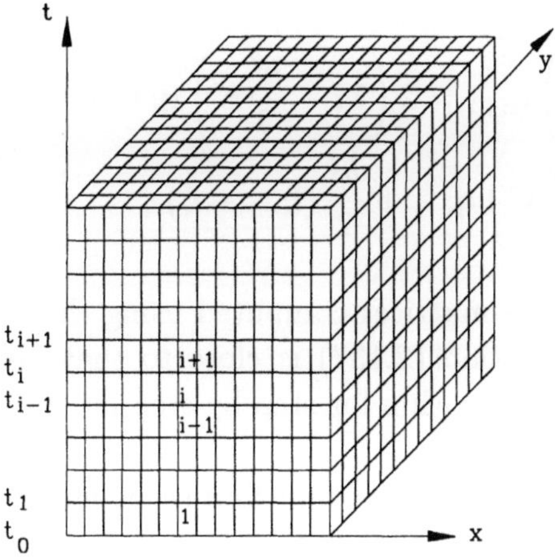

FIGURE 2.19. Space-time model of a rectangular plate using rectangular elements arranged in regular layers in time.

ith layer and the subscript in vector $\mathbf{q}_i$ refers to time t_i and not to the ith layer.

Once the matrices for the layers are assembled, the solution can be obtained starting from the first layer, in which the initial conditions are usually stated, and proceeding layer by layer in a way is similar to conventional step-by-step time integration. The equation for the ith layer can then be written in the form

$$\mathbf{B}_{i+1}\mathbf{q}_{i+1} = -\left(\mathbf{D}_i + \mathbf{A}_{i+1}\right)\mathbf{q}_i - \mathbf{C}_i\mathbf{q}_{i-1} + \frac{1}{2}\left(\mathbf{f}_{2_{i-1}} + \mathbf{f}_{1_i}\right) . \qquad (2.135)$$

Equation (2.135) directly yields the values of the generalized coordinates at time t_{i+1} from the values they take at time t_i and t_{i-1}.

The simplest type of shape functions for elements that are rectangular in time is

$$\mathbf{N}(x,y,z,t) = \left[\ \mathbf{N}_1(x,y,z)(1-\tau)\quad \mathbf{N}_1(x,y,z)\tau\ \right], \qquad (2.136)$$

where $\tau = t/h$ is the nondimensional time obtained by dividing the time t by the time step h and $\mathbf{N}_1$ is the matrix of the shape functions of the conventional FEM approach. With simple computations, it is possible to show that the integration in space can thus be performed independently from the integration in time. By also taking into account the presence of a possible Rayleigh dissipation function, equation (2.135) reduces to

$$\left(\mathbf{M} + \frac{h^2}{6}\mathbf{K} + \frac{h}{2}\mathbf{C}\right)\mathbf{q}_{i+1} = 2\left(\mathbf{M} - \frac{h^2}{3}\mathbf{K}\right)\mathbf{q}_i +$$
$$-\left(\mathbf{M} + \frac{h^2}{6}\mathbf{K} - \frac{h}{2}\mathbf{C}\right)\mathbf{q}_{i-1} + \frac{h^2}{2}\left(\mathbf{f}_{i-1} + \mathbf{f}_i\right), \tag{2.137}$$

where $\mathbf{M}$, $\mathbf{C}$, and $\mathbf{K}$ are the matrices computed using the conventional FEM approach. Equation (2.137) is close to that obtained from the central difference algorithm for step-by-step integration in time (see Section A.5).

However, as already stated, the space-time finite element approach gives better results if the elements are not rectangular in time and the shape functions cannot be expressed as the product of a function of space by a function of time. There are some limitations on the ratio between the space and time dimensions of the elements, which should approach the propagation velocity of the elastic waves in the material[8]. This consideration can cause a large proliferation of the number of the degrees of freedom.

2.8 Reduction of the number of degrees of freedom

2.8.1 General considerations

As already stated, when performing a dynamic analysis, there is a great advantage in reducing the size of the problem, particularly when using the FEM, which usually yields models with many degrees of freedom. It is not uncommon to use models with thousands or tens of thousands of degrees of freedom; when studying static problems, that does not constitute a problem for modern computers, but the solution of an eigenproblem of that size can still be a formidable problem. Note that because the FEM is a displacement method, i.e., first solves the displacements and then computes stresses and strains as derivatives of the displacements, the precision with which

---

[8]Z. Kaczkowski, "The method of finite space-time elements in dynamics of structures", *J. Tech. Phys.*, Vol. 16, No. 1, 69-84, 1975. See also C. Bajer, C. Bonthoux, "State-of-the-art in the space-time element method", *The Shock and Vibration Digest*, Vol. 23, 3-9, May 1991.

displacements, and all other entities directly linked with displacements including mode shapes and natural frequencies, are obtained is far greater, for a given mesh, than that achievable for stresses and strains. Conversely, this means that the mesh needs to be much finer when solving the stress field, which is typical of static problems, than when searching natural frequencies and mode shapes. Because it is often expedient to use the same mesh for both or a finer mesh is required to model a complex shape, a reduction of the number of degrees of freedom for dynamic solution is useful, particularly when only a limited number of natural frequencies are required.

Two approaches can be used: reducing the size of the model or leaving the model as is and using algorithms, such as the subspace iteration method, that search only the lowest natural frequencies. Although the two are more or less equivalent, the first leaves the choice of which degrees of freedom to retain to the user, and the second operates automatically. As a consequence, a skilled operator can use advantageously reduction techniques, which allow very good results with very few degrees of freedom. A general-purpose code for routine computations, which is sometimes used by analysts who are not very well trained, on the contrary, can advantageously use the second approach. Note that before computers were available, remarkable results were obtained using models with very few (often a single) degrees of freedom, but this required great computational ability and physical insight.

In this section, techniques aimed at reducing the size of the model, of both the condensation (or reduction or substructuring) and the component mode synthesis types, are described. Subspace iteration technique will be dealt with in Appendix 1.

2.8.2 Reduction techniques

Static reduction is based on the subdivision of the generalized coordinates $\mathbf{q}$ of the model into two types: master degrees of freedom $\mathbf{q}_1$ and slave degrees of freedom $\mathbf{q}_2$. When used to solve a static problem it yields exact results, i.e., the same results that would be obtained from the complete model. The stiffness matrix and the nodal force vector can be partitioned accordingly, and the equation expressing the static problem becomes

$$\begin{bmatrix} \mathbf{K}_{11} & \mathbf{K}_{12} \\ \mathbf{K}_{21} & \mathbf{K}_{22} \end{bmatrix} \left\{ \begin{array}{c} \mathbf{q}_1 \\ \mathbf{q}_2 \end{array} \right\} = \left\{ \begin{array}{c} \mathbf{f}_1 \\ \mathbf{f}_2 \end{array} \right\}. \tag{2.138}$$

Note that matrices $\mathbf{K}_{11}$ and $\mathbf{K}_{22}$ are symmetrical, while $\mathbf{K}_{12} = \mathbf{K}_{21}^T$ are neither symmetrical nor square. Solving the second set of equations (2.138) in $\mathbf{q}_2$, the following relationship linking the slave to the master coordinates is obtained:

$$\mathbf{q}_2 = -\mathbf{K}_{22}^{-1}\mathbf{K}_{21}\mathbf{q}_1 + \mathbf{K}_{22}^{-1}\mathbf{f}_2. \tag{2.139}$$

Introducing equation (2.139) into equation (2.138), the latter yields

$$\mathbf{K}_{cond}\mathbf{q}_1 = \mathbf{f}_{cond}, \tag{2.140}$$

where

$$\left\{ \begin{array}{l} \mathbf{K}_{cond} = \mathbf{K}_{11} - \mathbf{K}_{12}\mathbf{K}_{22}^{-1}\mathbf{K}_{12}^T, \\ \mathbf{f}_{cond} = \mathbf{f}_1 - \mathbf{K}_{12}\mathbf{K}_{22}^{-1}\mathbf{f}_2. \end{array} \right.$$

Equation (2.140) yields the master generalized displacements $\mathbf{q}_1$. The slave displacements can be obtained directly from equation (2.139) simply by multiplying some matrices.

The subdivision of the degrees of freedom between vectors $\mathbf{q}_1$ and $\mathbf{q}_2$ can be based on different criteria. The master degrees of freedom can simply be those in which the user is directly interested. Another type of choice can be that of physically subdividing the structure in two parts. The second practice, which can be generalized by subdividing the generalized coordinates into many subsets, is generally known as *solution by substructures* or *substructuring*. In particular, it can be expedient when the structure can be divided into many parts that are all connected to a single frame. If the generalized displacements of the connecting structure or frame are listed in vector $\mathbf{q}_0$ and those of the various substructures are included in vectors $\mathbf{q}_i$, the equation for the static solution of the complete structure has the form

$$\begin{bmatrix} \mathbf{K}_{00} & \mathbf{K}_{01} & \mathbf{K}_{02} & \cdots \\ & \mathbf{K}_{11} & \mathbf{0} & \cdots \\ & & \mathbf{K}_{22} & \cdots \\ \text{symm.} & & & \cdots \end{bmatrix} \left\{ \begin{array}{c} \mathbf{q}_0 \\ \mathbf{q}_1 \\ \mathbf{q}_2 \\ \cdots \end{array} \right\} = \left\{ \begin{array}{c} \mathbf{f}_0 \\ \mathbf{f}_1 \\ \mathbf{f}_2 \\ \cdots \end{array} \right\}. \tag{2.141}$$

The equations related to the ith substructure can be solved as

$$\mathbf{q}_i = -\mathbf{K}_{ii}^{-1}\mathbf{K}_{i0}\mathbf{q}_0 + \mathbf{K}_{ii}^{-1}\mathbf{f}_i . \tag{2.142}$$

The generalized displacements of the frame can be obtained using an equation of the type of equation (2.140) where the condensed matrices are

$$\begin{cases} \mathbf{K}_{cond} = \mathbf{K}_{00} - \displaystyle\sum_{\forall i} \mathbf{K}_{0i}\mathbf{K}_{ii}^{-1}\mathbf{K}_{0i}^T , \\[2mm] \mathbf{f}_{cond} = \mathbf{f}_0 - \displaystyle\sum_{\forall i} \mathbf{K}_{0i}\mathbf{K}_{ii}^{-1}\mathbf{f}_i . \end{cases} \tag{2.143}$$

As already stated, static reduction does not introduce any further approximation into the model. A similar reduction can be used in dynamic analysis without introducing approximations only if no generalized inertia is associated with the slave degrees of freedom. In this case, static reduction is advisable because the mass matrix of the original system is singular and the condensation procedure allows removal of the singularity. When using the lumped-parameters approach with beam elements and the moments of inertia of the cross sections are neglected, no inertia is associated with half of the degrees of freedom related to bending. Static reduction allows the removal of all of them and the obtaining of a nonsingular mass matrix. Generally speaking, however, the mass matrix is not singular and it is not possible to just neglect the inertia linked with some degrees of freedom. The so-called *Guyan reduction* is based on the assumption that the slave generalized displacements $\mathbf{q}_2$ can be computed directly from master displacements $\mathbf{q}_1$, neglecting inertia forces and external forces $\mathbf{f}_2$. In this case, equation (2.139), without the last term, can also be used for dynamic solution. By partitioning the mass matrix in the same way seen for the stiffness matrix, the kinetic energy of the structure can be expressed as

$$\mathcal{T} = \frac{1}{2}\left\{ \begin{array}{c} \dot{\mathbf{q}}_1 \\ -\mathbf{K}_{22}^{-1}\mathbf{K}_{21}\dot{\mathbf{q}}_1 \end{array} \right\}^T \begin{bmatrix} \mathbf{M}_{11} & \mathbf{M}_{12} \\ \mathbf{M}_{21} & \mathbf{M}_{22} \end{bmatrix} \left\{ \begin{array}{c} \dot{\mathbf{q}}_1 \\ -\mathbf{K}_{22}^{-1}\mathbf{K}_{21}\dot{\mathbf{q}}_1 \end{array} \right\} . \tag{2.144}$$

The kinetic energy is then $\mathcal{T} = \frac{1}{2}\dot{\mathbf{q}}_1^T\mathbf{M}_{cond}\dot{\mathbf{q}}_1$, where the condensed mass matrix is

$$\mathbf{M}_{cond} = \mathbf{M}_{11} - \mathbf{M}_{12}\mathbf{K}_{22}^{-1}\mathbf{K}_{12}^{T} - \left[\mathbf{M}_{12}\mathbf{K}_{22}^{-1}\mathbf{K}_{12}^{T}\right]^{T} + \\ +\mathbf{K}_{12}\mathbf{K}_{22}^{-1}\mathbf{M}_{22}\mathbf{K}_{22}^{-1}\mathbf{K}_{12}^{T} . \qquad (2.145)$$

Guyan reduction is not much more demanding from a computational viewpoint than static reduction because the only matrix inversion is that of $\mathbf{K}_{22}$, which has already been performed for the computation of the condensed stiffness matrix. If matrix $\mathbf{M}$ is diagonal, two of the terms of equation (2.145) vanish. Although approximate, it introduces errors that are usually very small, at least if the choice of the slave degrees of freedom is appropriate. Inertia forces related to slave degrees of freedom are actually not neglected, but their contribution to the kinetic energy is computed from a deformed configuration obtained on the basis of the master degrees of freedom alone. If the relevant mode shapes are only slightly influenced by the presence of some of the generalized masses or if some parts of the structure are so stiff that their deflected shape can be determined by a few coordinates, the results can be very good, even when very few master degrees of freedom are used.

In a way similar to that seen for the mass matrix, viscous or structural damping matrices can be reduced using equation (2.145) in which $\mathbf{M}$ has been substituted with $\mathbf{C}$ and $\mathbf{K}''$, respectively. Also, the reduction of damping matrices introduces errors that depend on the choice of the slave degrees of freedom but are usually very small when the degrees of freedom in which viscous dampers are applied or, in the case of hysteretic damping, where the loss factor of the material changes, are not eliminated. Alternatively, these degrees of freedom can be neglected when their displacement is well determined by some master displacement, as in the case of very stiff parts of the structure.

Example 2-9
Compute the first natural frequency of the shaft of Example 2-7 through Guyan reduction.

If the stiffness of the two shaft elements is much greater than that of the central support, the first mode shape is made by two parts of straight lines. In this case the results obtainable by condensing the model to the translational degree of freedom of node 2 are very close to those that can be obtained from the complete model, where the first natural frequency is concerned. By reordering the matrices, and choosing the mentioned degree of freedom as master and all others as slave, it follows that

$$\mathbf{K} = \frac{EI_y}{l^3} \left[\begin{array}{c|cccc} 24 + k^* & -6l & -6l & 6l & 6l \\ \hline -6l & 4l^2 & 2l^2 & 0 & 0 \\ -6l & 2l^2 & 4l^2 & 0 & 0 \\ 6l & 0 & 0 & 4l^2 & 2l^2 \\ 6l & 0 & 0 & 2l^2 & 4l^2 \end{array} \right] ,$$

$$\mathbf{M} = \frac{\rho A l}{420} \left[\begin{array}{c|cccc} 312 + m^* & 13l & -22l & 22l & -13l \\ \hline 13l & 4l^2 & -3l^2 & 0 & 0 \\ -22l & -3l^2 & 4l^2 & 0 & 0 \\ 22l & 0 & 0 & 4l^2 & -3l^2 \\ -13l & 0 & 0 & -3l^2 & 4l^2 \end{array} \right] .$$

The inversion of matrix $\mathbf{K}_{22}$ is straightforward, and Guyan reduction can be performed without difficulty:

$$\mathbf{K}_{22}^{-1} = \frac{l}{12EI_y} \left[\begin{array}{cccc} 4 & -2 & 0 & 0 \\ -2 & 4 & 0 & 0 \\ 0 & 0 & 4 & -2 \\ 0 & 0 & -2 & 4 \end{array} \right] ,$$

$$k_{cond} = k_G , \qquad m_{cond} = \frac{2}{3}\rho A l + m_G , \qquad \lambda = \sqrt{\frac{3k_G}{2\rho A l + 3m_G}}.$$

Note that in the condensed stiffness matrix, which reduces to a single element because only one master degree of freedom was considered, only the elastic characteristics of the central support are included. This could be easily predicted because the system becomes a *mechanism* (an underconstrained system capable of rigid-body motion) if the support on node 2 is removed.

The reduction can be operated directly on the dynamic stiffness matrix (dynamic reduction). This procedure does not introduce approximations, but the frequency must be stated before performing it. It will be dealt with when studying multi-degree-of-freedom nonlinear systems.

2.8.3 Component mode synthesis

When substructuring is used, the degrees of freedom of each structure can be divided into two sets: internal degrees of freedom and boundary degrees of freedom. The latter are all degrees of freedom that the substructure has in common with other parts of the structure. They are often referred to as *constraint degrees of freedom* because they express how the substructure is constrained to the rest of the system. *Internal degrees of freedom* are those belonging only to the relevant substructure. The largest possible reduction scheme is that in which all internal degrees of freedom are considered slave coordinates and all boundary degrees of freedom are considered master coordinates. In this way, however, the approximation of all modes in which the motion of the internal points of the substructure with respect to the motion of its boundary is important, can be quite rough.

A simple way to avoid this drawback is to also consider as master coordinates, together with the boundary degrees of freedom, some of the modal coordinates of the substructure constrained at its boundary. This procedure would obviously lead to exact results if all modes were retained, but because the total number of modes is equal to the number of internal degrees of freedom, the model obtained has as many degrees of freedom as the original one. As usual with modal practices, the computational advantages grow, together with the number of modes that can be neglected. The relevant matrices are partitioned as seen for reduction techniques, with subscript 1 referring to the boundary degrees of freedom and subscript 2 to the internal degrees of freedom. The displacement vector $\mathbf{q}_2$ can be assumed to be equal to the sum of the constrained modes $\mathbf{q}_2'$, i.e., the deformation pattern due to the displacements $\mathbf{q}_1$ when no force acts on the substructure, plus the constrained normal modes $\mathbf{q}_2''$, i.e., the natural modes of free vibration of the substructure when the boundary generalized displacements $\mathbf{q}_1$ are equal to zero.

The constrained modes $\mathbf{q}_2'$ can be expressed by equation (2.139) once the force vector $\mathbf{f}_2$ is set equal to zero. The constrained normal modes can easily be computed by solving the eigenproblem

$$\left(-\lambda^2 \mathbf{M}_{22} + \mathbf{K}_{22}\right) \mathbf{q}_2'' = \mathbf{0} \,.$$

Once the eigenproblem has been solved, the matrix of the eigenvectors $\boldsymbol{\Phi}$ can be used to perform the modal transformation $\mathbf{q}_2'' = \boldsymbol{\Phi}\boldsymbol{\eta}_2$. The generalized coordinates of the substructure can thus be expressed as

$$\left\{ \begin{array}{c} \mathbf{q}_1 \\ \mathbf{q}_2 \end{array} \right\} = \left\{ \begin{array}{c} \mathbf{q}_1 \\ -\mathbf{K}_{22}^{-1}\mathbf{K}_{21}\mathbf{q}_1 + \boldsymbol{\Phi}\boldsymbol{\eta}_2 \end{array} \right\} =$$

$$= \left[\begin{array}{cc} \mathbf{I} & \mathbf{0} \\ -\mathbf{K}_{22}^{-1}\mathbf{K}_{21} & \boldsymbol{\Phi} \end{array} \right] \left\{ \begin{array}{c} \mathbf{q}_1 \\ \boldsymbol{\eta}_2 \end{array} \right\} = \boldsymbol{\Psi} \left\{ \begin{array}{c} \mathbf{q}_1 \\ \boldsymbol{\eta}_2 \end{array} \right\} . \tag{2.146}$$

Equation (2.146) represents a coordinate transformation, allowing the expression of the deformation of the internal part of the substructure in terms of constrained and normal modes. Matrix $\boldsymbol{\Psi}$ expressing this transformation can be used to compute the new mass, stiffness, and, where needed, damping matrices and the force vector

$$\mathbf{M}^* = \boldsymbol{\Psi}^T\mathbf{M}\boldsymbol{\Psi}, \quad \mathbf{K}^* = \boldsymbol{\Psi}^T\mathbf{K}\boldsymbol{\Psi},$$
$$\mathbf{C}^* = \boldsymbol{\Psi}^T\mathbf{C}\boldsymbol{\Psi}, \quad \mathbf{f}^* = \boldsymbol{\Psi}^T\mathbf{f}. \tag{2.147}$$

If there are m constrained coordinates and n internal coordinates and if only k constrained normal modes are considered $(k < n)$, then the size of the original matrices $\mathbf{M}$, $\mathbf{K}$, and so on is $m + n$, while that of matrices $\mathbf{M}^*$, $\mathbf{K}^*$, and so on is $m + k$.

Once the transformation of coordinates of the substructures has been performed, they can be assembled in the same way already seen for elements: The boundary coordinates are common between the substructures and are actually assembled while the modal coordinates are typical of only one substructure at a time, in the same way as the coordinates of internal nodes of elements that had nodes of this type. Actually each substructure can be regarded as a large element, sometimes referred to as a *superelement* and the relevant procedures do not differ from those that are standard in the FEM.

The main advantage of substructuring is that of allowing the construction of the model and the analysis of the various parts of a large structure in an independent way. The results can then be assembled and the behaviour of the structure can be assessed from that of its parts. If this is done, however, the connecting nodes must be defined in such a way that the same boundary degrees of freedom are considered in the analysis of the various parts. It is, however, possible to use algorithms allowing the connecting of otherwise incompatible meshes.

All the methods discussed in this section, which are closely related to each other and are found in the literature in a variety of versions,

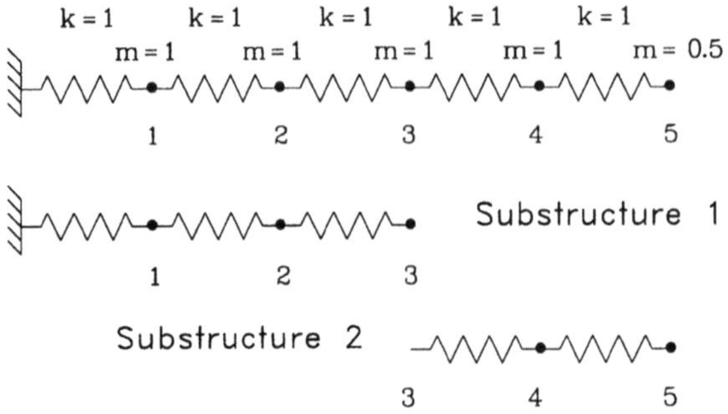

k = 1   k = 1   k = 1   k = 1   k = 1
m = 1   m = 1   m = 1   m = 1   m = 0.5

1   2   3   4   5

Substructure 1

1   2   3

Substructure 2   3   4   5

FIGURE 2.20. Sketch of the system and values of the relevant parameters.

are general for discrete systems and can also be used outside the FEM even if they became popular only with the use of the latter owing to the large number of degrees of freedom it yields.

Example 2-10
Consider the discrete system sketched in Figure 2.20. Study its dynamic behaviour and compare the results obtained using component-mode synthesis retaining different numbers of modes.
The total number of degrees of freedom of the system is five and the complete mass and stiffness matrices are

$$
\mathbf{K} = \begin{bmatrix} 2 & -1 & 0 & 0 & 0 \\ -1 & 2 & -1 & 0 & 0 \\ 0 & -1 & 2 & -1 & 0 \\ 0 & 0 & -1 & 2 & -1 \\ 0 & 0 & 0 & -1 & 1 \end{bmatrix} , \quad \mathbf{M} = \begin{bmatrix} 1 & 0 & 0 & 0 & 0 \\ 0 & 1 & 0 & 0 & 0 \\ 0 & 0 & 1 & 0 & 0 \\ 0 & 0 & 0 & 1 & 0 \\ 0 & 0 & 0 & 0 & 0.5 \end{bmatrix} .
$$

By directly solving the eigenproblem, the following matrix of the eigenvalues is obtained:

$$
[\lambda^2] = \mathrm{diag} \begin{bmatrix} 0.0979 & 0.8244 & 2.000 & 3.176 & 3.902 \end{bmatrix} .
$$

The structure is then subdivided into two substructures and the analysis is accordingly performed.

Substructure 1. Substructure 1 includes nodes 1, 2, and 3 with the masses located on them. The displacements at nodes 1 and 2 are internal coordinates, while the displacement at node 3 is a boundary coordinate. The mass and stiffness matrix of the substructure, partitioned with the boundary degree of freedom first and then the internal ones, are

$$\mathbf{K} = \begin{bmatrix} 1 & -1 & 0 \\ \hline -1 & 2 & -1 \\ 0 & -1 & 2 \end{bmatrix}, \qquad \mathbf{M} = \begin{bmatrix} 1 & 0 & 0 \\ \hline 0 & 1 & 0 \\ 0 & 0 & 1 \end{bmatrix}.$$

The matrix of the eigenvectors for the internal normal modes can be easily obtained by solving the eigenproblem related to matrices with subscript 22 and, by retaining all modes, matrices $\mathbf{K}^*$ and $\mathbf{M}^*$ of the first substructure can be computed:

$$\boldsymbol{\Phi} = \begin{bmatrix} \dfrac{\sqrt{2}}{2} & \dfrac{-\sqrt{2}}{2} \\ \dfrac{\sqrt{2}}{2} & \dfrac{\sqrt{2}}{2} \end{bmatrix}, \qquad \mathbf{K}^* = \begin{bmatrix} 0.3333 & 0 & 0 \\ \hline 0 & 1 & 0 \\ 0 & 0 & 3 \end{bmatrix},$$

$$\mathbf{M}^* = \begin{bmatrix} 1.556 & 0.7071 & -0.2357 \\ \hline 0.7071 & 1 & 0 \\ -0.2357 & 0 & 1 \end{bmatrix}.$$

Substructure 2. The second substructure includes nodes 3, 4, and 5 with the masses located on nodes 4 and 5. The mass located on node 3 has already been taken into account in the first substructure and must not be considered again. The displacements at nodes 4 and 5 are internal coordinates, while the displacement at node 3 is a boundary coordinate. The mass and stiffness matrix of the substructure, partitioned with the boundary degree of freedom first and then the internal ones, are

$$\mathbf{K} = \begin{bmatrix} 1 & -1 & 0 \\ \hline -1 & 2 & -1 \\ 0 & -1 & 1 \end{bmatrix}, \qquad \mathbf{M} = \begin{bmatrix} 0 & 0 & 0 \\ \hline 0 & 1 & 0 \\ 0 & 0 & 0.5 \end{bmatrix}.$$

Operating as seen for the first substructure, it follows

$$\boldsymbol{\Phi} = \begin{bmatrix} \dfrac{\sqrt{2}}{2} & \dfrac{-\sqrt{2}}{2} \\ 1 & 1 \end{bmatrix}, \qquad \mathbf{K}^* = \begin{bmatrix} 0 & 0 & 0 \\ \hline 0 & 0.5858 & 0 \\ 0 & 0 & 3.4142 \end{bmatrix},$$

$$\mathbf{M}^* = \begin{bmatrix} 1.5 & 1.2071 & -0.2071 \\ \hline 1.2071 & 1 & 0 \\ -0.2071 & 0 & 1 \end{bmatrix}.$$

The substructures can be assembled in the same way as the elements. The following map can be written

Subst.	d.o.f.	1	2	3		
1	type	boundary	modal	modal		
Subst.	d.o.f.	1			2	3
2	type	boundary			modal	modal
Global	d.o.f.	1	2	3	4	5

yielding the following global stiffness and mass matrices

$$\mathbf{K}^* = \left[\begin{array}{c|cccc} 0.3333 & 0 & 0 & 0 & 0 \\ \hline 0 & 1 & 0 & 0 & 0 \\ 0 & 0 & 3 & 0 & 0 \\ 0 & 0 & 0 & 0.5858 & 0 \\ 0 & 0 & 0 & 0 & 3.4142 \end{array}\right] ,$$

$$\mathbf{M}^* = \left[\begin{array}{c|cccc} 1.5 & 0.7071 & -0.2357 & 1.2071 & -0.2071 \\ \hline 0.7071 & 1 & 0 & 0 & 0 \\ -0.2357 & 0 & 1 & 0 & 0 \\ 1.2071 & 0 & 0 & 1 & 0 \\ -0.2071 & 0 & 0 & 0 & 1 \end{array}\right] .$$

The matrices have been partitioned in such a way to separate the boundary displacement degree of freedom from the modal degrees of freedom. If no modal coordinate is considered, the component-mode synthesis coincides with Guyan reduction, with only one master degree of freedom. If the third and fifth rows and columns are cancelled, only one internal normal mode is taken into account for each substructure. If the matrices are taken into account in complete form, all modes are considered and the result must coincide, except for computing approximations, with the exact ones. The results obtained in terms of the square of the natural frequency are

Size of matrices	5 (exact)	1 (Guyan red.)	3 (1 mode)	5 (2 modes)
Mode 1	0.0979	0.1091	0.0979	0.0979
Mode 2	0.8244	–	0.8245	0.8244
Mode 3	2.000	–	2.215	2.000
Mode 4	3.176	–	–	3.176
Mode 5	3.902	–	–	3.902

2.9 Exercises

Exercise 2-1

A transmission shaft is made of a circular steel tube and has two joints at the ends. Without considering the mass of the joints, compute the first five flexural natural frequencies of the shaft neglecting shear deformations. Repeat the computations, taking into account shear deformations and compare the results. Data: inner diameter = 80 mm, outer diameter = 100 mm, length = 1.5 m, $E = 2.1 \times 10^{11}$ N/m^2, $\nu = 0.3$, $\rho = 7{,}810$ kg/m^3.

Exercise 2-2

A cylindrical chimney is 40 m tall, has an inner diameter of 500 mm, and has a wall thickness of 15 mm. Compute the flexural natural frequencies and mode shapes (neglecting the influence of compressive stresses due to self-weight).

To perform a dynamic test, a small rocket attached to the top of the chimney and directed horizontally is fired. The thrust of the rocket instantaneously reaches the value of 100 N, lasts for 5 s, and then instantaneously drops to zero. Compute the time history of the various modes and the maximum stresses at the base of the stack. If necessary, assume that the damping is of structural type, with a loss factor $\eta = 0.02$. Use Duhamel's integral or numerical integration; if possible, also use an impulsive approach. Data: $E = 2.1 \times 10^{11}$ N/m^2, $\nu = 0.3$, $\rho = 7{,}810$ kg/m^3.

Exercise 2-3

Consider the chimney of Exercise 2-2. Compute the modal participation factors of the first five modes for an excitation provided by the horizontal motion of the supporting ground. Assume that the chimney is excited by a horizontal random motion of the ground, whose power spectral density is constant at 0.003 g^2/Hz between 3 and 30 Hz, increases at 9 dB/oct between 0.1 and 3 Hz, and drops at -12 dB/oct between 30 and 100 Hz. Compute the spectrum of the response of the first five modes and the r.m.s. values of the displacement at the top and of the stresses at the base.

Exercise 2-4

A connecting rod has an annular cross section and is hinged at the ends through two parallel cylindrical hinges. Compute the natural frequencies of the flexural vibrations in the plane containing the axes of the hinges and in a plane perpendicular to it. Repeat the computation when a compressive axial force of 10,000 N acts on the rod. Data: inner diameter $= 30$ mm, outer diameter $= 40$ mm, length $= 400$ mm, $E = 2.1 \times 10^{11}$ N/m^2, $\nu = 0.3$, $\rho = 7{,}810$ kg/m^3.

Exercise 2-5

Consider the taut string of Example 2-2. Compute the shift of the fundamental frequency due to a change of the tension equal to 10%. Compute the response of the string when it is plucked at a quarter of its length (assume the original value of the tension). Assume that the point at 1/4 of the length of the string is displaced of 10 mm and released, and compute the actual amplitude of the various modes and the maximum value of the kinetic energy related to them. How is the energy distributed among the modes?

Exercise 2-6

A chisel has a square cross section with side $a = 10$ mm and length $l = 500$ mm. Compute the time needed for the pressure wave caused by a hammer striking the end of the chisel to reach the bit. Assume that the bit of the chisel is pressed against a rigid surface and can be considered perfectly clamped. If the blow of the hammer causes a stress wave whose profile can be assumed to be a trapezium, plot the profile of the incident wave and of the wave reflected back at different times.

Exercise 2-7

Compute the length of a pipe with closed ends in such a way that the air contained into it can vibrate with a fundamental frequency corresponding with the central A (nominal frequency 440 Hz). Compute the other natural frequencies and state whether they are multiples of the fundamental frequency. Data: $\gamma \approx 7/5$, $p_0 = 1.007 \times 10^5$ Pa, $\rho_0 = 1.29$ kg/m^3.

Exercise 2-8

Consider the transmission shaft of Exercise 2-1. Compute the first flexural natural frequencies using the Myklestadt method. Lump the masss of the shaft in two, four, and eight stations.

Exercise 2-9

Repeat the computations of the preceding exercise using the FEM. Use three, five, and seven Timoshenko beam elements.

Exercise 2-10

Compute the field transfer matrix for a Timoshenko beam (equation 2.82) from the stiffness matrix of an element of the same type.

Exercise 2-11

Compute the natural frequencies of the connecting rod of Exercise 2-4 using the FEM. Use the geometric matrix to take into account the compressive axial force. First use a single beam element and then repeat the computation using a finer subdivision.

Exercise 2-12

Consider the transmission shaft of Example 2-7. Compute the numerical value of the natural frequency using the Rayleigh method. Compute the first three natural frequencies and the mode shapes of the flexural vibrations using a four-element subdivision and compare the result obtained for the first mode with that previously obtained. Data: inner diameter = 70 mm, outer diameter = 90 mm, length of each part of the shaft = 500 mm, $E = 2.1 \times 10^{11}$ N/m^2, $\nu = 0.3$, $\rho = 7{,}810$ kg/m^3, $m_G = 2$ kg, $k_G = 10^5$ N/m.

Exercise 2-13

Repeat the dynamic analysis of the previous exercise by using component-mode synthesis. Subdivide the system into two substructures coinciding with two spans of the shaft and take into account the first mode of each part.

Exercise 2-14

Consider a frame made by three beams arranged as an equilateral triangle. The frame is clamped at one of the vertices and carries two concentrated masses at the other vertices. Perform the dynamic analysis of the frame using one beam element for each side. Data: annular cross section with inner diameter = 50 mm, outer diameter = 70 mm, length of each side = 500 mm, $E = 2.1 \times 10^{11}$ N/m^2, $\nu = 0.3$, concentrated masses at the vertices $m_c = 10$ kg.

Exercise 2-15

Add two dampers with a damping coefficient $c = 0.1$ to the system of Figure 2.18 between nodes 2 and 3 and nodes 4 and 5. Compute the complex frequencies both usingh the complete model and resorting to the component-mode synthesis.

3

Nonlinear Systems

3.1   Linear versus nonlinear systems

In the preceding sections, only linear systems were studied. The real world is, however, generally nonlinear, and the use of linear models is always a source of approximation that is often justified by a number of considerations, such as:

1. The study of linear systems is much easier than that of non-linear systems. General information on the behaviour of the system can be drawn in the first case, while the results obtained studying nonlinear systems are usually applicable only to a given set of conditions.

2. Many actual systems show a behaviour that can be modeled as linear, at least in a broad range of conditions.

3. Even if a system is essentially nonlinear, it is possible, with a few exceptions, to linearize its behaviour for very small variations of its state in the vicinity of given conditions. In such a case the advantages in (1) can be exploited and a general solution for the behaviour *in the small* can be obtained.

In the case of structural dynamics, the aforementioned observations can be better specified as follows:

1. The study of linear systems can be performed using a set of second-order linear differential equations that can be solved by superimposing the solution of the homogeneous equations (defining the free behaviour of the system) to a particular solution of the complete equations (defining the forced behaviour). Above all, it is possible to resort to modal analysis and other techniques directly linked with the linearity of the model.

2. The behaviour of structures is well approximated using linear models if:

   - stresses and strains are low enough not to exceed the proportionality limit of the material, and displacements are small enough not to introduce geometrical nonlinearities,

   - no inherently nonlinear element is introduced into the system, or, at least, influences its behaviour in a significant way. Among the more common nonlinear elements, rolling element bearings (particularly ball bearings), elastomeric springs and dampers, and nonsymmetric shock absorbers can be mentioned,

   - no inherently nonlinear mechanism considerably influences the behaviour of the system. The main sources of nonlinearity are play, Coulomb friction, and dependence of structural damping of most materials on the amplitude of the stress cycle.

3. The linearization of the equations of motion allows one to obtain interesting information on the behaviour of nonlinear systems. The behaviour of the system about an equilibrium position, or, as is usually said, its behaviour in the small, allows assessment of the stability of the equilibrium. Even in cases in which no linearization is, strictly speaking, possible, as for systems whose behaviour is strongly influenced by the presence of nonsymmetrical dampers, a linearized analysis yields important results[1].

It must be stressed, however, that nonlinear systems can show a behaviour that is qualitatively different from that of linear sys-

---

[1]G. Genta, P. Campanile, "An approximated approach to the study of motor vehicle suspensions with nonlinear shock absorbers", *Meccanica*, 24, 1989, 47–57.

tems, and the behaviour in the large can differ significantly from the behaviour in the small. At least a qualitative knowledge of their behaviour is needed by the structural analyst, even if he plans to use, at most, linearized analysis: The differences between the actual behaviour of a structure or machine and the predicted one is often explained by the departure from linearity of some of its components.

One of the phenomena that can occur in nonlinear systems is that of chaotic vibration, which has been experimentally observed and numerically simulated in many mechanical systems. Their occurrence is usually linked with situations in which small changes in the initial conditions lead to large variations of the behaviour of the system. Their study is usually based on the numerical integration in time of the equations of motion or on the use of experimental techniques. This field is still mainly studied by theoretical mechanicists whose work have found, until now, limited practical application in structural analysis; only a short account on chaotic vibration will then be given here[2].

3.2 Equation of motion

A simple model of a nonlinear system with a single degree of freedom is shown in Figure 3.1a. It is similar to the linear system shown in Figure 1.1a, except that the force exerted by the nonlinear spring is proportional to neither the displacement x nor the velocity $\dot{x}$ but can be expressed by the general function $f(x,\dot{x})$, or, better, $f(x - x_A,\dot{x} - \dot{x}_A)$. The symbol used in the figure, although common, is not widely used as the corresponding symbol for linear springs. The symbol shown in Figure 3.1b is sometimes used for the case in which the force is also an explicit function of time. In the latter case, if the mass is assumed to be constant in time, the equation of motion of the system is

$$m\ddot{x} + f(x - x_A,\dot{x} - \dot{x}_A,t) = F(t) - mg \, . \qquad (3.1)$$

In addition to being nonlinear, the system is nonautonomous, because time appears explicitly in the equation through the forcing

---

[2] A textbook that can be used to start the study of chaotic vibration and that contains a large bibliography is F. C. Moon, *Chaotic vibrations, Wiley, New York, 1987.*

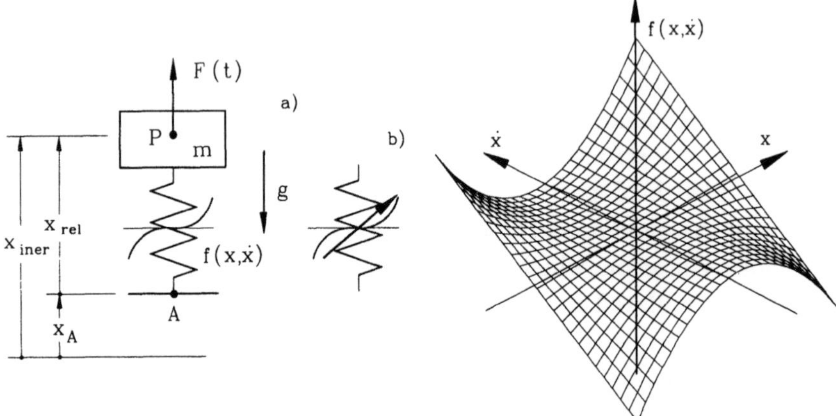

FIGURE 3.1. Nonlinear system with a single degree of freedom.

FIGURE 3.2. "State force mapping" for a nonlinear element characterized by the function $f(x,\dot{x}) = x - \mu\dot{x}(1-x^2)$, with $\mu = 0{,}2$ (Van der Pol's equation).

function $F(t)$ and function $f(x,\dot{x},t)$. Also the equation for the study of free vibrations is then nonautonomous. However, no system with time-dependent characteristics will be studied in the following sections, except when explicitly noted, and the reaction of the nonlinear spring will be expressed by a function $f(x,\dot{x})$ that is not an explicit function of time. This leads to an autonomous equation in the case of free behaviour of the system.

In the latter case, a simple way to represent the characteristics of the system is to plot a tridimensional diagram of the type shown in Figure 3.2 for the so-called Van der Pol oscillator. This type of representation is often referred to as *state force mapping*. Because the term expressing inertia forces is always linear, if the supporting point A can move, there is a definite advantage to resorting to a noninertial x_{rel} coordinate (Figure 3.1). The absolute acceleration to which the inertia force is proportional is $\ddot{x}_{iner} = \ddot{x}_{rel} + \ddot{x}_A$, and equation (3.1), with f not an explicit function of time, written in terms of relative displacements reduces to

$$m\ddot{x} + f(x,\dot{x}) = F(t) - mg - m\ddot{x}_A . \tag{3.2}$$

In this case, the system is not insensitive to a translation of the system of reference, which is independent from time and, consequently, the constant terms of equation (3.2) cannot be neglected or studied separately.

There are cases in which function $f(x,\dot{x})$ can be subdivided into a linear part, $c\dot{x} + kx$, and a nonlinear part, $f'(x,\dot{x})$. The equation of motion (3.2) in this case can be written in the form

$$m\ddot{x} + c\dot{x} + kx + f'(x,\dot{x}) = F(t) - mg - m\ddot{x}_A \,. \tag{3.3}$$

The aforementioned equations can be extended to multi-degree-of-freedom systems. By separating the nonlinear part of the equations from the linear autonomous part, the equations of motion of a general nonlinear system can be written in the following matrix form:

$$\mathbf{M}\ddot{\mathbf{x}} + \mathbf{C}\dot{\mathbf{x}} + \mathbf{K}\mathbf{x} + \mu\mathbf{g} = \mathbf{f}(t) \,. \tag{3.4}$$

Matrices $\mathbf{M}$, $\mathbf{C}$, and $\mathbf{K}$, all of order $n \times n$, define the behaviour of the linear system that can be associated to the original system and are defined accordingly with what seen in Chapter 1. Vector $\mathbf{g}$ contains the nonlinear part of the system. It consists of n functions g_i, which, in the most general case, are $g_i = g_i(\mathbf{x},\dot{\mathbf{x}},\mu,t)$.

In many cases, however, functions g_i do not depend explicitly on time, and the free behaviour can be studied using an autonomous equation. Parameter μ is a number that controls the relative size of the nonlinear and linear parts of the system. With increasing μ, the behaviour of the system gets less linear and, generally speaking, the solution of the equations of motion becomes increasingly difficult.

It is possible to resort to the modal transformation using the modal coordinates of the linearized system also in the case of nonlinear systems. The modal coordinates of the linearized system allow the location of any possible point in the configuration space and, hence, the description of any possible deformed shape of the nonlinear system. Equation (3.4) can thus be written in the form

$$\overline{\mathbf{M}}\ddot{\boldsymbol{\eta}} + \overline{\mathbf{C}}\dot{\boldsymbol{\eta}} + \overline{\mathbf{K}}\boldsymbol{\eta} + \mu\boldsymbol{\Phi}^T\mathbf{g} = \bar{\mathbf{f}}(t) \,. \tag{3.5}$$

However equations of motion (3.5) are not uncoupled, for two reasons. First, the modal damping matrix is not diagonal, except in the case of proportional damping. As shown in Chapter 1, in many cases the modal coupling due to damping can be neglected. A second cause of coupling is the nonlinearity of the system. In each equation, all functions g_i, which are now functions of the modal coordinates η_i and velocities $\dot{\eta}_i$, can be present. If the system is only weakly nonlinear, this coupling can also be neglected and the behaviour of the system can be approximated with that of a number of nonlinear systems with a single degree of freedom.

It must, however, be remembered that to uncouple the various modes is only an approximation in the case of a nonlinear system, even if no damping is present. Moreover, there is no way to estimate the approximation linked to the uncoupling of the modes. In most cases the modal coupling due to nonlinearities makes the modal formulation (3.5) of the equation of motion practically useless.

In the following sections, systems with a single degree of freedom will mainly be dealt with.

Also, in the case of nonlinear systems, the motion can be studied with reference to the state space. Operating in the same way seen in Section 1.3, introducing the state vector

$$\mathbf{z} = \left\{ \begin{matrix} \dot{\mathbf{x}} \\ \mathbf{x} \end{matrix} \right\}$$

and including all forcing functions into the excitation $\mathbf{f}$, the equation of motion (3.2) of a system with a single degree of freedom can be written in the form

$$\dot{\mathbf{z}} = \left\{ \begin{matrix} \dfrac{1}{m} f(x,\dot{x},t) \\ \dot{x} \end{matrix} \right\} . \tag{3.6}$$

If equation (3.4) can be used, the state space equation can be written in the form

$$\dot{\mathbf{z}} = \boldsymbol{A}\mathbf{z} + \left\{ \begin{matrix} -\mu \mathbf{M}^{-1}\mathbf{g} \\ 0 \end{matrix} \right\} + \boldsymbol{B}\mathbf{u}(t) . \tag{3.7}$$

where the dynamic matrix, the input gain matrix, and the input vector are defined in Section 1.4.

If the input acting on the system is constant in time, a static equilibrium position $\mathbf{x}_0$ can be easily obtained by assuming that both velocity and acceleration are equal to 0. In the case of a system with a single degree of freedom (equation 3.2), it follows that

$$f(x_0,0) = F_0 - mg , \tag{3.8}$$

where F_0 is a constant value of the force $F(t)$ and x_0 is the equilibrium position. Note that there may be many solutions of the nonlinear equation (3.8), i.e., the system may have several equilibrium positions.

If function $f(x,\dot{x})$ can be differentiated with respect to x and $\dot{x}$ in the point of the state plane with coordinates $x = x_0$, $\dot{x} = 0$, the motion in the vicinity of an equilibrium position or motion in the small can be studied through the following linearized equation:

$$m\ddot{x} + \left(\frac{\partial f}{\partial \dot{x}}\right)_{\substack{x = x_0 \\ \dot{x} = 0}} \dot{x} + \left(\frac{\partial f}{\partial x}\right)_{\substack{x = x_0 \\ \dot{x} = 0}} (x - x_0) = F(t) - mg - m\ddot{x}_A \,.$$

$$(3.9)$$

The linearized equation (3.9) allows the study of the stability of the static equilibrium position: If the derivative $\partial f/\partial x$ is positive, the equilibrium is statically stable, i.e., when point P is displaced from that position, a force that opposes this displacement is exerted on it. If it is negative, the equilibrium position is unstable; if it vanishes, the knowledge of the higher-order derivatives is needed to analyze the behaviour of the system. If the derivative $\partial f/\partial \dot{x}$ is positive, the system shows a damped behaviour, and if the position is statically stable, dynamic stability is also assured. If it is negative, self-excited divergent oscillations are to be expected, but if it vanishes an undamped behaviour can be predicted.

In the case of systems with many degrees of freedom, the state equation (3.7) can be linearized as

$$\dot{\mathbf{z}} = \left[\begin{array}{cc} -\mathbf{M}^{-1}\left(\mathbf{C} + \mu\left[\frac{\partial g_i}{\partial \dot{x}_j}\right]\right) & -\mathbf{M}^{-1}\left(\mathbf{K} + \mu\left[\frac{\partial g_i}{\partial x_j}\right]\right) \\ \mathbf{I} & 0 \end{array} \right] \mathbf{z} + \boldsymbol{B}\mathbf{u}(t) \,,$$

$$(3.10)$$

where $(\partial g_i/\partial \dot{x}_j)$ and $(\partial g_i/\partial x_j)$ are jacobian matrices containing the derivatives of functions $\mathbf{g}$ with respect to $\dot{\mathbf{x}}$ and $\mathbf{x}$.

The stability in the small can be studied as usual by computing the eigenvalues of the dynamic matrix of the system.

Apart from the study of the stability in the small, i.e., in the vicinity of an equilibrium point, in the case of nonlinear systems, it is possible to define a stability in the large. The stability of nonlinear systems will be dealt with in Section 5.5.4.

3.3   Free oscillations of the undamped system

3.3.1   Direct integration of the equations of motion

If function $f(x,\dot{x})$ does not depend on the velocity, i.e., is $f(x)$ only, the system is conservative and no energy dissipation can take place in it: It can then be said to be *undamped*. Assume that x_0 is an equilibrium position and, without loss of generality, that $f(x_0) = 0$; the free oscillations about the position x_0, i.e., the oscillations that can take place when no force function of time acts on the system, can be studied using the equation

$$m\ddot{x} + f(x) = 0 \,. \tag{3.11}$$

Equation (3.11) is nonlinear but autonomous. If function $f(x)$ is substituted by its Taylor series about point x_0, as $f(x_0) = 0$, the equation of motion becomes

$$m\ddot{x} + (x - x_0)\left(\frac{\partial f}{\partial x}\right)_{x=x_0} + \frac{(x - x_0)^2}{2!}\left(\frac{\partial^2 f}{\partial x^2}\right)_{x=x_0} +$$

$$+ \frac{(x - x_0)^3}{3!}\left(\frac{\partial^3 f}{\partial x^3}\right)_{x=x_0} + \ldots = 0 \,. \tag{3.12}$$

There is no general closed-form solution of equation (3.11), so the study of the dynamic behaviour of nonlinear systems is often performed by resorting to numerical experiments. Numerical simulation is a very powerful tool without which many phenomena could not even be discovered. However, it has the drawback of supplying general information on the behaviour of the system only at the cost of performing a large quantity of experiments. In this, it is very similar to physical experimentation. In many cases it is possible to resort to techniques that yield results that, although only being approximated, are valid in general, at least from a qualitative viewpoint. To get a physical insight to the relevant phenomena, greater stress will be placed on analytical techniques, even when the equations they yield need, at any rate, to be solved numerically. The reader must, however, be aware that only the simultaneous use of analytical and numerical techniques can yield useful information on the behaviour of most nonlinear systems.

Consider a system whose behaviour is modeled by equation (3.11). Assume that $f(x)$ is an odd function of $x - x_0$, positive when $x - x_0$

is positive, and set the origin of the x-coordinate in the static equilibrium position ($x_0 = 0$). Under this assumption, force $-f(x)$ is a symmetrical restoring force trying to keep mass m in the static equilibrium position. The motion of the system then consists of undamped oscillations centered about the position of static equilibrium. In these conditions, because function $f(x)$ is an odd function of x, only the derivatives whose order is odd are present in equation (3.12).

The acceleration can be written as $\ddot{x} = 1/2 d(\dot{x}^2)/dx$, so equation (3.11) can be rearranged in the form

$$\frac{d(\dot{x}^2)}{dx} = -\frac{2}{m} f(x) . \tag{3.13}$$

The physical meaning of equation (3.13) is clear: The variation of the kinetic energy of the system is, at any instant, equal to the work $d\mathcal{L}$ performed by the force $-f(x)$ in the displacement dx. By separating the variables and integrating, the following expression is obtained:

$$\dot{x}^2 = -\frac{2}{m} \int f(x)dx + C . \tag{3.14}$$

As the system is symmetrical with respect to point $x = 0$, the undamped free oscillations will be symmetrical about this point. During oscillation, the velocity vanishes when the extreme position x_m is reached: $\dot{x} = 0$ for $x = x_m$.

Consequently, the constant of integration in equation (3.14) and the limits of integration can be computed. Equation (3.14) can then be integrated, again obtaining

$$t = \sqrt{\frac{m}{2}} \int_0^x \frac{du}{\sqrt{\int_u^{x_m} f(u')du'}} , \tag{3.15}$$

where the dummy variables u and u' have been introduced. Equation (3.15), which has been obtained assuming that at time $t = 0$ point P passes through the equilibrium position, gives the time at which the same point passes through the point whose coordinate is x and can be considered the law of motion of the system. Instead of being expressed as $x = x(t)$, it is in the form $t = t(x)$ and obviously applies only for one-quarter of the period of oscillation, with x included between 0 and x_m. Because the system is symmetrical, it describes completely the motion as all quarters of the period are identical.

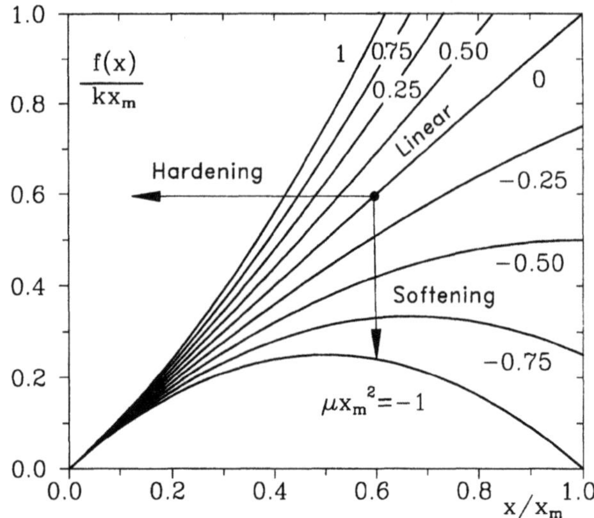

FIGURE 3.3. Law $f(x)$ expressed as a third-degree polynomial with two terms; nondimensional plot for both hardening and softening systems.

The period of oscillation is four times the time needed to reach the maximum amplitude x_m and can be easily computed

$$T = 4\sqrt{\frac{m}{2}} \int_0^{x_m} \frac{dx}{\sqrt{\int_x^{x_m} f(u)\,du}} . \tag{3.16}$$

If the velocity at any position is known, equation (3.14) can be used to compute the value of the amplitude of the motion. As an example, if the initial velocity $(\dot{x})_0$ is known, the value of x_m, i.e., the amplitude of the oscillation, can be obtained from the equation

$$(\dot{x})_0^2 = \frac{2}{m} \int_0^{x_m} f(x)\,dx . \tag{3.17}$$

Equations (3.15) to (3.17) allow the complete study of the free motion of an undamped nonlinear autonomous system. Unfortunately, the integral of equation (3.15) can be solved in closed form only in a few particular cases and, generally speaking, numerical integration is needed. The direct integration of the equation of motion shown here is, however, much simpler than numerical integration of the equation of motion, because only a simple integral has to be performed numerically.

It must be noted that the period is a function of the amplitude of oscillation: This is a general characteristic of nonlinear systems.

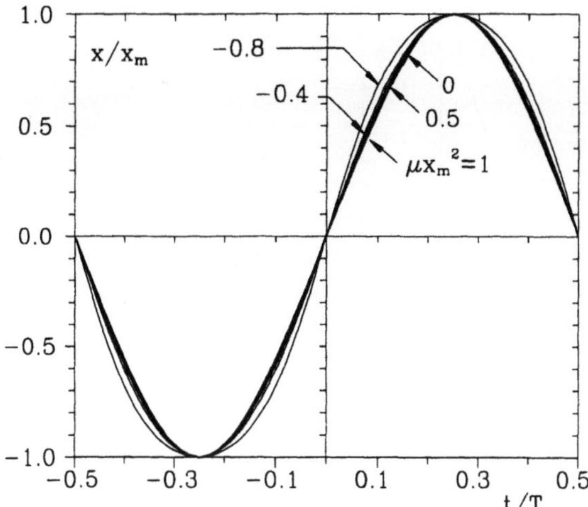

FIGURE 3.4. Time history of the free oscillations of a system with nonlinear restoring force of the type shown in Figure 3.3.

Consider a system in which the nonlinear spring has a characteristic that can be expressed by the following law $f(x)$:

$$f(x) = kx(1 + \mu x^2).\tag{3.18}$$

Equation (3.18) can be considered the expression of equation (3.12) for a system with a restoring force symmetrical with respect to the static equilibrium position (which makes even powers disappear), truncated at the second term. For a general nonlinear spring with a symmetrical characteristic, equation (3.18) can then be regarded as a second approximation, where the first approximation is linearization. Parameter μ has the dimensions of a length at the power -2.

Constant k is usually positive, a condition necessary to lead to a stable static equilibrium position at $x = 0$. If μ is positive, the spring is said to be of the *hardening* type, because its stiffness increases with the displacement. On the contrary, if μ is negative, the spring is said to be of the *softening* type. Equation (3.18) is reported in nondimensional form in Figure 3.3. An expression of the same type of equation (3.18), but with both k and μ negative, has been widely assumed for studies on the chaotic behaviour of mechanical systems.

The maximum elongation in free vibration can be correlated to the speed at which the system passes the equilibrium position through equation (3.19), which, in this case, can be solved directly, yielding

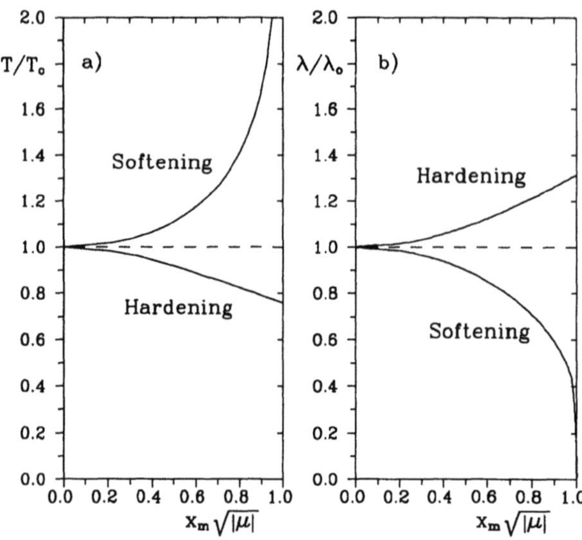

FIGURE 3.5. Frequency of the fundamental harmonic as a function of the amplitude of the free oscillations of a system with a law $f(x)$ of the type shown in Figure 3.3. The period and frequency are made nondimensional by dividing by the values characterizing the linear system.

$$x_m = \sqrt{\frac{1 - \sqrt{1 + 2\frac{m}{k}\mu(\dot{x})_0^2}}{|\mu|}}. \tag{3.19}$$

From equation (3.19), it is clear that in the case of softening systems, oscillatory motion is possible only if the amplitude is not greater than $\sqrt{1/|\mu|}$. If this value is reached, the force is not a restoring force anymore, because it changes its sign and causes point P to move indefinitely away from the static equilibrium position. All the equations reported thus far must be used with care in the context of softening systems, always verifying that the displacements are small enough, or, which is the same, that the total energy of the system is not higher than the maximum potential energy it can store.

The waveform of the free oscillations, obtained by numerically integrating equation (3.15), is reported in nondimensional form in Figure 3.4 for different values of parameter μ, or, better, of the nondimensional parameter μx_m^2. From Figure 3.4, it is clear that all curves are quite close to the sine wave characterizing the behaviour of the linear system ($\mu = 0$). Only in the case of softening systems oscillating with an amplitude close to the maximum possible amplitude, is the waveform far from that of a harmonic oscillation. The period,

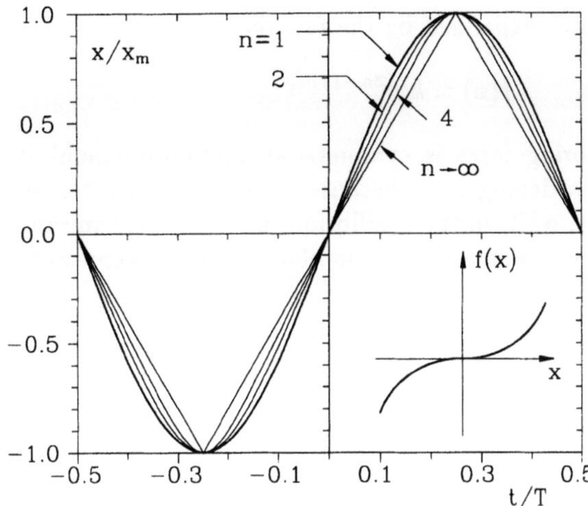

FIGURE 3.6. Time history of the free oscillations of a system with nonlinear restoring force of the type $f'(x) = kx^{2n-1}$.

obtained by numerical integration of equation (3.16), is plotted as a function of $x_m\sqrt{|\mu|}$ in Figure 3.5a. In Figure 3.5b the frequency of the fundamental harmonic has been plotted. Period and frequency are reported in nondimensional form, by dividing by, respectively, the period and the frequency of the linear system obtained by setting $\mu=0$. As is clear from the figure, the frequency increases with the amplitude (i.e., the period of oscillation gets shorter) in the case of hardening systems, while decreases in the case of softening ones. The dependence of the period on the amplitude of oscillation is a general rule for nonlinear systems.

The free response $x(t)$ is not harmonic, because of the nonlinearity of function $f(x)$. It is, however, periodic with period T and can easily be transformed into a series of harmonic terms. Because function $f(x)$ was assumed to be odd, it is easy to demonstrate that only terms whose circular frequency is an odd multiple of the fundamental frequency $2\pi/T$ are present. If at time $t = 0$ the system passes through the origin, the law $x(t)$ can be expressed as

$$x = x_m\left[a_1 \sin\left(\frac{2\pi}{T}t\right) + a_3 \sin\left(\frac{6\pi}{T}t\right) + \ldots + a_i \sin\left(\frac{2i\pi}{T}t\right) + \ldots\right].$$

$$(3.20)$$

Example 3-1

Consider a restoring force expressed by the function

$$f(x) = kx^{(2n-1)}$$

with $n \geq 1$. The restoring force is symmetrical about the equilibrium position and is of the hardening type, because the stiffness increases with increasing displacement, while in the equilibrium position it is vanishingly small. The time history and the maximum elongation are expressed by the equations

$$t = \sqrt{n}\sqrt{\frac{m}{2}} \int_0^x \frac{du}{\sqrt{x_m^{2n} - u^{2n}}},$$

$$x_m = \left[(\dot{x})_0 \sqrt{n}\sqrt{\frac{m}{2}}\right]^{1/n}.$$

If $n = 1$ the system behaves linearly. It is easy to verify that equation (3.15) yields a harmonic law $x(t)$. If $n \to \infty$ the system can be assimilated to two rigid walls in the points of coordinates $x = 1$ and $x = -1$. Point P bounces between the walls, traveling at constant speed in the space between them. If $n \neq 1$, no closed-form integration of the equation of motion is possible. When $n = 2$, i.e., the restoring force follows a cubic law in the displacement, the value of the period is

$$T = \frac{4}{x_m}\sqrt{\frac{2m}{k}} \int_0^1 \frac{d\chi}{\sqrt{1 - \chi^4}},$$

where $\chi = x/x_m$.

By substituting for the elliptical integral its approximate value, the period can be expressed by the formula

$$T = \frac{7.4164}{x_m}\sqrt{\frac{m}{k}}.$$

As usual for nonlinear systems, the period is a function of the amplitude, and in this case it is simply proportional to the reciprocal of the latter. The law $x(t)$ for different values of n is plotted in nondimensional form for different values of n in Figure 3.6.

The values of the first five coefficients of the series for the response are reported in Table 3.1. Because the integral expressing the waveform has to be solved by numerical methods, the coefficients of the series (3.20) also must be computed numerically. All coefficients are subject to numerical errors, which can be quite significant in the higher-order harmonics, except for the case with $n \to \infty$, in which they can be computed in closed form.

	a_1	a_3	a_5	a_7	a_9
$n = 1$ (linear)	1	0	0	0	0
$n = 2$	0.9522	-0.0440	0.0025	-0.0004	0.0002
$n = 3$	0.9238	-0.0641	0.0094	-0.0015	0.0004
$n = 4$	0.9045	-0.0745	0.0152	-0.0036	0.0010
$n = 5$	0.8907	-0.0804	0.0195	-0.0058	0.0019
$n \to \infty$	0.8106	-0.0901	0.0324	-0.0165	0.0100

TABLE 3.1. Coefficients of the first five harmonics of the series (3.20) for the response with different values of n. The coefficients were computed by numerically integrating equation (3.15) (Gauss method) in 400 steps in each quarter of period, and then using a discrete Fourier algorithm (1,600 points in the period). For $n \to \infty$ the coefficients were computed using the formula: $a_i = 8/(i\pi)^2$

From Table 3.1, an interesting consideration can be drawn: Even if the nonlinearity is strong, the response of the system is not far from being harmonic, i.e., the contribution of all harmonics except the first is quite small. This result is linked with the fact that function $f(x)$ was assumed to be odd. If no symmetry was assumed, higher-order harmonics would have been more important, and all of them, including an average deflection or zero-order harmonic, could have been present.

3.3.2   Series solutions: Harmonic balance

To avoid the difficulties linked with the integration of equation (3.15), sometimes a few of the terms of the series (3.20) are approximately evaluated, and often only the first term is computed. In the case of free oscillations of the undamped system, it gives an approximate relationship between the amplitude and the value of the fundamental frequency of the motion.

The simplest technique is the so-called *harmonic balance*. It consists simply of the introduction of a series solution of the type of equation (3.20) into the equation of motion. Because all harmonics of the motion must be balanced, a set of nonlinear algebraic equations in the amplitudes of the various harmonics is obtained. The procedure is approximate for two reasons: The series (3.20) is truncated, usually after few terms, if not at the first one, and some of the harmonic terms cannot be balanced, because the number of equations obtained is greater than the number of unknowns. Although it may not seem rigourous mathematically and not approximate enough in some cases, if properly used with sound engineering common sense, this is a powerful tool for solving many nonlinear problems.

Example 3-2

Find an approximate solution for the fundamental harmonic of the free response of an undamped system whose nonlinear spring has a law expressed by equation (3.18). If only the fundamental harmonic is retained, the solution can be expressed in the form

$$x = x_m \sin(\lambda t).$$

By introducing the assumed time history of the solution into the equation of motion, it follows

$$(-m\lambda^2 + k)x_m \sin(\lambda t) + k\mu x_m^3 \sin^3(\lambda t) = 0.$$

By remembering some trigonometric identities and introducing the natural frequency of the linearized system $\lambda_0 = \sqrt{k/m}$, the latter equation yields

$$\left[1 - \left(\frac{\lambda}{\lambda_0}\right)^2\right] x_m \sin(\lambda t) + \frac{3}{4}\mu x_m^3 \sin(\lambda t) - \frac{1}{4}\mu x_m^3 \sin^3(3\lambda t) = 0.$$

To satisfy this equation for all values of time, the coefficients of both $\sin(\lambda t)$ and $\sin(3\lambda t)$ must vanish. The first condition leads to the equation

$$\left[1 - \left(\frac{\lambda}{\lambda_0}\right)^2\right] + \frac{3}{4}\mu x_m^3 = 0,$$

which is the required relationship linking the frequency of the fundamental harmonic with the amplitude.

Balancing the third harmonic would lead to another equation in the unknown x_m that cannot be satisfied. This fact is linked with the intrinsic approximations of the method. The condition for balancing the fundamental harmonic yields the following relationship between the amplitude and the frequency

$$\frac{\lambda}{\lambda_0} = \sqrt{1 + \frac{3}{4}\mu x_m^3},$$

which has been plotted in nondimensional form in Figure 3.7, together with the solution already reported in Figure 3.5b. The precision obtained through the harmonic balance technique is very good in the case of oscillations with small amplitude and mildly nonlinear systems, i.e., for small values of parameter $|\mu|x_m^2$.

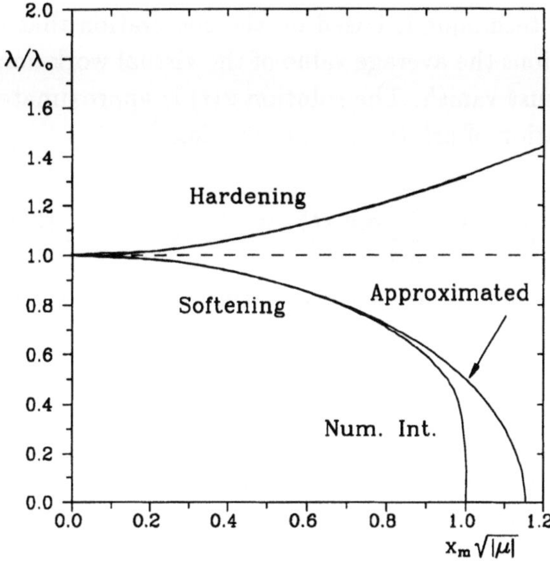

FIGURE 3.7. Frequency of the first harmonic as a function of the amplitude of the free oscillations of a system with a law $f(x)$ of the type shown in Figure 3.3; approximate values obtained using the harmonic balance procedure compared with the results obtained through numerical integration of the equations of motion. For hardening systems, the two curves are completely superimposed.

With increasing values of $|\mu|x_m^2$, the solution loses precision, particularly in the case of softening systems. In particular, the value of the amplitude at which the natural frequency vanishes, i.e., the period tends to infinity or the system no longer has an oscillatory behaviour, is $\sqrt{4/3|\mu|}$, and the correct value is $\sqrt{1/|\mu|}$.

A better approximation can be obtained by assuming a solution in which two harmonics are included:

$$x = a_1 \sin(\lambda t) + a_2 \sin(3\lambda t).$$

There are now two unknowns and two harmonics can be balanced. Note that a better evaluation of the frequency of the fundamental harmonic is obtained in this way. The computations involved are, however, quite complex because the solution of two coupled nonlinear equations must be obtained.

3.3.3 Ritz averaging technique

Other solution methods, which can also be labeled series solutions, are the so-called averaging techniques. Into this category falls, among others, the method of Kirilov and Bogoljubov and that usually referred to as the Ritz averaging method, which will be described in detail. Practically, they are based on the substitution of some nonlinear functions with their average over one period, thus obtaining simplified expressions yielding approximate results. Usually the averaging is based on energy consideration.

The Ritz averaging technique is based on the observation that in free undamped vibrations the average value of the virtual work done in a complete cycle must vanish. The solution $x(t)$ is approximated with a linear combination of arbitrary time functions[3]

$$x(t) = \sum_{i=1}^{n} a_i \phi_i(t) \,, \qquad (3.21)$$

where a_i are constants and ϕ_i are arbitrary time functions. If $f(x)$ is an odd function, functions ϕ_i can be the harmonic terms of the series (3.20).

When looking for a first approximation of the fundamental harmonic of the time history of the motion, only one function ϕ_i is used. As in the case of the harmonic balance technique, constants a_i can be computed by introducing the solution (3.21) into the equation of motion. The latter clearly cannot be satisfied exactly, because only an approximation of the true time history has been assumed. Instead of requiring that the equation of motion be satisfied at each instant, which is impossible, the equation of motion is satisfied as an average over one period of the motion.

The equation of motion (3.11) can be considered an equation of dynamic equilibrium, stating that elastic and inertial forces must always be in equilibrium. If a virtual displacement δx is imposed on the system starting from a dynamic equilibrium condition characterized by given values of the position and velocity, the virtual work done by the forces applied to it must vanish, i.e.,

$$[m\ddot{x} + f(x)]\,\delta x = 0 \,. \qquad (3.22)$$

As already said, this condition cannot be satisfied exactly. The

---

[3]Equation (3.21) must not be confused with expression (2.74) used in connection with the assumed-modes method. In the latter method, the unknowns $a_i(t)$ are functions of time, and the assumed functions are functions of the space coordinates. The problem is that of obtaining the deformed shape of the system, continuous or discrete, which is approximated by arbitrary functions, while the time history is not difficult to compute (in the case of free oscillations, it is harmonic). In this case, however, the system is nonlinear; then the difficulty lies in the determination of the time history, which is approximated by the arbitrary time functions. In the case of systems with a single degree of freedom, the unknowns are constants, and they are functions of space coordinates if the method is applied to continuous systems or vectors for multi-degree-of-freedom discrete systems.

averaged approach consists of stating that, although not vanishing in any instant, the virtual work must be equal to zero as an average over a period. This means that the integral of equation (3.22) over one period is equal to zero. If a set of n virtual displacements are stated using the same functions ϕ_i assumed in the definition of the time history, it follows that

$$\delta x = \sum_{i=1}^{n} \delta a_i \phi_i(t) . \tag{3.23}$$

The averaged expression of the virtual work is then

$$\delta \mathcal{L} = \sum_{i=1}^{n} \int_{0}^{T} \left[\ddot{x} + \frac{f(x)}{m} \right] \delta a_i \phi_i(t) dt = 0 . \tag{3.24}$$

Because the virtual displacements δa_i are arbitrary, each term of the sum must vanish in order to satisfy equation (3.24)

$$\int_{0}^{T} \left[\ddot{x} + \frac{f(x)}{m} \right] \phi_i(t) dt = 0, \qquad \text{for } i = 1,2,\ldots,n . \tag{3.25}$$

The n equations (3.25) are a set of equations in the n unknowns a_i. Because such equations are generally nonlinear, their solution can require the use of numerical techniques and may constitute a difficult mathematical problem.

Example 3-3
Repeat the computation of Example 3-2 by resorting to the Ritz averaging technique.
The linear combination expressing the time history will be assumed to reduce to a single harmonic term and to coincide with that used in Example 3-2. There is only one function of time $\phi(t) = \sin(\lambda t)$ and, correspondingly, only one unknown coefficient, $a = x_m$. Equation (3.25), reduces to

$$\int_0^T \left[x_m \left(\frac{k}{m} - \lambda^2 \right) \sin(\lambda t) + \frac{k}{m} \mu x_m^3 \sin^3(\lambda t) \right] \sin(\lambda t) dt = 0 .$$

Remembering that

$$\int_0^T \sin^2(\lambda t) dt = \frac{\pi}{\lambda} , \qquad \int_0^T \sin^4(\lambda t) dt = \frac{3\pi}{4\lambda} ,$$

and introducing the natural frequency of the linearized system $\lambda_0 = \sqrt{k/m}$, the equation linking the fundamental frequency of the free vibration with the amplitude is

$$\left[1 - \left(\frac{\lambda}{\lambda_0} \right)^2 \right] + \frac{3}{4} \mu x_m^3 = 0 ,$$

This result is coincident with that obtained trough the harmonic balance technique. In this case, a second approximation solution could also be searched by using two arbitrary functions instead of one, but the complexity of the relevant computations would be overwhelming.

3.3.4 Iterative and perturbation techniques

Perturbation and iterative methods are restricted to weakly nonlinear and nonautonomous systems (or quasi-linear systems), i.e., systems whose equation of motion can be separated into a part containing only linear terms and a second one, relatively small with respect to the first, containing the nonlinear or nonautonomous terms. What *weakly* actually means is difficult to assess, and the applicability of this approach depends on many factors, including the method used and the nature of the problem.

In the iterative approach, a zero-order approximation solution is stated and then introduced into the equation of motion in order to obtain a new solution.

The process is repeated until the required precision is obtained. In perturbation techniques the solution is assumed to be a power series of a small parameter linked with the nonlinear part of the system, e.g., parameter μ of equation (3.4). As a generating solution (i.e., the part of the solution remaining when the small parameter is set to 0) the solution of the linear equation is assumed. Both approaches are, however, plagued by the presence of the so-called secular terms, i.e., terms that increase in time to infinity, which are introduced by the solution algorithms and must be eliminated at each step of

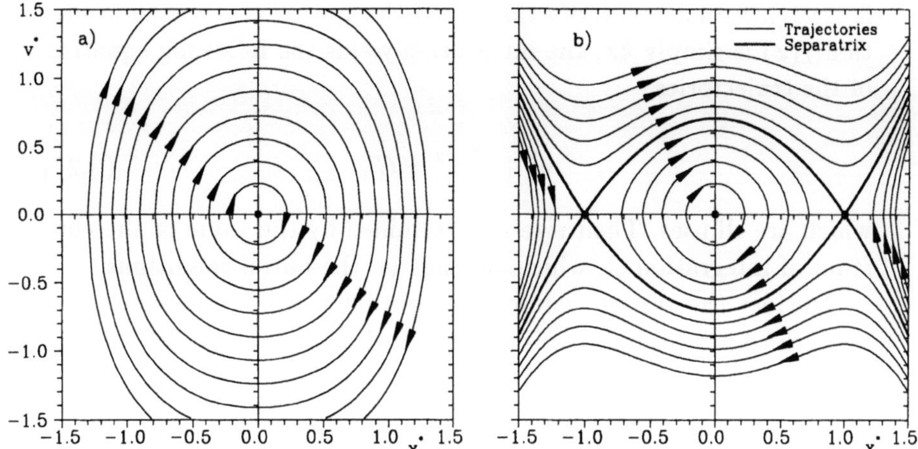

FIGURE 3.8. State portrait of a system whose restoring force is expressed by equation (3.18). Nondimensional velocity $v^* = v\sqrt{|\mu|m/k}$ as a function of nondimensional displacement $x^* = x\sqrt{|\mu|}$ for (a) hardening and (b) softening systems.

the solution. The term *secular* derives from the astronomical field, in which the perturbation technique was first applied. A detailed explanation of these methods is beyond the scope of this book; the reader can refer to many texts[4].

3.3.5 *Solution in the state plane*

By rewriting the state equation (3.6) for the free behaviour of an undamped system and explicitly introducing the velocity $v = \dot{x}$, the following equation is obtained

$$\begin{cases} \dot{v} = -\dfrac{1}{m}f(x)\,, \\ \dot{x} = v\,, \end{cases} \tag{3.26}$$

or

$$\frac{dv}{dx} = -\frac{f(x)}{m}\,.$$

The second equation has been obtained by eliminating time between the two equations. It directly yields the slope of the trajectories of the system in the state plane and can be integrated without problems, even if in many cases numerical integration is required. It

---

[4]For example, L. Meirovitch, *Methods of analytical dynamics*, McGraw-Hill, New York, 1970.

is straightforward to verify that when the system is linear and function $f(x)$ is simply kx, the integration yields the following equation of the trajectories

$$v^2 + \frac{k}{m}x^2 = C \,, \tag{3.27}$$

which are ellipses. Their size is determined by the value of the constant of integration C, which depends on the initial conditions.

Consider the system studied in Figure 3.3, whose nonlinear part has a third power characteristic. Introducing equation (3.18) into the second equation (3.26) and integrating, it follows that

$$v^2 + \frac{k}{m}x^2 \left(1 + \frac{\mu}{2}x^2\right) = C \,. \tag{3.28}$$

Equation (3.28) is plotted in Figure 3.8 in nondimensional form for various values of constant C. Stating a nondimensional velocity $v^* = v\sqrt{|\mu|m/k}$ and a nondimensional coordinate $x^* = x\sqrt{|\mu|}$, the behaviour of all possible systems whose function $f(x)$ is of the type of equation (3.18) can be summarized in just two plots; one for hardening systems (Figure 3.8a) and one for softening systems (Figure 3.8b)

Note that all trajectories run from left to right in the upper half-plane, where the velocity is positive, and from right to left in the lower half-plane.

If the system is hardening there is a single equilibrium position and the trajectories are very similar to ellipses (circles, in the nondimensional plot) for small amplitudes. With growing amplitudes, the curves are elongated on the velocity axis and they lose the shape of ellipses. The equilibrium point is what is called a *singular point* of the center type. No trajectory can pass through that point, but all of them orbit around it.

In the case of softening systems, as already said, there are three equilibrium points; one stable (the origin) and two unstable. The state portrait is divided into five domains or basins of attraction by a separatrix. The first domain of attraction is centered in the origin and is the basin of the stable solution. In it, the motion is periodic, with trajectories of elliptical (circular) form, if the amplitude is small, and an increasingly elongated form (along the displacement axis) with growing amplitude. In the two basins of attraction that lie above and below the first one, the motion is nonperiodic and characterized

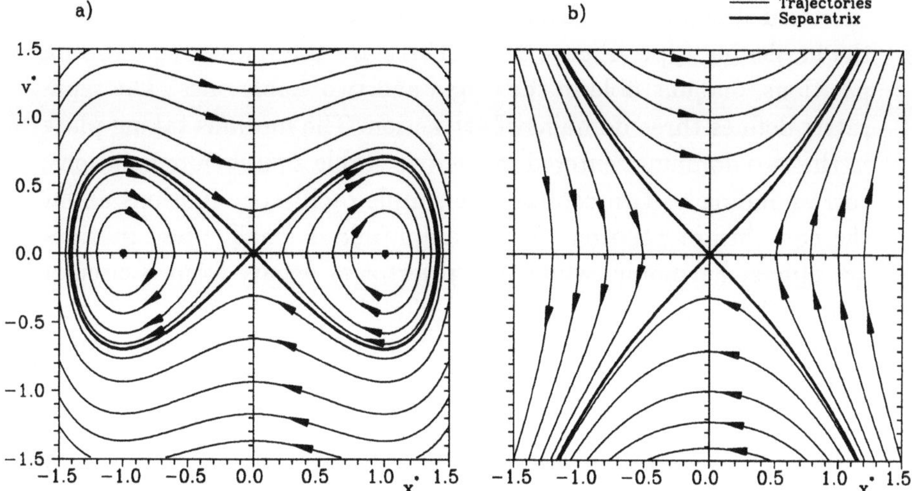

FIGURE 3.9. State portrait of a system whose restoring force is expressed by equation (3.18), as in Figure 3.8, but with a negative value of the stiffness k. Nondimensional velocity $v^* = v\sqrt{|\mu|m/k}$ as a function of nondimensional displacement $x^* = x\sqrt{|\mu|}$ for (a) hardening (in this case with negative value of μ) and (b) softening systems.

by high speed. Point P comes from the right (lower basin) or left (upper basin) and slows down, feeling a repulsive force unless the force changes sign in correspondence with the unstable equilibrium position. It is then caught by the central force field of the origin, accelerates again, and dashes past it. It then slows down again until the force changes sign at the other unstable equilibrium position and then accelerates away under the effect of the repulsive force field.

The other two domains of attraction, on the right and on the left, are characterized by large displacement and relatively small speed. The motion is again nonperiodic, and point P comes toward the origin, but the repulsive force prevents it from reaching the unstable equilibrium position and drives it back.

The separatrices pass through the unstable equilibrium points: They represent the trajectory of a point that has just enough energy to reach the unstable equilibrium position but gets there with vanishingly small speed. The two unstable equilibrium points are again singular, this time of the saddle type.

If the linear term is repulsive instead of attractive, i.e., the stiffness k is negative, the state portraits are of the type shown in Figure 3.9. When μ is negative, i.e., the coefficient of the term containing the third power of the displacement is positive, the behaviour is of the hardening type (Figure 3.9a). There are three static equilibrium positions, one unstable in the origin and two stable ones. The separatrix defines three domains of attraction. The motions taking place in the two domains centered in the two stable equilibrium positions, corresponding to two centers, are simple oscillatory motions about the equilibrium positions. If the amplitude is small, these motions are almost harmonic, while the trajectories depart from a circular shape when the amplitude grows. The domain of attraction lying outside the separatrix is that of oscillatory motions of large amplitude that cross all three equilibrium positions and are centered about the origin. The origin is a saddle point.

In the case of softening systems with repulsive linear stiffness, no oscillatory motion is possible and nonperiodic motions occur in all four domains of attraction visible in the state portrait, and the origin is again a saddle point.

Note that all the approximate methods seen in the previous sections are applicable only within the domains of attraction of the stable equilibrium positions, while the current solution yields exact results in the whole state plane. A more detailed analysis of the state-plane approach and a review of the possible singular points will be seen when dealing with damped systems.

3.4   Forced oscillations of the undamped system

3.4.1   Approximate evaluation of the response to a harmonic forcing function using the Ritz averaging technique

The forced oscillations of the undamped system can be studied by using the nonautonomous equation obtained by adding a forcing function to equation (3.11). By separating the linear part of the restoring force $f(x)$ from the nonlinear part as in equation (3.3), the following nonlinear and nonautonomous equation is obtained:

$$m\ddot{x} + kx + f'(x) = F(t)\,. \tag{3.29}$$

The solution of equation (3.29) is, from the mathematical viewpoint, a formidable task and no general solution can be easily obtained. The only general approach is the numerical integration of the equation of motion, but, as already stated, it does not yield general results, and it corresponds to performing a numerical experiment.

Among the approximate methods, the Ritz averaging technique can be used simply, taking into account the virtual work due to the forcing function $F(t)$, i.e., by introducing the latter into equation (3.25), which becomes

$$\int_0^T \left[\ddot{x} + \frac{k}{m}x + \frac{f'(x)}{m} + \frac{f(x)}{m} - \frac{F(t)}{m} \right] \phi_i(t)dt = 0 , \qquad (3.30)$$

for $i = 1,2,\ldots,n$.

If the number of functions $\phi_i(t)$ is high (it is usually enough when it is greater than 1) the computational difficulties can become overwhelming. Harmonic balance techniques can also be used, obtaining results that are usually identical to those achievable through averaging techniques. In this case, the complexity of the analysis grows very rapidly with the number of harmonics considered.

If both the nonlinear part $f'(x)$ of $f(x)$ and the nonautonomous part of $F(t)$ are small, iterative or perturbation techniques can be used. Unfortunately, little general insight of the relevant phenomena can be gained through solutions of this type, due to the impossibility of resorting to the superimposition technique. The advantages of these complex analytical techniques over the direct numerical integration of the equation of motion are then questionable.

If the forcing function is harmonic in time with frequency λ, assuming that time $t = 0$ coincides with the instant at which the forcing function vanishes while increasing from negative values, the expression for $F(t)$ is $f_0 \sin(\lambda t)$. If the Ritz averaging technique is used, it is not necessary to assume that the nonlinear part of the restoring force $f'(x)$ is small. The accuracy of the results is, however, better in the case of mildly nonlinear systems.

Even in the case of the simplest expressions of function $f'(x)$, it is impossible to reach a general solution of the equation of motion. The steady-state response will be assumed to be periodic, with a period equal to the period of the forcing function, and an approximate evaluation of the fundamental harmonic of the response will be searched. Because no damping has been introduced into the system, it is possible to assume that the excitation and the response are in phase with each other. Reducing the response to its fundamental harmonic, a time history of the type

$$x = x_m \sin(\lambda t) \tag{3.31}$$

can be assumed. Note that if function $f'(x)$ is not odd, equation (3.32) can lead to poor results. Because even harmonics are also present in the response, the presence of a constant term of the displacement can be expected and must be included in the response, even when only an approximation of the fundamental harmonic is searched.

By applying the Ritz averaging technique in the form of equation (3.30), it follows that

$$\int_0^T \left[m\ddot{x} + kx + f'(x) - f_0 \sin(\lambda t) \right] \sin(\lambda t) dt = 0 . \tag{3.32}$$

By defining a nondimensional time variable $\tau = \lambda t$, whose values span from 0 to 2π in a period, introducing equation (3.31) into equation (3.32), and integrating, it follows that

$$x_m \left[1 - \left(\frac{\lambda}{\lambda_0} \right) \right]^2 - \frac{f_0}{k} + \frac{1}{k\pi} \int_0^{2\pi} f'(x) \sin(\tau) d\tau = 0 . \tag{3.33}$$

Equation (3.33) allows the computation of the amplitude of the response x_m as a function of the amplitude and the frequency of the forcing function. Once that function $f'(x)$ has been stated, it can easily be transformed through equation (3.31) into a function of time and then integrated. In the worst cases, the integration can be performed using numerical techniques. The frequency response, i.e., the amplitude of the response as a function of the forcing frequency, can thus be computed easily. Note that, because of the nonlinearity of the system, the frequency response also depends on the amplitude of the excitation.

Often, it is simpler to obtain the frequency response in the form of a function $\lambda(x_m)$ instead of in the more common form $x_m(\lambda)$ by solving equation (3.33) in the frequency

$$\frac{\lambda}{\lambda_0} = \sqrt{ 1 - \frac{f_0}{kx_m} + \frac{1}{k\pi x_m} \int_0^{2\pi} f'(x) \sin(\tau) d\tau } . \tag{3.34}$$

3.4.2 Undamped Duffing's equation: First-approximation solution

Consider a system whose restoring force is that studied in Figure 3.3, with a nonlinear part following a third-power law, on which a harmonic forcing function is acting. The relevant equation of motion,

$$m\ddot{x} + kx(1 + \mu x^2) = f_0 \sin(\lambda t)\,, \tag{3.35}$$

is often referred to as *undamped Duffing's equation*, because its general properties were first studied in detail by G. Duffing in 1918[5].

The study of such equations is particularly interesting because it yields a deep insight to many properties peculiar to nonlinear systems. As usual with nonlinear systems, there is no general solution to equation (3.35), but different techniques can be used to obtain approximate solutions. As already stated, the solution of the equation of motion of a nonlinear system is, generally speaking, nonharmonic, even if the forcing function is harmonic. However, a good approximation, at least of the fundamental harmonic of the response, can be found using the Ritz averaging technique and assuming a monoharmonic time history.

By introducing the nonlinear function $f'(x) = k\mu x^3$ into equation (3.34), it follows that

$$\frac{\lambda}{\lambda_0} = \sqrt{1 - \frac{f_0}{kx_m} + \frac{3}{4}\mu x_m^2} = \sqrt{1 - \left(\frac{x_m k}{f_0}\right)^{-1} + \frac{3}{4}\frac{\mu f_0^2}{k^2}\left(\frac{x_m k}{f_0}\right)^2}\,. \tag{3.36}$$

The second form of equation (3.36) evidences two nondimensional parameters, namely, $x_m k/f_0$ e $\sqrt{|\mu|}f_0/k$. The first is nothing other than the magnification factor defined in the way already seen for linear systems. It can be plotted as a function of the nondimensional frequency of the forcing function λ/λ_0. The second defines the nonlinearity of the system. If such a parameter vanishes, the response of the linear system is found. Otherwise, the dependence of the magnification factor on $\sqrt{|\mu|}f_0/k$ shows that it is a function of the amplitude of the excitation as well as of the parameters of the system. As already stated, this is typical of nonlinear systems.

---

[5]G. Duffing, *Erzwungene Schwingungen bei veränderlicher Eigenfrequenz*, F. Vieweg u. Sohn, Braunschweig, Germany, 1918.

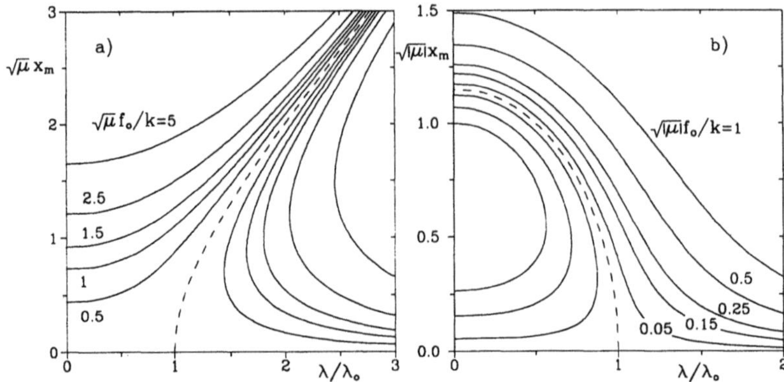

FIGURE 3.10. Response of an undamped system governed by Duffing's equation; amplitude of the response as a function of the frequency for various values of the amplitude of the harmonic forcing function; plot for (a) hardening and (b) softening systems in nondimensional form.

The amplitude of the response x_m is plotted as a function of the forcing frequency, using the amplitude of the forcing function f_0/k as a parameter, in Figures 3.10a and 3.10b, for the cases of hardening and softening systems, respectively. Note that the solution obtained here is only an approximation that gets worse when the value of the nondimensional frequency λ/λ_0 is far from unity. The approximation is particularly bad in static conditions, as can be immediately assessed by comparing the expression for the amplitude obtainable from equation (3.36) with ($\lambda = 0$) and the expression obtainable directly from the equation of motion (3.35) rewritten for static conditions (constant forcing function, $\ddot{x} = 0$).

The dashed curves have been obtained from equation (3.36) by setting to zero the amplitude of the forcing function. They yield the frequency of the free oscillations as a function of their amplitude and coincide with the curves of Figure 3.7. Such curves are often referred to as the *backbone* or *skeleton* of the response and are very important in the evaluation of the response to harmonic excitation of nonlinear systems, both undamped and damped. The backbone curve expresses the conditions for the resonance of a nonlinear system, assuming that such a term has a meaning in this case.

Also, in the case of the nonlinear system, three zones can be identified in the frequency response:

1 At low frequency the behaviour of the system is dominated by the stiffness, which, in this case, is nonlinear. The magnification factor depends on the force f_0; the phase is 0.

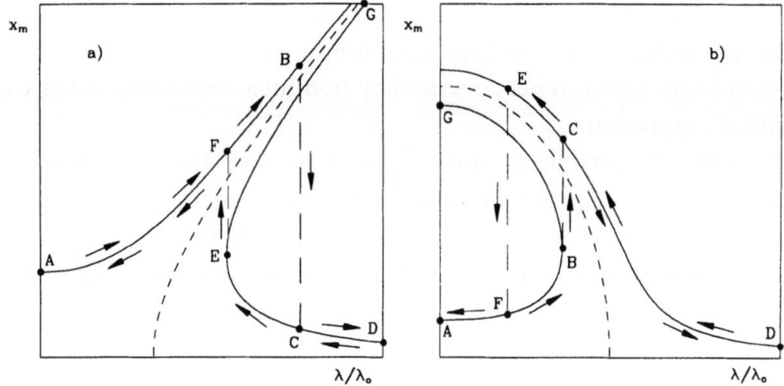

FIGURE 3.11. Jump phenomenon in (a) hardening and (b) softening systems.

2 At high frequency the behaviour is dominated by the inertia of
  mass m. Because no nonlinearity is involved in inertia forces,
  the response is very similar to that of the linear system; the
  phase is 180°.

3 When conditions approach those typical of free oscillations, i.e.,
  when the curve $x_m(\lambda/\lambda_0)$ approaches the backbone, a sort of
  resonance occurs and the behaviour of the system is dominated
  by damping. Little can be said by using an undamped model.

In the nonlinear case, however, the resonance peak is not straight,
but rather it slopes to the right for hardening systems and to the
left for softening ones. Consequently, there is no finite value of the
frequency for which the amplitude of the response grows to infinitely
large values when damping tends to zero.

Consider the hardening system with an excitation with very low
frequency. The response corresponds to point A in Figure 3.11a. With
increasing frequency, the amplitude of the response grows, following
the line AFB, always remaining in phase with the excitation. After
reaching point F, however, the solution is no longer the only possible
one, and, at a certain frequency, mainly defined by the damping of
the system (point B), the amplitude drops at the value corresponding
to point C in the figure, and the phase lag shifts to 180°. At higher
frequencies, the line CD of the plot is followed.

If, on the contrary, the frequency is decreased from that corresponding to point D, the line DCE will be followed, with a phase of
180°. At frequencies lower than that of point E, no solution with a
phase of 180° exists, and an increase in the amplitude of the response
and a change of phase to 0 occur (line EF). The amplitude then de-

creases following line FA. This behaviour is often referred to as the *jump phenomenon* and is typical of nonlinear systems. It also holds, qualitatively, for systems governed by nonlinear equations other than Duffing's equation.

A similar behaviour is shown by softening systems, as illustrated in Figure 3.11b. The path followed when the frequency of the excitation increases is AFBCD, with a jump at BC. When the frequency is reduced, the path is DCEFA, with a downward jump at EF. While the downward jumps (BC in hardening systems and EF in softening systems) occur at a frequency that is often not well determined, depending mainly on damping but also on external disturbances, the frequency at which the upward jumps occur (BC and EF, respectively) can be computed, at least within the first-approximation solution.

It is easy to verify that the frequency at which point E is located in Figure 3.11a or point B is located in Figure 3.11b is

$$\frac{\lambda}{\lambda_0} = \sqrt{1 \pm \sqrt[3]{\frac{81}{16}|\mu|\left(\frac{f_0}{k}\right)^2}}, \tag{3.37}$$

where the upper sign holds for hardening systems and the lower one for softening systems.

It is possible to demonstrate that when three values of the amplitude are found, the greatest and the smallest correspond to stable conditions of the system, while the intermediate one is unstable and cannot actually be obtained. The unstable branch is the one indicated as EG in Figure 3.11a (hardening system) and BG in Figure 3.11b (softening system). The attraction domains of the solutions will be studied in detail when dealing with the damped Duffing's equation.

The solution obtained earlier is only a first-approximation estimate of the fundamental harmonic of the solution, in which harmonics with frequencies 3λ, 5λ, ..., etc., are also present. There are only odd-order harmonics because function $f(x)$ is odd. Although their amplitude is usually not great, their presence is a symptom of the presence of nonlinearities, as the sloping resonance peak. A detailed study of the harmonic content of the solution of Duffing's equation, based on the perturbation technique, is reported in the already mentioned book by L. Meirovitch, where the interested reader can find all the details. Only a short summary of the results will be reported here.

A harmonic motion is used as a generating solution. The amplitude of the response is computed by stating that the secular terms, which are in the equation yielding the second term of the power series $x^{(1)}$, must vanish and the same value of the amplitude obtained using the Ritz averaging technique is obtained. The term $x^{(1)}$, which contains a term with frequency λ and a second one with frequency 3λ, is then obtained. Their amplitude is computed by stating that the secular terms that appear in the solution for $x^{(2)}$ vanish. Note that while obtaining the solution for the third harmonic, a new estimate for the amplitude of the fundamental harmonic, which, however, is usually not much different from that previously found, is obtained. The analysis can continue obtaining new harmonics and corrections for the already computed results, but the relevant computations become very involving.

Harmonics whose frequency is a submultiple of the forcing function (i.e., subharmonics) can also be excited. In linear systems, subharmonics can be present when the frequency of the forcing function is a multiple of the natural frequency; these subharmonic oscillations are, however, seldom observed as the damping of the system makes them disappear quite soon. The presence of sustained subharmonic oscillations is then another characteristic feature of nonlinear systems. In the case of a system governed by Duffing's equation, only odd subharmonics, i.e., subharmonics with frequencies $\lambda/3$, $\lambda/5$, $\lambda/7$, ..., are found, because the restoring force is an odd function of the displacement.

When a linear system is excited by a polyharmonic forcing function (i.e., a forcing function that can be considered the sum of harmonic components), the response is also polyharmonic and can be computed by adding the responses to each component of the excitation. In the case of nonlinear systems, the response is much more complex, and if a Fourier analysis is performed, it contains harmonics that are not present in the excitation. This could be expected, because the effect of distorting the waveform, which is peculiar to nonlinear systems, is exactly that of introducing new harmonics.

Consider a system governed by Duffing's equation excited by a forcing function consisting of the sum of two harmonic components λ_1 and λ_2. By using perturbation techniques, a first approximation solution containing harmonic components with frequencies λ_1, λ_2,

$2\lambda_1 + \lambda_2$, $2\lambda_1 - \lambda_2$, $\lambda_1 + 2\lambda_2$, $\lambda_1 - 2\lambda_2$, $3\lambda_1$, and $3\lambda_2$ is obtained. The fundamental harmonics with frequencies λ_1 and λ_2, which would have been in the response of a linear system, are then present together with higher-order harmonics (third-order, in this case) and combination tones. If the solution is carried on to higher-order approximations, other harmonics are found. Note that the particular higher-order harmonics and combination tones found are typical of systems governed by Duffing's equation and that other frequencies would have been obtained if a different restoring force was assumed.

Example 3-4

Evaluate the amplitude of the fundamental harmonic of the response to a harmonic forcing function of a piecewise linear system, i.e., a system in which the restoring force has a characteristic that can be represented by a number of straight lines. Assume that the restoring force is expressed by an odd function of the displacement and that there are only two values of the stiffness (Figure 3.12a).

By identifying the stiffness of the first part of the characteristics k_1 with the stiffness k of the linearized system, function $f'(x)$ expressing the nonlinear part of the restoring force is

$$\begin{cases} f'(x) = k(\alpha - 1)(x + x_1) & \text{per } x < -x_1 \,, \\ f'(x) = 0 & \text{per } -x_1 < x < x_1 \,, \\ f'(x) = k(\alpha - 1)(x - x_1) & \text{per } x > x_1 \,, \end{cases}$$

where $\alpha = k_2/k_1$.

Function $f'(x)$ must be integrated separately in each of the fields identified in Figure 3.12b. The first expression for $f'(x)$ holds between τ_3 and τ_4, while the third equation holds between τ_1 and τ_2. Function $f'(x)$ vanishes between 0 and τ_1, τ_2 and τ_3, and τ_4 and 2π. The integral to be introduced into equation (3.34) is simply

$$\int_0^{2\pi} f'(x) \sin(\tau) d\tau = k(\alpha - 1) \left\{ \int_{\tau_1}^{\pi - \tau_1} \left[x_m \sin^2(\tau) - x_1 \sin(\tau) \right] d\tau + \right.$$

$$\left. + \int_{\pi + \tau_1}^{2\pi - \tau_1} \left[x_m \sin^2(\tau) - x_1 \sin(\tau) \right] d\tau \right\} ,$$

where the value of τ_1 is $\tau_1 = \arcsin(x_1/x_m)$.

The response of the system can thus be easily computed through equation (3.34). By performing the integration and writing the response in nondimensional form, it follows that

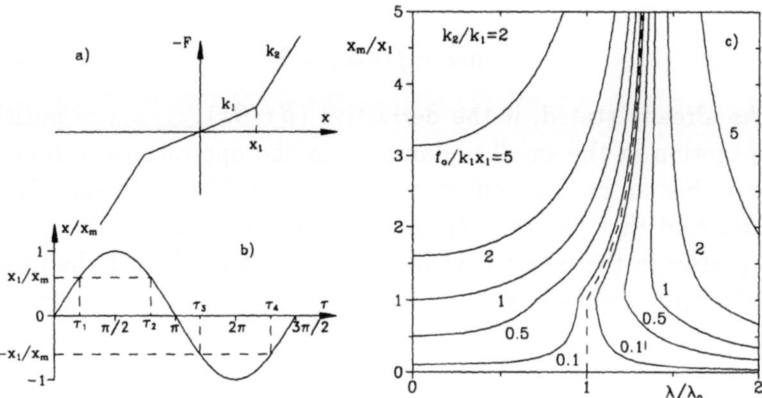

FIGURE 3.12. System with a piecewise linear characteristic: (a) force-displacement curve; (b) law $x(\tau)$ (only the fundamental harmonic is considered); (c) approximate response in nondimensional form.

$$\frac{\lambda}{\lambda_0} = \sqrt{1 - \frac{f_0}{kx_1}\frac{x_1}{x_m} + (\alpha - 1)\left[1 + \frac{\sin(2\tau_1) - 2\tau_1}{\pi} - \frac{4x_1}{\pi x_m}\cos(\tau_1)\right]}\,.$$

This expression holds only if $x_m > x_1$, otherwise the system behaves as linear. The response is plotted in Figure 3.12c in nondimensional form, for the case in which $\alpha = 2$. If $\alpha > 1$, as in the case shown, the behaviour is typical of hardening systems, with a resonance curve leaning toward the high values of the frequency.

It must be stressed that the solution so obtained is only an approximation, yielding results as close to the correct ones as the response is similar to a harmonic function. In the case studied, this happens if the amplitude is either only slightly higher than x_1 or very much higher than the same value. The limiting case with $x_m \gg x_1$ is that of a linear system with stiffness k_2, whose natural frequency is $\lambda = \lambda_0\sqrt{\alpha}$.

3.5 Free oscillations of the damped system

3.5.1 Direct integration of the equation of motion

Consider a system of the type shown in Figure 3.1, in which the restoring force is a function not only of the position of point P, but also of its velocity. If $f(x_0,0)$ vanishes, the free oscillations about an equilibrium position can be studied using a nonlinear autonomous equation that is very similar to equation (3.11):

$$m\ddot{x} + f(x,\dot{x}) = 0 \,. \tag{3.38}$$

As already stated, if the derivative $(\partial f/\partial \dot{x})_{x=x_0,\dot{x}=0}$ is positive, the motion in the small is damped. In the opposite case, function $f(x,\dot{x})$ has an exciting effect on the motion. Generally speaking, it is not said that function $f(x,\dot{x})$ maintains its damping or exciting properties in the whole field in which it is defined, and the motion in the large can be complex. It can have, for example, a damping effect for large values of x and an exciting nature for values of x smaller than a given quantity. In this case, the motion in the large is self-excited at small amplitudes and damped in the case of large motions, and the existence of a limit cycle in the phase plane $x,\dot{x}$, to which all oscillations tend with time, can usually be demonstrated.

There are cases in which function $f(x,\dot{x})$ can be considered as the sum of a function of the displacement x and a function of the velocity $\dot{x}$, i.e., in which the restoring force and the damping force act in a separate way. Using the symbol $f(x)$ for the former and $\beta(\dot{x})$ for the latter, equation (3.38) reduces to

$$m\ddot{x} + \beta(\dot{x}) + f(x) = 0 \,. \tag{3.39}$$

As in the case of undamped systems, equation (3.38) can be rewritten as a first-order differential equation:

$$m\dot{x}\frac{\dot{x}}{dx} = -f(x,\dot{x}) \,. \tag{3.40}$$

In general, however, no separation of the variables is possible, and equation (3.19) cannot be integrated in the way seen for the undamped systems in Section 3.3.1. Each case has to be studied in a particular way, often yielding only qualitative results. To obtain quantitative results, it is usually necessary to resort to approximate methods, like iterative or perturbation procedures, or to the numerical integration of the equations of motion. Note that the motion is, strictly speaking, not periodic; the amplitude decreases in time. The Ritz averaging technique cannot be used unless the system is lightly damped, and then the reduction of amplitude in each period is small enough to be neglected.

Example 3-5

Consider a nonlinear system in which a damper following the Coulomb model is added to a device providing a restoring force having no damping component. The term Coulomb damping is commonly used for a drag force that is independent from the speed. It has important practical applications since it can be used to model dry friction.

Equation (3.39) can be used, and function $\beta(\dot{x})$ can be expressed in the form $\beta(\dot{x}) = F\dot{x}/|\dot{x}|$, where F is the absolute value of the drag force. By introducing the aforementioned expression for $\beta(\dot{x})$ into equation (3.40), the following equation can be obtained:

$$\frac{1}{2}\frac{d(\dot{x})^2}{dx} = \frac{-f(x) \mp F}{m},$$

where the upper sign holds when the velocity is positive. It can be integrated as

$$(\dot{x})^2 = \frac{2}{m}\int [-f(x) \mp F] + C,$$

with the obvious limitation that when the sign of the velocity changes, the integration must be interrupted and resumed with a different sign and a different value of the constant C. By separation of the variables and further integration, the following relationship between time t and coordinate x can be obtained

$$t = \pm\sqrt{\frac{m}{2}}\int \frac{du}{\sqrt{\int [-f(x) \mp F]\,dx + C}} + C_1.$$

This equation can be studied in a way not much different from the one seen for undamped systems in Section 3.3.1. The case in which force $f(x)$ is linear with displacement x ($f(x) = kx$) will be studied in detail. If at time $t=0$ the mass is displaced in the position $x = x_0$ (with positive value of x_0) and then is released with vanishingly small velocity, the lower sign in the equation must be chosen at the beginning, because the velocity is negative, at least for a while.

This clearly holds only if the restoring force is strong enough to overcome dry friction, i.e., $x_0 > F/k$; otherwise no motion results. Actually a greater value of the friction is to be expected when the mass m is at standstill, but this effect will be neglected here. It is easy to perform the integration and to compute the value of constant C, obtaining

$$\begin{cases} (\dot{x})^2 = \dfrac{2}{m}\left(-\dfrac{kx^2}{2} + Fx + \dfrac{kx_0^2}{2} - Fx_0\right), \\[2ex] t = \pm\sqrt{\dfrac{m}{2}}\displaystyle\int \dfrac{du}{\sqrt{x_0^2 - 2\frac{F}{k}x_0 + 2\frac{F}{k}u - u^2}} + C_1. \end{cases}$$

where constant C_1 and the sign must be chosen in such a way that $x = x_0$ for $t=0$ and that x decreases with increasing time. By performing the integration it follows:

$$t = \pm \sqrt{\frac{m}{2}} \, \arcsin \left(\frac{kx - F}{kx_0 - F} \right) + C_1 \,,$$

or, expressing x as a function of time and expressing the constant of integration as a phase angle Φ,

$$x = \left(x_0 - \frac{F}{k} \right) \sin \left(\sqrt{\frac{k}{m}} t + \Phi \right) + \frac{F}{k} = \left(x_0 - \frac{F}{k} \right) \cos \left(\sqrt{\frac{k}{m}} t \right) + \frac{F}{k} \,.$$

The last expression is obtained considering that at time $t=0$ the mass m is in x_0 and the phase angle Φ is equal to $\pi/2$. It holds for half a period $(0 < t < \pi\sqrt{m/k})$.

The motion in the first half-period follows a harmonic law, with the only difference with respect to the undamped case of being centered about the value F/k instead of being centered on the time axis. The study can then proceed in the same way for the second half-period, remembering that the sign of the velocity is changed and that the starting value of x is that obtained at the end of the first half-period.

At the end of a full period the position of mass m is $x_2 = x_0 - 4F/k$. The amplitude decreases by the quantity $4F/k$ in a period. The motion then proceeds, as shown in Figure 3.13, with half-periods of a sine law, all with the same frequency (coinciding with that of the undamped system), each one with an amplitude reduced by the amount $2F/k$. Eventually the mass stops within the band with half-width F/k and all motion is extinguished. The motion so obtained has some similarities with that obtained for viscous damping, in particular where the frequency is concerned, but with the important difference that the law of motion is included, at least within the usual approximations, between two straight lines of equation

$$x = \pm \left(x_0 - \frac{2F}{\pi\sqrt{mk}} t \right)$$

and that the motion extinguishes in a finite time in a position that is different from the equilibrium position of the spring. The number n of half-periods needed to stop the system is

$$n = \mathrm{int} \left(\frac{x_0 k}{2F} + \frac{1}{2} \right) \,.$$

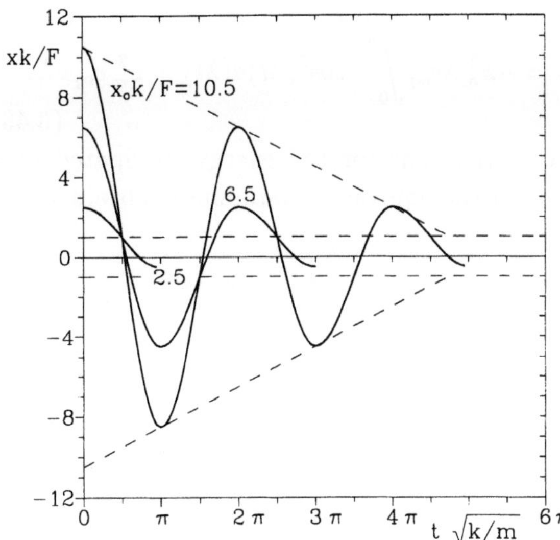

FIGURE 3.13. Free oscillations of a linear system with dry friction (Coulomb damping).

3.5.2 Equivalent damping

When damping is small, it is possible to approximate the behaviour of the system by adding a suitable equivalent linear damping to the undamped system. The linear damping can be chosen in such a way that the energy it dissipates in a cycle is the same dissipated by the actual damping. If damping is small enough to allow neglect of the decrease of amplitude that occurs in a cycle and neglecting also preloading, the energy dissipated in the period T can be computed as

$$\mathcal{L} = \int_0^T -f(x,\dot{x})\dot{x}\,dt\,. \tag{3.41}$$

Assuming that the time history of the motion is substantially harmonic, the law $x(t)$ can be expressed by equation (3.31), obtained by neglecting all harmonics except the fundamental. The expression of the energy dissipated, then, is

$$\mathcal{L} = -x_m \int_0^{2\pi} -f(x,\dot{x}) \cos(\lambda t)d(\lambda t)\,. \tag{3.42}$$

The energy dissipated in a cycle by the equivalent linear (viscous) damping is simply

$$\mathcal{L}_{eq} = -\int_0^T c_{eq}\dot{x}^2\,dt = -x_m^2\lambda c_{eq}\int_0^{2\pi}\cos^2(\lambda t)d(\lambda t) = x_m^2 c_{eq}\pi\lambda\,.$$

$$(3.43)$$

By equating the two expressions for the energy dissipated in a cycle, the following value of the equivalent damping is obtained

$$c_{eq} = \frac{1}{x_m\pi\lambda}\int_0^{2\pi} f(x,\dot{x})\cos(\lambda t)d(\lambda t) = \frac{1}{x_m\pi\lambda}\int_0^{2\pi}\beta(\dot{x})\cos(\lambda t)d(\lambda t)\,;$$

$$(3.44)$$

the second expression holds if equation (3.39) can be used, i.e., if the restoring force and the damping force are independent of each other. The integral that appears in equation (3.44) can be easily computed, at least in a numerical way, once that law $f(x,\dot{x})$ (or $\beta(\dot{x})$) has been defined. Note that the equivalent damping depends on both the amplitude and the frequency of the motion. This definition of the equivalent damping is consistent with that given in Section 1.8 for structural damping. In that case, however, the equivalent damping was independent from the amplitude of the motion, because structural damping is a form of linear damping.

Example 3-6
If the Reynolds number is high enough, aerodynamic damping is proportional to the square of the velocity. Compute the value of the equivalent damping.
Law $\beta(\dot{x})$ can be expressed in this case as $\beta(\dot{x}) = c\dot{x}|\dot{x}|$. By introducing it into equation (3.44), it follows

$$c_{eq} = \frac{cx_m\lambda}{\pi}\left[\int_{-\pi/2}^{\pi/2}\cos^3(\lambda t)d(\lambda t) + \int_{\pi/2}^{3\pi/2} -\cos^3(\lambda t)d(\lambda t)\right] = \frac{8cx_m\lambda}{3\pi}\,.$$

3.5.3 Solution in the state plane

The equation yielding the trajectories in the state plane is still equation (3.26), where function f now depends on both position and velocity. In general, it is impossible to integrate it in closed form, but there is no difficulty in obtaining the state portrait by numerical integration. Explicitly introducing the velocity $v = \dot{x}$ into the state equation for a system with restoring force expressed by equation (3.18), linear damping, and no excitation, the following equation yielding the trajectories in the state plane is obtained:

$$\frac{dv}{dx} = -\frac{kx(1 + \mu x^2) + cv}{mv} .$$

(3.45)

Stating a nondimensional velocity $v^* = v\sqrt{|\mu|}m/k$ and a nondimensional coordinate $x^* = x\sqrt{|\mu|}$, the behaviour of the system depends on a single nondimensional parameter, the damping ratio $\zeta = c/2\sqrt{km}$. The state portrait is plotted in Figure 3.14 in nondimensional form for a value of $\zeta = 0.1$. The two plots are related to hardening systems (Figure 3.14a) and softening systems (Figure 3.14b)

Note that the state portraits are qualitatively different from those related to the undamped system reported in Figure 3.8. If the system is hardening there is a single equilibrium position and the trajectories are spirals that wind inward, toward the equilibrium point in the origin. Because all trajectories tend to the singular point, this is an attractor and its basin of attraction extends to the whole state plane. A singular point of this type is said to be a *stable focus*.

The state portrait for softening systems (Figure 3.14b) is more complex. In this case there are three singular points, corresponding to the equilibrium positions. The one in the centre is again a stable focus, but now its basin of attraction does not encompass all the state plane, but only the zone that lies within the two separatrices.

The other two singular points are saddle points and are repellors because the state trajectories tend to depart from them, even if they can be initially attracted. The parts of the plane outside the basin of attraction of the stable equilibrium positions are characterized by nonperiodic motions, which are essentially not different from the ones seen for the undamped system.

If the linear term is repulsive instead of attractive, i.e., the stiffness k is negative, the state portraits are of the type shown in Figure 3.15. When μ is also negative, i.e., the coefficient of the term containing the third power of the displacement is positive, the behaviour is of the hardening type (Figure 3.15a). There are three static equilibrium positions; one unstable in the origin and two stable ones. The corresponding singular points are a saddle point and two stable foci.

The separatrix defines two domains of attractions, whose shape in this case is quite intricate. The domain of the focus on the right

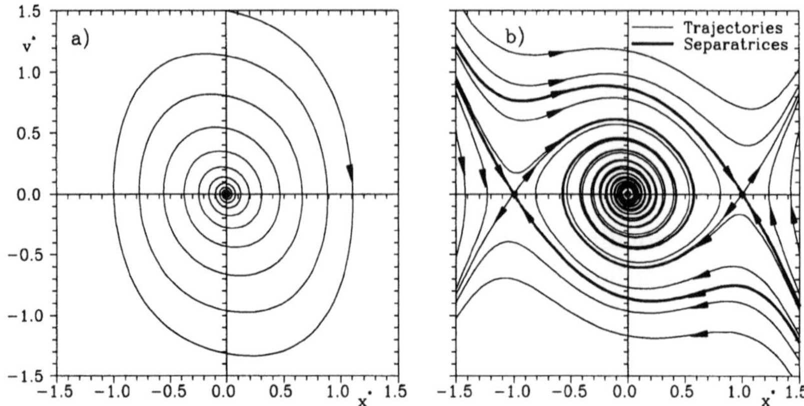

FIGURE 3.14. State portrait of a system whose restoring force is expressed by equation (3.18). Nondimensional velocity $v^* = v\sqrt{|\mu|m/k}$ as a function of nondimensional displacement $x^* = x\sqrt{|\mu|}$ for (a) hardening and (b) softening systems. Nondimensional damping $\zeta = c/2\sqrt{km} = 0.1$.

has been hatched in the figure, in order to make the state portrait easier to understand. All motions end up in either of the foci, but the interwoven nature of the domains of attraction makes the final result quite sensitive on the initial condition. Consider, for example, a motion starting from rest with point P on the negative side of the x-coordinate. If the initial value of x is small, the motion is attracted toward the equilibrium position at the left. With increasing absolute value of x, however, motions that are attracted alternatively by the two different equilibrium positions are found.

In the case of softening systems with repulsive linear stiffness, no oscillatory motion is possible and nonperiodic motions occur in all domains of attraction visible in the state portrait. The origin is again a saddle point.

The types of singular points seen up to this point (centers, foci, and saddle points) are not the only possible ones. A *singular point* is an equilibrium position of the system; it can easily be found from equation (3.8), which, without loss of generality, can be written as $f(x_0,0) = 0$, as static forces acting on the system can be included in function $f(x,\dot{x})$.

The free motion in the small in the vicinity of the singular point can easily be studied using the homogeneous linear equation associated with equation (3.9). The usual practical definitions of static stability, when the stiffness is positive and point P is attracted toward the equilibrium position when displaced, and of dynamic stability,

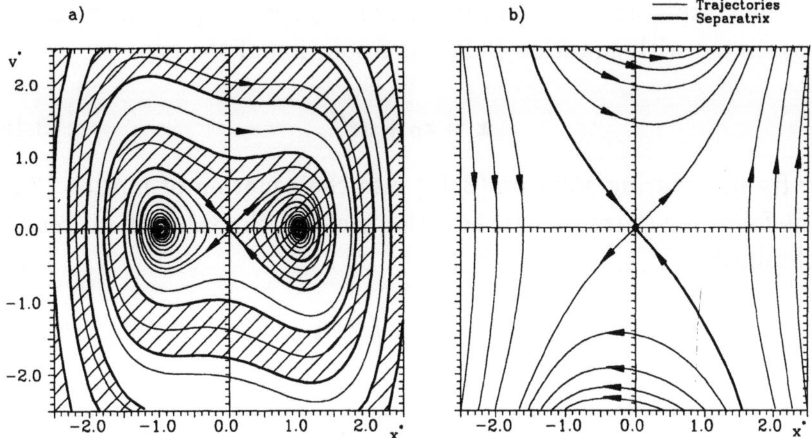

FIGURE 3.15. State portrait of a system whose restoring force is expressed by equation (3.18), as in Figure 3.14, but with a negative value of the stiffness k. Nondimensional velocity $v^* = v\sqrt{|\mu|}m/k$ as a function of nondimensional displacement $x^* = x\sqrt{|\mu|}$ for (a) hardening (in this case with negative value of μ) and (b) softening systems. Nondimensional damping $\zeta = c/2\sqrt{|k|m} = 0.1$.

when the damping is positive and the time history actually reaches the equilibrium position, at least asymptotically, can be used.

More elaborate definitions for stability are needed, however. Consider a vector $\vec{X}(t)$ in the state plane, whose components are $x(t)$ and $v(t)$, and use the symbol $|\vec{X}(t)|$ for its euclidean norm. Following Liapunov, an equilibrium position defined by $\vec{X}_0 = \{x_0,0\}$ is stable if, for any arbitrarily small positive quantity ϵ, there exists a positive quantity δ such that the inequality

$$|\vec{X}(t) - \vec{X}_0| < \epsilon \qquad \text{for } 0 \le t < \infty \qquad (3.46)$$

holds if $|\vec{X}(0) - \vec{X}_0| < \delta$.

This means that any trajectory starting within a circle of radius δ centered in the equilibrium point in the state plane remains within a circle of radius ϵ for all values of time. If to this condition it is added that

$$\lim_{t \to \infty} |\vec{X}(t) - \vec{X}_0| = 0 \,, \qquad (3.47)$$

then the equilibrium position is asymptotically stable. This definition also holds for systems with many degrees of freedom, provided that vector $\vec{X}(t)$ has as many components as the number of dimensions of the state space. The linearized homogeneous equation of motion associated with equation (3.9) has a solution of the type

$$x = x_0 + ae^{st} \, . \tag{3.48}$$

By introducing solution (3.48) into the equation of motion (3.9), the following characteristic equation yielding the value of s is readily obtained:

$$s^2 + \left(\frac{\partial f}{\partial \dot{x}}\right)_{\dot{x}=0} s + \left(\frac{\partial f}{\partial x}\right)_{x=x_0} = 0 \, . \tag{3.49}$$

If the roots of the equation are real and have the same sign, the singular point is a node. If they are positive the node is unstable; if they are negative the node is stable. An example of a stable node is the equilibrium position of an overdamped linear system. If the roots of the equation are real but their signs are different, the singular point is a saddle point. An example was shown in Figure 3.8b. The motion in the vicinity of a node or saddle point is nonperiodic. If the roots of the equation are imaginary and obviously have opposite signs, i.e., they must be conjugate, the singular point is a center and the motion is an undamped oscillation. An example of a center is the equilibrium position of Figure 3.7a. If the roots of the equation are complex conjugate, the singular point is a focus. If their real part is positive, the focus is unstable; if it is negative, the focus is stable. An example of a stable focus is the equilibrium position of Figure 3.14a. The motion is a damped or self-excited oscillation.

A stable node or stable focus is strictly an attractors; centers can also be called attractors because all state trajectories lying within a certain basin orbit about them.

The roots of equation (3.49) can be reported on the Argand plane, obtaining a plot of the same type as those shown in Table 3.2. As already stated for linear systems, this type of representation is very useful, particularly when the stability of a system with many degrees of freedom has to be studied: If any root lies at the right of the imaginary axis, i.e., it has a positive real part, the system is unstable. If the influence of the variation of a parameter on the stability of the system has to be studied, the plot can be drawn for different values of the relevant parameter. Each root then describes a curve on the Argand plane (the root locus), and the onset of instability corresponds to the value of the parameter that causes one of the

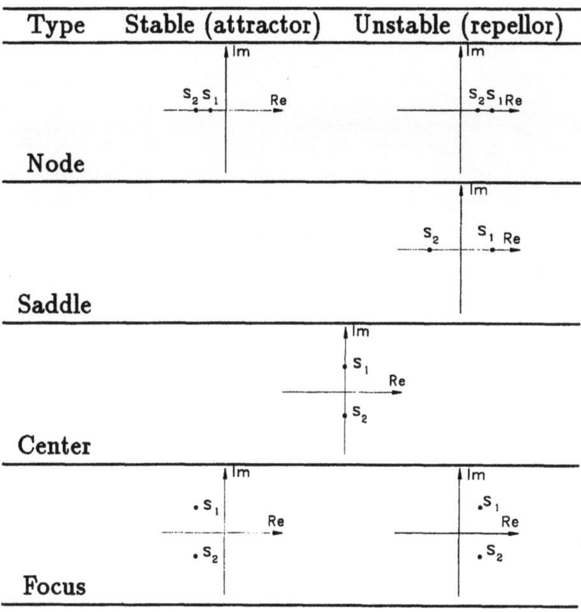

TABLE 3.2. Various types of singular points for a system with a single degree of freedom together with the corresponding position of the roots of equation (3.49) on the Argand plane. Note that the center has been considered as an intermediate situation between stable and unstable singular points.

curves to cross the imaginary axis. Also, if stability in the large is considered, apart from singular points, there are attractors and repellors of another type: the limit cycles. Consider, for example, the so-called Van der Pol oscillator, described by the equation

$$\ddot{x} + \zeta(x^2 - 1)\dot{x} + x = 0, \qquad \text{with } \zeta > 0. \tag{3.50}$$

The state portrait, obtained by numerical integration with $\zeta = 0.2$, is reported in Figure 3.16. The central singular point is an unstable focus. All trajectories tend to a stable limit cycle, which has a shape tending to a circle when $\zeta \to 0$. Within the limit cycle the behaviour of the system is self-excited; outside, it is damped.

If a single singular point is surrounded by several limit cycles nested within each other, they are alternatively stable and unstable, with the unstable limit cycles acting as separatrices between the attractors represented by the stable ones. Sometimes two limit cycles, a stable and an unstable one, can merge into one another and give way to a semistable limit cycle, which acts as an attractor on one side and as a repellor, or separatrix, on the other side.

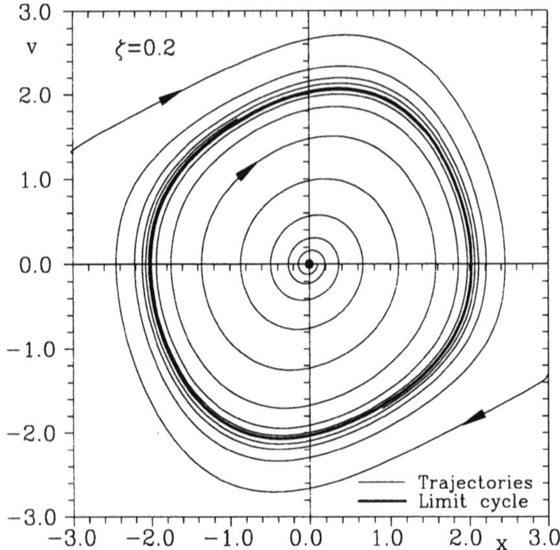

FIGURE 3.16. State portrait of the Van der Pol oscillator.

If the number of dimensions of the state space is greater than two, other types of singular points exist, and other types of attractors, as toroidal attractors and strange attractors, can be present.

3.6   Forced oscillations of the damped system

3.6.1   First approximation of the response to a harmonic forcing function

Consider a nonlinear system excited by a harmonic forcing function. If the equation of motion is written separating the linear part of the restoring and damping forces from the nonlinear part expressed by function $f'(x,\dot{x})$ and introducing the natural frequency of the linearized system λ_0, the equation of motion is

$$\ddot{x} + 2\zeta\lambda_0\dot{x} + \lambda_0^2 x + \frac{\lambda_0^2}{k}f(x,\dot{x}) = \frac{\lambda_0^2}{k}f_0\sin(\lambda t)\,. \qquad (3.51)$$

Due to damping, the response will not be in phase with the excitation, and when searching a first approximation of the fundamental harmonic, two unknowns must be stated: The amplitude and the phase of the response, or the amplitude in phase and the amplitude in quadrature. Instead of working in phase with the excitation, it is simpler to assume time $t = 0$ in the instant in which the response vanishes, i.e., to work in phase with the response. The response can then be assumed to be of the type of equation (3.31), and the unknown phase can be included in the expression of the forcing function, which becomes $f_0 \sin(\lambda t + \Phi)$. Using the Ritz averaging technique, it follows that

$$\int_0^T \left[\ddot{x} + 2\zeta\lambda_0\dot{x} + \lambda_0^2 x + \frac{\lambda_0^2}{k}f(x,\dot{x}) - \frac{\lambda_0^2}{k}f_0 \sin(\lambda t + \Phi) \right] \phi(t)dt = 0 ,$$

$$(3.52)$$

where the arbitrary function $\phi(t)$ has been assumed as $\phi(t) = \sin(\lambda t)$, and the unknown a is the amplitude x_m. By remembering that

$$\int_0^{2\pi} \sin(\alpha)\cos(\alpha)d\alpha = 0$$

and introducing the nondimensional time $\tau = \lambda t$, it follows that

$$\left[1 - \left(\frac{\lambda}{\lambda_0}\right)^2 \right] x_m - \frac{f_0}{k}\cos(\Phi) + \frac{1}{k\pi}\int_0^{2\pi} f'(x,\dot{x})\sin(\tau)d\tau = 0 . \quad (3.53)$$

Equation (3.53) contains the two unknowns, x_m and Φ, and another equation, which can be readily obtained by observing that the energy dissipated by the system must be equal to the work performed by the excitation, is needed. Again, instead of equating the relevant quantities in each instant, the equation is written as an average in a cycle. The energy dissipated in a cycle is expressed by the integral

$$\mathcal{L}_d = \int_0^T \left[\ddot{x} + 2\zeta\lambda_0\dot{x} + \lambda_0^2 x + \frac{\lambda_0^2}{k}f'(x,\dot{x}) \right] \dot{x}dt . \quad (3.54)$$

By introducing into equation (3.54) the expression of the velocity $\dot{x}$ and integrating, it follows that

$$\mathcal{L}_d = 2\pi\zeta\lambda_0\lambda^2 x_m + \frac{\lambda\lambda_0^2 x_m}{k}\left[\int_0^{2\pi} f'(x,\dot{x})\cos(\tau)d\tau \right] . \quad (3.55)$$

The work performed by the forcing function in a cycle is

$$\mathcal{L}_f = \frac{\lambda_0^2}{k}\int_0^T f_0 \sin(\lambda t - \Phi)\dot{x}dt = \frac{\pi x_m \lambda\lambda_0^2}{k}f_0 \sin(\Phi) . \quad (3.56)$$

Equating expressions (3.55) and (3.56), the required equation is obtained:

$$2\zeta\frac{\lambda}{\lambda_0}x_m + \frac{1}{k\pi}\int_0^{2\pi} f'(x,\dot{x})\cos(\tau)d\tau - \frac{f_0}{k}\sin(\Phi) = 0\,. \qquad (3.57)$$

Equations (3.53) and (3.57) allow the computation of the phase and the amplitude of the response, once function $f'(x,\dot{x})$ has been stated. Often, however, the problem can be solved in a simpler way. Assume that the nonlinear restoring force can be expressed by function $f'(x)$ while the nonlinear damping force can be expressed by function $\beta'(\dot{x})$ and that $f'(x)$ and $\beta'(\dot{x})$ are odd functions of x and $\dot{x}$, respectively. This means that $\beta'(-\dot{x}) = -\beta'(\dot{x})$ and $f'(-x) = -f'(x)$. Displacement $x(t)$ expressed by equation (3.31) is itself an odd function of time, while its derivative, the velocity, is an even function of time. From this, it follows that functions $\beta'(\tau)$ and $f(\tau)$ are even and odd, respectively. A consequence of this consideration is that

$$\int_0^{2\pi} \beta'(\dot{x})\sin(\tau)d\tau = 0\,; \qquad \int_0^{2\pi} f'(x)\cos(\tau)d\tau = 0\,. \qquad (3.58)$$

Note that if functions $\beta'(\dot{x})$ and $f'(x)$ are not odd, the center of the forced oscillations does not coincide with the static equilibrium position, and the assumption that the response can be approximated with a simple harmonic motion of the type of equation (3.31) becomes unacceptable. At least a constant term of the displacement must be included, and the whole procedure shown here must be modified. Actually, if equation (3.58) is not satisfied, the restoring force $f'(x)$ would enter equation (3.57), i.e., would participate to the energy dissipation, which is clearly not the case.

By substituting function $f'(x,\dot{x})$ with functions $\beta'(\dot{x})$ and $f'(x)$ and remembering equation (3.58), equations (3.53) and (3.57) reduce to

$$\begin{cases} \left[1-\left(\dfrac{\lambda}{\lambda_0}\right)^2\right]x_m + \dfrac{1}{k\pi}\displaystyle\int_0^{2\pi} f'(x)\sin(\tau)d\tau = \dfrac{f_0}{k}\cos(\Phi)\,, \\[2ex] 2x_m\zeta\dfrac{\lambda}{\lambda_0} + \dfrac{1}{k\pi}\displaystyle\int_0^{2\pi} \beta'(\dot{x})\cos(\tau)d\tau = \dfrac{f_0}{k}\sin(\Phi)\,. \end{cases} \qquad (3.59)$$

By adding the squares of the two equations (3.59), a relationship that allows the computation of the amplitude of the response as a function of the driving frequency is readily obtained:

$$\left\{ \left[1 - \left(\frac{\lambda}{\lambda_0} \right)^2 \right] x_m + \frac{1}{k\pi} \int_0^{2\pi} f'(x) \sin(\tau)d\tau \right\}^2 +$$
$$+ \left[2x_m \zeta \frac{\lambda}{\lambda_0} + \frac{1}{k\pi} \int_0^{2\pi} \beta'(\dot{x}) \cos(\tau)d\tau \right]^2 = \left(\frac{f_0}{k} \right)^2 . \tag{3.60}$$

The phase can be easily computed by dividing the second equation (3.59) by the first:

$$\Phi = \arctan \left\{ \frac{2x_m \zeta \dfrac{\lambda}{\lambda_0} + \dfrac{1}{k\pi} \displaystyle\int_0^{2\pi} \beta'(\dot{x}) \cos(\tau)d\tau}{\left[1 - \left(\dfrac{\lambda}{\lambda_0} \right)^2 \right] x_m + \dfrac{1}{k\pi} \displaystyle\int_0^{2\pi} f'(x) \sin(\tau)d\tau} \right\} . \tag{3.61}$$

3.6.2 Duffing's equation with viscous damping

Consider, for example, a nonlinear system governed by Duffing's equation, to which a linear viscous damper has been added. The equation of motion is

$$m\ddot{x} + c\dot{x} + kx(1 + \mu x^2) = f_0 \sin(\lambda t) . \tag{3.62}$$

The procedure seen in the preceding section can be used provided that

$$f'(x) = k\mu x^3 ; \qquad \beta'(\dot{x}) = 0 . \tag{3.63}$$

Remembering that

$$\int_0^{2\pi} f'(x) \sin(\tau)d\tau = k\mu \int_0^{2\pi} [x_m \sin(\tau)]^3 \sin(\tau)d\tau = \frac{3\pi}{4} k\mu x_m^3 , \tag{3.64}$$

equations (3.60) and (3.61) allow the computation of the amplitude and the phase of the response

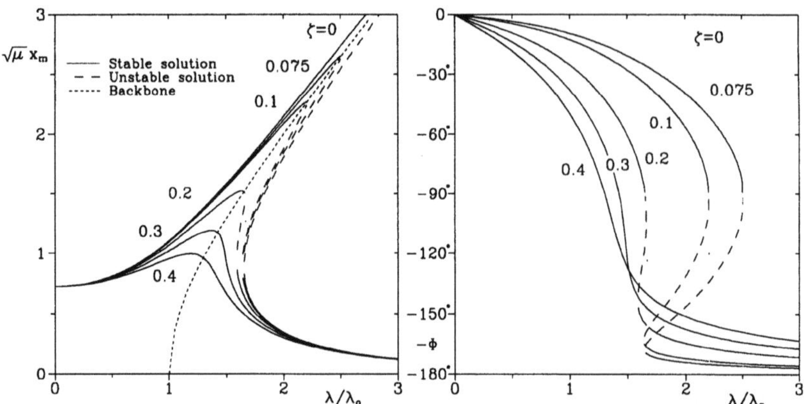

FIGURE 3.17. Amplitude and phase of the response of a system governed by Duffing's equation (hardening spring) with viscous damping under the effect of a harmonic excitation; $\sqrt{\mu}f_0/k=1$. Phase $-\varPhi$ of the displacement with respect to the force, which is negative, as in the case of linear systems, has been plotted.

$$\begin{cases} \left[(k-m\lambda^2)x_m+\dfrac{3}{4}k\mu x_m^3\right]^2+x_m^2c^2\lambda^2=f_0^2 \\[2mm] \varPhi=\arctan\left(\dfrac{c\lambda}{k-m\lambda^2+\frac{3}{4}k\mu x_m^2}\right). \end{cases} \tag{3.65}$$

The phase so obtained is that of the exciting force with respect to the displacement and is positive, i.e., the force leads the displacement. With simple modifications, the first equation (3.65) can be expressed in the following nondimensional form:

$$\begin{aligned} &\left(\frac{\lambda}{\lambda_0}\right)^4-2\left(\frac{\lambda}{\lambda_0}\right)^2\left[1+\tfrac{3}{4}\left(\sqrt{\mu}x_m\right)^2-2\zeta^2\right]+ \\ &+\left[1+\tfrac{3}{4}\left(\sqrt{\mu}x_m\right)^2\right]^2-\left(\frac{\sqrt{\mu}f_0}{k}\frac{1}{\sqrt{\mu}x_m}\right)^2=0, \end{aligned} \tag{3.66}$$

which expresses the nondimensional amplitude $\sqrt{\mu}x_m$ as a function of the nondimensional frequency λ/λ_0 with parameters $\sqrt{\mu}f_0/k$ and ζ. For computational convenience, equation (3.66) can be solved in the frequency and function $\sqrt{\mu}x_m(\lambda/\lambda_0)$ can then be obtained from function $\lambda/\lambda_0(\sqrt{\mu}x_m)$. Unfortunately, because such a plot is a function of two parameters, namely, excitation and damping, it is not possible to summarize the general behaviour of the system in a single chart.

The amplitude and phase, computed with a single value of the excitation $\sqrt{\mu}f_0/k=1$ and some different values of ζ, are plotted in Figure 3.17. The plot refers to hardening systems. For $\zeta=0$ the so-

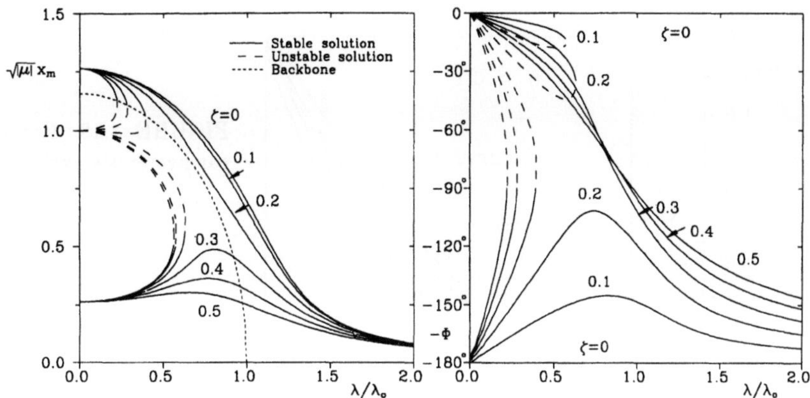

FIGURE 3.18. Same as Figure 3.17, but for softening systems; $\sqrt{\mu}f_0/k = 0.25$.

lution of the undamped system is obtained. With increasing values of the damping, the resonance peak gets smaller. The jump phenomenon is present only for small values of the damping, as can be easily inferred from the consideration that at high values of ζ only one possible value of the amplitude can be found at any value of the frequency. The same consideration can be drawn from the plot of the phase against the frequency. At high damping, the phase changes continuously from $0°$ to $180°$ with increasing frequency. If the damping is small enough, an abrupt variation of phase takes place when the system jumps from one configuration to the other. It is interesting to note that the frequency at which the phase takes the value of $90°$ depends on the damping. Such a value of the phase occurs when the amplitude curve crosses the resonance line, i.e., the backbone of the response.

The case of a softening system is shown in Figure 3.18. Note that at high values of damping the curves split into two parts; the lower one is that actually followed by the system and the upper one is merely a theoretical result, affected by strong approximations. A similar pattern is shown by the plot of the phase, in which the curves are constituted by two branches at low values of the damping, when the jump phenomenon is present. The phase takes a value of 0 at low frequency, to decrease slowly following the upper branch until the jump occurs. The phase then shifts beyond $-90°$, to decrease in absolute value slightly and then increase up to $-180°$ at very high frequency. At high values of damping, when no jump occurs, the phase follows the line on the right.

As already stated, the results reported here are approximate and

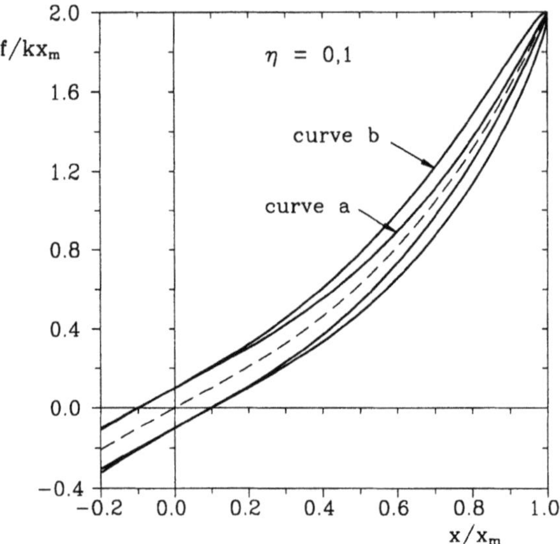

FIGURE 3.19. Nonlinear hysteresis cycles.

refer only to the fundamental harmonic of the response. For a more complete evaluation of the response, it is possible to resort to perturbation techniques, which lead to involved computations, or to numerically integrate the equations of motion.

3.6.3   Duffing's equation with structural damping

It is also possible to use the structural damping model in the case of nonlinear systems. Consider, for example, a system with a single degree of freedom with a hardening spring whose characteristic is of the type seen in Duffing's equation, and assume that the response $x(t)$ can be approximated with a harmonic law. To introduce structural damping, use the complex stiffness model with regard to the linear part of the restoring force of the system

$$f(x) = kx_m \sin(\lambda t + \eta) + \mu k x_m^3 \sin^3(\lambda t) . \qquad (3.67)$$

The linear part of the restoring force is out of phase with respect to the displacement, and a hysteresis cycle of the type shown in Figure 3.19, Curve a, is obtained (note that the phase is directly assumed to be equal to the loss factor η). If the whole reaction $f(x)$ is assumed to lag the displacement (i.e., $\lambda t + \eta$ is introduced also in the term in $\sin^3$), a hysteresis cycle of the type shown in Figure 3.19, Curve b, is obtained. The hysteresis cycles shown in Figure 3.19 allow the

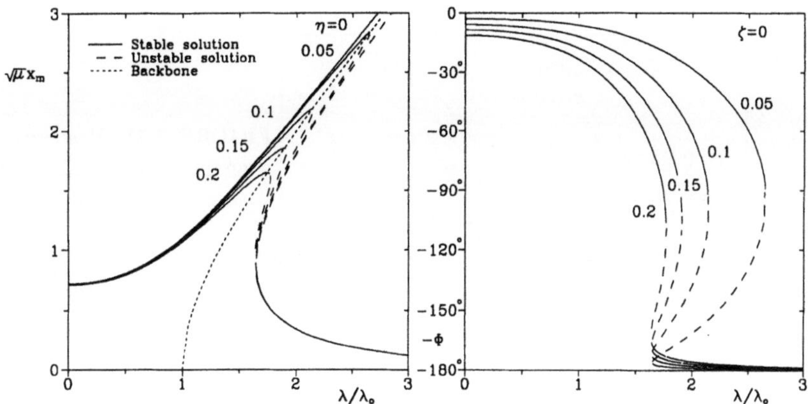

FIGURE 3.20. Amplitude and phase of the response of a system governed by Duffing's equation (hardening spring) with structural damping as defined by equation (3.67) under the effect of a harmonic excitation; $\sqrt{\mu}f_0/k=1$.

modeling of the actual behaviour of many actual machine elements, as elastomeric springs and vibration insulators.

The response of a system governed by Duffing's equation with structural damping of the type of equation (3.67) can be easily computed

$$
\begin{cases}
\left(\dfrac{\lambda}{\lambda_0}\right)^4 - 2\left(\dfrac{\lambda}{\lambda_0}\right)^2 \left[1 + \dfrac{3}{4}(\sqrt{\mu}x_m)^2\right] + (1+\eta^2)\left[1 + \dfrac{3}{4}(\sqrt{\mu}x_m)^2\right]^2 + \\[2ex]
\qquad - \left(\dfrac{\sqrt{\mu}f_0}{k}\dfrac{1}{\sqrt{\mu}x_m}\right)^2 = 0 \,, \\[3ex]
\Phi = \arctan\left[\eta\dfrac{1 + \frac{3}{4}\mu x_m^2}{1 - \left(\frac{\lambda}{\lambda_0}\right)^2 + \frac{3}{4}\mu x_m^2}\right] .
\end{cases}
\tag{3.68}
$$

The response is plotted in Figure 3.20, for $\sqrt{\mu}f_0/k=1$ and various values of the loss factor η.

3.6.4 Backbone and limit envelope

Also in the case of damped systems it is possible to define the backbone of the response. If the amplitude of the forcing function vanishes, the first equation (3.59) yields the backbone of the response

$$
\frac{\lambda}{\lambda_0} = \sqrt{1 + \frac{1}{k\pi x_m}\int_0^{2\pi} f'(x)\sin(\tau)d\tau} \,.
\tag{3.69}
$$

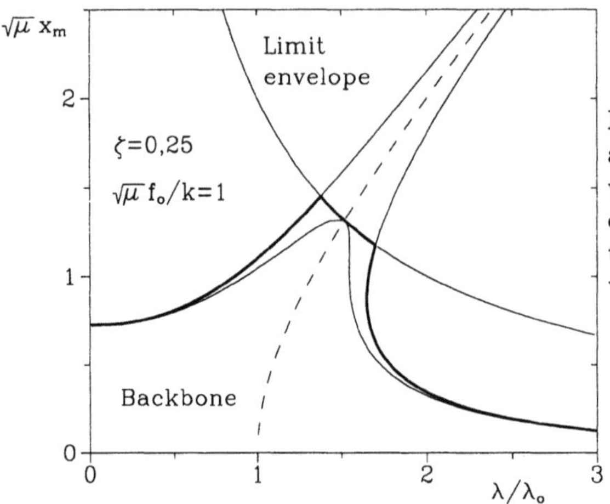

FIGURE 3.21. Response of a nonlinear damped system with a single degree of freedom approximated by cutting the undamped response with the limit envelope.

The backbone depends on neither damping nor the forcing function. Actually it is a feature of the conservative part of the system and is more correlated to its free behaviour than to the response to a given excitation. Its role is the same that the natural frequency has for linear systems; on the backbone, inertia forces balance exactly nonlinear restoring forces and the response lags the excitation by 90°. The value of the amplitude in such resonant conditions can be computed by stating that the phase lag is 90°. From the second equation (3.59), it follows that

$$\frac{\lambda}{\lambda_0} = \frac{f_0}{2\zeta k x_m} - \frac{1}{2\zeta k \pi x_m} \int_0^{2\pi} \beta'(\dot{x}) \cos(\tau) d\tau,$$
$$\frac{\lambda}{\lambda_0} = \frac{f_0}{2\zeta k x_m},$$

(3.70)

where the last expression has been obtained for Duffing's equation. Equations (3.70) define a curve in the $x_m(\lambda)$ plane that is usually referred to as a *limit envelope*. It divides the plane in two regions: The one spanning under the curve includes all possible working conditions, as the forcing function can supply all the energy needed to sustain the oscillations. In the zone above the limit envelope, the energy dissipated by damping is greater than that supplied by the excitation and then no steady oscillation is possible.

While the backbone defines the conditions in which there is equilibrium between inertia forces and restoring forces (resonance conditions), the limit envelope states the conditions for equilibrium between damping and external forces. The latter does not depend on

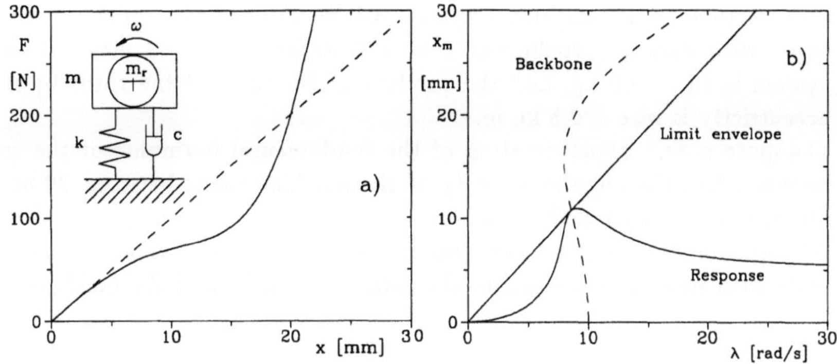

FIGURE 3.22. (a) Scheme of the system and force-displacement characteristics of the elastomeric springs; (b) first approximation of the fundamental harmonic of the response, backbone, and limit envelope.

the mass of the system or on the type of restoring force. In particular, it does not change whether the restoring force is linear, hardening, or softening, provided that damping and the forcing function are not changed.

The response of a damped nonlinear system with a single degree of freedom can easily be approximated by cutting the undamped response using the limit envelope (Figure 3.21). This procedure is the generalization of that shown in Figure 1.9 for linear systems. The properties of the backbone and the limit envelope are valid for any type of functions $f'(x)$ and $\beta'(\dot{x})$ (provided they are odd) and any law linking the amplitude of the forcing function f_0 to the frequency λ. In some cases the limit envelope can be completely above the backbone, with no intersection between them. The amplitude in this case grows indefinitely, with no downward jump with increasing frequency, as the forcing function is powerful enough to sustain the oscillation with the larger amplitude at any value of the frequency.

Example 3-7
A rotating machine is suspended through a system of elastomeric springs whose force-displacement characteristics (Figure 3.22a) can be approximated by the equation $F = k_1 x - k_2 x^3 + k_3 x^5$, where $k_1 = 10^4$ N/m, $k_2 = 4 \times 10^7$ N/m^3, and $k_3 = 10^{11}$ N/m^5.

The damping of the suspension system will be assumed to be of the viscous type, with damping coefficient $c = 400$ Ns/m. The total mass of the system is $m = 100$ kg, and the product of the mass of the rotor by its eccentricity is $m_r \epsilon = 0.5$ kg m.

Compute a first approximation of the fundamental harmonic of the response when the angular velocity of the machine spans between 20 and 300 rpm (between 2 and 30 rad/s).

The unbalance causes a force whose vertical component varies harmonically with frequency λ equal to the rotational speed ω of the machine

$$F = m_r \epsilon \lambda^2 \sin(\lambda t).$$

The equation of motion of the system is

$$m\ddot{x} + c\dot{x} + k_1 x - k_2 x^3 + k_3 x^5 = m_r \epsilon \lambda^2 \sin(\lambda t).$$

The natural frequency and the damping ratio of the linearized system are

$$\lambda_0 = \sqrt{\frac{k_1}{m}} = 10 \text{ rad/s}; \qquad \zeta = \frac{c}{2\sqrt{k_1 m}} = 0.2.$$

The backbone and the limit envelope of the response can be obtained directly from equations (3.69) and (3.70)

$$\frac{\lambda}{\lambda_0} = \sqrt{1 - \frac{3k_2}{4k_1}x_m^2 + \frac{15k_3}{24k_1}x_m^4}; \qquad \frac{\lambda}{\lambda_0} = \frac{m_r \epsilon \lambda^2}{2\zeta k_1 x_m}.$$

The amplitude of the fundamental harmonic of the response can be computed using equation (3.60). By performing the relevant integrations and rewriting the equation as a quadratic equation in λ^2, it follows that

$$\left(\frac{\lambda}{\lambda_0}\right)^4 \left[1 - \left(\frac{m_r \epsilon}{m x_m}\right)^2\right] - 2\left(\frac{\lambda}{\lambda_0}\right)^2 \left(1 - \frac{3k_2}{4k_1}x_m^2 + \frac{15k_3}{24k_1}x_m^4 - 2\zeta^2\right) +$$
$$+ \left(1 - \frac{3k_2}{4k_1}x_m^2 + \frac{15k_3}{24k_1}x_m^4\right)^2 = 0.$$

The frequency response of the system is plotted, together with the backbone and the limit envelope, in Figure 3.22b.

Example 3-8

Consider a linear system with dry friction, modeled as shown in Example 3-5 using the Coulomb damping model, excited by a harmonic forcing function. The equation of motion is

$$m\ddot{x} + F\frac{\dot{x}}{|\dot{x}|} + kx = f_0 \sin(\lambda t).$$

The integral defining the equivalent damping must be computed separately for the parts of the cycle in which the velocity is positive $(-\pi/2 < \lambda t < \pi/2)$ or negative $(\pi/2 < \lambda t < 3\pi/2)$

$$c_{eq} = \frac{4F}{x_m \pi \lambda}; \qquad \zeta_{eq} = \frac{2F\lambda_0}{x_m \pi \lambda k}.$$

The amplitude of the response of the linear system with equivlent damping can be computed using equations (1.73). Note that in this case the damping depends on the amplitude and a further rearranging of the equations is required to obtain the following frequency response

$$x_m = \frac{f_0}{k}\sqrt{\frac{1 - \left(\dfrac{4F}{\pi f_0}\right)^2}{1 - \left(\dfrac{\lambda}{\lambda_0}\right)^2}}; \qquad \Phi = \arctan\left[\frac{-1}{\sqrt{\left(\dfrac{\pi f_0}{4F}\right)^2 - 1}}\right].$$

An apparently inconsistent result is thus obtained, because at resonance the amplitude tends to infinity, in spite of damping, at least if $F/f_0 < \pi/4$. This can, however, be easily explained by the consideration that both the energy dissipated by damping and the energy input due to the exciting force is proportional to the amplitude. Thus, if dry friction is small enough to allow the system to oscillate, it cannot avoid the building up of an infinite amplitude at resonance. The case with viscous damping was different, as the drag force was proportional to the amplitude, and consequently, the energy dissipated was proportional to its square. No matter how low the damping of the system was, there was, a finite value of the amplitude for which the energy dissipated by damping was equal to the work performed by the excitation.

The same conclusion can be immediately drawn from the study of the limit envelope. By introducing the equivalent damping into the equation of the limit envelope, it follows that $F/f_0 = \pi/4$. This expression does not represent a line on the λ, x_m-plane, i.e., no limit envelope actually exists, but it is nevertheless the condition for which motion is possible, as stated earlier.

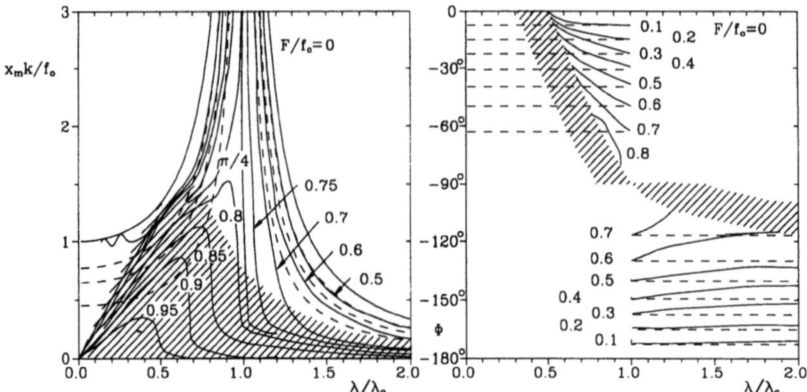

FIGURE 3.23. Response of a system with linear restoring force and Coulomb damping to a harmonic excitation; exact solution (full lines) against the current approximate solution (dashed lines). In the hatched region the motion is not continuous; in the zone in the lower left corner more than one stop occurs in each cycle.

The results here obtained are only an approximation, which is close to the correct results only if the motion of the system is not far from being a harmonic motion. An exact solution to the problem was obtained by Den Hartog (J. P. Den Hartog, "Forced vibrations with combined coulomb and viscous damping", *Trans. ASME*, Vol. 53, 1931) and is reported in Figure 3.23. In the part of Figure 3.23 lying within the hatched region, the motion is not continuous but contains at least one stop at each cycle. In the part in the lower-left corner, the motion stops at least twice each cycle, and no solution was found. The dashed lines are the solutions obtained from the current model using the equivalent damping concept. In the vicinity of the resonance peak, the approximate curves are very close to those obtained in the aforementioned paper, as could be expected. On the contrary, when the motion is intermittent, or, generally speaking, when damping is not small, the approximate solution is quite far from the correct one. Where motion is intermittent no phase has been computed, because the very concept of phase loses its meaning. Also, in cases where the approximate solution leads to the conclusion that no motion is possible, an exact solution exists.

3.6.5 Van der Pol method: Stability of the steady-state solution

The approach seen in Section 3.6.1 allows a first approximation of the fundamental harmonic of the steady-state response to a harmonic forcing function to be obtained. The so-called Van der Pol method relies on similar approximations but also allows the study of solutions that are not restricted to steady-state motions; it is a powerful tool in the study of the stability of the motion. Consider a nonlinear nonautonomous equation of motion of the type

$$m\ddot{x} + f(x,\dot{x}) = f_0 \cos(\lambda t) \tag{3.71}$$

and assume that the solution has the form

$$x(t) = x_1(t)\cos(\lambda t) + x_2(t)\sin(\lambda t) + \bar{x}, \tag{3.72}$$

where x_1 and x_2 are slowly varying functions of time and a constant displacement $\bar{x}$ is taken into account for considering the case of nonsymmetrical damping or restoring force. The motion is then a harmonic motion with slowly varying amplitude. x_1 and x_2 are the in-phase and in-quadrature amplitudes, respectively, while $\bar{x}$ is usually vanishingly small because, in many cases, function $f(x,\dot{x})$ is odd in both x and $\dot{x}$.

It is possible to show that by introducing the solution (3.72) into the equation (3.71) and neglecting the second derivatives of functions x_1 and x_2 with respect to time, which are negligible due to the assumption that the amplitude varies slowly in time, the problem reduces to

$$\begin{cases} \dfrac{dx_1}{dt} = F_1(x_1,x_2,\bar{x})\,, \\ \dfrac{dx_2}{dt} = F_2(x_1,x_2,\bar{x})\,, \\ 0 = F_3(x_1,x_2,\bar{x})\,. \end{cases} \tag{3.73}$$

In the case where $\bar{x}$ is vanishingly small, the last equation is not present. Only this case will be studied in detail here. Equation (3.73) can be dealt with using the same methods as for the study of equation (3.26), and the motion can be studied in the (x_1,x_2)–plane. This plane is sometimes referred to as the Van der Pol plane, but the terms phase and state plane are also common. The author thinks that the latter terms are better reserved to the $(x,\dot{x})$–plane, to avoid confusion. Neglecting $\bar{x}$, equation (3.73) can be reduced to

$$\frac{dx_2}{dx_1} = \frac{F_2(x_1,x_2)}{F_1(x_1,x_2)}, \tag{3.74}$$

which allows plotting of the trajectories of the system in the Van der Pol plane and the obtaining of the steady-state solution as fixed

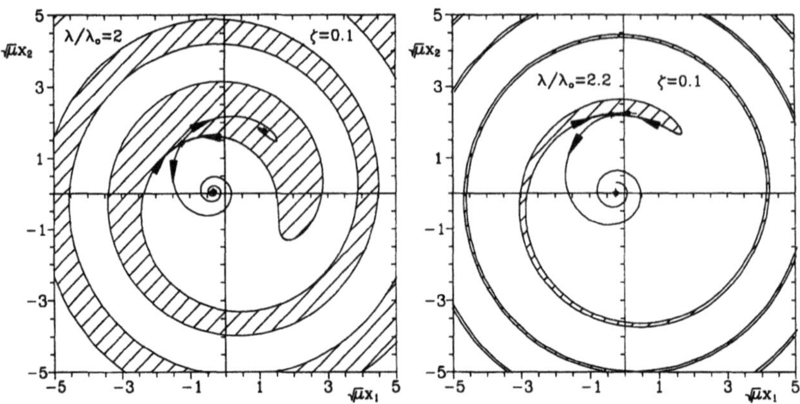

FIGURE 3.24. Trajectories in the Van der Pol plane for a system governed by Duffing's equation, with $\zeta=0.1$ and $f_0\sqrt{\mu}/k=1$; nondimensional plots drawn for $\lambda/\lambda_0 = 2.0$ and 2.2, respectively. The hatched zone is the basin of attraction of the stable solution with larger amplitude.

points in the same plane. The stability of the steady-state solutions can be studied in the same way seen for the stability of equilibrium positions in the phase plane.

Consider, for example, Duffing's equation. By introducing the solution (3.72) into the equation of motion and neglecting the terms containing the second derivatives of x_1 and x_2 with respect to time, those containing the product of their derivatives for the damping coefficient, and those containing the trigonometric functions of $3\lambda t$ (as seen for the harmonic balance method) and then equating the coefficients of the sine and cosine of λt, it follows that

$$\begin{cases} \dfrac{dx_1}{dt} = \dfrac{1}{2\lambda}\left[(\lambda_0^2 - \lambda^2)x_2 - 2\zeta\lambda\lambda_0 x_1 + \dfrac{3}{4}\mu\lambda_0^2 x_2(x_1^2 + x_2^2)\right], \\[2ex] \dfrac{dx_2}{dt} = \dfrac{1}{2\lambda}\left[-(\lambda_0^2 - \lambda^2)x_1 - 2\zeta\lambda\lambda_0 x_2 + \dfrac{f_0}{k}\lambda_0^2 - \dfrac{3}{4}\mu\lambda_0^2 x_1(x_1^2 + x_2^2)\right]. \end{cases}$$
$$(3.75)$$

Equation (3.75) is just equation (3.73) with the right-hand sides defining functions F_1 and F_2. The trajectories obtained by numerically integrating equation (3.74) for $\zeta = 0.1$, $f_0\sqrt{\mu}/k = 1$ and two values of λ/λ_0 are plotted in nondimensional form in Figure 3.24.

The three singular points are the steady-state solutions that coincide with those obtained using the Ritz averaging technique. The solutions with the maximum and minimum amplitudes correspond to stable foci, while the intermediate one is a saddle point. The stability of the former two solutions and the instability of the last one is so demonstrated. The basins of attraction of the two stable solutions have a complex shape and are quite interwoven. This means that starting from a certain position in the Van der Pol plane (i.e., starting from an oscillation with a certain amplitude and phase), the system settles to the oscillation with larger amplitude; starting from an oscillation with larger amplitude, the steady-state condition with smaller amplitude is obtained. Also the phase can be important in deciding which solution the system will tend toward, although the importance of phase decreases with growing amplitude, as the separatrices tend to assume a more circular form.

Consider a system governed by Duffing's equation, excited by a forcing function with a frequency lower than that at which three solutions can exist. There is only one solution, which in the Van der Pol plane is a stable focus. If the forcing frequency is increased, at a certain moment two more solutions appear and the plane splits into two domains of attraction. The domain of the larger solution is very large and that of the smaller one is very narrow. With increasing frequency the smaller domain of attraction grows while the other one reduces. In the center of the frequency range in which three solutions exist, the two domains have comparable extensions (left-hand side of Figure 3.24). With increasing frequency the domain of the larger solution reduces its size, to disappear at the frequency at which the jump takes place and three solutions no longer exist. This explains why it can be difficult to sustain an oscillation with the larger amplitude near the jump frequency: When the domain of attraction of this solution is very small, as in the right-hand side of Figure 3.24, and the solution is very near the separatrix, any small casual change of amplitude makes the system shift toward the other solution.

To study the stability of a singular point x_{1_0}, x_{2_0}, equation (3.73) can be easily linearized about it. Because it is a singular point, functions F_1 and F_2 vanish and the linearized equation reduces to

$$\begin{cases} \dfrac{dx_1}{dt} = \left(\dfrac{\partial F_1}{\partial x_1}\right)_{x_{1_0}, x_{2_0}} x_1 + \left(\dfrac{\partial F_1}{\partial x_2}\right)_{x_{1_0}, x_{2_0}} x_2\,; \\[4mm] \dfrac{dx_2}{dt} = \left(\dfrac{\partial F_2}{\partial x_1}\right)_{x_{1_0}, x_{2_0}} x_1 + \left(\dfrac{\partial F_2}{\partial x_2}\right)_{x_{1_0}, x_{2_0}} x_2\,. \end{cases} \tag{3.76}$$

By assuming a solution of the type $\mathbf{x} = \mathbf{a}e^{st}$, the following characteristic equation is obtained

$$
\det \left[\begin{array}{cc} \left(\dfrac{\partial F_1}{\partial x_1}\right)_{x_{1_0},x_{2_0}} - s & \left(\dfrac{\partial F_1}{\partial x_2}\right)_{x_{1_0},x_{2_0}} \\[2ex] \left(\dfrac{\partial F_2}{\partial x_1}\right)_{x_{1_0},x_{2_0}} & \left(\dfrac{\partial F_2}{\partial x_2}\right)_{x_{1_0},x_{2_0}} - s \end{array} \right] = 0 , \qquad (3.77)
$$

i.e.,

$$
s^2 + \beta s + \gamma = 0 ,
$$

where

$$
\left\{ \begin{array}{l} \beta = - \left(\dfrac{\partial F_1}{\partial x_1}\right)_{x_{1_0},x_{2_0}} - \left(\dfrac{\partial F_2}{\partial x_2}\right)_{x_{1_0},x_{2_0}} , \\[2ex] \gamma = \left(\dfrac{\partial F_1}{\partial x_1}\right)_{x_{1_0},x_{2_0}} \left(\dfrac{\partial F_2}{\partial x_2}\right)_{x_{1_0},x_{2_0}} - \left(\dfrac{\partial F_1}{\partial x_2}\right)_{x_{1_0},x_{2_0}} \left(\dfrac{\partial F_2}{\partial x_1}\right)_{x_{1_0},x_{2_0}} . \end{array} \right.
$$

From the signs of the real and imaginary parts of the solutions s it is possible to state whether the point is a saddle, a node, a focus, or a center and to study its stability, following the same rules seen in Section 3.5.3. Note that the Van der Pol method is approximate as are the harmonic balance method and Ritz averaging technique, the latter used with only the fundamental harmonic: It is practically an application of harmonic balance to slowly varying nonstationary motions. Its main advantage is that of eliminating time by working in a plane characterized by a reference frame rotating with angular velocity equal to the forcing frequency λ in the Argand plane of Figure 1.7, the only difference being that the direction of the ordinate axis is reversed. In that plane, the equation of motion is autonomous, and a bidimensional state plane can be used. In this sense, the Van der Pol plane is a state plane.

3.6.6 Poincaré mapping

As already stated in Section 1.3, in the case of nonautonomous systems the state space must have one added dimension: time. The trajectories related to a system with a single degree of freedom are thus tridimensional curves. The equation of motion of the system can be easily written in the form of a set of three autonomous equations in the variables x, v, and t

$$
\left\{ \begin{array}{l} \dot{v} = F(x,v,t) , \\ \dot{x} = v , \\ \dot{t} = 1 , \end{array} \right. \qquad (3.78)
$$

where function $F(x,v,t)$ also includes the forcing function. In the case of periodic motion, the trajectories wind around the time axis, assuming the shape of a helix whose pitch is equal to the period of the response.

Consider, for example, a linear damped system with $\zeta = 0.2$, excited at a frequency slightly smaller than the natural frequency ($\lambda_n/\lambda = 1.5$). Assume that at time $t = 0$ the system is at a standstill and the forcing function is at its maximum positive value. The tridimensional phase trajectory, computed using equation (1.77) is plotted in Figure 3.25a. The choice of a linear system for this example is due to the fact that an exact solution for the time history exists. The time history is plotted in Figure 3.25b. Clearly, it is the projection of the phase trajectory on the (x,t)–plane. The state projection, i.e., the projection of the tridimensional trajectory on the (x,v)–plane, is plotted in Figure 3.25c; the projected trajectory intersects at many points. The dots in the figure are the positions at time $t = 0$ and after a number of periods of the forcing function.

The Van der Pol plane was seen in Section 3.6.5 with the Van der Pol approximate method. Actually, it can be defined by the following coordinate transformation

$$\begin{cases} x(t) = x_1(t)\cos(\lambda t) + x_2(t)\sin(\lambda t)\,, \\ \frac{1}{\lambda}v(t) = -x_1(t)\sin(\lambda t) + x_2(t)\cos(\lambda t)\,. \end{cases} \qquad (3.79)$$

The first equation (3.79) coincides with equation (3.72) when the constant term in the latter has been neglected: Coordinate transformation (3.79) defines the Van der Pol plane as well as equation (3.72). The inverse transformation is obviously

$$\begin{cases} x_1(t) = x(t)\cos(\lambda t) - \frac{1}{\lambda}v(t)\sin(\lambda t)\,, \\ x_2(t) = x(t)\sin(\lambda t) + \frac{1}{\lambda}v(t)\cos(\lambda t)\,. \end{cases} \qquad (3.80)$$

No assumption on functions $x_1(t)$ and $x_2(t)$, such as they are slowly varying in time, has been made and, consequently, the exact trajectory can be represented. If λ is the fundamental frequency of the response, the Van der Pol plane can be considered a plane that travels along the time axis while rotating about the origin with angular velocity λ. The trajectories then unwind in the Van der Pol plane, yielding a sort of state portrait with nonintersecting lines.

The trajectory in the Van der Pol plane for the system of Figure 3.25 has been plotted in Figure 3.25d. In the same figure the tra-

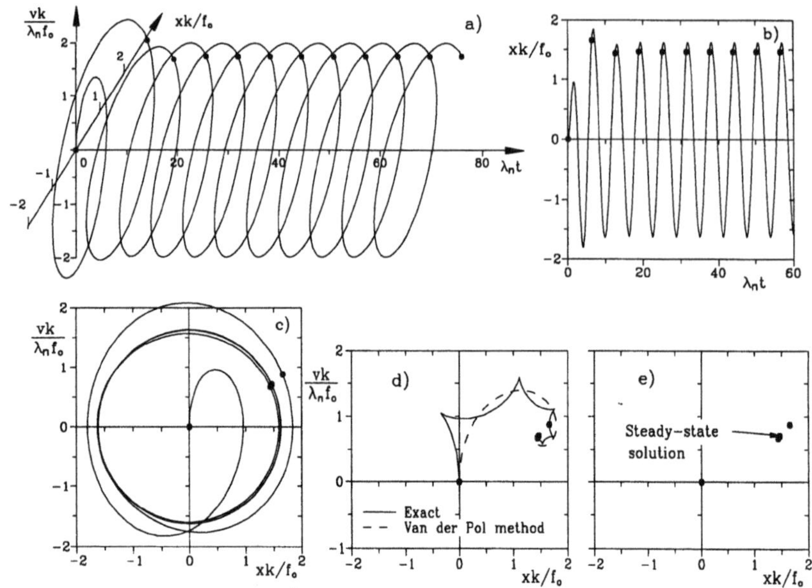

FIGURE 3.25. Behaviour of a linear damped system excited by a harmonic forcing function in the state space. The dots are the positions at time $t = 0$ and after a number of periods of the forcing function; (a) tridimensional state trajectories; (b) time history as the projection of the state trajectory on (x,t)–plane; (c) state projection on (x,v)–plane; (d) trajectories in the Van der Pol plane; (e) Poincaré map. Nondimensional plot with $\zeta = 0.2$ and $\lambda_n/\lambda = 1.5$.

jectory obtained using the Van der Pol method, i.e., assuming that functions $x_1(t)$ and $x_2(t)$ are slowly varying in time, has been plotted with a dashed line. Note that the approximations linked with the Van der Pol solution in this case do not affect the steady-state solution, because it is exactly harmonic, only the way in which it is reached.

A more common way of representing the tridimensional trajectory on a plane is the so-called *Poincaré mapping* or *Poincaré section*. A set of points on the tridimensional trajectory is sampled with an interval equal to the period of the forcing function and then projected on the (x,v)–plane. When studying the steady-state motion with a period equal to the period of the forcing function, all points on the Poincaré map are superimposed on each other. If the system starts from a nonstationary motion, a set of points that get nearer to the one related to the stationary condition is obtained. The attractor on the Poincaré map for a system that undergoes a steady-state oscillation with a period equal to the period of the forcing function is then made of a single point. The Poincaré mapping of the tridimensional trajectory shown in Figure 3.25a is plotted in Figure 3.25e.

A nonlinear system excited by a harmonic forcing function with frequency λ can, in certain cases, exhibit what is usually called a *subharmonic oscillation*, i.e., an oscillatory motion whose fundamental frequency is smaller than that of the forcing function, usually a submultiple of it. A subharmonic of order n is a solution with fundamental frequency equal to λ/n, i.e., with a period equal to n times the period of the forcing function. Because the points used for the construction of the Poincaré mapping are sampled with a period equal to that of the excitation, the map for steady-state oscillations is made of n distinct points. The attractor is then a set of n points, usually referred to as *periodic points*, which are cyclically touched at regular intervals by the tridimensional state trajectory.

In the Poincaré map different attractors can exist, separated by lines that define their basins of attraction. Also, simple point attractors can coexist with n-point attractors: This simply means that, depending on the initial conditions, i.e., depending on the basin of attraction in which the first point is, the motion can evolve into a periodic motion with fundamental frequency λ or into a subharmonic motion with frequency λ/n. To show the basins of attraction in the Poincaré map, some plots obtained by Hayashi from Duffing's equation with vanishing linear stiffness, using a specially designed analog computer are reported in Figures 3.26 and 3.27. The domains of attraction reported in Figure 3.26 refer to a particular set of parameters for which two stable steady-state solutions exist (points 2 and 3). Point 1 is a saddle point corresponding to an unstable solution and lies on the separatrix of the domains of attraction.

The whole picture is not very different from those reported in Figure 3.24, although the latter was only approximate, having been obtained with the Van der Pol method. The solution reported in Figure 3.26 is exact, at least within the approximations allowed by the analog computer used.

A far more complicated situation is reported in Figure 3.27. Working with the same equation, but with different values of the parameters, subharmonic solutions of the third-order have also been obtained, and the domains of attraction become very intricate. Again solutions 2 and 3 correspond to periodic attractors with fundamental frequency equal to the frequency of the forcing function, and solu-

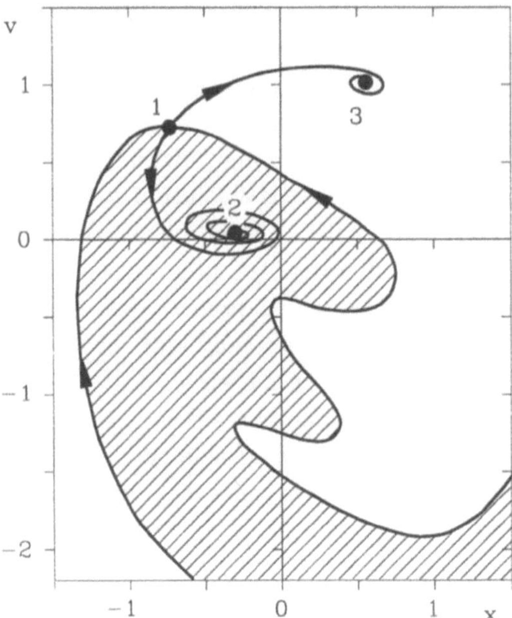

FIGURE 3.26. Domains of attraction in the Poincaré map for Duffing's equation. From C. Hayashi, *Nonlinear oscillations in physical systems*, Princeton University Press, Princeton, N.J., 1985.

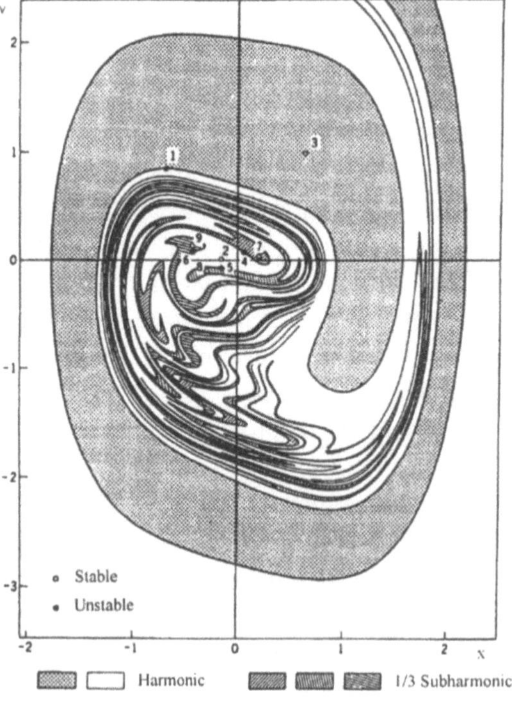

FIGURE 3.27. Domains of attraction in the Poincaré map for Duffing's equation, as in Figure 3.26, but with different values of the parameters. From C. Hayashi, *Nonlinear oscillations in physical systems*, Princeton University Press, Princeton, N.J., 1985.

tion 1 is a saddle point, as in Figure 3.26. In this case, however, two subharmonic solutions of the third order also exist. The one represented by periodic points 4, 5, and 6 is unstable and the points lay on the separatrix, while the solution represented by points 7, 8, and 9 is stable. Note that all domains of attraction of the subharmonic solution lie within the domain of attraction of solution 2.

3.6.7 Systems with many degrees of freedom

Consider a system with many degrees of freedom modeled by equation (3.4), with functions g_i not depending explicitly on time. In many cases, some of the functions vanish, a case of practical interest being that of a linear system that is constrained by a number of supports having nonlinear behaviour. The degrees of freedom can be reordered and divided into two groups. The first group contains the degrees of freedom corresponding to equations containing a nonlinear part. The second group contains all other degrees of freedom. The equations of motion can thus be written in the form

$$\begin{bmatrix} \mathbf{M}_{11} & \mathbf{M}_{12} \\ \mathbf{M}_{21} & \mathbf{M}_{22} \end{bmatrix} \begin{Bmatrix} \ddot{\mathbf{x}}_1 \\ \ddot{\mathbf{x}}_2 \end{Bmatrix} + \begin{bmatrix} \mathbf{C}_{11} & \mathbf{C}_{12} \\ \mathbf{C}_{21} & \mathbf{C}_{22} \end{bmatrix} \begin{Bmatrix} \dot{\mathbf{x}}_1 \\ \dot{\mathbf{x}}_2 \end{Bmatrix} + $$
$$+ \begin{bmatrix} \mathbf{K}_{11} & \mathbf{K}_{12} \\ \mathbf{K}_{21} & \mathbf{K}_{22} \end{bmatrix} \begin{Bmatrix} \mathbf{x}_1 \\ \mathbf{x}_2 \end{Bmatrix} + \mu \begin{Bmatrix} \mathbf{g} \\ 0 \end{Bmatrix} = \begin{Bmatrix} \mathbf{f}_1(t) \\ \mathbf{f}_2(t) \end{Bmatrix} . \tag{3.81}$$

If the nonlinear degrees of freedom are regarded as master degrees of freedom while the others are considered slave, the application of the usual Guyan reduction technique allows the elimination of all linear equations, yielding a set of equations that are fewer in number than the original ones. Functions g_i can also contain slave coordinates, but this does not introduce difficulties because they are easily computed from the master coordinates through equation (2.139). As usual with Guyan reduction, the procedure yields only approximate results; if a large-scale condensation leading to the suppression of all linear degrees of freedom is performed, these approximations may well lead to unacceptable results.

If only a first approximation of the fundamental harmonic of the response to a harmonic forcing function is to be obtained, a solution of the type $\{x\} = \{x_{0_1}\} \sin(\lambda t) + \{x_{0_2}\} \cos(\lambda t)$ can be assumed. A similar expression can be used for the forcing function. Because the motion is harmonic, it is possible to reduce the linear part of the system to a single matrix, the dynamic stiffness matrix

$$\mathbf{K}_{dyn} = \mathbf{K} - \lambda^2 \mathbf{M} + i\lambda \mathbf{C},$$

which is generally complex.

If not all degrees of freedom are related to the nonlinear behaviour of the system, the dynamic stiffness matrix can be partitioned in the same way for the stiffness matrix, and the usual procedure of static reduction can be performed on the dynamic stiffness matrix without any difficulty and without introducing approximations other than those implicit in the assumption that the time history of the response is harmonic. This reduction technique is usually referred to as *dynamic reduction* and can be used in the case of linear systems. Note that the dynamic stiffness matrix is a function of the frequency of the forcing function λ and then, when computing the frequency response of the system, the reduction procedure has to be repeated for each value of the frequency. Moreover, no result can be obtained at those frequencies for which matrix $\mathbf{K}_{dyn_{22}}(\lambda)$ is singular. The fact that when the system is damped the dynamic stiffness matrix is complex does not complicate the relevant computations.

This technique allows the reduction of the number of nonlinear equations to be solved, but the computation of the response, although easier than that of the original system, is still a formidable problem. The Ritz averaging technique can effectively be used to transform the set of second-order nonlinear differential equations into a set of algebraic nonlinear equations.

Consider an undamped system (i.e., with $\mathbf{C} = 0$ and functions g_i depending only on the coordinates) and assume that all forcing functions are in phase. Owing also to the absence of damping, all the responses, reduced to their fundamental harmonic, are in phase with each other and with the excitation, and their time histories can be written in the form $\mathbf{f}(t) = \mathbf{f}_0 \sin(\lambda t)$, $\mathbf{x}(t) = \mathbf{x}_0 \sin(\lambda t)$. The equations of motion of the reduced system can be written in the form

$$(\mathbf{K}_{dyn_{cond}} \mathbf{x}_0 - \mathbf{f}_{dyn_0}) \sin(\lambda t) + \mathbf{g}. \tag{3.82}$$

The Ritz averaging technique can be directly applied to equation (3.82), yielding

$$\mathbf{K}_{dyn_{cond}} \mathbf{x}_0 - \mathbf{f}_{dyn_0} + \frac{1}{\pi \lambda} \int_0^{2\pi} \mathbf{g} \sin(\lambda t) d(\lambda t) = 0, \tag{3.83}$$

which is the required set of algebraic nonlinear equations. Generally speaking, the integration of equation (3.83) can be performed without much difficulty. The problem of solving the set of nonlinear algebraic equations can, however, be difficult, particularly in the fields of frequency where more than one solution exists. There is a number of computation procedures that can be used, but the Newton-Raphson method is usually the best choice, although it cannot guarantee convergence in all cases.

If damping has not been neglected or the excitation is not synchronized, both time histories of the response and excitation can be expressed as the sum of a sine and a cosine function, and the application of the Ritz averaging technique yields a set of $2n$ equations, where n is the number of master degrees of freedom.

Also in the case of systems with many degrees of freedom, it is possible to plot the backbone of the response, which is usually made of a number of separate branches equal to the number of natural frequencies of the linearized system. The backbone can be obtained simply by computing the response to a vanishingly small excitation. On the contrary, the limit envelope, in general, does not exist, except in particular cases. However, when it exists, it is a very powerful tool for understanding the behaviour of the system.

Example 3-9

Consider a structure whose behaviour is linear, supported on a number m of elastic constraints with a force-elongation characteristic of the type used in Duffing's equation: $F_i = -k_i x_i (1 + \mu_i x_i^2)$. Assume that the system is discretized in such a way that the generalized displacements of the supporting points are chosen as the first m generalized coordinates and the linear part of the characteristic of the supports is accounted for in the stiffness matrix of the system. Note that each function g_i contains only the generalized coordinate x_i. It is then possible to separate the nonlinear part of the problem from the linear part, eliminating the last $n - m$ equations by dynamic reduction. By using the Ritz averaging technique, the following set of m algebraic equations is obtained

$$\mathbf{K}_{dyn_{cond}} \mathbf{x}_0 - \mathbf{f}_{dyn_0} + \frac{3}{4} \{ k_i \mu_i x_{0_i}^3 \} = 0 \,,$$

where the condensed matrix and vector must be computed for each value of the frequency λ. A set of m cubic equations is so obtained.

In the particular case of a system with two degrees of freedom, a simple solution can be found by solving one of the unknowns from the equation in which it is at first power, and then substituting it into the other equation. From the first equation the following value for x_2 can be obtained

$$x_2 = \frac{\delta - m_{11}x_1\lambda^2}{\lambda^2 m_2 - k_{12}},$$

where

$$\delta = x_1(k_{11} + k_1\mu_1 x_1^2) - f_1\,.$$

By introducing this value of x_2 into the second equation of motion, it yields an equation of the fourth degree in λ^2, which can be solved directly.

Example 3-10
Consider the rigid beam on nonlinear springs sketched in Figure 3.28a, excited by a harmonic forcing function applied to the center of mass. With simple computations and introducing the radius of inertia $r = \sqrt{J/m}$, the mass and stiffness matrices and the force vector can be shown to take the values

$$\mathbf{M} = \frac{m}{l^2}\left[\begin{array}{cc} a^2 + r^2 & ab - r^2 \\ ab - r^2 & b^2 + r^2 \end{array}\right], \qquad \mathbf{K} = \left[\begin{array}{cc} k_1 & 0 \\ 0 & k_2 \end{array}\right],$$

$$\mathbf{f}_0 = \frac{f_0}{l}\left\{\begin{array}{c} a \\ b \end{array}\right\}.$$

To simplify the computations without loss of generality, the characteristics of the two supports will be considered equal: $k_1 = k_2 = k$, $\mu_1 = \mu_2 = \mu$. The results computed for the values of the parameters shown in Figure 3.28 are reported in the same figure with those related to the linear system ($\mu = 0$). The displacements x_1 and x_2 at the supporting points are reported in Figure 3.28a. The same results are converted to displacement at the center of gravity x_G and rotation θ in Figure 3.28b. The last plot allows the conclusion that the second mode of the linear system consists mainly of a rotation about the center of mass and that the nonlinear effects are not great. The value of $\mu = 8,000$ m^{-2} is, however, not high, as the nondimensional parameter $\sqrt{\mu}x_0$ reaches a value of only 0.36 when the amplitude of motion is as high as 0.004 m.

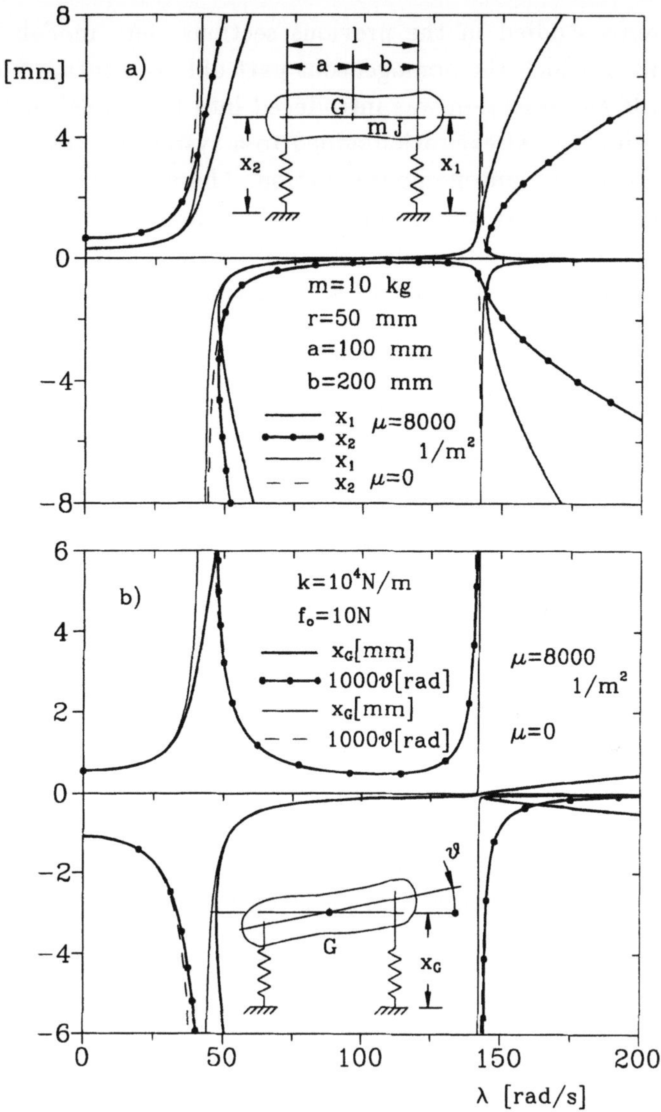

FIGURE 3.28. Response of the system with two degrees of freedom sketched in the figure, excited by a harmonic forcing function applied in the center of gravity; (a) amplitude of the displacements of the supporting points; (b) amplitude of the displacement of the center of gravity x_G and of rotation θ.

3.7 Parametrically excited systems

3.7.1 General considerations

All systems studied in the previous sections were modeled using equations in which the homogeneous part did not contain functions of time. If an excitation was introduced into the model, it had the form of an external excitation added to a system that was in itself modeled using an autonomous equation. There are, however, many systems of practical importance for which models of this type do not hold. The simplest of such systems are linear systems in which the mass, damping, or stiffness are functions of time. The general equation of motion is

$$\ddot{x} + p_1(t)\dot{x} + p_2(t)x = f(t) \,. \tag{3.84}$$

Even if the external excitation is set to zero, i.e., if $f(t) = 0$, the presence of the functions of time in the homogeneous equation can act as an excitation. Because this type of excitation acts from within the parameters of the system, it is usually referred to as *parametric excitation*. As in the case of equation (3.84), parametric excitation can coexist with external excitation. Although the detailed study of systems with parametric excitation is beyond the scope of this book, a survey of the Hill and Mathieu equation will be reported here. In Chapters 4 and 5, some forms of parametric excitation, linked with the dynamic behaviour of unsymmetrical rotors and reciprocating machinery, will be shown.

Equation (3.84) is linear, although its coefficients are not constant, and its general solution can be obtained by adding a particular solution of the complete equation to the general solution of the homogeneous equation. Moreover, if $x_1(t)$ and $x_2(t)$ are two independent solutions of the homogeneous equation, its general solution can be obtained by the linear combination $x(t) = C_1 x_1(t) + C_2 x_2(t)$.

Consider a system modeled with a linear second-order differential equation of the type of equation (3.84) but without external excitation and with functions $p_1(t)$ and $p_2(t)$, which are periodic in time with period T. The study of equations of this type was published by Floquet in 1883, and, hence, is usually referred to as *Floquet theory*. Under the only added condition that $p_1(t)$ is differentiable with respect to time, it is possible to demonstrate that the equation of motion can be reduced to the form

$$\ddot{x}^* + p(t)x^* = 0\,, \qquad \text{with } p(t) = p(t+T)\,, \qquad (3.85)$$

where

$$x = x^* e^{-\frac{1}{2}\int p_1(t)dt}\,, \qquad p(t) = p_2 - \frac{1}{4}p_1^2 - \frac{1}{2}\dot{p}_1\,.$$

This means that the free behaviour of the damped system can be obtained from that of an undamped system by multiplying the time history of the latter by an appropriate decaying factor and slightly modifying the frequency by a change of the stiffness. This clearly holds also for linear systems with constant parameters. Equation (3.85) is usually referred to as *Hill's equation*, because it was first studied by Hill is 1886 in the determination of the perigee of lunar orbit. From Floquet theory, it follows that the two independent solutions $x_i(t)$, whose linear combinations yield all possible solutions of the homogeneous equation, can be written as

$$x_i(t) = e^{\alpha_i t}\phi_i(t) \qquad\qquad i = 1,2\,, \qquad (3.86)$$

where functions $\phi_i(t)$ are periodic with period T and constants α_i can be complex. From equation (3.86) it immediately follows that if the real parts of constants α_i are positive, the solution is unbounded and the behaviour of the system is unstable. If the imaginary part of the same constants is not vanishingly small, the result is the product of a function with period T for another function whose period can be different, and then the solution can be an oscillation, damped or self-excited, whose period can be different from that of function $p(t)$.

A particular form of Hill's equation is the Mathieu equation:

$$\ddot{x} + [\delta + 2\epsilon\cos(\lambda t)]x = 0\,. \qquad (3.87)$$

3.7.2 *Pendulum on a moving support: Mathieu equation*

Consider a pendulum suspended to a point A that can move with a prescribed law (Figure 3.29). To obtain the equation of motion through the Lagrange equation, the position and the velocity of point P must be computed

$$\left\{\begin{array}{c} x \\ y \end{array}\right\}_P = \left\{\begin{array}{c} x_A + l\sin(\theta) \\ y_A - l\cos(\theta) \end{array}\right\} \qquad \left\{\begin{array}{c} \dot{x} \\ \dot{y} \end{array}\right\}_P = \left\{\begin{array}{c} \dot{x}_A + l\dot{\theta}\cos(\theta) \\ \dot{y}_A + l\dot{\theta}\sin(\theta) \end{array}\right\}.$$

$$(3.88)$$

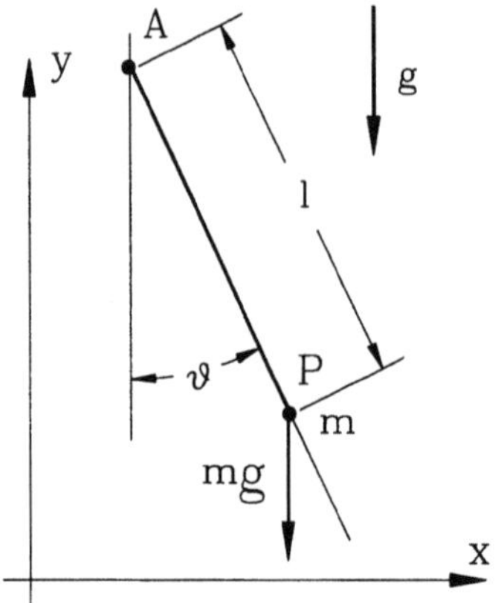

FIGURE 3.29. Pendulum on a moving support; sketch of the system.

The kinetic and potential energies are immediately obtained

$$
\mathcal{T} = \tfrac{1}{2}m\left\{\dot{x}_A^2 + \dot{y}_A^2 + l^2\dot{\theta}^2 + 2\dot{\theta}l[\dot{x}_A\cos(\theta) + \dot{y}_A\sin(\theta)]\right\},
$$
$$
\mathcal{U} = mg[y_A - l\cos(\theta)].
$$

(3.89)

By introducing equation (3-89) into the Lagrange equation and performing all relevant derivatives, the following equation of motion can be obtained

$$
\ddot{\theta} + \frac{\ddot{x}_A}{l}\cos(\theta) + \frac{g + \ddot{y}_A}{l}\sin(\theta) = 0.
$$

(3.90)

Equation (3.90) is a nonlinear equation with parametric excitation. If the amplitude of the motion is small, it can be linearized, obtaining

$$
\ddot{\theta} + \frac{g + \ddot{y}_A}{l}\theta = -\frac{\ddot{x}_A}{l}.
$$

(3.91)

The motion of the supporting point in the x-direction acts as an external excitation, and its effects can be easily computed using the techniques seen in Chapter 1. The motion in the y-direction, on the contrary, provides a parametric excitation. The homogeneous equation obtained by neglecting the motion in the x-direction is Hill's equation. If the motion in the y-direction of the supporting point is harmonic with amplitude A and frequency λ, the relevant equation of motion becomes

$$
\frac{d^2\theta}{d\tau^2} + [\delta + 2\epsilon\cos(2\tau)]\theta = 0,
$$

(3.92)

where $\tau = \lambda/2t$, $\delta = 4g/l\lambda^2$, and $\epsilon = -2A/l$. It is a Mathieu equation in its standard form (3.87), and its solution is of the type of equation (3.86) where function $\phi(\tau)$ is periodic with period equal to π and, consequently, can be expressed by the following Fourier series

$$\phi(\tau) = \sum_{k=-\infty}^{\infty} \phi_k e^{2ik\tau}. \tag{3.93}$$

Remembering that $2\cos(2\tau) = e^{2i\tau} - e^{-2i\tau}$ and substituting the series expressing the solution into the equation of motion (3.92), the latter yields

$$\sum_{k=-\infty}^{\infty} \phi_k[(\alpha + 2ik)^2 + \delta]e^{(\alpha+2ik)\tau} + $$
$$+\epsilon \sum_{k=-\infty}^{\infty} \phi_k\{e^{[\alpha+2i(k+1)]\tau} + e^{[\alpha+2i(k-1)]\tau}\} = 0. \tag{3.94}$$

By separately equating the various terms of equation (3.94), a set made of an infinity of equations is obtained

$$\begin{bmatrix} \cdots & \cdots & \cdots & \cdots & \cdots & \cdots & \cdots \\ \cdots & (\alpha-4i)^2+\delta & \epsilon & 0 & 0 & 0 & \cdots \\ \cdots & \epsilon & (\alpha-2i)^2+\delta & \epsilon & 0 & 0 & \cdots \\ \cdots & 0 & \epsilon & \alpha^2+\delta & \epsilon & 0 & \cdots \\ \cdots & 0 & 0 & \epsilon & (\alpha+2i)^2+\delta & \epsilon & \cdots \\ \cdots & 0 & 0 & 0 & \epsilon & (\alpha+4i)^2+\delta & \cdots \\ \cdots & \cdots & \cdots & \cdots & \cdots & \cdots & \cdots \end{bmatrix} \times$$

$$\times \begin{Bmatrix} \cdots \\ \phi_{-2} \\ \phi_{-1} \\ \phi_0 \\ \phi_1 \\ \phi_2 \\ \cdots \end{Bmatrix} = \{0\}. \tag{3.95}$$

Equation (3.95) can be used to compute the exponent α of equation (3.86) by equating to zero the determinant of the matrix of the coefficients. Such a determinant, which has an infinity of rows and columns, is usually referred to as Hill's infinite determinant. Although is impossible to obtain an exact solution of this eigenproblem, approximate solutions are readily obtainable by only considering the central part of the determinant. For example, if only the central part with dimension 3×3 is retained, the constant term and the fundamental harmonic of the series in equation (3.93) are computed. This will be referred to as *first approximation*. The second approximation, allowing the computation of the second harmonic of the series, comes from the 5×5 determinant obtained considering all terms written explicitly in equation (3.95).

Hill's determinant can be also used to assess the stability of the motion. The motion is unstable, i.e., the amplitude grows indefinitely in time, if the real part of exponent α is positive. The motion can be stable or unstable depending on the values assumed by parameters δ and ϵ: It is possible to plot on the (δ,ϵ)–plane the boundary of the regions in which an unstable behaviour is present simply by setting to zero the real part of α. In particular, when $\alpha = 0$ an undamped oscillation with nondimensional period equal to π is obtained, while when $\alpha = \pm i$ the undamped motion has a period of 2π.

By introducing these values of α, the equation stating that Hill's determinant is equal to zero becomes an eigenproblem in δ, if ϵ is stated, or in ϵ if δ is stated. The hatched regions in Figure 3.30a are the regions of instability in the (δ,ϵ)–plane, obtained from the fourth-order approximation (determinant whose size is 9×9). These results are obviously only approximate, because a reduced form of Hill's determinant has been used.

Note that when δ is negative the motion is almost always unstable, except for a few combinations of δ and ϵ. A negative δ physically means that the acceleration field is directed toward the suspension point, i.e., the pendulum is inverted and the values of ϵ that lead to stability in this case correspond to values of the parametric excitation causing the pendulum to be stable in its inverted position. When δ is positive, i.e., the pendulum is in its normal position, the fields of instability are very narrow unless the parametric excitation ϵ is quite large.

If damping is not neglected, the linearized homogeneous equation of motion of the pendulum in nondimensional form is

$$\frac{d^2\theta}{d\tau^2} + 2\zeta\frac{d\theta}{d\tau} + [\delta + 2\epsilon\cos(2\tau)]\theta = 0 , \qquad (3.96)$$

where $\zeta = c/m\lambda$. By using the transformation seen for equation (3.85), equation (3.96) can be reduced to a Mathieu equation in standard form. The presence of damping reduces the growth rate of

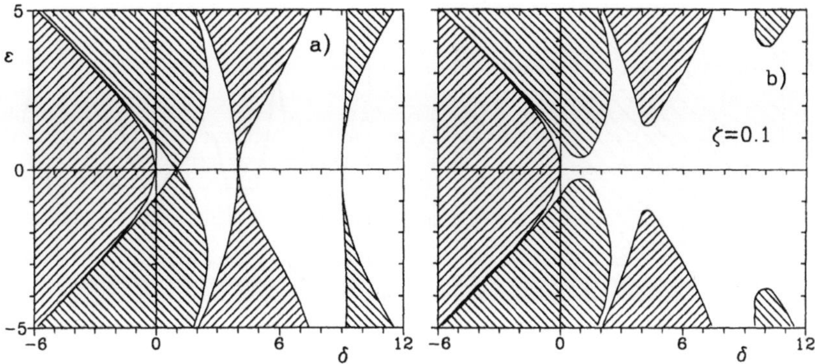

FIGURE 3.30. Regions of instability in the (δ,ϵ)-plane for Mathieu equation; (a) without damping; (b) with damping.

the oscillations from α to $\alpha - \zeta$ and modifies the natural frequency from $\sqrt{\delta}$ to $\sqrt{\delta - \zeta^2}$. The regions of instability, computed using the fourth approximation of Hill's determinant, are plotted in Figure 3.30b for a value of $\zeta = 0.1$. The effect of damping is that of reducing the instability regions, particularly for high values of δ.

There are many different approaches, all approximate, to the study of the Mathieu and Hill equations. Only that based on Hill's determinant is, however, presented here, because it is a very suitable tool for the study of systems with many degrees of freedom and is very easily implemented on digital computers.

Example 3-11

Consider a pendulum whose length is 0.5 m connected to a support that can move in a vertical direction (along the y-axis in Figure 3.29). Using the results obtained from the Mathieu equation, choose the values of the frequency and the amplitude of the harmonic time history of the displacement of the supporting point so that the inverted vertical position of the pendulum is stable. Verify this stability by numerically integrating the equation of motion of the pendulum with initial conditions corresponding to a displacement from the vertical of 0.05 rad (about 2.86°).

At first the relevant portion of the stability map of Figure 3.30a is plotted in Figure 3.31a. A point lying within the stability zone is chosen (point A, with $\delta = -0.1$, $\epsilon = 0.75$). The frequency and the amplitude of the motion of the supporting point are immediately computed (note that $g = -9.81$ m/s^2, as the pendulum is inverted)

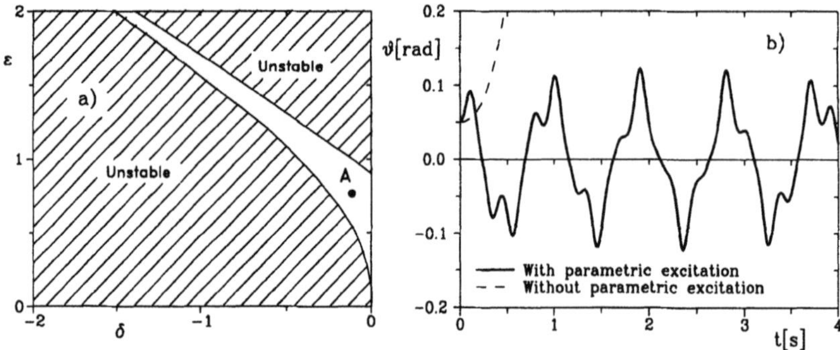

FIGURE 3.31. (a) Boundary of the stability zone for the inverted pendulum in the (δ,ϵ)–plane (Figure 3.30a); (b) time histories of the response of the inverted pendulum with and without parametric excitation supplied by the vertical motion of the supporting point.

$$\lambda = \sqrt{\frac{4g}{l\delta}} = 28 \text{ rad/s} = 4{,}45 \text{ Hz}\,; \qquad |A| = \frac{l\epsilon}{2} = 0{,}188 \text{ m}\,.$$

The equation of motion of the pendulum, without making any small-oscillations assumption, is then

$$\ddot{\theta} - \left[19{,}62 + 294{,}78\cos(28t)\right]\sin(\theta) = 0\,.$$

The result of the numerical integration of the equation of motion of the system, with initial conditions $\theta = 0.05$, $\dot{\theta} = 0$, is reported in Figure 3.31b, with the time history computed without taking into account parametric excitation. From the figure, it is clear that the prescribed motion of the supporting point allows the inverted pendulum to oscillate about the vertical position and it falls off if the supporting point is fixed.

3.8   An outline of chaotic vibrations

The possibility of performing numerical experiments on nonlinear differential equations that cannot be integrated analytically allowed the discovery that a system modeled using a standard differential equation, and hence a fully deterministic model, can give way to a seemingly random motion. Actually, the possibility of obtaining irregular time histories from a deterministic system excited by a regular forcing function had been observed for a long time, but was usually ascribed to noise or the influence of some external disturbances rather than the intrinsic behaviour of the system. However, it is clear that such irregular behaviour may take place in many systems due to their own characteristics and that very simple models can also simulate this behaviour.

The study of the forced response of the Van der Pol equation did show that in the cases in which two stable attractors exist, when starting the motion from certain positions of the state space the initial transient can hesitate to choose the attractor to converge to. An arbitrarily long transient shifting from the region around one attractor to the region around the other can be present before the motion eventually settles out. This behaviour is now known as *transient chaos*.

Another form of chaotic behaviour, perhaps the true one, is *persistent chaos*. When a system undergoes such a process, the seemingly random motion will not settle out in time, and no attractor, in the conventional sense, is present. Because chaotic motion does not die out, it is possible to identify a different sort of attractor, typical of chaotic motion, which is called a *strange* or *chaotic attractor*.

To give way to chaotic motion, the state space must have at least three dimensions, and the system must be quite sensitive to the initial conditions. The latter condition seems to be the key factor for chaotic behaviour. The first deep study on systems that are very sensitive to initial conditions was performed by Lorentz in 1963 on a three-equation model of atmospheric dynamics for weather predictions. He found that very slight changes in some parameters lead, after time, to large variations in the results.

Consider, for example, the following damped Duffing equation with vanishing linear stiffness:

$$\ddot{x} + 0.05\dot{x} + x^3 = 7.5\cos(t).$$

This very equation has been studied by Ueda[6], and the results have also been reported by many other authors. To show how much the time history of the system can depend on the initial conditions, two time histories starting from the conditions $x = 3$, $v = 4$ and $x = 3.1$, $v = 4.1$ for $t = 0$ are plotted in Figure 3.32a. The time histories are initially very similar but, as times goes on, the differences quickly build up, and after time they do not look related anymore. Of course, the system is fully deterministic, and every time a com-

---

[6]Y. Ueda, "Steady motion exhibited by Duffing's equation: A picture book of regular and chaotic motions", *New approaches in nonlinear problems in dynamics*, P.J. Holmes (Editor), SIAM, Philadelphia, 1980, 311–322.

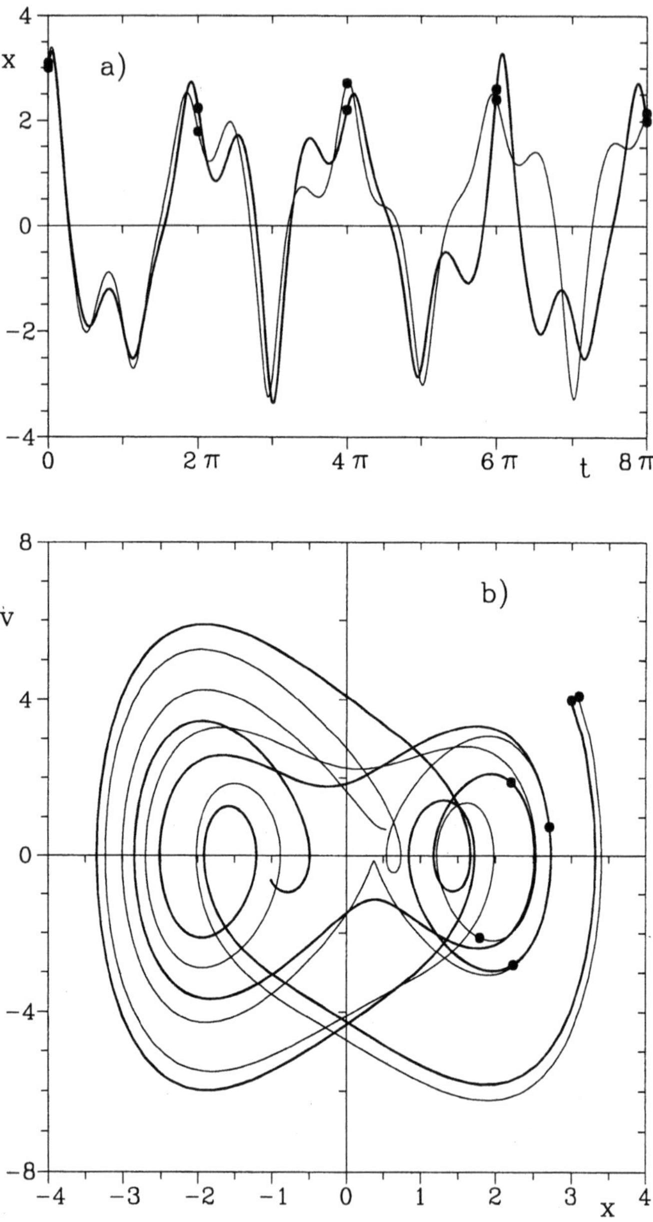

FIGURE 3.32. (a) Time history and (b) state projection for a system modeled by the equation $\ddot{x} + 0.05x + x^3 = 7.5\cos(t)$ with two different initial conditions.

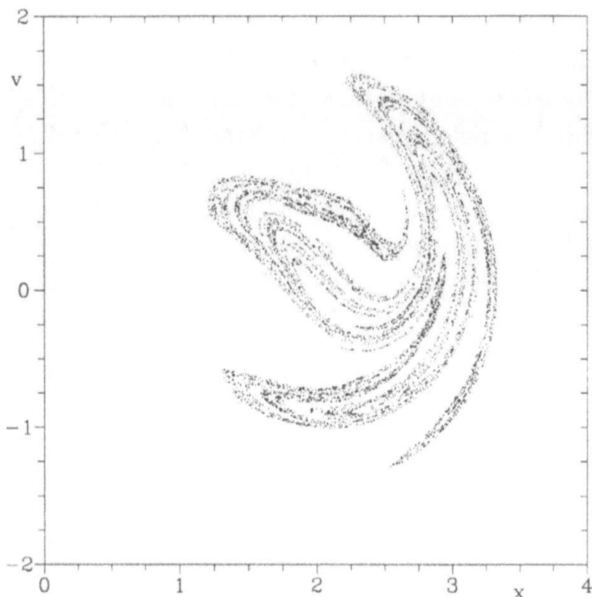

FIGURE 3.33. Ueda attractor represented as a Poincaré map; picture obtained following the system through 8,000 cycles.

putation is started with exactly the same initial conditions, identical results are found. However, if the initial conditions are known in an approximate way, completely different results are obtained depending on the approximation chosen.

A system that shows this behaviour is said to be structurally unstable. To define structural stability, it is possible to verify whether two trajectories starting in the state space within a sphere of small radius δ, tend to remain close to each other within a sphere with a limited radius or separate. To show the same behaviour in the state projection, the projection of the state trajectories on the (x,v)–plane of the system studied in Figure 3.33a are reported in Figure 3.32b. The heavy dots denote the position at the ends of each cycle, i.e., the points that enter the Poincaré map.

The time history and state projection seem quite irregular, but if the system is followed for a longer period of time, a particular form of regularity emerges. After the transient dies out, although the time history does not show any regularity or periodicity, all points of the Poincaré map settle on a very complicated geometrical figure that is the attractor of the system. The attractor resulting from the aforementioned equation, usually referred to as the *Ueda attractor*, is shown in Figure 3.33.

The Ueda attractor is one of the better-studied chaotic attractors. Its shape is quite complex, and its structure follows a fractal geometry. This is a general feature of strange attractors. To show the fractal structure, a very detailed representation must be obtained, which means that the numerical integration must be carried on for many cycles with sufficient precision. A great deal of computer time is needed to perform detailed studies on chaotic vibrations.

In addition to a strange attractor, another characteristic feature of chaotic motion is the spectrum of the time history. Although the response of a linear system to a harmonic excitation is monoharmonic and that of a nonlinear nonchaotic system contains a number of harmonics (usually very few), when chaotic motion is present a continuous spectrum, like those encountered in random vibrations, is found.

Obviously the phenomena are different, because there is nothing random in chaotic motion. If the same numerical experiment is repeated several times with exactly the same initial conditions, the same outputs are obtained. And even if the initial conditions are different, once steady-state chaotic motion is achieved, the same attractor is found. The presence of a well-defined attractor makes it clear that under the random-like appearance, there is an underlying order.

The power spectral density of the law $x(t)$ shown in Figure 3.32a is shown in Figure 3.34. The motion was followed for many periods, and then a fast Fourier algorithm was applied on a set of 8,192 points. The fundamental frequency and some higher harmonics are clearly visible, but the spectrum is continuous as if broadband noise were present, as is typical for chaotic vibrations.

The Duffing equation can give way to chaotic response to harmonic excitation not only with the values of the various parameters considered here. Actually, chaotic motion has been observed with both hardening and softening systems, with positive, negative, or vanishing linear parts. The presence of multiple equilibrium positions is then not needed to enable the occurrence of chaotic vibrations. On the contrary, for chaotic motion, strong nonlinearity is needed. The more experimental work, physical and numerical, performed on chaotic vibration, the more this type of behaviour seems to be a common possibility for heavily nonlinear systems.

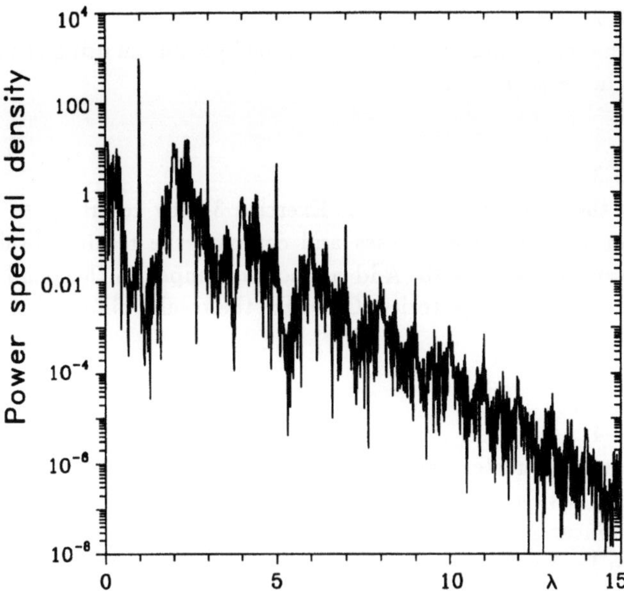

FIGURE 3.34. Power spectral density of the law $x(t)$ shown in Figure 3.32a obtained using a fast Fourier algorithm on a set of 8,192 points and Hanning windowing.

Much theoretical and experimental work is still needed before chaotic behaviour of mechanical systems can be fully understood and design methods based on it developed. For now, the study of chaotic systems is mostly a theoretical issue, pursued by mathematicians and mechanicists with strong theoretical interests. The author, however, thinks that these studies can, in the future, supply the basis for applicative studies that will influence design and structural analysis methods.

The interested reader can find more details on this topics in the many specialized texts that have recently been published. Some of them are listed in the bibliography.

3.9 Exercises

Exercise 3-1

Plot the state force surface for the restoring force acting on a damped pendulum without assuming that the displacement from the vertical position is small.

Exercise 3-2

Study the free oscillations of an undamped pendulum using the technique described in Section 3.3.1.

Exercise 3-3

Repeat the study performed in Exercise 3-2 by resorting to the state plane. Plot the state trajectories and compare the results obtained with those shown in Figure 3.8b. Add a viscous damper to the pendulum and again plot the state trajectories. Compare the results obtained with Figure 3.14b.

Exercise 3-4

Consider the quarter car model studied in Section 1.9. Substitute the linear spring with a spring with cubic characteristics with the same stiffness and a nonlinear parameter $\mu = 400$ m^{-2}. Plot the frequency response of the system for amplitudes of the excitation equal to 10 mm and 25 mm.

Exercise 3-5

Consider the system of Example 2.8 and substitute for the linear support with stiffness k_G a nonlinear support with a hardening characteristic of the same type as seen for Duffing's equation. Compute the backbone of the response. Assume that a harmonic excitation with variable frequency and amplitude 20 N acts on the central joint. Compute the fundamental harmonic of the response of the system in the range of frequencies between 1 and 100 Hz (only a first approximation is required).

Data: inner diameter = 70 mm, outer diameter = 90 mm, length of each part of the shaft = 500 mm, E = 2.1 $\times 10^{11}$ N/m^2, $\nu = 0.3$, $\rho = 7,810$ kg/m^3, mass of the joint 2 kg, linear stiffness of the central support 105 N/m, $\mu = 5 \times 10^7$ 1/m^2.

Exercise 3-6

Add to the system of the previous exercise a linear viscous damper with damping coefficient $c = 200$ Ns/m located on the central support. Plot the limit envelope and compute the forced response of the system, using the same excitation of Exercise 3-5. State whether the system shows the jump phenomenon.

Exercise 3-7

Consider the system of Example 1.2 and neglect damping. Substitute a nonlinear element with a cubic hardening characteristic to the shaft connecting the first moment of inertia to point A. The linear part of the stiffness coincides with the stiffness in Example 1.2, while $\mu = 30 \ 1/\text{rad}^2$. Compute a first approximation of the fundamental harmonic of the frequency response of the system when a moment with an amplitude of 0.5 Nm and frequency variable between 0 and 6 rad/s acts on the third inertia. Compare the results obtained with those shown in Figure 1.17. Do not resort to modal coordinates, but use dynamic reduction to reduce the nonlinear problem to a single equation.

Exercise 3-8

Repeat the computations of the preceding example by resorting to modal coordinates. State whether the coupling between the modes due to the nonlinear behaviour of the system is large.

Exercise 3-9

Repeat the computations of Example 3.7 without neglecting damping.

Exercise 3-10

Consider the double pendulum of Exercise 1-15. Write the nonlinear equations of motion and those obtained by retaining two terms of the series for the trigonometric functions of the generalized coordinates. Study by numerical integration of the equations of motion the free motion starting from a configuration with only angle θ_2 different from 0 and all velocities equal to 0. Perform the computation for two cases with $\theta_{2_0} = 2°$ and $60°$, respectively. Compare the results with those obtainable from the linearized system (analytical solution).

4
Dynamic Behaviour of Rotating Machinery

4.1   Rotors and structures

The preceding chapters have been devoted to the study of the dynamic behaviour of structures, i.e., of mechanical systems that are stationary with respect to an inertial frame of reference, apart from the vibratory motion that is the object of the study. Many machine elements, however, do not comply with this definition because, due to their rotational motion, it is not possible to define an inertial system of reference in which the element is stationary.

This section will be devoted to the study of the dynamic behaviour of rotors. Following the ISO definition, a *rotor* is a body suspended through a set of cylindrical hinges or bearings that allow it to rotate freely about an axis fixed in space. Transmission shafts, parts of reciprocating machines that have only rotational motion, and many other rotating machine elements can thus be considered rotors. If no reference is made to the type of supports or their existence, a space vehicle or a celestial body that rotates about an axis whose direction in space is constant can also be regarded as a rotor, at least in some features of their behaviour. However, although in rotors on fixed bearings the spin speed is usually imposed, isolated rotors are governed by the conservation of linear and angular momenta. The parts of the machine that do not rotate will be referred to with the general definition of *stator*.

Some simplifying assumptions allowing the building of mathematical models that are not too complex for the dynamic study of rotors must be stated. The equations that describe the motion of even a simple rigid body in the tridimensional space are actually quite complex, particularly when dealing with the rotational degrees of freedom, and do not allow the use of any linearized model. Once the reference frames xyz and $\xi\eta\zeta$ (the first inertial and the second fixed to the rigid body and coinciding with its principal axes of inertia) are stated, the six equations of motion under the action of the generic force $\vec{F}$ and moment $\vec{M}$ can be written in the form

$$
\begin{cases}
m\ddot{x} = F_x\,, \\
m\ddot{y} = F_y\,, \\
m\ddot{z} = F_z\,,
\end{cases}
\qquad
\begin{cases}
M_\xi = \dot{\Omega}_\xi J_\xi + \Omega_\eta \Omega_\zeta (J_\zeta - J_\eta)\,, \\
M_\eta = \dot{\Omega}_\eta J_\eta + \Omega_\xi \Omega_\zeta (J_\xi - J_\zeta)\,, \\
M_\zeta = \dot{\Omega}_\zeta J_\zeta + \Omega_\xi \Omega_\eta (J_\eta - J_\xi)\,.
\end{cases}
\qquad (4.1)
$$

The last three equations for the rotational degrees of freedom, which are the well-known Euler equations, are clearly nonlinear in the angular velocity $\vec{\Omega}$. However, some simplifications that allow the linearization of the equations related to the rotational degrees of freedom can be made for the case of rotors. The rotor has, in the undeformed configuration, a well-defined and fixed rotation axis, which coincides with one of the baricentrical principal axes of inertia, if the rotor is perfectly balanced. Actually, this is only approximately true, but the *unbalance*, i.e., the deviation from this ideal condition, is usually small. The displacement of the rotational axis from its nominal position due to the deformations of the system is also assumed to be small. The two assumptions of small unbalance and small displacement allow the linearization of the equations of motion in a way consistent with what has been seen for the case of the dynamics of structures where similar small-displacement assumptions were required to obtain linear equations of motion.

Some cases that, strictly speaking, could not be studied using the aforementioned assumptions, can still be dealt with in the same way. Consider, for example, the rotor of an aircraft gas turbine during maneuvered flight. The direction of the axis of the rotor changes continuously in time, and no small-angle assumption can be considered for this motion. However, the motion of the rotor can be studied in a reference frame that is fixed to the aircraft, provided that the motion of the latter can be considered independent of the dynamic behaviour of the first and the related inertia forces are added. This way of separating the problem into its dynamic and quasi-static parts is possible if the characteristic times of the different phenomena under study are widely different. In the example given earlier, this is clearly the case if the frequencies that characterize the motion of the rotor of the turbine with respect to the aircraft are of several Hertz (periods of fractions of seconds), while the rotations of the airframe have characteristic times of the order of several seconds. On the contrary, the seismic actions on the rotor of a machine in a building can have frequencies of the same order of magnitude as those that characterize the rotor itself, and the problem may have to be studied without any uncoupling being possible.

Another common assumption is that of axial symmetry of the rotor. If this assumption holds, the dynamic study is greatly simplified and a nonrotating reference frame is usually also chosen for the study of rotor motions. If, on the contrary, the rotor cannot be considered axially symmetrical, the study becomes very complicated, unless an axial symmetry assumption can be made on the nonrotating parts of the system. In the latter case, a reference frame that rotates at the angular velocity of the rotor can be used and simplified equations can be obtained. If both stator and rotor are isotropic with respect to the rotation axis, very simple models can be devised.

4.1.1 Vibrations of rotors: The Campbell diagram

As already stated, a rotor is a body that is free to rotate about a well-defined axis. Usually the dynamic study of rotors is performed under the assumption that the angular velocity about the axis of rotation, usually referred to as *spin speed*, is constant, at least in its average value. Because the natural frequencies of a deformable rotor, or more generally, of a machine that contains a rotor, can depend on the spin speed, the dynamic behaviour of such systems is usually summarized by a plot of the natural frequencies as a function of the rotational speed. Because in many cases the frequencies of the exciting forces also depend on the speed, they can be reported on the same plot, obtaining what is generally known as a *Campbell diagram*. If the dynamic behaviour of the system can be described in terms of complex frequencies, a second plot, in which the decay rates are reported as functions of the spin speed, can be obtained with the Campbell diagram. Alternatively, the frequency can be plotted

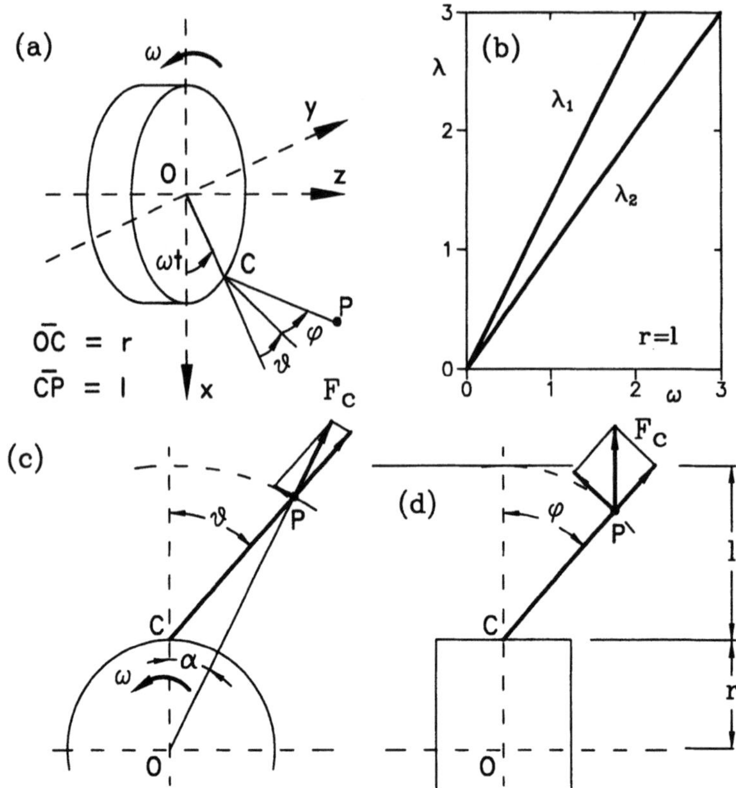

FIGURE 4.1. Rotating pendulum. (a) Sketch of the system and generalized co-ordinates, (b) Campbell diagram, (c) situation in the xy plane, (d) situation in a plane containing the z-axis.

against the decay rate, obtaining what is generally referred to as a *roots locus*.

There are cases in which, in spite of what was said earlier, the natural frequencies are constant with respect to the rotational speed: The Campbell diagram in this case is made of horizontal straight lines.

Note that the Campbell diagram can be plotted only in the case of linear systems, because only in this case does the very concept of natural frequencies apply. However, in the case of nonlinear systems, the Campbell diagram of the linearized system may yield important information on the behaviour of the system.

Example 4-1

A case in which the natural frequencies of the system are strongly influenced by the spin speed is that of the rotating pendulum, which can be described as a pendulum attached to the outer radius of a disc that rotates at a constant angular velocity (Figure 4.1a).

As the angular velocity ω of the disc is imposed, the system only has two degrees of freedom, and angles θ (between the projection of line PC on the plane of the disc and radius OC) and ϕ (between line PC and the mentioned plane), can be assumed to be generalized coordinates. All other force fields except the centrifugal field are neglected. The position of point P is

$$(\overline{\text{P-O}}) = \left\{ \begin{array}{c} r\cos(\omega t) + l\cos(\phi)\cos(\omega t + \theta) \\ r\sin(\omega t) + l\cos(\phi)\sin(\omega t + \theta) \\ l\sin(\phi) \end{array} \right\}.$$

By differentiating the expressions of the coordinates with respect to time, the velocity of point P is readily obtained

$$\vec{V}_{\text{P}} = \left\{ \begin{array}{c} -\omega r\sin(\omega t) - \dot{\phi}l\sin(\phi)\cos(\omega t + \theta) - l(\omega + \dot{\theta})\cos(\phi)\sin(\omega t + \theta) \\ \omega r\cos(\omega t) - \dot{\phi}l\sin(\phi)\sin(\omega t + \theta) + l(\omega + \dot{\theta})\cos(\phi)\cos(\omega t + \theta) \\ \dot{\phi}l\cos(\phi) \end{array} \right\}.$$

The kinetic energy of the mass located in point P is simply

$$T = \frac{1}{2}m|\vec{V}_{\text{P}}|^2 = \frac{1}{2}m\big[\omega^2 r^2 + \dot{\phi}^2 l^2 + l^2(\omega + \dot{\theta})^2\cos^2(\phi) +$$

$$-2\omega r l\dot{\phi}\sin(\phi)\sin(\theta) + 2\omega r l(\omega + \dot{\theta})\cos(\phi)\cos(\theta)\big].$$

The equations of motion can be easily obtained by resorting to Lagrange equations. They are clearly nonlinear, due to the presence of trigonometric functions of the generalized coordinates. With simple computations, the following equations of motion are obtained:

$$\left\{ \begin{array}{l} l\ddot{\theta}\cos^2(\phi) - 2l(\omega + \dot{\theta})\dot{\phi}\cos(\phi)\sin(\phi) + \omega^2 r\cos(\phi)\sin(\theta) = 0 \\ l\ddot{\phi} + l(\omega + \dot{\theta})^2\cos(\phi)\sin(\phi) + \omega^2 r\sin(\phi)\cos(\theta) = 0. \end{array} \right.$$

The equations of motion can be linearized in the study of the small oscillations of the pendulum about the static equilibrium position

$$\left\{ \begin{array}{l} l\ddot{\theta} + \omega^2 r\theta = 0 \\ l\ddot{\phi} + \omega^2(r + l)\phi = 0. \end{array} \right.$$

The linearized equations can also be obtained directly from an expression of the kinetic energy truncated after quadratic terms. By introducing the series for the sine and cosine or the generalized coordinates in which terms of order greater than two are discarded, the kinetic energy can be written as

$$T = \frac{1}{2}m\left[\omega^2(r+l)^2 + \dot{\phi}^2l^2 + \dot{\theta}^2l^2 - \omega^2l(r+l)\phi^2 - \omega^2rl\theta^2 + 2\omega l(r+l)\dot{\theta}\right].$$

Remembering equation (1.8), the expression of the kinetic energy can be subdivided into three terms:

$$T_0 = m\left[\omega^2(r+l)^2 - \omega^2l(r+l)\phi^2 - \omega^2rl\theta^2\right]/2$$

is independent from the generalized velocities. Its role is similar to that of the potential energy and is usually referred to as *dynamic potential*. Apart from a constant term, whose derivatives are nil, it yields the so-called geometric stiffness terms in the equation of motion. As usual in rotating systems, they constitute a centrifugal stiffening and are proportional to the square of the spin speed.

$$T_1 = m\omega l(r+l)\dot{\theta}$$

is linear in the generalized velocities. However, this term is independent of the displacements, and its derivatives in the equations of motion are nil: There is no gyroscopic term in the equations of the rotating pendulum.

$$T_2 = m\left[\dot{\phi}^2l^2 + \dot{\theta}^2l^2\right]/2$$

is quadratic in the generalized velocities and yields the inertial terms of the equations of motion, which are also present in natural systems.

Because the linearized equations of motion are decoupled, the motion in the rotation plane xy is uncoupled, within the validity of the linearization of the equations of motion, from the motion in axial direction z. The former is the motion of a pendulum whose length is l within a constant force field whose acceleration is $r\omega^2$, and the latter is the motion of the same pendulum within a constant force field whose acceleration is $(r+l)\omega^2$. The natural frequencies of the motions outside and within the rotation planes are, respectively,

$$\lambda_1 = \omega\sqrt{1+\frac{r}{l}}, \qquad \lambda_2 = \omega\sqrt{\frac{r}{l}}.$$

The Campbell diagram, shown in Figure 4.1b for the case in which $r = l$, is then made of two straight lines. A simple explanation of the different behaviour of the system in a plane containing the axis of rotation and a plane perpendicular to it (xy-plane) is shown in Figures 4.1c and d. In the former, the restoring force acting on the pendulum in a direction perpendicular to line PC is

$$F_c \sin(\phi) \approx m\omega^2 (r + l)\phi.$$

In the xy-plane the restoring force is

$$F_c \sin(\theta - \alpha) \approx m\omega^2 (r + l)(\theta - \alpha).$$

As $\alpha \approx \theta l/(r + l)$, the restoring force is $\approx m\omega^2 r\theta$.

The analysis reported here shows a simple case in which the natural frequency of the system depends on the angular velocity. It also shows how the cylindrical symmetry of the centrifugal field induces a sort of anisotropy into the system: The behaviour in the xy-plane is different from that in any plane containing the axis of rotation. The rotating pendulum also has some important technological applications and will be studied again when dealing with damping of torsional vibration.

4.1.2 Forced vibrations of rotors: Critical speeds

Often rotors are subjected to forces that vary in time, and sometimes their time history is harmonic. This is the case, for example, of forces due to the unbalance of the rotor itself, which can be described as a vector rotating with the same angular speed as the rotor and whose components in the fixed reference frame vary harmonically in time with circular frequency equal to the rotational speed ω. In other cases the time history is less regular but, if it is periodical, it can always be represented as the sum of harmonic components. In the case of rotors there are also, however, instances in which the forcing functions can be described only in statistical terms.

In the first two cases the frequency of the forcing function or of its harmonic components is often linked with the spin speed of the rotor and can be plotted on the Campbell diagram. In the case of the excitation due to unbalance, for example, the forcing frequency can be represented on the ($\lambda\omega$)-plane by the straight line $\lambda = \omega$, i.e., by the bisector of the first quadrant. In this case, the excitation is said to be *synchronous*. The relationship linking the frequency of the forcing function to the spin speed is often of simple proportionality and can be represented on the Campbell diagram by a straight line

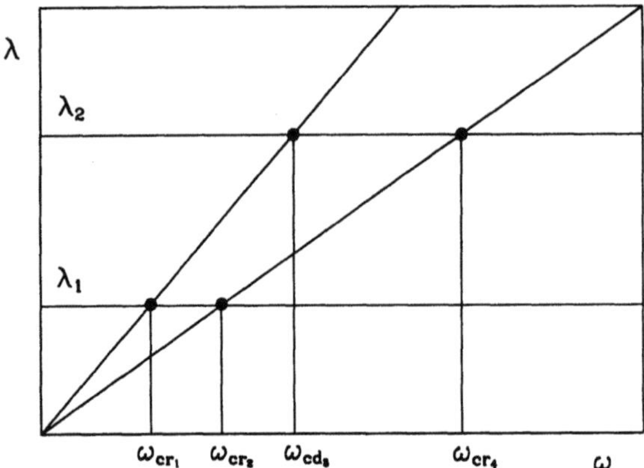

FIGURE 4.2. Graphical representation of the critical speeds in the case in which the forcing frequencies are proportional to ω while the natural frequencies are constant.

through the origin.

The spin speeds at which one of the forcing functions has a frequency coinciding with one of the natural frequencies of the system are usually referred to as *critical speeds* and can be identified on the Campbell diagram by the intersections of the curves related to the natural frequencies with those related to the forcing frequencies. A case in which the natural frequencies are independent of the speed and the forcing frequencies are proportional to ω is reported in Figure 4.2.

Not all the intersections on the Campbell diagram are equally dangerous. If the frequency of a forcing function coincides with the natural frequency of a mode that is completely uncoupled with it (or, better, if the modal force corresponding to the forcing function and the resonant mode is vanishingly small), no resonance actually occurs. For example, if the frequency of the driving torque (i.e., of the torsional moment on the rotor) is coincident with a flexural natural frequency of the rotor and torsional and flexural behaviour are completely uncoupled, no resonance takes place. In other cases the resonance can be very weak and the damping of the system can be sufficient to avoid any measurable effect.

There are, however, cases in which a very strong resonance takes place and the rotor cannot operate at or near a critical speed without incurring strong vibrations and even a catastrophic failure. In particular, the resonances due to the coincidence of one of the flexural natural frequencies with the spin speed are particularly dangerous. They can be detected on the Campbell diagram by the intersection of the curves related to the natural frequencies with the straight line

$\lambda = \omega$. They are usually referred to as *flexural critical speeds*, without further indications, while other critical speeds related to bending behaviour, which are usually less dangerous, are often said to be *secondary critical speeds*.

When a rotor operates at a critical speed, with the meaning just given, the amplitude of the vibration grows linearly in time and only the damping of the stator and the supports (as will be shown later, the damping of the rotor is completely ineffective in this case) and the unavoidable nonlinearities, which show up when the amplitude grows, can prevent the failure of the rotor. Actually, it is possible to design rotating machinery in such a way that operation at a critical speed is possible for a limited period of time, but it is, at any rate, necessary that the normal operating range be either below or above the critical speeds and that sustained operation at critical speed be avoided. An example of a machine operating above the first critical speed is that of the common domestic washing machine: In the transition between the washing and the spinning modes the crossing of a critical speed is generally easily detected from strong vibrations, even without instruments.

The speed range spanning from zero to the first critical speed is usually referred to as the *subcritical range*; above the first critical speed the *supercritical range* starts. A growing number of machines work in the supercritical range, and then at least one of the critical speeds must be crossed during start-up and shut-down procedures.

If the Campbell diagram related to flexural vibrations is made by straight lines parallel to the ω-axis, i.e., if the natural frequencies are independent of the speed, the numerical values of the critical speeds coincide with those of the natural frequencies at standstill, as can be seen from Figure 4.2. Some confusion between the concepts of critical speed and natural frequency that can still be found can probably be ascribed to this. Even if the numerical values are coincident, the two physical phenomena are different, particularly where the stressing of the rotor is concerned.

The force due to unbalance, which can be expressed in a stationary reference frame as a vector rotating with angular velocity ω, has a constant direction if seen in a reference frame fixed to the rotor. If the machine is axially symmetrical, as will be seen in the following sections, at the critical speed the rotor rotates in a deflected configuration but does not vibrate, in the sense that there is no cyclic stressing. The *flexural critical speed* can, in fact, be defined as the speed at which the centrifugal forces due to the bending of the rotor are in indifferent equilibrium with the elastic restoring forces, and from this point of view the situation is more similar to that characterizing elastic instability than that typical of vibratory phenomena. A rotor operating at a critical speed is then not subject to vibrations of any type but is a source of periodic excitations that can cause vibrations, often very strong, in the nonrotating parts of the machine.

In addition to flexural critical speeds, torsional critical speeds can also be very dangerous, particularly in the case of reciprocating machinery. Many devices whose aim is to reduce the amplitude of the vibrations induced by critical speeds have been developed.

The very concept of critical speed has been defined with reference to a linear system, and it is impossible to define critical speeds in this sense in the case of nonlinear rotors. However, a more general definition of critical speed, as a spin speed in which strong vibrations are encountered, is often used. This definition, which also holds in the case of nonlinear rotors, has a certain degree of arbitrarity, because the amplitude of the vibration depends on the cause that produces it. In the case of nonlinear rotors the speed at which the maximum amplitude is reached, i.e., the critical speed following the last definition, also depends on the entity of the exciting causes (for example, the unbalance in the case of flexural critical speeds). The critical speeds of linear systems are, on the contrary, characteristics of the system and are independent of the excitations.

4.2   Fields of instability

Rotors can develop an unstable behaviour in well-defined velocity ranges. The velocities at which this unstable behaviour occurs must not, however, be confused with the critical speeds of the rotor because the two phenomena are completely different. The term *unstable* can have several meanings, and different definitions of stability exist, one of the most common being that introduced by Liapunov and reported in Section 3.5.3.

However, this theoretical definition of stability can be difficult to apply in many engineering situations, and a technical definition of stability can be used: The behaviour of a machine is considered *stable*

when the amplitude of vibration in normal operation does not exceed
a value considered acceptable. For rotating machinery, was stated by
A. Muszynska:[1]

> a rotating machine is stable if its rotor performs a pure
> rotational motion around an appropriate axis at a re-
> quired rotational speed and this motion is not accom-
> panied by other modes of vibrations of the rotor, its
> elements or other stationary parts of the machine, or,
> if such vibrations take place, their amplitudes do not
> exceed admitted, acceptable values. The stable rotating
> machine is immune to external perturbing forces, i.e.,
> any random perturbation cannot drastically change its
> behaviour. Such a perturbation causes only a transient
> decaying process leading to a previous regime of perfor-
> mance, or to a new one, which is included in the accept-
> able limits.

When studying the behaviour of damped linear systems, the amplitude of free vibration was seen to decay exponentially in time,
due to the energy dissipation due to damping. In the case of rotors,
however, there is a source of energy, the centrifugal field, that may
in some cases cause an unbounded growth in time of the amplitude
of free vibrations. The frequency ranges in which this growth occurs,
i.e., in which self-excited vibrations can develop, are usually called
instability fields or *instability ranges*, and the speed at which the first
such field starts is the *threshold of instability*.

Instability ranges must not be confused with critical speeds: Critical speeds are a sort of resonance between a natural frequency and a
forcing function acting on the rotor, while in instability ranges true
self-excited vibrations occur. They need the presence of some source
of energy to sustain the vibration with increasing amplitude, and
in this case the energy can be supplied by the kinetic energy linked
with rotation at the spin speed ω. It is easy to verify that the kinetic
energy stored in the rotor is greater by some orders of magnitude
than the elastic potential energy the rotor can store without failure.

Consider, for example, a thin ring with radius r and material density ρ rotating at speed ω. As the hoop stress in the ring is simply
$\sigma_h = \rho\omega^2 r^2$, the relationship linking the kinetic energy with the

---

[1] A. Muszynska, *Rotor Instability*, Senior Mechanical Engineering Seminar,
Carson City, June 1984.

stress is $T = \frac{1}{2}mr^2\omega^2 = \frac{1}{2}m\sigma_c/\rho$.

The maximum potential energy the ring can store in an axisymmetrical tensile deformation is $\mathcal{U} = \frac{1}{2}V\sigma_U^2/E = \frac{1}{2}m\sigma_U^2/E\rho$, where V and σ_U are the material volume and the ultimate strength, respectively. The ratio between the kinetic energy and the potential energy corresponding to a deformation causing the failure of the ring (at failure $\sigma_h = \sigma_U$) is $T/\mathcal{U} = E/\sigma_U$. By introducing the expression of the hoop stress into the last equation, it is easy to see that ratio E/σ_U is nothing other than the ratio v_s/V_U between the speed of sound in the material and the peripheral velocity at failure of the ring, whose value is usually far greater than 10. The kinetic energy stored in the rotor is then at least one or two orders of magnitude greater than the energy needed to deform the rotor until failure occurs. Similar considerations would hold for other geometrical configurations or deformation patterns. It is then sufficient that a small portion of the kinetic energy of the rotor is transformed into deformation potential energy to cause failure. This situation is typical of structural elements that are in close contact with an energy source that can excite and sustain vibration, another case being that of aeroelastic vibrations.

It is easy, at least from a theoretical point of view, to predict the onset of unstable working conditions in a linear system. If the time history of the system is expressed in the form

$$\mathbf{x} = \mathbf{x}_0 e^{i\lambda t}, \tag{4.2}$$

the sign of the decay rate, i.e., of the imaginary part of the complex frequency λ gives directly the stability condition: If it is positive the amplitude decays exponentially in time (stable condition), while unstable operation is characterized by the negative imaginary part of the complex frequency. The plot of the decay rate as a function of the spin speed must then be drawn with the Campbell diagram. At vanishingly small speeds all decay rates are obviously positive, because there is no external source of energy that can excite vibration. With increasing speed the decay rate of some modes can decrease, showing a reduction of stability. If at a certain value of the spin speed ω one of them vanishes and then becomes negative, that speed is the threshold of instability of the system.

It is clear that when the time history is written in the form

$$\mathbf{x} = \mathbf{x}_0 e^{st} , \qquad (4.3)$$

the system is stable if the real parts of all complex eigenvalues are negative.

Actually, it can be very difficult to evaluate the decay rate with enough precision, because it is influenced by many factors that are very difficult to evaluate, one of them being damping. Often, it is only possible to perform a first-approximation theoretical or numerical study, whose results must be verified experimentally.

As a general rule, if the conditions for uncoupling of flexural, axial and torsional behaviour are met, only the first can give way to self-excited vibration. There are many mechanisms that can cause the establishment of unstable conditions, including internal rotor damping due to material damping and friction between the various components assembled by bolting, riveting, shrink fitting, and so on; rubbing between stator and rotor; and fluid forces in journal bearings and seals. All the mentioned mechanisms are potentially dangerous, but they do not necessarily always cause instability. The better known unstabilizing effects are those due to the material damping in the rotor and lubricated journal bearings. The latter can cause the well-known oil whip phenomenon, which consists of very strong vibrations starting from a speed that is, in many cases, almost twice the first flexural critical speed.

To make it easier to distinguish between critical speeds and fields of instability, the following features can be listed:

Critical speeds

- They occur at well-defined values of the spin speed.
- The amplitude grows linearly in time if no damping is present. It can be maintained within reasonable limits and, as a consequence, a critical speed can be passed.
- The value of the speed is fixed, but that of the maximum amplitude depends on the amplitude of the perturbation that causes it. In particular, the main flexural critical speeds do not depend on the amount of unbalance, but the amplitude increases with increasing unbalance.

Fields of instability

- Their span is usually quite large. Often all speeds in excess of the threshold of instability give way to unstable behaviour.

- The threshold of instability, if it exists, is usually located in the supercritical range.

- The amplitude grows exponentially in time. It grows in an uncontrollable way, and then working above the threshold of instability is impossible. When it falls within the working range, the system must be modified to raise it well above the maximum operating speed. Only possible nonlinearities of the system can maintain the amplitude within a limit, giving way to a limit cycle.

4.3   The linear Jeffcott rotor

4.3.1   Equations of motion of the undamped system

The simplest model that can be used to study the flexural behaviour of rotors is the so-called *Jeffcott rotor*, consisting of a point mass attached to a massless shaft. The only force acting on the mass m is that due to the elastic restoring force of the shaft. The weight of the rotor is then neglected, but this does not detract much from the validity of the model. The stiffness k providing the restoring force can be considered the stiffness of the shaft, the supporting structure, or a combination of the two. The two schemes sketched in Figure 4.3 yield the same results, as long as the system is undamped and axially symmetrical.

Point P, in which mass m is fixed, is always contained in the xy–plane. This statement is justified by the uncoupling between axial and radial motions and relies on the small-displacement assumptions that are at the base of linear structural analysis. In the study of the flexural behaviour a model with two degrees of freedom can then be used. With x and y being the coordinates of point P at the generic time t, the equations of motion of mass m are simply

$$\begin{cases} m\ddot{x} + kx = 0 \\ m\ddot{y} + ky = 0 \, . \end{cases} \tag{4.4}$$

The equations of motion along each axis are coincident with the equation of the free motion of a system with a single degree of freedom and their solution is a harmonic motion with frequency $\lambda_n = \sqrt{k/m}$. The motion of point P can thus be thought of as the combination of two harmonic motions taking place along axes x and

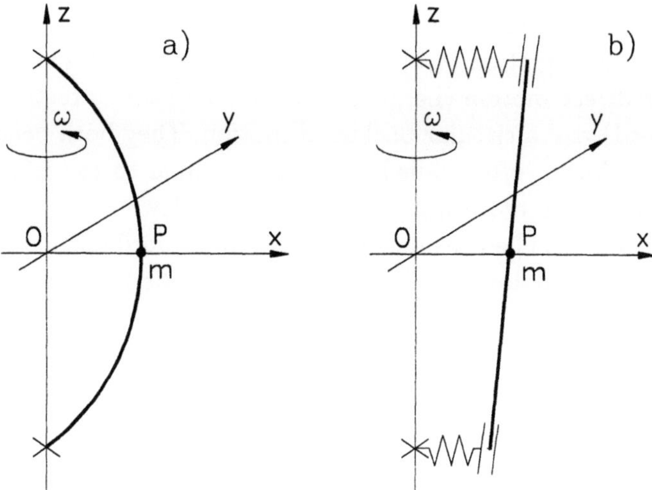

FIGURE 4.3. Sketch of a Jeffcott rotor. The model is sketched in its deformed configuration at the time in which point P crosses the xz–plane; (a) flexible shaft on stiff supports, (b) stiff shaft on compliant supports.

y with the same frequency λ, coinciding with the natural frequency of the nonrotating shaft. They can add to each other giving way to a trajectory of point P that can be circular, elliptical, or straight in any direction in the xy-plane, depending on the initial conditions.

The same result could be obtained using the complex coordinate

$$z = x + iy \tag{4.5}$$

in the xy-plane. By multiplying the second equation (4.4) by the imaginary unit i and adding the two equations, it follows that

$$m\ddot{z} + kz = 0\,. \tag{4.6}$$

The solution of this homogeneous differential equation is $z = z_0 e^{i\lambda t}$, where z_0 is, generally speaking, a complex number $z_0 = x_0 + iy_0$. By introducing this solution into equation (4.6), the latter yields a homogeneous algebraic equation that has solutions other than the trivial solution $z_0 = 0$ only if $\lambda = \pm\sqrt{k/m}$.

The general solution of equation (4.6) is then

$$z = Z_1 e^{i\sqrt{\frac{k}{m}}t} + Z_2 e^{-i\sqrt{\frac{k}{m}}t}\,. \tag{4.7}$$

The physical meaning of equation (4.7) is obvious: z is a vector that rotates in the xy-plane with angular velocity λ. If the amplitude z_0 is real, point P crosses the x-axis at time $t = 0$. The motion expressed by equation (4.7) is the superimposition of a circular forward or direct motion (i.e., occurring in the same direction as the spin speed) and a circular backward motion. They both occur at an angular velocity, often called *whirl speed*, equal to the natural frequency of the nonrotating system. In the following study, the spin speed will always be considered positive: A forward motion will then be characterized by positive whirl speed λ (first quadrant of the Campbell diagram), while a backward motion is characterized by a negative value of λ (fourth quadrant of the Campbell diagram). The result of the superimposition of the two motions depends on the initial conditions, i.e., on the values of complex constants Z_1 and Z_2. If, for example, Z_2 is equal to 0 a circular forward whirling occurs, while if the two constants are equal and real a harmonic vibration along the x-axis takes place.

The result obtained obviously does not depend on which solution is used, (either that based on the study of the motion in the xz- and yz-planes (equation 4.5) or that based on the use of complex coordinates (equation 4.6)). However, they enlighten different aspects of the phenomenon, and their results are not exactly equivalent. In the first, λ is the frequency of two harmonic motions in two planes and its sign has no physical meaning. Vector $e^{i\lambda t}$, which can be used to express the motion in the form $x = x_0 e^{i\lambda t}$, $y = y_0 e^{i\lambda t}$, is a rotating vector in the Argand plane and only the real components of $x_0 e^{i\lambda t}$ and $y_0 e^{i\lambda t}$ have a physical meaning.

In the solution of equation (4.6), λ is, however, a true angular velocity, and the relevant vector rotates in the physical space and not in the Argand plane. The deflected shape rotates about the undeformed configuration with angular velocity λ, although no rotation of a material object with that speed takes place. As a consequence, its sign states the direction of rotation of the deflected configuration.

At any rate, the natural frequency of the rotor or the whirl speed does not depend on the spin speed ω: The Campbell diagram of a Jeffcott rotor is then made of straight lines, as shown in Figure 4.2. The flexural critical speed, defined as the speed at which the natural frequency of the system is coincident with the frequency of rotation, coincides with the natural frequency of the nonrotating system:

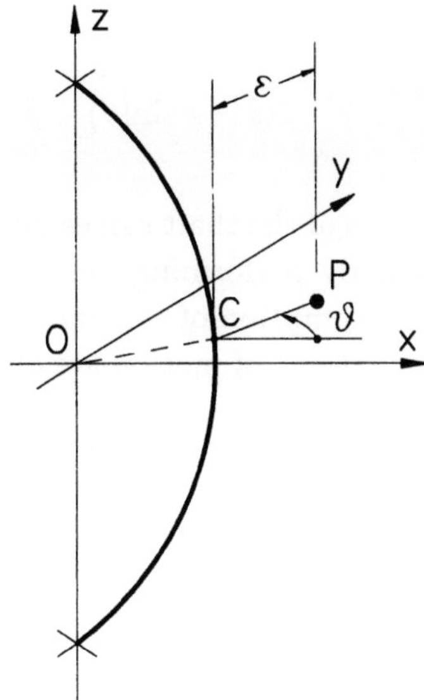

FIGURE 4.4. Unbalanced Jeffcott rotor sketched at time t.

$$\omega_{cr} = \sqrt{\frac{k}{m}}. \tag{4.8}$$

In real life, the center of gravity P of mass m cannot be exactly coincident with the center C of the cross section of the shaft. Although the distance between the two points can be small, the presence of the eccentricity ϵ (Figure 4.4) causes a static unbalance $m\epsilon$ that can strongly affect the behaviour of the system.

If the angle between line CP and the x-axis is imposed by the driving system, i.e., function $\theta(t)$ can be regarded as a known function of time, the number of degrees of freedom of the system is again equal to 2. Assuming the displacements x and y of point C as generalized coordinates, the position and velocity of point P and the Lagrangian function are, respectively,

$$(\overline{P\text{-}O}) = \left\{ \begin{array}{c} x_P \\ y_P \end{array} \right\} = \left\{ \begin{array}{c} x + \epsilon\cos(\theta) \\ y + \epsilon\sin(\theta) \end{array} \right\}$$

$$V_P = \left\{ \begin{array}{c} \dot{x}_P \\ \dot{y}_P \end{array} \right\} = \left\{ \begin{array}{c} \dot{x} - \epsilon\dot{\theta}\sin(\theta) \\ \dot{y} + \epsilon\dot{\theta}\cos(\theta) \end{array} \right\}$$

$$T - U = \frac{1}{2}m\left\{ \dot{x}^2 + \dot{y}^2 + \epsilon^2\dot{\theta}^2 + 2\epsilon\dot{\theta}\left[-\dot{x}\sin(\theta) + \dot{y}\cos(\theta) \right] \right\} - \frac{1}{2}k\left(x^2 + y^2 \right).$$
$$(4.9)$$

The equation of motion of the shaft can easily be obtained through the Lagrange equation. By performing the relevant derivatives and remembering that no assumption of constant spin speed $\dot{\theta}$ has been made, the following equations of motion are obtained:

$$\begin{cases} m\left[\ddot{x} - \epsilon\dot{\theta}^2\cos(\theta) - \epsilon\ddot{\theta}\sin(\theta) \right] + kx = 0 \\ m\left[\ddot{y} - \epsilon\dot{\theta}^2\sin(\theta) + \epsilon\ddot{\theta}\cos(\theta) \right] + ky = 0. \end{cases} \qquad (4.10)$$

By multiplying the second equation of motion (4.10) by the imaginary unit i and adding the two equations, the following equation of motion in terms of the complex coordinate z is readily obtained

$$m\ddot{z} + kz = m\epsilon\left(\dot{\theta}^2 - i\ddot{\theta} \right)e^{i\theta}. \qquad (4.11)$$

The homogeneous equation is the same equation (4.6) already studied and does not contain the angular velocity of the rotor.

4.3.2 Unbalance response at constant speed

If the angular velocity of the rotor $\omega = \dot{\theta}$ is assumed to be constant, angle θ can be expressed as ωt, and equation (4.11) reduces to

$$m\ddot{z} + kz = m\epsilon\omega^2 e^{i\omega t}, \qquad (4.12)$$

whose particular integral is simply $z = z_0 e^{i\omega t}$. By introducing it into equation (4.12) the latter transforms into the algebraic equation

$$(-m\omega^2 + k)z_0 = m\epsilon\omega^2,$$

which yields

$$z_0 = \epsilon\frac{\omega^2}{\omega_{cr}^2 - \omega^2}, \qquad (4.13)$$

where ω_{cr} is the critical speed defined by equation (4.8). The value of the amplitude z_0 so obtained is real and, as a consequence, vector $(\overline{\text{C-O}})$, i.e., z, rotates with velocity ω in the xy-plane remaining in line with vector $(\overline{\text{P-C}})$. The value of ω that causes the denominator of the

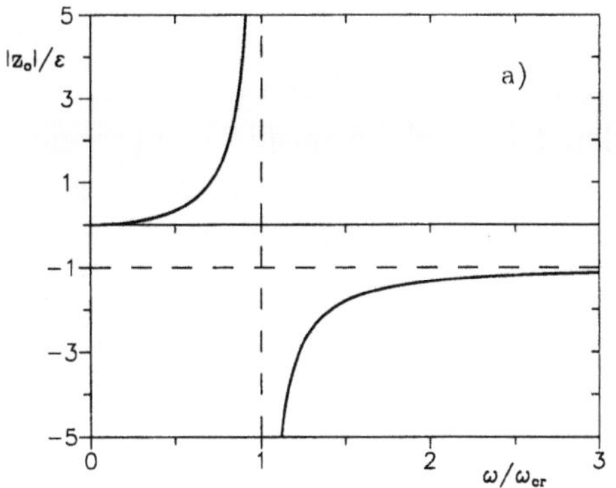
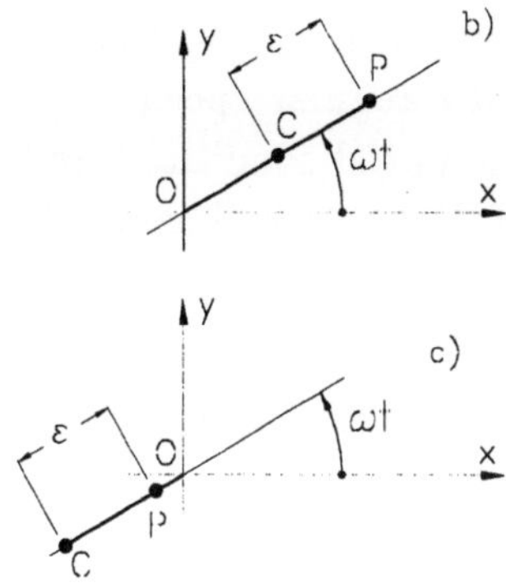

FIGURE 4.5. (a) Unbalance response of an undamped Jeffcott rotor. Nondimensional amplitude as a function of the nondimensional spin speed. Configuration of the system in (b) subcritical conditions and (c) supercritical conditions.

expression for z_0 in equation (4.13) to vanish, i.e., which causes the amplitude to reach an infinite value, is coincident with the flexural critical speed of the rotor.

The amplitude of motion of point C due to the presence of the unbalance $m\epsilon$, i.e., the unbalance response, is reported as a function of the speed in the nondimensional plot of Figure 4.5a. In the subcritical range, the amplitude grows from zero to a value tending to infinity at the critical speed, always remaining positive. In the supercritical range, however, the value of the amplitude z_0 is negative, and its absolute value decreases monotonically with the speed. When the speed tends to infinity the amplitude tends to $-\epsilon$. The sign of the solution determines the equilibrium configurations as shown in Figures 4.5b and c. When the solution is positive, in the subcritical field, points O, C, and P are aligned in the mentioned order and the center of gravity of the rotor lies outside the deformed configuration of the shaft. In the supercritical field, however, point P lies between points C and O, and when the speed tends to infinity the amplitude z_0 tends to $-\epsilon$, or point P tends to point O. This phenomenon is usually referred to as *self-centering* because the rotor tends to rotate about its center of mass instead of its geometrical center. When the behaviour is controlled by the stiffness, rotation takes place about a point near the geometrical center, while when it is dominated by the inertia it occurs about a point near the mass center.

The motion of point C can then be expressed as the superimposition of a free motion that can be circular, elliptical, or even rectilinear occurring with frequency $\lambda = \sqrt{k/m}$ and a circular motion with angular speed ω. The amplitude of the first is controlled by the initial conditions and is independent of the speed. The presence of damping will cause the free motion to decay in time or, in some cases, to increase in time as will be shown in the following section. The amplitude of the second is strongly influenced by the speed and is constant in time. The unbalance response has been computed assuming constant spin speed: Figure 4.5 gives the amplitude of the motion at the various speeds and cannot be used to describe the time history of the amplitude during the acceleration of the rotor.

The Jeffcott rotor is obviously an oversimplification of real-world rotors, but nevertheless it allows an understanding and modeling, at least qualitative, of some of the most important phenomena typical of rotor dynamics, the presence of critical speeds and self-centering. In the next sections damping will be added to the model to extend its applicability and the stability of the self-centered configuration will be studied. As a last remark, it must be clearly stated that the coincidence of the critical speed, computed from the unbalance response or from the free behaviour of the system, with the natural frequency of the undamped system is a peculiar characteristic of the Jeffcott model or, more generally, of all those rotors in which the natural frequency does not depend on the speed and must not be considered a general feature.

4.3.3   System with viscous damping: Free behaviour

When considering a damped rotor it is very important to distinguish between the damping effects that can be associated to the stationary parts of the machine, usually referred to as nonrotating damping, and those directly associated with the rotor, or rotating damping. The former usually has a stabilizing effect that the designer can use to achieve the required stability in the whole working range of the machine. The latter, on the contrary, can usually reduce the amplitude of vibration in subcritical conditions but shows unstabilizing effects in the supercritical range. Designers must then be very careful when studying machines operating in the supercritical range, taking into account that all mechanisms increasing energy dissipation within the rotor, such as material damping; friction in threaded, riveted, shrink-fitted connections in built-up rotors; splined shafts;

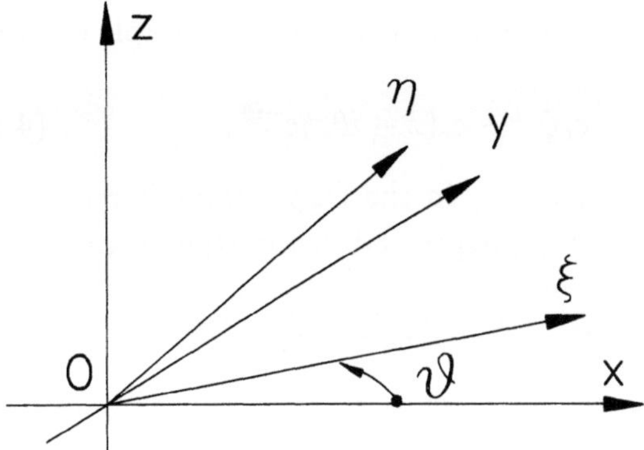

FIGURE 4.6. Reference frames $Oxyz$ and $O\xi\eta z$ at time t.

intershaft dampers in multishaft machines; and so on; can cause severe instability problems.

The model of Figure 4.3 can be extended to the damped system simply by adding to the right-hand side of equation (4.11) the generalized forces due to damping. Because the generalized coordinates chosen are simply the displacements in the x- and y-directions of point C, such forces are the components in the directions of the x- and y-axes of damping forces applied by the shaft to the point mass in point C. If nonrotating damping is considered, assuming a viscous damping model and using the complex notation, the force is

$$\vec{F}_n = F_{n_x} + iF_{n_y} = -c_n\dot{x} - ic_n\dot{y} = -c_n\dot{z}, \qquad (4.14)$$

where c_n is the nonrotating damping coefficient.

For the study of rotating damping, a rotating reference frame $O\xi\eta z$ (Figure 4.6) must be introduced. The origin and z-axis of the rotating frame are the same as those of the fixed reference frame $Oxyz$ of Figure 4.3, but axes ξ and η rotate in the xy-plane at the same speed of the rotor. When the rotational speed is constant, the angle between the two reference frames is simply given by ωt. Let a complex coordinate ζ be defined in the $\xi\eta$-plane

$$\zeta = \xi + i\eta = ze^{-i\theta}. \qquad (4.15)$$

From equation (4.15), the derivative of the complex coordinate ζ is readily obtained:

$$\dot{\zeta} = (\dot{z} - i\dot{\theta}z)e^{-i\theta}.$$

The force due to a rotating viscous damping, with damping coefficient c_r, can be expressed in the $O\xi\eta z$ frame by the simple equation

$$\vec{F}_{r_{\xi\eta}} = -c_r\dot{\zeta} = -c_r(\dot{z} - i\dot{\theta}z)e^{-i\theta}\,. \tag{4.16}$$

The expression of the force $F_{r_{xy}}$ in the $Oxyz$ frame is readily obtained from that of vector $F_{r_{\xi\eta}}$, expressed in the $O\xi\eta z$ frame

$$\vec{F}_{r_{xy}} = \vec{F}_{r_{\xi\eta}}e^{i\theta} = -c_r(\dot{z} - i\dot{\theta}z)\,. \tag{4.17}$$

Introducing the expressions (4.14) and (4.17) for the forces due to nonrotating and rotating damping on the right-hand side of the equation of motion (4.11), it follows that

$$m\ddot{z} + (c_r + c_n)\dot{z} + (k - ic_r\dot{\theta})z = m\epsilon(\dot{\theta}^2 - i\ddot{\theta})e^{i\theta}\,. \tag{4.18}$$

If the spin speed is constant, equation (4.18) is easily modified by neglecting the term in $\ddot{\theta}$. In the latter case, the equation of motion is formally identical to the equation of motion of a mass m suspended on a spring with complex stiffness $k - i\omega c_r$ and a viscous damper with damping coefficient $c_n + c_r$ on which a force with harmonic time history with frequency $\omega = \dot{\theta}$ and amplitude $m\epsilon\omega^2$ is acting.

Equation (4.18) can be solved, as usual, by adding a particular integral to the complementary function. The solution of the homogeneous equation yielding the behaviour of a perfectly balanced rotor is the usual one, $z = z_0 e^{i\lambda t}$, where both the amplitude z_0 and the frequency λ are expressed by complex numbers. By introducing this solution into equation (4.18), the following characteristic equation is obtained

$$m\lambda^2 + i(c_r + c_n)\lambda + k - i\omega c_r = 0\,. \tag{4.19}$$

The roots of this quadratic equation with complex coefficients are

$$\lambda = i\frac{c_r + c_n}{2m} \pm \sqrt{\frac{-(c_r + c_n)^2 + 4m(k - i\omega c_r)}{4m^2}}\,. \tag{4.20}$$

The real and imaginary parts of the complex frequency can be easily separated by resorting to the well-known formula

$$\sqrt{a \pm ib} = \sqrt{\frac{\sqrt{a^2 + b^2} + a}{2}} \pm i\sqrt{\frac{\sqrt{a^2 + b^2} - a}{2}}\,, \tag{4.21}$$

which yields

$$\lambda = \pm \frac{1}{\sqrt{2}} \sqrt{\Gamma + \sqrt{\Gamma^2 + \left(\frac{\omega c_r}{m}\right)^2}} + i\left[\frac{c_r + c_n}{2m} \mp \frac{1}{\sqrt{2}}\sqrt{-\Gamma + \sqrt{\Gamma^2 + \left(\frac{\omega c_r}{m}\right)^2}}\right]$$

$$(4.22)$$

where

$$\Gamma = \frac{k}{m} - \frac{(c_r + c_n)^2}{4m^2}. \tag{4.23}$$

Two values of the complex whirl frequency or whirl speed λ can be found for each value of the spin speed ω. They are not conjugate, because equation (4.19) has complex coefficients. Without loss of generality, because it is sufficient to assume that at time $t = 0$ point C crosses the x-axis, the amplitude z_0 can be assumed to be real. Separating the real part λ_R of the complex frequency λ from the imaginary part λ_I, the time history can then be written as

$$\begin{cases} x = z_0 e^{-\lambda_I t} \cos(\lambda_R t) \\ y = z_0 e^{-\lambda_I t} \sin(\lambda_R t). \end{cases} \tag{4.24}$$

A logarithmic spiral is then obtained as the result of two damped or amplified harmonic motions. The real part of λ has the meaning of a true angular velocity: It is the angular velocity at which the deflected shape rotates about the undeformed configuration. The imaginary part is a decay rate: If it is positive the amplitude decays in time and point C tends to point O. The rotor has a stable behaviour as whirl motion tends to reduce its amplitude. If, however, it is negative, the amplitude grows exponentially in time. The motion is unstable, as any small perturbation can trigger this self-excited whirling.

The first of the two values of λ obtained in equation (4.22) (the one with the upper signs) has a positive real part and an imaginary part that can be positive or negative. It corresponds to a forward whirl mode, which can be either damped or self-excited, depending on the sign of λ_I. With simple computations, the condition for stability can be shown to be

$$\omega < \sqrt{\frac{k}{m}\left(1 + \frac{c_n}{c_r}\right)}. \tag{4.25}$$

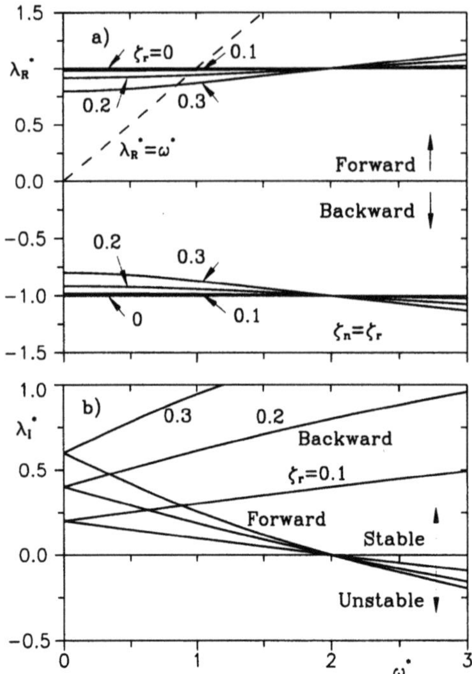

FIGURE 4.7. (a) Real and (b) imaginary parts of the complex whirl frequency as functions of the spin speed in nondimensional form. The curves for four different values of damping have been plotted. Rotating and nonrotating damping have been assumed to be equal: The threshold of instability is then twice the critical speed.

The rotor is then stable in the subcritical range, i.e., when the speed is lower than the critical speed $\sqrt{k/m}$ of the undamped system. In the supercritical field, stability depends on the value of ratio c_n/c_r between the nonrotating and the rotating damping. If there is no nonrotating damping, the motion is unstable in the whole supercritical range. Increasing nonrotating damping, the threshold of instability becomes higher.

The second value of λ (lower signs in equation (4.22)) has a negative real part and a positive imaginary part. It corresponds to a damped backward whirl mode that usually damps out quite fast and has very little practical interest. Equation (4.22) can be rewritten in the following nondimensional form

$$\begin{cases} \lambda_R^* = \pm\sqrt{\Gamma^* + \sqrt{\Gamma^{*2} + \omega^{*2}\zeta_r^2}} \\ \lambda_I^* = \zeta_r + \zeta_n \mp \sqrt{-\Gamma^* + \sqrt{\Gamma^{*2} + \omega^{*2}\zeta_r^2}}\,, \end{cases} \tag{4.26}$$

where $\lambda^* = \lambda/\sqrt{k/m}$, $\omega^* = \omega/\sqrt{k/m}$ $\zeta_r = c_r/2\sqrt{km}$, $\zeta_n = c_n/2\sqrt{km}$ and $\Gamma^* = [1 - (\zeta_n + \zeta_r)^2]/2$.

The nondimensional whirl frequency λ^* is then a function of the

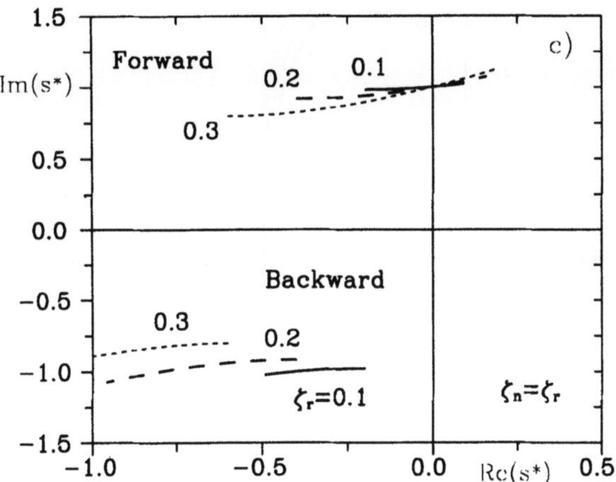

FIGURE 4.8. Nondimensional roots locus for the same system as Figure 4.7

spin speed ω^* and of only two parameters ζ_n and ζ_r. The real and imaginary parts of λ^* are plotted as functions of the spin speed in Figure 4.7. Four values of ζ_r have been considered, and rotating and nonrotating damping have been assumed to be equal ($\zeta_n = \zeta_r$). It is clear that the real part of λ is little affected by the presence of damping: A value of ζ_n equal to 0.1 is already quite high and, in most cases, the curve $\lambda_R(\omega)$ cannot be distinguished from that of the undamped case. The first and fourth quadrants of the (λ_R,ω)-plane have been shown in the Campbell diagram. When using the complex notation, the sign of λ_R has an important physical meaning and it is advisable to plot at least two quadrants of the λ_R,ω-plane: the first and second, or, as in the figure, the first and fourth.

Another way of representing the real and imaginary parts of the complex whirl frequency is the roots locus (Figure 4.8), in which the imaginary parts of eigenvalues s are plotted as functions of the real parts. As usual for root loci, a solution of the type of equation (4.3) is assumed: The imaginary part of s has the meaning of whirl speed, while the real part is the decay rate changed in sign.

A difference between roots loci in general dynamics (e.g., that of Figure 1.5) and those encountered in rotordynamics is that the latter are not symmetrical with respect to the real axis. This is because the eigenvalues are not conjugate or, in physical terms, the forward and backward whirl speeds are not equal.

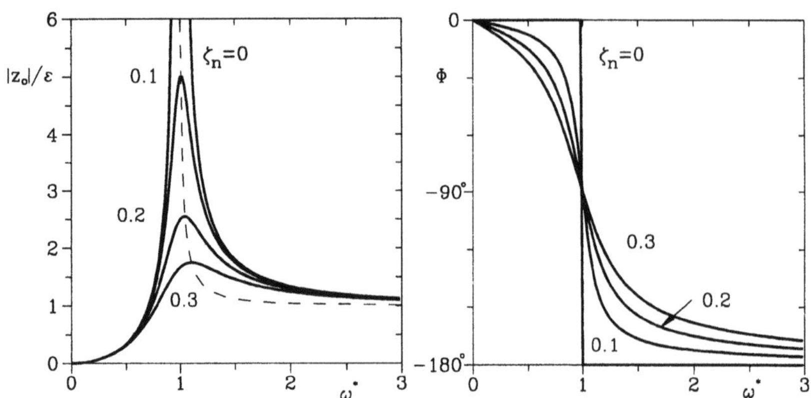

FIGURE 4.9. Nondimensional amplitude and phase of the unbalance response for three different values of nonrotating damping. Full line indicates curves for different values of damping; dashed line indicates line connecting the peaks.

4.3.4 System with viscous damping: Unbalance response

If the rotor is not perfectly balanced, it is necessary to resort to the nonhomogeneous equation (4.18). If the angular acceleration is neglected, the complementary function and the particular integral are, respectively,

$$z = Z_1 e^{i\lambda_1 t} + Z_2 e^{i\lambda_2 t}; \qquad z = z_0 e^{i\omega t}. \qquad (4.27)$$

The first allows the description of the motion of the perfectly balanced rotor, while the second yields the response to the static unbalance $m\epsilon$. The amplitude of the unbalance response is then obtained by introducing the particular integral into the equation of motion, obtaining

$$z_0(-m\omega^2 + i\omega c_n + k) = m\epsilon\omega^2. \qquad (4.28)$$

As it was easily predictable, rotating damping does not enter equation (4.28): Unbalance produces a synchronous excitation, i.e., an excitation that rotates in the xy-plane at the same speed ω as the rotor, and the latter rotates in the deflected configuration but is not subject to deformations that change in time.

The amplitude z_0 is obviously expressed by a complex number. By separating the real from the imaginary part, it follows that

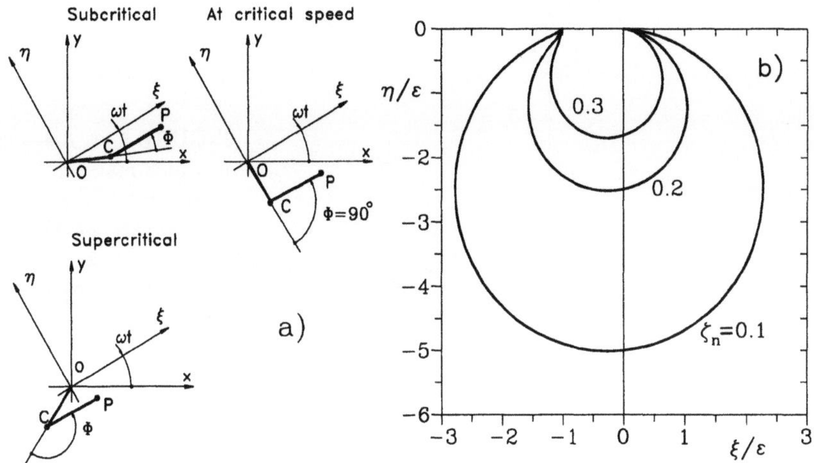

FIGURE 4.10. (a) Situation in the xy-plane in subcritical conditions, at the critical speed, and in supercritical conditions; (b) trajectories of point C in the $\xi\eta$-plane expressed in nondimensional form.

$$\Re(z_0) = \frac{m\epsilon\omega^2(k - m\omega^2)}{(k - m\omega^2)^2 + \omega^2 c_n^2} = \epsilon\frac{\omega^{*2}(1 - \omega^{*2})}{(1 - \omega^{*2})^2 + 4\zeta_n^2\omega^{*2}},$$

$$\Im(z_0) = -\frac{m\epsilon\omega^3 c_n}{(k - m\omega^2)^2 + \omega^2 c_n^2} = -\epsilon\frac{2\omega^{*3}\zeta_n}{(1 - \omega^{*2})^2 + 4\zeta_n^2\omega^{*2}},$$

$$|z_0| = \epsilon\frac{\omega^{*2}}{\sqrt{(1 - \omega^{*2})^2 + 4\zeta_n^2\omega^{*2}}},$$

$$\Phi = \arctan\left(\frac{-2\omega^*\zeta_n}{1 - \omega^{*2}}\right). \tag{4.29}$$

The amplitude and phase of z_0 are plotted in nondimensional form as functions of the speed in Figure 4.9. The different curves have been obtained with different values of the nonrotating damping ζ_n. The equation yielding the unbalance response is identical to that yielding the amplitude of the response of a vibrating system with a single degree of freedom to a harmonic excitation whose amplitude is proportional to the square of the frequency. Figure 4.9 is then identical to Figure 1.14b. The damped resonance peak lies at the right of the undamped resonance, as was the case for Figure 1.14b, and not on the left, as in Figure 1.8.

The situation in the xy-plane is shown in Figure 4.10b. Since angle Φ is always negative, point C always lags the line forming an angle of ωt with the x-axis, i.e., ξ-axis. The delay is exactly 90° at the crit-

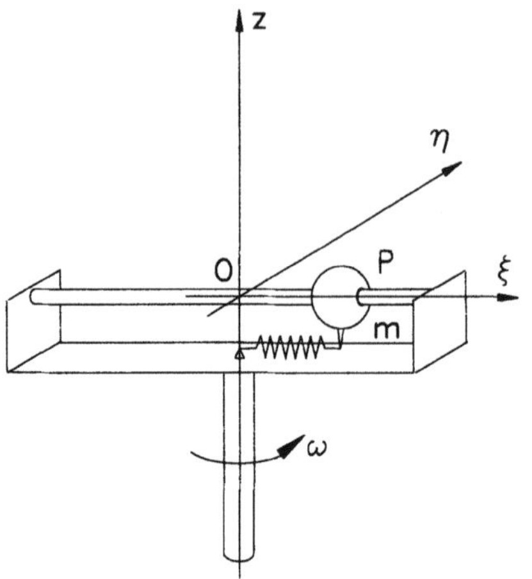

FIGURE 4.11. A system in which mass m is constrained to move along the xi-axis of the rotating reference frame. No self-centering is possible.

ical speed. The plot of the real and imaginary parts of the complex amplitude z_0 yields directly the trajectory of point C in the $\xi\eta$-plane (Figure 4.10b). By adding vector ϵ to vector z_0, the trajectory of point P is obtained.

The plot of the trajectory of C is very similar to a Nyquist diagram, with the important difference that the latter is plotted in the Argand plane, while the former gives the actual position of point C in the $\xi\eta$-plane in the physical space. The term *trajectory* has been used with an extended meaning: All these figures were obtained neglecting the angular acceleration, and the points of the trajectories refer to steady-state conditions at different speeds and are not successive positions during an acceleration of the rotor. However, if the acceleration is very slow and dynamic effects linked to it can be neglected, the curves of Figure 4.10b can be assumed to be at least a good approximation for the actual trajectories in the $\xi\eta$-plane. From them it is clear that self-centering is strictly linked with an increase of phase Φ from $0°$ to $-180°$, i.e., to a rotation of point C in the $\xi\eta$-plane.

If point C were constrained to remain on the ξ-axis, no self-centering would occur, as in the case of the system sketched in Figure 4.11, where mass m is allowed to move along a rotating guide and is constrained to the center of rotation O by a spring of stiffness k. The system is only apparently similar to the Jeffcott rotor of Figure 4.3: If the length of the guide were infinite, it would behave similarly to a Jeffcott rotor in the subcritical field, but in supercritical conditions point P would not come back to the self-centered position and would remain at infinity (or, in real life, at the end of the guide).

In the damped Jeffcott rotor of Figure 4.3, the motion of point C can be considered as the superimposition of a backward inward-spiral motion, which usually decays very quickly in time, a forward spiral motion, whose amplitude can be decreasing, constant, or increasing in time depending on the stability of the system, and a circular synchronous motion, with constant amplitude. The amplitude of the latter depends on the spin speed and the eccentricity ϵ.

4.3.5 System with structural damping

Rotating and nonrotating damping have been supposed to be of the viscous type. This can be a realistic model for nonrotating damping, particularly when dampers of the squeeze-film type are used, but for the rotating damping, a structural damping model is usually better suited. There is no difficulty in introducing the complex stiffness model described in Section 1.9. Equation (1.89) yielding the equivalent damping must, however, be modified into $c_{eq} = \eta k/\lambda_m$, where λ_m is the frequency at which the material goes through the hysteresis cycle. It coincides with $|\lambda_R|$, modulus of the real part of the whirl frequency λ, in the case of nonrotating damping, while it takes the value $|\lambda_R - \omega|$ for rotating damping. If they are both of the structural type, by neglecting the imaginary part of frequency λ in the computation of λ_m, equation (4.19) for the study of free whirling becomes

$$- m\lambda^2 + k + i \left(\frac{\lambda_R}{|\lambda_R|} \eta_n k_n + \frac{\lambda_R - \omega}{|\lambda_R - \omega|} \eta_r k_r \right) = 0 . \qquad (4.30)$$

The terms neglected operating in this way $-\eta_n k_n \lambda_I/|\lambda_R|$ $-\eta_r k_r \lambda_I/|\lambda_R - \omega|$ are very small with respect to k, at least if the loss factor is small and λ_R is different enough from ω. Note that at the threshold of instability, the imaginary part of the whirl frequency vanishes and then equation (4.30) yields exact results. Equation (4.30) can be written in the form

$$- m\lambda^2 + k + i(\pm \eta_n k_n \pm \eta_r k_r) = 0 . \qquad (4.31)$$

In the case of forward subcritical motion, the positive signs hold; in supercritical conditions the sign of η_n is positive and that of η_r

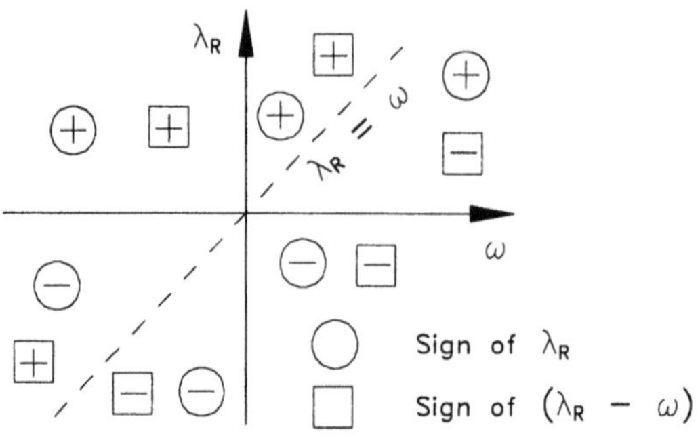

FIGURE 4.12. Signs of λ_R and $\lambda_R - \omega$ in the various zones of the Campbell diagram.

is negative. They are both negative in the case of backward motion (Figure 4.12).

The characteristic equation can be solved in the same way as the case of viscous damping. The real and imaginary parts of the whirl frequency are

$$\begin{cases} \lambda_R = \pm \dfrac{1}{\sqrt{2m}} \sqrt{k + \sqrt{k^2 + (k_n \eta_n \pm k_r \eta_r)^2}} \, , \\[2mm] \lambda_I = \dfrac{1}{\sqrt{2m}} \sqrt{-k + \sqrt{k^2 + (k_n \eta_n + k_r \eta_r)^2}} \, , \end{cases} \tag{4.32}$$

in the case of backward or of forward subcritical whirl; in the super-critical field, the imaginary part is

$$\begin{cases} \lambda_I = \dfrac{1}{\sqrt{2m}} \sqrt{-k + \sqrt{k^2 + (k_n \eta_n - k_r \eta_r)^2}} & \text{if } \eta_n k_n > \eta_r k_r \, , \\[2mm] \lambda_I = -\dfrac{1}{\sqrt{2m}} \sqrt{-k + \sqrt{k^2 + (k_n \eta_n - k_r \eta_r)^2}} & \text{if } \eta_n k_n < \eta_r k_r \, . \end{cases} \tag{4.33}$$

The condition for stability in the supercritical field is then $\eta_n k_n > \eta_r k_r$. If this condition is satisfied, no threshold of instability exists and the system is stable at any speed. If, on the contrary, it is not satisfied, the threshold of instability coincides with the critical speed, and no supercritical running is possible.

The case in which there are different types of damping can be studied in the same way. If rotating damping is of the structural type and nonrotating damping is viscous, the condition for stability is

$$c_n > \eta_r k_r \sqrt{\frac{m}{k}} \, .$$

In the case of the unbalance response, rotating damping has no influence on the behaviour of the system. If nonrotating damping is of the structural type, the amplitude and phase of the unbalance response are

$$|z_0| = \epsilon \frac{\omega^{*2}}{\sqrt{(1 - \omega^{*2})^2 + \eta_n^2}} \,,$$

$$\Phi = \arctan\left(\frac{-\eta_n}{1 - \omega^{*2}}\right) . \tag{4.34}$$

Note that the assumption of constant loss factor is only a rough approximation. If the dependence of the loss factor from the frequency is known, the free whirling frequencies can be computed iteratively. Because the whirl speed is little influenced by the value of the damping, the procedure can converge quickly. For the computation of the unbalance response, however, the dependence of damping from the frequency introduces no computational difficulty. Other effects, such as the dependence of damping from the value of the maximum stress, introduce nonlinearities into the model and make the solution very difficult.

4.3.6 Equations of motion in real coordinates

Equation (4.18) can be written using real coordinates x and y

$$\begin{bmatrix} m & 0 \\ 0 & m \end{bmatrix} \begin{Bmatrix} \ddot{x} \\ \ddot{y} \end{Bmatrix} + \begin{bmatrix} c_n + c_r & 0 \\ 0 & c_n + c_r \end{bmatrix} \begin{Bmatrix} \dot{x} \\ \dot{y} \end{Bmatrix} +$$

$$+ \begin{bmatrix} k & \omega c_r \\ -\omega c_r & k \end{bmatrix} \begin{Bmatrix} x \\ y \end{Bmatrix} = m\epsilon \begin{Bmatrix} \dot{\theta}^2 \cos(\theta) + \ddot{\theta} \sin(\theta) \\ \dot{\theta}^2 \sin(\theta) - \ddot{\theta} \cos(\theta) \end{Bmatrix} . \tag{4.35}$$

Equation (4.35) has the form of equation (1.5), where the skew-symmetric gyroscopic matrix $\mathbf{G}$ vanishes while the skew-symmetric circulatory matrix $\mathbf{H}$ is present and contains rotating damping. As it will be shown better for gyroscopic systems, imaginary terms in the equation based on complex coordinates correspond to skew symmetric terms when real coordinates are used. The presence of a circulatory matrix is linked with unstabilizing effects, as in the current case due to rotating damping.

4.3.7 Stability in the supercritical field

The steady-state unbalance response of an undamped or damped Jeffcott rotor was computed in the preceding sections. However, no conclusion about the stability of the equilibrium position of the system was reached. A simple way to state the stability of the equilibrium position is by observing that the complete solution of the equation of motion can be obtained by adding the solution for free whirling to the unbalance response. When the first one leads to a stable behaviour, the overall behaviour of the system is stable. The equilibrium position is then stable in the whole supercritical range up to the previously computed threshold of instability.

However, better evidence can be obtained by writing the equation of motion with reference to the rotating reference frame $O\xi\eta$ shown in Figure 4.6. If the angular velocity ω is constant, the position, velocity, and acceleration of point C can then be expressed as functions of complex coordinate ζ

$$\begin{cases} z = \zeta e^{i\omega t}, \\ \dot{z} = \left(\dot{\zeta} + i\omega\zeta\right) e^{i\omega t}, \\ \ddot{z} = \left(\ddot{\zeta} + 2i\omega\dot{\zeta} - \omega^2\zeta\right) e^{i\omega t}. \end{cases} \qquad (4.36)$$

By introducing equation (4.36) into the equation of motion of the damped system (4.18) written for constant speed and rearranging the various terms, it follows that

$$m\ddot{\zeta} + (c_n + c_r + 2im\omega)\dot{\zeta} + (k - m\omega^2 + i\omega c_n)\zeta = m\epsilon\omega^2. \qquad (4.37)$$

Note that the main differences between equation (4.18) written in the inertial frame and equation (4.37) are the presence of the terms $2im\omega\dot{\zeta}$ linked with Coriolis acceleration and $m\omega^2\zeta$ due to centrifugal forces and the fact that in the rotating frame unbalance forces are constant, in both direction and modulus. From equation (4.37), the static equilibrium equation is readily obtained. By equating to zero all derivatives of the generalized coordinates with respect to time, the equilibrium position ζ_0 is obtained:

$$\zeta_0 = \frac{m\epsilon\omega^2}{k - m\omega^2 + i\omega c_n}. \qquad (4.38)$$

Note that equation (4.38) coincides with equation (4.29). The stability of the equilibrium position is easily studied by assuming that the motion of point C occurs in the vicinity of the position expressed by equation (4.38)

$$\zeta(t) = \zeta_1(t) + \zeta_0 \,. \tag{4.39}$$

By introducing equation (4.39) into the equation of motion (4.37) and remembering the expression for the equilibrium position (4.38), the following equation for the motion about the equilibrium position is obtained

$$m\ddot{\zeta}_1 + (c_n + c_r + 2im\omega)\dot{\zeta}_1 + (k - m\omega^2 + i\omega c_n)\zeta_1 = 0 \,. \tag{4.40}$$

The solution of equation (4.40) is of the usual type $\zeta_1 = \zeta_{1_0}e^{i\lambda't}$, yielding a spiral motion about the equilibrium position whose amplitude can be either decreasing or increasing in time depending on the sign of the imaginary part of the complex frequency λ'. By introducing this solution into equation (4.40), the following characteristic equation allowing the computation of the complex frequency is readily obtained:

$$-m\lambda'^2 + \left[i(c_n + c_r) - 2m\omega\right]\lambda' + k - m\omega^2 + i\omega c_n = 0 \,. \tag{4.41}$$

The solution of the characteristic equation is

$$\lambda' = -\omega + i\frac{c_r + c_n}{2m} \pm \sqrt{\frac{-(c_r + c_n)^2 + 4m(k - i\omega c_r)}{4m^2}} \,. \tag{4.42}$$

By comparing equation (4.42) with equation (4.20), it is immediately clear that the imaginary parts of the two expressions are equal and the real parts differ by a term equal to $-\omega$. This result is expected because the former is expressed in a frame of reference that rotates at an angular velocity equal to ω with respect to the one in which equation (4.20) is expressed. Because the imaginary part of the complex frequency is the same as that given by equation (4.20), when condition (4.25) for stability is satisfied, the motion about the equilibrium position is, in the rotating reference frame, a decaying spiral and the equilibrium position is stable.

The equation of motion in the rotating frame (4.37) can be written using the real coordinates ξ and η instead of the complex coordinate ζ, obtaining

$$\begin{bmatrix} m & 0 \\ 0 & m \end{bmatrix} \begin{Bmatrix} \ddot{\xi} \\ \ddot{\eta} \end{Bmatrix} + \begin{bmatrix} c_n + c_r & -2m\omega \\ 2m\omega & c_n + c_r \end{bmatrix} \begin{Bmatrix} \dot{\xi} \\ \dot{\eta} \end{Bmatrix} + \qquad (4.43)$$

$$+ \begin{bmatrix} k - m\omega^2 & -\omega c_n \\ \omega c_n & k - m\omega^2 \end{bmatrix} \begin{Bmatrix} \xi \\ \eta \end{Bmatrix} = \begin{Bmatrix} m\epsilon\omega^2 \\ 0 \end{Bmatrix}.$$

In this case both the gyroscopic $\mathbf{G}$ and the gyroscopic circulatory matrix $\mathbf{H}$ are present, the first is due to Coriolis forces due to the fact that the reference frame is noninertial. A sort of centrifugal stiffening term (the terms $-m\omega^2$ in $\mathbf{K}$) is present but is again due to the noninertial reference frame rather than a true centrifugal stiffening.

4.3.8 Acceleration through the critical speed

If the angular velocity of the rotor is not constant, the equation of motion for the lateral behaviour of the damped system is equation (4.18). A third equation must be added to express the dependence of the angular displacement θ from the driving torque M_z. Because the third degree of freedom of the system is linked with rotation about the z-axis, the polar moment of inertia J_z cannot be neglected. No torsional elasticity of the shaft is taken into account, because the driving torque is assumed to be directly applied to a torsionally stiff rotor and the Lagrangian of the system is expressed by equation (4.9), to which a term $J_z \dot{\theta}^2/2$ must be added.

By performing the relevant derivatives of the Lagrangian with respect to the angular coordinate θ and its derivative, the equation of motion for rotations about the z-axis is

$$(J_z + m\epsilon^2)\ddot{\theta} + m\epsilon\left[-\ddot{x}\sin(\theta) + \ddot{y}\cos(\theta)\right] = M_z. \qquad (4.44)$$

Equation (4.44), which can be expressed in terms of the complex coordinate z, and equation (4.18) are the equations of motion of the system

$$\begin{cases} m\ddot{z} + (c_r + c_n)\dot{z} + (k - i\dot{\theta}c_r)z = m\epsilon(\dot{\theta}^2 - i\ddot{\theta})e^{i\theta}, \\ (J_p + m\epsilon^2)\ddot{\theta} + m\epsilon\Im\left(\ddot{z}e^{-i\theta}\right) = M_z. \end{cases} \qquad (4.45)$$

The same equations can be written in a rotor-fixed reference frame, which, in this case does not rotate at constant speed. Equations

(4.36) expressing the absolute position, velocity, and acceleration as functions of the same characteristics expressed in the rotating frame must be modified to take into account the fact that the rotational speed is not constant

$$
\begin{cases}
z = \zeta e^{i\theta}, \\
\dot{z} = \left(\dot{\zeta} + i\dot{\theta}\zeta\right) e^{i\theta}, \\
\ddot{z} = \left(\ddot{\zeta} + 2i\dot{\theta}\dot{\zeta} + i\ddot{\theta}\zeta - \dot{\theta}^2\zeta\right) e^{i\theta}.
\end{cases}
\tag{4.46}
$$

The equations of motion written with reference to the rotating frame are then

$$
\begin{cases}
m\ddot{\zeta} + (c_n + c_r + 2im\omega)\dot{\zeta} + (k - m\omega^2 + i\omega c_n + im\ddot{\theta})\zeta = m\epsilon(\dot{\theta}^2 - i\ddot{\theta}), \\
(J_p + m\epsilon^2)\ddot{\theta} + m\epsilon\,\Im\left(\ddot{\zeta} + 2i\dot{\theta}\dot{\zeta} + i\ddot{\theta}\zeta - \dot{\theta}^2\zeta\right) = M_z,
\end{cases}
\tag{4.47}
$$

or, using real coordinates,

$$
\begin{bmatrix} m & 0 \\ 0 & m \end{bmatrix}
\begin{Bmatrix} \ddot{\xi} \\ \ddot{\eta} \end{Bmatrix}
+
\begin{bmatrix} c_n + c_r & -2m\dot{\theta} \\ 2m\dot{\theta} & c_n + c_r \end{bmatrix}
\begin{Bmatrix} \dot{\xi} \\ \dot{\eta} \end{Bmatrix}
+
$$

$$
+
\begin{bmatrix} k - m\dot{\theta}^2 & -\dot{\theta}c_n - m\ddot{\theta} \\ \dot{\theta}c_n + m\ddot{\theta} & k - m\dot{\theta}^2 \end{bmatrix}
\begin{Bmatrix} \xi \\ \eta \end{Bmatrix}
= m\epsilon
\begin{Bmatrix} \dot{\theta}^2 \\ -\ddot{\theta} \end{Bmatrix},
\tag{4.48}
$$

$$
(J_p + m\epsilon^2)\ddot{\theta} + m\epsilon\left(\ddot{\eta} + 2\dot{\theta}\dot{\xi} + \ddot{\theta}\xi - \dot{\theta}^2\eta\right) = M_z.
$$

The study of an accelerating Jeffcott rotor can be performed following two different schemes: Either the time history of the driving torque or that of the angular displacement can be stated. In the first case the two equations are coupled and the solution can get very complicated. In the second case, however, they uncouple and the first equation directly yields the flexural behaviour of the system, while the second allows the computation of the time history of the driving torque needed to follow the assumed acceleration law. The laws $\theta(t)$, $\omega(t) = \dot{\theta}(t)$, and $a(t) = \ddot{\theta}(t)$ are known and there is no difficulty in numerically integrating the equations of motion.

Even if in the literature some solutions for the constant acceleration case can be found,[2] they are so complicated that today it is

---

[2] See, for example, F.M. Dimentberg, *Flexural vibrations of rotating shafts*, Butterworths, London, 1961.

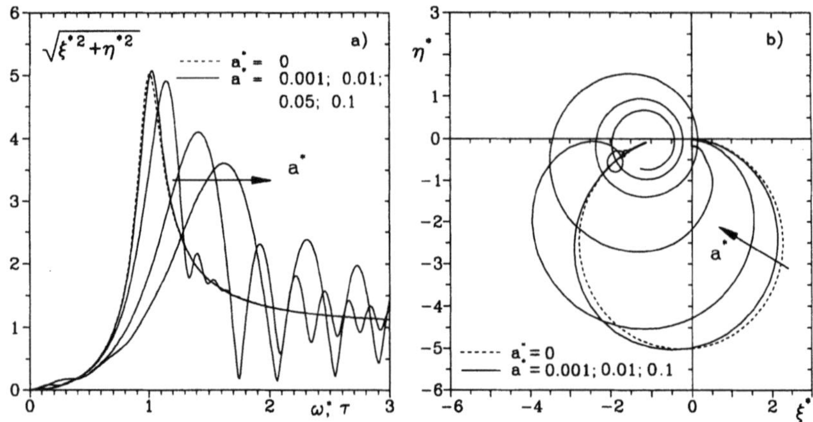

FIGURE 4.13. Motion of a Jeffcott rotor with $\zeta_n = 0.1$ and $\zeta_r = 0.01$ crossing a critical speed with constant acceleration. Nondimensional plot for some values of the angular acceleration; (a) time history of the amplitude; (b) trajectory in the rotating plane $\xi^* \eta^*$.

easier to perform the numerical integration of the equations of motion. The amplitude of the motion of a Jeffcott rotor with $\zeta_n = 0.1$ and $\zeta_r = 0.01$ accelerating from standstill to a speed equal to three times the critical speed with constant acceleration is plotted in Figure 4.13a. The plot has been obtained in nondimensional form, using the following nondimensional parameters: $\xi^* = \xi/\epsilon$, $\eta^* = \eta/\epsilon$, $\tau = t/t_1 = ta/\omega_{cr}$, and $a^* = a/\omega_{cr}^2$. Note that the nondimensional time τ and the nondimensional velocity ω^* coincide, because time has been made nondimensional with reference to time t_1 needed to reach the critical speed.

With very low values of the acceleration, the motion follows a pattern very similar to that seen for the steady-state case, as was easily predicted. With increasing values of the acceleration, the peak amplitude is reduced and the self-centered conditions are reached after some oscillations are damped out. The oscillations become stronger with increasing acceleration.

The trajectory of point C in the rotating plane is shown in Figure 4.13b. From this plot, it is clear that the oscillations are actually the result of a spiral motion of the system taking place about the self-centered position. Note that the threshold of instability is never exceeded and the motion is stable. When the speed becomes higher than the threshold of instability, the spiral stops decaying and starts growing as unstable behaviour develops.

The driving torque needed to perform the acceleration is shown in

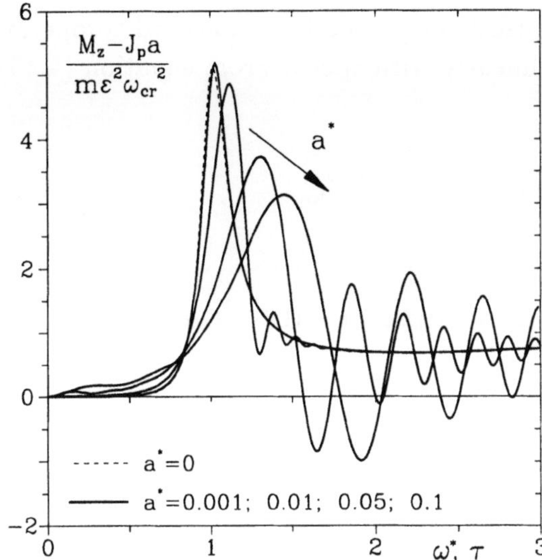

FIGURE 4.14. Nondimensional driving torque needed to perform the acceleration through the critical speed. Same system as Figure 4.13.

Figure 4.14. Actually, the nondimensional torque plotted is only the amount of torque that exceeds the value aJ_z needed to accelerate a perfectly balanced rotor. Also, strong oscillations are visible here in the supercritical field. In Figure 4.14 the curve related to a vanishingly small acceleration has also been plotted. The equation of this curve can be obtained in closed form by imposing that the displacement ζ is constant in time at the value expressed by equation (4.38) and stating $a = 0$ in the second equation (4.47)

$$M_z = m\varepsilon \Im(-\zeta\omega^2) = \frac{2k\epsilon^2 \omega^{*^5} \zeta_n}{(1 - \omega^{*^2})^2 + 4\zeta_n^2 \omega^{*^2}}. \qquad (4.49)$$

Equation (4.49) yields the torque needed to drive the rotor at constant speed against the drag provided by nonrotating damping due to the unbalance of the rotor. Note that this drag has a peak at the critical speed, whose value is

$$M_{z_{max}} = \frac{k\epsilon^2}{2\zeta_n}. \qquad (4.50)$$

If the driving torque is smaller than the value given by equation (4.50), the rotor stalls, i.e., it fails to accelerate beyond the critical speed. All the power supplied by the driving system is dissipated by nonrotating damping, and acceleration is no longer possible. Obviously, the value so computed must be added to that required to overcome all other forms of drag, such as aerodynamic or bearing drag. In the high supercritical range, the torque needed to operate at constant speed grows linearly with speed. From equation (4.49), it follows that

$$\lim_{\omega \to \infty} M_z = 2k\epsilon^2 \omega^* \zeta_n \,. \tag{4.51}$$

4.4   Model with four degrees of freedom: Gyroscopic effect

4.4.1   Generalized coordinates and equations of motion

In the Jeffcott model the rotor was assumed to be a point mass. When studying the acceleration through the critical speed, the moment of inertia about the axis of rotation was introduced, but it had no effect on the flexural behaviour. Actually, the rotor's moments of inertia can considerably influence its dynamic behaviour as they are responsible for the gyroscopic moments that cause the natural frequencies of bending modes to depend on the spin speed and the Campbell diagram to be different from a number of straight lines running in a horizontal direction. The simplest model to evaluate this effect is still that in Figure 4.3, where instead of a point mass, a rigid body is located in P, with nonvanishing moments of inertia and one of the principal axes of inertia coinciding, in the undeformed position, with the z-axis. Its ellipsoid of inertia has axial symmetry with respect to the same axis. The principal moments of inertia of the rigid body will be referred to as the *polar* moment of inertia J_p about the rotation axis and *transversal* moment of inertia J_t about any axis in the rotation plane.

If $J_p > J_t$, the body is usually referred to as a *disc*; the limiting case is that of an infinitely thin disc in which $J_p = 2J_t$. If $J_p = J_t$, the inertia ellipsoid degenerates into a sphere and the situation is, in many aspects, similar to that of the Jeffcott rotor. If $J_p < J_t$, the rotor is usually referred to as a *long rotor*.

Assume also that owing to small errors, the position of point P, in which the center of gravity of the rigid body is located, does not coincide with that of point C, the center of the shaft. The distance between the two points is the eccentricity ϵ. Moreover, the axis of symmetry of the rigid body does not coincide exactly with the rota-

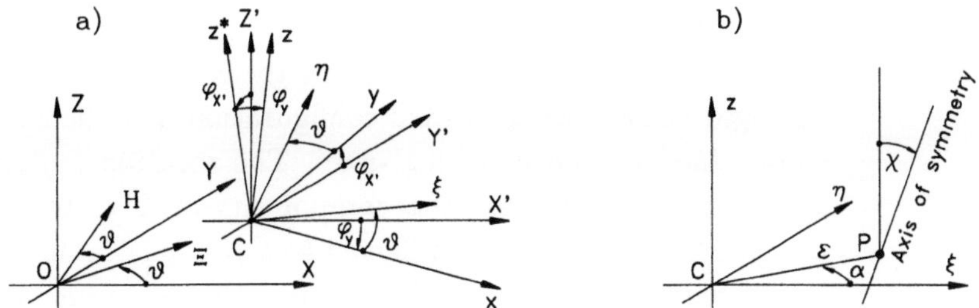

FIGURE 4.15. (a) Reference frames; (b) unbalance condition in the $C\xi\eta z$ reference frame.

tion axis. The angle between them is the angular error χ. These two errors, which are assumed to be small, cause a static unbalance and a couple unbalance, respectively.

Strictly speaking, the system has six degrees of freedom, and six generalized coordinates must be defined for the study of its dynamic behaviour. The uncoupling between axial, flexural, and torsional behaviour seen for beams will be assumed to hold for the elastic part of the system. In this section it will be shown that the same uncoupling also occurs for the inertial part of the model and that a model with four degrees of freedom is sufficient for the study of the flexural behaviour at constant speed, at least under wide simplifying assumptions.

The generalized coordinates will be defined with reference to the frames shown in Figure 4.15a.

- Frame $OXYZ$: Inertial frame, with the origin in O and the Z-axis coinciding with the rotation axis of the rotor.

- Frame $O\Xi HZ$ with the origin in O and the Z-axis coinciding with that of the preceding frame: Axes Ξ, H rotate in the XY-plane with angular velocity ω, in the case of constant speed operation. In the case of variable speed, angle θ between the two systems coincides with the same angle defined in Figure 4.4. It will be referred to as a *rotating frame*.

- Frame $CX'Y'Z'$ with origin in C Its axes remain parallel with those of frame $OXYZ$. The $X'Y'$-plane remains parallel to the XY-plane.

- Frame $Cxyz$ with origin in C: The z-axis coincides with the rotation axis of the rigid body in the deformed position and the x- and y-axes are defined by the following rotations:

- Rotate the axes of $CX'Y'Z'$ frame about the X'-axis of an angle $\phi_{X'}$ until the Y'-axis enters the rotation plane of the rigid body in its deformed configuration. Let the axes so obtained be the y- and z^*-axes. The rotation matrix allowing one to express the components of a vector in $CX'yz^*$ frame from those referred to $CX'Y'Z'$ frame (or to the inertial frame, as the directions of the axes coincide) is

$$
\mathbf{R}_1 = \begin{bmatrix} 1 & 0 & 0 \\ 0 & \cos(\phi_{X'}) & \sin(\phi_{X'}) \\ 0 & -\sin(\phi_{X'}) & \cos(\phi_{X'}) \end{bmatrix} .
$$

- Rotate the frame obtained after the mentioned rotation, about the y-axis until the X'-axis also enters the rotation plane of the rigid body in its deformed configuration. Let the axis so obtained be the x-axis and the rotation angle be ϕ_y. After the two mentioned rotations, the z-axis coincides, apart from the angular error χ, with the symmetry axis of the rigid body in its deformed configuration. Frame $Cxyz$ is centered in the center of the shaft of the rigid body and follows it in its whirling motion. However, it does not rotate with the spin speed ω. It will be referred to as *whirling frame*. Let the matrix expressing this second rotation be

$$
\mathbf{R}_2 = \begin{bmatrix} \cos(\phi_y) & 0 & -\sin(\phi_y) \\ 0 & 1 & 0 \\ \sin(\phi_y) & 0 & \cos(\phi_y) \end{bmatrix} .
$$

- Frame $C\xi\eta z$ with origin in C: It is obtained from frame $Cxyz$ by rotating the x- and y-axes in the xy-plane of an angle equal to rotation angle θ of the rotor corresponding to the spin speed. If rotation occurs with constant spin speed ω, angle θ is equal to ωt. Frame $C\xi\eta z$ is actually fixed to the rigid body, although not being centered in its center of gravity owing to the eccentricity ϵ, and not being principal of inertia due to the angular error χ. It will be referred to as *rotating and whirling frame*. The matrix allowing one to express a vector in the $C\xi\eta z$-frame from the components in the $Cxyz$-frame is

$$\mathbf{R}_3 = \begin{bmatrix} \cos(\theta) & \sin(\theta) & 0 \\ -\sin(\theta) & \cos(\theta) & 0 \\ 0 & 0 & 1 \end{bmatrix}.$$

As already stated, the rotor is assumed to be slightly unbalanced. As the position of the rotor in the $C\xi\eta z$-frame is immaterial, the principal axis of inertia corresponding to the moment of inertia J_p will be assumed to be parallel to the ξz-plane. As the static unbalance cannot be assumed to lie in the same plane as the couple unbalance, the eccentricity cannot be assumed to lie along the ξ-axis, as was the case of the Jeffcott rotor. The conditions of unbalance are summarized in Figure 4.15b, where the static unbalance is shown to lead the couple unbalance of a phase angle α. A fourth rotation matrix $\mathbf{R}_4$ allowing passage from the rotor system of reference $C\xi\eta z$ to the principal axes of the rigid body is thus defined

$$\mathbf{R}_4 = \begin{bmatrix} \cos(\chi) & 0 & -\sin(\chi) \\ 0 & 1 & 0 \\ \sin(\chi) & 0 & \cos(\chi) \end{bmatrix}.$$

Take the X-, Y- and Z-coordinates of point C and angles $\phi_{X'}$, ϕ_y, and θ as generalized coordinates of the rigid body. A small displacement assumption on coordinates X, Y, Z, $\phi_{X'}$, and ϕ_y will allow great simplification of the problem. Coordinate θ, on the contrary, cannot be considered small.

To compute the kinetic energy of the rigid body, the velocity of the center of gravity (point P) and the angular velocity expressed in the principal system of inertia must be computed. The position of point P is easily obtained

$$\overline{(P-O)} = \left\{ \begin{array}{c} X \\ Y \\ Z \end{array} \right\} + \mathbf{R}_1^T \mathbf{R}_2^T \mathbf{R}_3'^{\,T} \left\{ \begin{array}{c} \epsilon \\ 0 \\ 0 \end{array} \right\}, \tag{4.52}$$

where matrix $\mathbf{R}_3'$ is a matrix identical to $\mathbf{R}_3$, but obtained substituting angle $\theta + \alpha$ to angle θ. The small-displacement assumption allows the linearization of the trigonometric functions of angles $\phi_{X'}$ and ϕ_y and the neglect of some terms, which are of the same order of magnitude as those that are neglected when truncating the series for the trigonometric functions after the first term[3]. Equation (4.52) then reduces to

$$(\overline{P-O}) = \left\{ \begin{array}{c} X + \epsilon \cos(\theta + \alpha) \\ Y + \epsilon \sin(\theta + \alpha) \\ Z + \epsilon \left[\phi_{X'} \sin(\theta + \alpha) - \phi_y \cos(\theta + \alpha) \right] \end{array} \right\}. \quad (4.53)$$

The velocity of the center of gravity or the rotor is easily computed by performing the derivatives of vector $(\overline{P - textO})$ with respect to time. As angle α is constant, it follows that

$$V_P = \left\{ \begin{array}{c} \dot{X} - \epsilon\dot{\theta} \sin(\theta + \alpha) \\ \dot{Y} + \epsilon\dot{\theta} \cos(\theta + \alpha) \\ \dot{Z} + \epsilon \left[(\dot{\theta}\phi_{X'} - \dot{\phi}_y) \cos(\theta + \alpha) + (\dot{\theta}\phi_y + \dot{\phi}_{X'}) \sin(\theta + \alpha) \right] \end{array} \right\}.$$

$$(4.54)$$

In the third line of equation (4.54) there are two terms: The first is the velocity in the axial direction due to the displacement of point C in the same direction, and the second is the velocity in the axial direction due to the eccentricity and rotations of the cross section of the shaft. It is easy to verify that the last term causes a coupling between bending and axial behaviour of the rotor; however, if the eccentricity is small, it is negligible compared to the first one. In the following developments, all terms containing the product of the eccentricity or the angular error by a small quantity will be neglected and no axial-flexural coupling will be obtained. The translational kinetic energy is then

$$T_t = \frac{1}{2} m V_P^2 = \frac{1}{2} m \left\{ \dot{X}^2 + \dot{Y}^2 + \dot{Z}^2 + \right. \quad (4.55)$$

$$\left. + \epsilon^2 \dot{\theta}^2 + \epsilon\dot{\theta} \left[-\dot{X} \sin(\theta + \alpha) + \dot{Y} \cos(\theta + \alpha) \right] \right\}.$$

---

[3] All terms of order higher than two in the expression of the kinetic energy give way to nonlinear terms in the equations of motion. By linearizing the expressions of the displacements and the velocities, the expression of the kinetic energy is then simplified in such a way to yield linearized equations of motion. The results so obtained were checked against those obtained by computing the kinetic energy using the complete expression of the displacements and then making the simplifications at the end of the computations. The two procedures yield, as expected, identical results.

For the computation of the rotational kinetic energy, the angular velocity must be expressed with reference to a frame coinciding with the principal axes of the rotor. The three components of the angular velocity can be considered vectors acting in different directions: $\dot{\phi}_{X'}$ along the X'-axis, $\dot{\phi}_y$ along the y-axis, and $\dot{\theta}$ along the z-axis. Using the relevant rotation matrices, the components of the angular velocity along the principal axes of inertia of the rotor $\mathbf{\Omega}$ are

$$\mathbf{\Omega} = \mathbf{R}_4\mathbf{R}_3\mathbf{R}_2 \left\{ \begin{array}{c} \dot{\phi}_{X'} \\ 0 \\ 0 \end{array} \right\} + \mathbf{R}_4\mathbf{R}_3 \left\{ \begin{array}{c} 0 \\ \dot{\phi}_y \\ 0 \end{array} \right\} + \mathbf{R}_4 \left\{ \begin{array}{c} 0 \\ 0 \\ \dot{\theta} \end{array} \right\} . \tag{4.56}$$

By remembering the small-displacement assumptions, equation (4.56) reduces to

$$\mathbf{\Omega} = \left\{ \begin{array}{c} \dot{\phi}_{X'}\cos(\theta) + \dot{\phi}_y\sin(\theta) - \chi\dot{\theta} \\ -\dot{\phi}_{X'}\sin(\theta) + \dot{\phi}_y\cos(\theta) \\ \dot{\phi}_{X'}[\chi\cos(\theta) + \phi_y] + \dot{\phi}_y\chi\sin(\theta) + \dot{\theta} \end{array} \right\} . \tag{4.57}$$

As the components of $\mathbf{\Omega}$ are referred to the principal axes of inertia, neglecting the small terms, the rotational kinetic energy can be easily computed as

$$\mathcal{T}_r = \frac{1}{2}\left\{ J_t(\dot{\phi}_{X'}^2 + \dot{\phi}_y^2 + \chi^2\dot{\theta}^2) + J_t(\dot{\theta}^2 + 2\dot{\theta}\dot{\phi}_{X'}\phi_y) + \right. \tag{4.58}$$

$$\left. +2\dot{\theta}\chi(J_p - J_t)\left[\dot{\phi}_{X'}\cos(\theta) + \dot{\phi}_y\sin(\theta)\right]\right\} .$$

The equations of motion can be obtained as usual through the Lagrange equation. In this case it is expedient to introduce into the Lagrange equations only the expression of the kinetic energy and then to add the elastic reaction of the shaft directly, as external generalized forces. The generalized forces Q_i that must be included in the equations of motion related to the translational degrees of freedom are simply the forces F_X, F_Y, and F_Z applied on the rigid body in the direction of the axes of the inertial frame. By performing the relevant derivatives, the first three equations of motion are obtained with simple computations

$$\begin{cases} m\ddot{X} = m\epsilon \left[\ddot{\theta}\sin(\theta+\alpha) + \dot{\theta}^2\cos(\theta+\alpha)\right] + F_X \,, \\ m\ddot{Y} = m\epsilon \left[-\ddot{\theta}\cos(\theta+\alpha) + \dot{\theta}^2\sin(\theta+\alpha)\right] + F_Y \,, \\ m\ddot{Z} = F_Z \,. \end{cases} \qquad (4.59)$$

The generalized forces to be introduced into the three equations of motion related to the rotational degrees of freedom are not exactly the moments about the axes of the $Cxyz$ frame because the first generalized coordinate for rotation is referred to the X'-axis. It is easy to show that the generalized forces are related to the components of the moment acting on the rigid body in the directions of the axes of $Cxyz$ frame by the equations

$$Q_{\phi_{x'}} = M_x\cos(\phi_y) + M_z\sin(\phi_y), \qquad Q_{\phi_y} = M_y, \qquad Q_\theta = M_z \,.$$
$$(4.60)$$

By using expressions (4.60) for the generalized forces, performing all the relevant derivatives and linearizing again, the following equations of motion can be obtained

$$\begin{cases} J_t\ddot{\phi}_{x'} + J_p\dot{\theta}\dot{\phi}_y + J_p\dot{\theta}\dot{\phi}_y = \chi(J_t - J_p)[\ddot{\theta}\cos(\theta) - \dot{\theta}^2\sin(\theta)]+ \\ \qquad + M_x + \phi_y M_z \,, \\ J_t\ddot{\phi}_y - J_p\dot{\theta}\dot{\phi}_{x'} = \chi(J_t - J_p)[\ddot{\theta}\sin(\theta) + \dot{\theta}^2\cos(\theta)] + M_y \,, \\ (J_p + J_t\chi^2 + m\epsilon^2)\ddot{\theta} + m\epsilon\left[\ddot{Y}\cos(\theta) - \ddot{X}\sin(\theta)\right] + \\ \qquad +\chi(J_p - J_t)[\ddot{\phi}_{x'}\cos(\theta) + \ddot{\phi}_y\sin(\theta)] = M_z \,. \end{cases}$$
$$(4.61)$$

Moment M_z can be obtained from the last equation and substituted into the first. As in the first equation, M_z is multiplied by the small quantity ϕ_y; only the term $J_p\dot{\theta}$ needs to be considered. By introducing the value of the moment about the z-axis into the first equation, it reduces to the form

$$J_t\ddot{\phi}_{x'} + J_p\dot{\theta}\dot{\phi}_y = \chi(J_t - J_p)[\ddot{\theta}\cos(\theta) - \dot{\theta}^2\sin(\theta)] + M_x \,. \qquad (4.62)$$

Equation (4.62) and the last two equations (4.61) are the three equations of motion for the rotational degrees of freedom.

4.4.2 Behaviour of the system at constant speed

The equation related to the translational degree of freedom along the z-axis uncouples from the others. If the angular velocity of the rotor is constant, the last equation also uncouples and the flexural behaviour can be studied separately from axial and torsional behaviour. By stating that $\theta = \omega t$, the four relevant equations of motion reduce to

$$\begin{cases} m\ddot{X} = m\epsilon\omega^2 \cos(\omega t + \alpha) + F_X\,, \\ m\ddot{Y} = m\epsilon\omega^2 \sin(\omega t + \alpha) + F_Y\,, \\ J_t\ddot{\phi}_{X'} + J_p\omega\dot{\phi}_y = -\chi\omega^2(J_t - J_p)\sin(\omega t) + M_x\,, \\ J_t\ddot{\phi}_y - J_p\omega\dot{\phi}_{X'} = \chi\omega^2(J_t - J_p)\cos(\omega t) + M_y\,. \end{cases} \tag{4.63}$$

The only forces and moments acting on the rigid body in P that will be considered are those due to the elastic reaction of the shaft. Because the behaviour of the shaft is assumed to be linear, they are linked to the generalized coordinates by the stiffness matrix of the shaft. The situation in the xz-plane is similar to that in the yz-plane, but if the same elements of the stiffness matrix are used, owing to the axial symmetry of the shaft, the different sign convention in the two coordinate planes compels the use of opposite signs for the elements with subscripts 12 and 21

$$\begin{aligned} \left\{ \begin{array}{c} F_X \\ M_y \end{array} \right\} &= - \left[\begin{array}{cc} K_{11} & K_{12} \\ K_{12} & K_{22} \end{array} \right] \left\{ \begin{array}{c} X \\ \phi_y \end{array} \right\}, \\ \left\{ \begin{array}{c} F_Y \\ M_x \end{array} \right\} &= - \left[\begin{array}{cc} K_{11} & -K_{12} \\ -K_{12} & K_{22} \end{array} \right] \left\{ \begin{array}{c} Y \\ \phi_{X'} \end{array} \right\}. \end{aligned} \tag{4.64}$$

Note that while forces are defined with reference to the undeflected configuration, as is typical of the linear theory of elasticity, moments are, on the contrary, referred to the xyz-axes, which follow the deflected configuration. This is essentially equivalent, because rotations are assumed to be small and can be configured as the releasing of one of the usual simplifications of the theory of elasticity. By introducing the expressions (4.64) of the forces and moments into the equations of motion (4.63), it follows that

$$\begin{cases} m\ddot{X} + K_{11}X + K_{12}\phi_y = m\epsilon\omega^2\cos(\omega t + \alpha)\,, \\ m\ddot{Y} + K_{11}Y - K_{12}\phi_{X'} = m\epsilon\omega^2\sin(\omega t + \alpha)\,, \\ J_t\ddot{\phi}_{X'} + J_p\omega\dot{\phi}_y - K_{12}Y + K_{22}\phi_{X'} = -\chi\omega^2(J_t - J_p)\sin(\omega t)\,, \\ J_t\ddot{\phi}_y - J_p\omega\dot{\phi}_{X'} + K_{12}X + K_{22}\phi_y = \chi\omega^2(J_t - J_p)\cos(\omega t)\,. \end{cases}$$

$$(4.65)$$

Equations (4.65) can be written in the following matrix form:

$$\begin{bmatrix} m & 0 & 0 & 0 \\ 0 & J_t & 0 & 0 \\ 0 & 0 & m & 0 \\ 0 & 0 & 0 & J_t \end{bmatrix} \begin{Bmatrix} \ddot{X} \\ \ddot{\phi}_y \\ \ddot{Y} \\ \ddot{\phi}_{X'} \end{Bmatrix} + \omega \begin{bmatrix} 0 & 0 & 0 & 0 \\ 0 & 0 & 0 & -J_p \\ 0 & 0 & 0 & 0 \\ 0 & J_p & 0 & 0 \end{bmatrix} \begin{Bmatrix} \dot{X} \\ \dot{\phi}_y \\ \dot{Y} \\ \dot{\phi}_{X'} \end{Bmatrix} +$$

$$+ \begin{bmatrix} K_{11} & K_{12} & 0 & 0 \\ K_{12} & K_{22} & 0 & 0 \\ 0 & 0 & K_{11} & -K_{12} \\ 0 & 0 & -K_{12} & K_{22} \end{bmatrix} \begin{Bmatrix} X \\ \phi_y \\ Y \\ \phi_{X'} \end{Bmatrix} = \omega^2 \begin{Bmatrix} m\epsilon\cos(\omega t + \alpha) \\ \chi(J_t - J_p)\cos(\omega t) \\ m\epsilon\sin(\omega t + \alpha) \\ -\chi(J_t - J_p)\sin(\omega t) \end{Bmatrix}\,.$$

$$(4.66)$$

If $-\phi_{X'}$ is used instead of $\phi_{X'}$ as the generalized coordinate for rotations about the X-axis, the stiffness matrix assumes a more regular pattern with all terms positive and the skew-symmetric gyroscopic matrix, containing the polar moments of inertia, is replaced by its transpose. Also, in the current case, it is possible to define a set of complex coordinates that allow the equations of motion to be written in a more compact form

$$\begin{cases} z = X + iY\,, \\ \phi = \phi_y - i\phi_{X'}\,. \end{cases}$$

$$(4.67)$$

Multiplying the second equation (4.66) by the imaginary unit i and adding it to the first one, and multiplying the third equation by $-i$ and adding it to the fourth one, reduces the equations of motion to

$$\begin{cases} m\ddot{z} + K_{11}z + K_{12}\phi = m\epsilon\omega^2 e^{i(\omega t + \alpha)}\,, \\ J_t\ddot{\phi} - i\omega J_p\dot{\phi} + K_{12}z + K_{22}\phi = \chi\omega^2(J_t - J_p)e^{i\omega t}\,, \end{cases}$$

$$(4.68)$$

or, in the more compact form,

$$\mathbf{M}\ddot{\mathbf{q}} - i\omega\mathbf{G}\dot{\mathbf{q}} + \mathbf{K}\mathbf{q} = \omega^2\mathbf{f}e^{i\omega t}\,.$$

$$(4.69)$$

The vector of complex coordinates, mass matrix, gyroscopic matrix, stiffness matrix, and vector of unbalances in equations (4.69) are, respectively

$$\mathbf{q} = \left\{ \begin{array}{c} z \\ \phi \end{array} \right\}, \qquad \mathbf{M} = \left[\begin{array}{cc} m & 0 \\ 0 & J_t \end{array} \right], \qquad \mathbf{G} = \left[\begin{array}{cc} 0 & 0 \\ 0 & J_p \end{array} \right],$$

$$\mathbf{K} = \left[\begin{array}{cc} K_{11} & K_{12} \\ K_{12} & K_{22} \end{array} \right], \qquad \mathbf{f} = \left\{ \begin{array}{c} m\epsilon e^{i\alpha} \\ \chi(J_t - J_p) \end{array} \right\}.$$

Note that all matrices are symmetric when using the complex coordinate notation, and when using real coordinates, the gyroscopic matrix is skew symmetric.

The presence of damping can be accounted for in a very easy way if the viscous or hysteretic damping models can be accepted. As in the case of the Jeffcott model, it is important to distinguish between nonrotating and rotating damping and to introduce separately the two damping matrices. In the case of viscous damping, the equation of motion of the damped system is

$$\mathbf{M}\ddot{\mathbf{q}} + (\mathbf{C}_n + \mathbf{C}_r - i\omega\mathbf{G})\dot{\mathbf{q}} + (\mathbf{K} - i\omega\mathbf{C}_r)\mathbf{q} = \omega^2\mathbf{f}e^{i\omega t}, \qquad (4.70)$$

where the structure of matrices $\mathbf{C}_n$ and $\mathbf{C}_r$ is similar to that of the stiffness matrix.

To conclude this section, it must be observed that by linearizing the equation of motion of a rotating gyroscopic system a linear equation very similar to that of a discrete vibrating system has been obtained. The main differences are the presence of the gyroscopic matrix and, in the case of damped systems, of a term $i\omega\mathbf{C}_r$ linked to rotating damping. Obviously, when the spin speed ω tends to zero, the equation of motion of a nonrotating system is obtained.

4.4.3 Free whirling of the undamped system

The free whirling of the undamped system can be studied using the homogeneous equation associated with equation (4.68). Introducing a solution of the type

$$z = z_0 e^{i\lambda t}, \qquad \phi = \phi_0 e^{i\lambda t}$$

into the equation of motion, the following algebraic linear equations are readily obtained

$$\begin{cases} z_0 \left(-m\lambda^2 + K_{11}\right) + \phi_0 K_{12} = 0\,, \\ z_0 K_{12} + \phi_0 \left(-J_t\lambda^2 + J_p\lambda\omega + K_{22}\right) = 0\,. \end{cases} \qquad (4.71)$$

The characteristic equation allowing computation of the whirl frequency is

$$\det \begin{bmatrix} -m\lambda^2 + K_{11} & K_{12} \\ K_{12} & -J_t\lambda^2 + J_p\lambda\omega + K_{22} \end{bmatrix} = 0\,, \qquad (4.72)$$

which yields

$$\lambda^4 - \omega\lambda^3\frac{J_p}{J_t} - \lambda^2\left(\frac{K_{11}}{m} + \frac{K_{22}}{J_t}\right) + \omega\lambda\frac{K_{11}J_p}{mJ_t} + \frac{K_{11}K_{22} - K_{12}^2}{mJ_t} = 0\,. \qquad (4.73)$$

Equation (4.73) has four real roots, two of which are positive. The Campbell diagram of the system is of the type shown in Figure 4.16: At each speed ω, four whirl modes, occurring at different frequencies, are possible. Two of them occur in the forward direction and two in the backward direction.

Because all solutions of equation (4.73) are real, the corresponding eigenvectors $\mathbf{q}_i$ are also real. At time $t = 0$, both Y and $\phi_{X'}$ vanish: the z-axis is contained in a plane also containing the Z-axis and rotating about the latter with a constant angular velocity equal to the whirl speed λ. The axis of the rotor describes a cone whose axis is the rotor axis in its undeformed position. Because the deflected shape is contained in a plane, the motion could have been studied in such a plane from the beginning, using a model with only two degrees of freedom. This model, however, cannot be used to study the damped system, and a more general model was preferred from the beginning. Actually, the free whirling of the system can be more complex because the motion is the combination of all four circular whirling motions that occur at different frequencies

$$\mathbf{q} = \sum_{k=1}^{4} \mathbf{q}_k e^{i\lambda_k t}\,. \qquad (4.74)$$

The gyroscopic terms couple the behaviour in the planes passing through the rotation axis and make it impossible for the system to

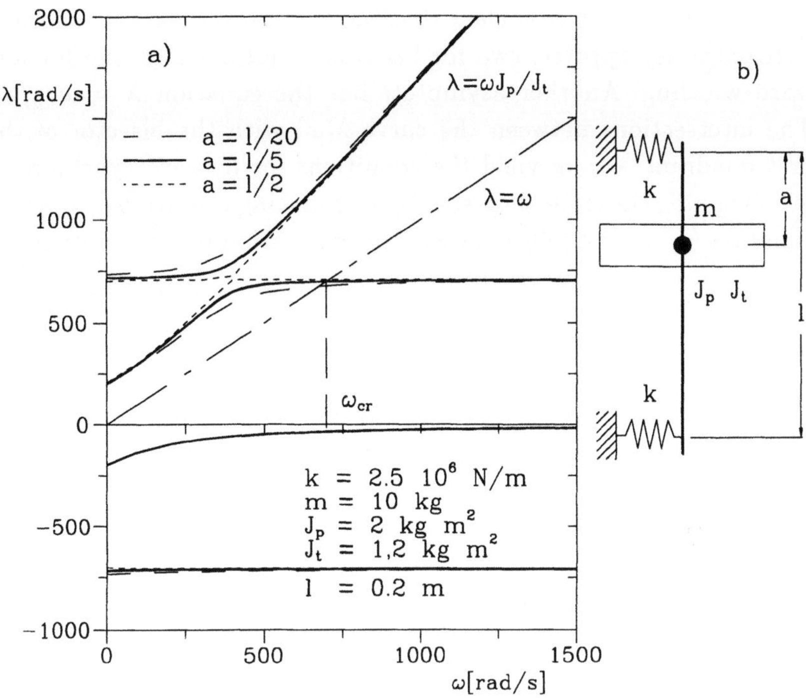

FIGURE 4.16. Campbell diagram of a system made by a rigid rotor on elastic supports.

perform elliptical or rectilinear motions: An equation of the type of equation (4.74) can yield elliptical motions only in the case in which two of the eigenfrequencies λ_k have the same modulus and opposite sign, as was the case of the Jeffcott rotor. The introduction of gyroscopic moments causes forward whirl frequencies to be different from backward whirl frequencies and all whirl motions to be circular. Obviously the four circular modes can add to each other, yielding Lissajous curves. Another effect of gyroscopic moments is that causing the whirl frequencies to be functions of the spin speed. Generally speaking, the frequency of forward modes increases with speed, while the frequency of backward modes decreases.

To avoid solving equation (4.73) in λ, the Campbell diagram can be obtained solving the same equation in ω. The equation is linear in this unknown and a closed-form solution can be easily found

$$\omega = \frac{mJ_t\lambda^4 - (J_tK_{11} + mK_{22})\lambda^2 + K_{11}K_{22} - K_{12}^2}{\lambda J_p(m\lambda^2 - K_{11})}. \tag{4.75}$$

From Figure 4.16 it is clear that the Campbell diagram has three horizontal asymptotes; two for backward motions and one for forward whirling. Another asymptote has the equation $\lambda = \omega J_p/J_t$. The intersections between the curve $\lambda(\omega)$ with the bisector of the first quadrant $\lambda = \omega$ yield the conditions for forward synchronous whirling, i.e., the critical speeds. By introducing condition $\omega = \lambda$ into equation (4.73) the following quadratic equation in ω^2 is obtained:

$$\omega^4 m(J_p - J_t) - \omega^2[(J_p - J_t)K_{11} - mK_{22}] - K_{11}K_{22} + K_{12}^2 = 0 . \quad (4.76)$$

By neglecting the negative solutions corresponding to the intersections of curve $\lambda(\omega)$ with the straight line $\lambda = \omega$ lying in the third quadrant of the $\lambda\omega$-plane, the following values of the critical speed are obtained:

$$\omega_{cr} = \sqrt{\frac{K_{11}(J_p - J_t) - mK_{22} \pm \sqrt{\left[K_{11}(J_p - J_t) + mK_{22}\right]^2 - 4m(J_p - J_t)K_{12}^2}}{2m(J_p - J_t)}} .$$
$$(4.77)$$

Introducing the ratio $\delta = (J_p - J_t)/m$, equation (4.77) can be written in the more compact form

$$\omega_{cr} = \frac{1}{\sqrt{2m}}\sqrt{K_{11} - \frac{K_{22}}{\delta} \pm \sqrt{\left(K_{11} + \frac{K_{22}}{\delta}\right)^2 - 4\frac{K_{12}^2}{\delta}}} . \quad (4.78)$$

If $J_p < J_t$ or $\delta < 0$, as happens in the case of long rotors, there are two real solutions and, as a consequence, two values of the critical speed. If, on the contrary, $J_p > J_t$ or $\delta > 0$, as happens in the case of discs, one of the solutions is imaginary and only one critical speed exists. If $K_{12} = 0$, i.e. if translational and rotational degrees of freedom are uncoupled, the two equations can be studied separately and the Campbell diagram degenerates into two straight lines with equations $\lambda = \pm\sqrt{K_{11}/m}$, as in the case of Jeffcott rotors, and in a curve that has for asymptotes the ω-axis in backward whirling, and the straight line $\lambda = \omega J_p/J_t$ in forward whirling. The straight lines and the curve can cross in the first quadrant, as in Figure 4.16, or in the fourth quadrant. The values of the critical speed are

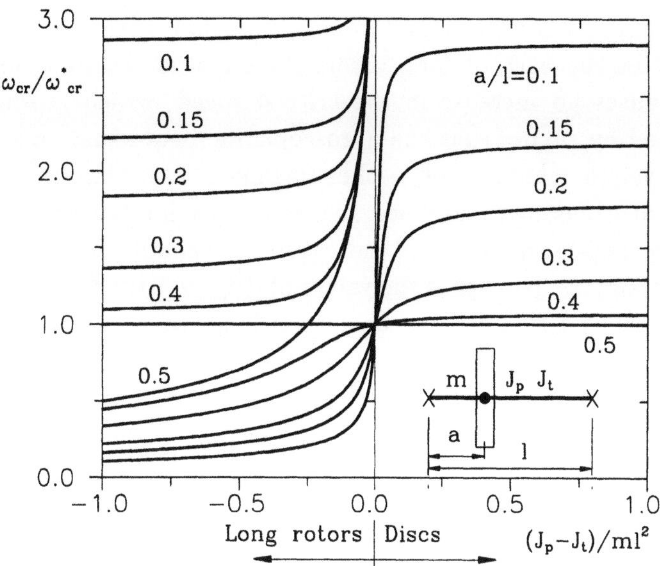

FIGURE 4.17. Influence of the gyroscopic moment on critical speeds.

$$\begin{cases} \omega_{cr\,I} = \sqrt{\dfrac{K_{11}}{m}}, \\[2ex] \omega_{cr\,II} = \sqrt{-\dfrac{K_{22}}{m\delta}} = \sqrt{-\dfrac{K_{22}}{J_p - J_t}}. \end{cases} \tag{4.79}$$

The second value is real only if $\delta < 0$.

Consider a rotor made by a rigid gyroscopic body attached to a flexible uniform shaft running on rigid bearings (Figure 4.17). The stiffness matrix of the system is

$$\mathbf{K} = \frac{3EIl}{a(l-a)} \begin{bmatrix} \dfrac{l^2 - 3al + 3a^2}{a^2(l-a)^2} & \dfrac{2a-l}{a(l-a)} \\[2ex] \dfrac{2a-l}{a(l-a)} & 1 \end{bmatrix}. \tag{4.80}$$

The influence of the gyroscopic moment on the critical speeds can be shown by plotting the values of the critical speed or, better, of ratio $\omega_{cr}/\omega^*_{cr}$, where ω^*_{cr} is the critical speed computed neglecting gyroscopic effects, as functions of δ or, better, of nondimensional parameter $\delta/l^2 = (J_p-J_t)/ml^2$. The graph, plotted for various values of a/l, is shown in Figure 4.17. If the rotor is at midspan, i.e., if $a/l = 0.5$, the gyroscopic moment has no effect on critical speeds in the case of the discs and causes a second critical speed to be present in the case of long rotors. In all other cases, a disc-type rotor causes an increase in the critical speed, which is sometimes explained by saying that the gyroscopic moment causes a stiffening of the system, which is only a phenomenological explanation because no actual stiffening takes place. In the case of a long rotor, the critical speed decreases and a second critical speed, usually quite higher than the first, occurs. Note that the values of $|(J_p - J_t)|/ml^2$ are in practice quite small, particularly in the case of long rotors, and Figure 4.17 holds only for a narrow zone about the ordinate axis. On the right side of this narrow zone the discs are too thin and they can no longer be considered rigid bodies, and on the left the rotor becomes too long to be considered rigid again. The considerations drawn from Figure 4.17, although obtained for a particular case, are, however, qualitatively applicable in general.

4.4.4 Unbalance response of the undamped system

The unbalance response can be easily studied starting from equation (4.69), whose particular integral can be expressed in the form $z = z_0 e^{i\omega t}$, $\phi = \phi_0 e^{i\omega t}$, where the amplitudes z_o and ϕ_o are, in general, complex numbers because static and couple unbalances do not lie in the same plane, i.e., the phase angle α is, in general, not equal to zero. The effects of the two types of unbalance can be studied separately, and then the relevant solutions can be added to each other.

In the case of a static unbalance with vanishing phase α, the amplitude of the response is

$$\begin{cases} z_0 = m\epsilon\omega^2 \dfrac{(J_p - J_t)\omega^2 + K_{22}}{\Delta}, \\[2mm] \phi_0 = -m\epsilon\omega^2 \dfrac{K_{12}}{\Delta}, \end{cases} \tag{4.81}$$

where

$$\Delta = -m(J_p - J_t)\omega^4 + [K_{11}(J_p - J_t) + mK_{22}]\omega^2 + K_{11}K_{22} - K_{12}^2.$$

In the case of couple unbalance, the response is

$$\begin{cases} z_0 = \chi\omega^2(J_p - J_t)\dfrac{K_{12}}{\Delta}, \\[2mm] \phi_0 = \chi\omega^2(J_p - J_t)\dfrac{m\omega^2 - K_{11}}{\Delta}. \end{cases} \tag{4.82}$$

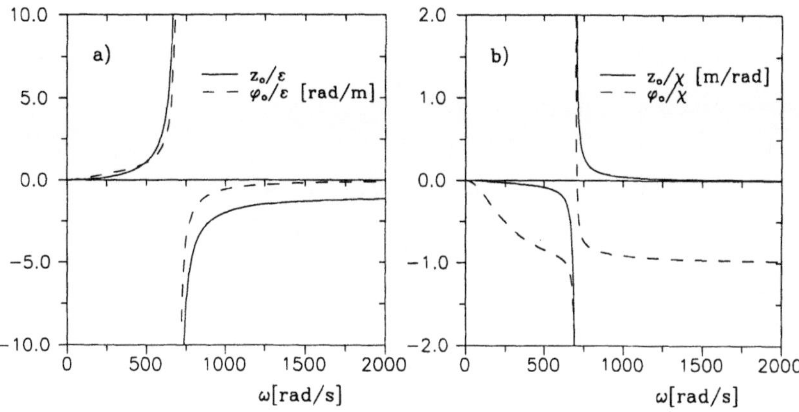

FIGURE 4.18. Amplitude of the unbalance response for the same rotor of Figure 4.16 with $l/a = 4$; (a) response to static unbalance; (b) response to couple unbalance.

Equations (4.81) and (4.82) are plotted in Figure 4.18 for the same rotor as Figure 4.16 with $l/a = 4$. As is clear from the figure, in the current case, self-centering occurs at speeds in excess of the critical speed in the case of static unbalance, while a certain self-centering also occurs in the subcritical field in the case of couple unbalance.

This is because the system has only one critical speed, because $J_p > J_t$, and the natural frequency at standstill related to the rotational mode is lower than that linked with translational mode.

Example 4-2
The schematic cross section of a turbomolecular pump is reported in Figure 4.19b, together with the few basic dimensions that are required. Knowing that the operating speed is 30,000 rpm and that the first critical speed must be lower than 15,000 rpm and the second must be above the working range, state the dimensions of the shaft.
The rotor of the pump can be modeled as a rigid body, with its center of gravity at the connection with the shaft, attached to a massless shaft. The inertial characteristics of the rigid body are $m = 6$ kg, $J_p = 0.035$ kg m^2, and $J_t = 0.055$ kg m^2. The material of the shaft has a Young's modulus $E = 2.1 \times 10^{11}$ N/m^2.
The mass of the shaft has been neglected, and the model with two complex degrees of freedom studied in Section 4.6 can be used. The model is sketched in Figure 4.19a. Choose as generalized coordinates the displacement and rotation at the center of mass of the rigid body, neglect the compliance of the bearings and of the stator of the machine, and assume that the shaft has constant diameter.

(a)

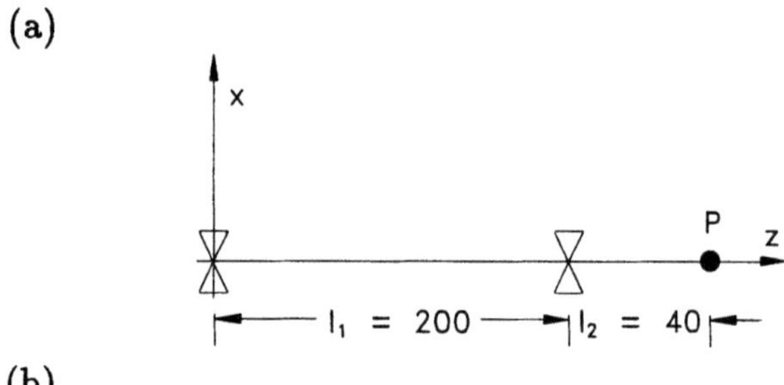

(b)

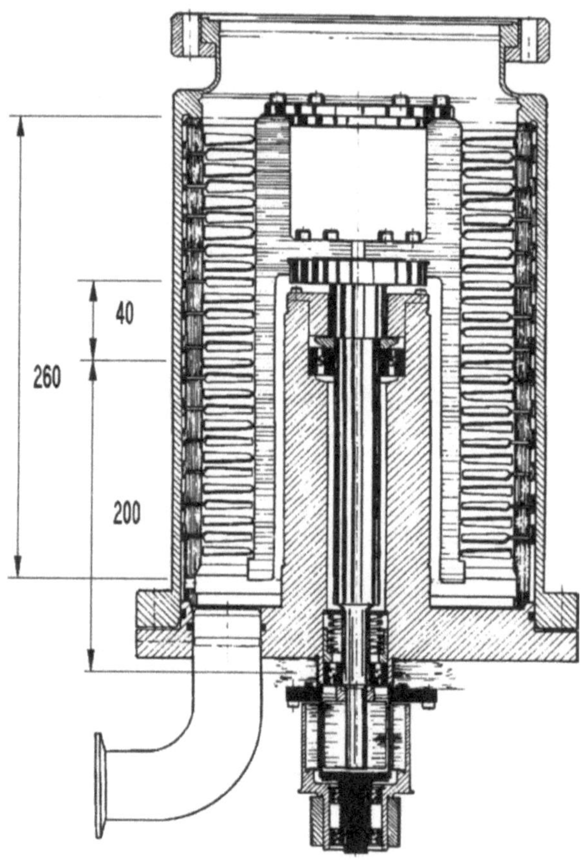

FIGURE 4.19. Turbomolecular pump; (a) model of the rotor; (b) schematic drawing (*Vacuum Technology; Its Foundation, Formulae and Tables*, Leybold-Heraeus, Köln, Germany). The drawing refers to a machine no longer in production.

The stiffness matrix can be computed using the FEM. However, due to the simple geometry, it is possible to immediately obtain the compliance matrix, which can then be inverted to yield the stiffness matrix. By applying a unit force and a unit moment in point P and computing the displacement and rotation in the same point (for simple geometries like the current one, the relevant formulas are reported in many handbooks), the elements of the compliance matrix β are obtained: $\beta_{11} = l_2^2(l_1 + l_2)/3EI$, $\beta_{12} = l_2(2l_1 + 3l_2)/6EI$, $\beta_{22} = (l_1 + 3l_2)/3EI$. The stiffness matrix $\mathbf{K}$ is obtained by inverting the compliance matrix β

$$\mathbf{K} = \frac{6EI}{l_2^3(4l_1 + 3l_2)} \begin{bmatrix} 2(l_1 + 3l_2) & -l_2(2l_1 + 3l_2) \\ -l_2(2l_1 + 3l_2) & 2l_2^2(l_1 + l_2) \end{bmatrix} =$$

$$= I \begin{bmatrix} 1.3696 & -0.0445 \\ -0.0445 & 0.001644 \end{bmatrix} \times 10^{16}.$$

The critical speeds can be computed using equation (4.77): $\omega_{cr_I} = 0.8611 \times 10^7\sqrt{I}$ rad/s $= 0.88223 \times 10^8\sqrt{I}$ rpm, $\omega_{cr_{II}} = 5.504 \times 10^7\sqrt{I}$ rad/s $= 5.2566 \times 10^8\sqrt{I}$ rpm.

The large difference between the first and second critical speeds allows the choice of a low value of the first critical speed without any danger that the second is located within the working range. By assuming that $\omega_{cr_I} = 10{,}000$ rpm, a value of the moment of inertia of the shaft $I = 1.479 \times 10^{-8}$ m^4 and then a shaft diameter of 23.4 mm is obtained. Choosing a value of 24 mm, the moment of inertia is $I = 1.628 \times 10^{-8}$ m^4 and the critical speeds are $\omega_{cr_I} = 10{,}490$ rpm, $\omega_{cr_{II}} = 67{,}080$ rpm. The second value is quite high, and it is likely that other critical speeds are located between the computed values. A more realistic model, which also takes into account the mass of the shaft and the compliance of the bearings, must, however, be used to obtain them.

4.5 Dynamic study of rotors with many degrees of freedom

4.5.1 General considerations

Two simplified models that allow at least a qualitative understanding of the dynamic behaviour of linear axisymmetrical rotors have been shown in the preceding sections. In them the rotor was modeled as a point mass or a rigid body connected to a massless elastic shaft running in massless elastic bearings. These models are, however, inadequate to supply quantitative information on the behaviour of complex rotating systems, and more accurate models are needed in the design stage. Also, in the case of rotors, both the continuum and the discretized approaches are possible, at least in theory.

Consider a rotor that can be modeled as a beam that is slender enough to use the standard beam theory. If the supports can be assumed to be rigid, there is no difficulty computing the natural frequencies at standstill. If the polar moment of inertia of the cross sections can be neglected, gyroscopic moments do not significantly affect the dynamic behaviour of the system when rotating. In this situation, the natural frequencies are independent of the spin speed, and the values computed for the nonrotating beam hold for any value of the speed ω. The Campbell diagram is made by straight lines parallel to the ω-axis, and the values of the critical speeds coincide with those of the natural frequencies.

Transmission shafts with a simple geometrical shape are usually modeled as cylindrical Euler-Bernoulli beams. There is, however, no difficulty studying the dynamic behaviour of a cylindrical shaft, taking into account the rotational inertia of the cross section and shear deformation. Also, the gyroscopic effect can be accounted for, and it is possible to show that the three effects are of the same order of magnitude, and if one of them is considered, it is advisable to include them all. In the case of rotors, the gyroscopic effect causes a stiffening of the system, which increases with increasing speed. In critical speed conditions, usually the combined effect of the three phenomena is that of increasing the critical speed, while at standstill they cause a decrease of the natural frequency.[4].

The continuum approach is, however, feasible only in the case of very simple geometrical shapes, such as shafts with constant cross section. Dealing with more complicated shapes using the continuum approach leads to analytical difficulties, and today the most popular methods are based on some discretization technique. In particular, methods based on transfer matrices were, and still are, widely used, even if they are now yielding to methods based on the FEM.

Very often, rotors are modeled using beam elements, with inertial properties modeled using the lumped or the consistent approach. Although there is no difficulty using other types of elements, the relevant gyroscopic matrices are usually not available to users of

---

[4]See, for example, F.M. Dimentberg, *Flexural vibrations of rotating shafts*, Butterworths, London, 1961; G. Genta, "Consistent matrices in rotor dynamics", *Meccanica*, Vol. 20, (1985), 235–248.

standard finite element codes.

4.5.2 Lumped-parameters models: Myklestadt-Prohl method

Among lumped-parameters methods, the transfer matrix approach, namely, the Myklestadt method was, and still is, widely used in rotor dynamics, mainly for critical speed prediction. The formulation seen in Section 2.6.4 holds with the only difference of adding the gyroscopic moment in the node transfer matrices expressed by equation (2.83). As it can be easily obtained from the second equation (4.71), the gyroscopic moment due to the ith rigid body located in the ith node can be accounted for by substituting $\lambda \omega J_p - \lambda^2 J_t$ to $-\lambda^2 J_t$ in the expression of the generalized inertia force related to the rotational degree of freedom. By performing the mentioned substitution in equation (2.83), the Myklestadt method is easily converted to the study of the dynamic behaviour of rotors. If the aim is only to evaluate the critical speeds, it is possible to introduce $\lambda = \omega = \omega_{cr}$ into the node transfer matrix, which becomes

$$
\mathbf{T}_{i_n} = \begin{bmatrix}
1 & 0 & 0 & 0 \\
0 & 1 & 0 & 0 \\
-m_i \omega_{cr}^2 + k_c & 0 & 1 & 0 \\
0 & -(J_{t_i} - J_{p_i})\omega_{cr}^2 + \chi_c & 0 & 1
\end{bmatrix}. \tag{4.83}
$$

The computation follows the guidelines already seen in Section 6.3.4 without modifications.

When using the stiffness or compliance approach, the whole structure is modeled as for nonrotating systems in Section 2.6.1. If complex coordinates are used, the relevant equations are either equation (4.69) or (4.70), depending on whether or not damping is neglected. The size of all matrices and vectors is equal to the number of complex degrees of freedom, which coincides with the number of degrees of freedom related to bending behaviour in the xz- and yz-planes. Mass, stiffness, and damping matrices $\mathbf{M}$, $\mathbf{K}$, and $\mathbf{C}$ are those related to the flexural behaviour of the nonrotating system in the xz-plane, the choice of the relevant plane depends on the way in which the complex coordinates were defined. This is obvious because the equation of motion must become the same that describes the free behaviour of the nonrotating system when the spin speed ω tends to zero. Also, in this case the stiffness matrix can be directly obtained using the FEM or from the matrix of the coefficients of influence.

If the complex coordinates are ordered as follows

$$\mathbf{q} = \begin{bmatrix} z_1 & \phi_1 & z_2 & \phi_2 & \ldots & z_n & \phi_n \end{bmatrix}^T , \qquad (4.84)$$

matrices $\mathbf{M}$ and $\mathbf{G}$ and vector $\mathbf{f}$ of the unbalances are, respectively,

$$\mathbf{M} = \mathrm{diag}\begin{bmatrix} m_1 & J_{t_1} & m_2 & J_{t_2} & \ldots & m_n & J_{t_n} \end{bmatrix} ,$$

$$\mathbf{G} = \mathrm{diag}\begin{bmatrix} 0 & J_{p_1} & 0 & J_{p_2} & \ldots & 0 & J_{p_n} \end{bmatrix} , \qquad (4.85)$$

$$\mathbf{f} = \begin{bmatrix} m_1\epsilon_1 e^{i\alpha_1} & (J_{t_1} - J_{p_1})\chi_1 e^{i\beta_1} & \ldots & m_n\epsilon_n e^{i\alpha_n} & (J_{t_n} - J_{p_n})\chi_n e^{i\beta_n} \end{bmatrix}^T .$$

The phases α_i e β_i are needed to take into account the possible different orientations in space of the vectors expressing static and couple unbalances. Rotating damping matrix $\mathbf{C}_r$ can be built the same way as for general damping matrices, taking into account only the contribution to the overall energy dissipation due to rotating parts of the machine.

4.5.3 Models with consistent inertial properties

There is no difficulty in adding the contribution of rotation to the kinetic energy of an element of any type into the FEM. The mass matrix that results from this approach is coincident with that related to the nonrotating model. Also, the stiffness and damping matrices are not affected by the fact that the system rotates, apart from a possible geometrical effect due to the stiffening of the element that can be ascribed to centrifugal stressing. This effect, which is usually neglected in the formulation of the beam elements used to model rotating shafts, must be considered when dealing with beam elements in a direction perpendicular to the rotation axis, such as those used to model turbine blades, propellers, and similar structural members. If centrifugal stiffening must be considered, it can be generally taken into account by adding a term of the type $\omega^2\mathbf{K}_\omega$ to the stiffness matrix, where $\mathbf{K}_\omega$ is a matrix of constants that can be computed at the element level and then assembled in the usual way.

The effect of axial forces acting on the rotor can be accounted for by using the standard formulation of the geometric matrix.

The greatest difference between a finite element model for a rotor and that for the corresponding nonrotating structure is the presence in the first of a gyroscopic term. It is possible to show with simple computations that the consistent gyroscopic matrix of a beam element is identical to the part of the mass matrix related to the rotational inertia of the cross sections in which the polar moment of the cross section I_z is substituted for the transversal moment of inertia I_y. As in case of an axisymmetrical beam $I_z = 2I_y$, the gyroscopic matrix is equal to twice the rotational contribution to the mass matrix. If there is damping in the system, the damping matrices of the nonrotating and rotating elements can be assembled separately, directly yielding the two matrices $\mathbf{C}_n$ and $\mathbf{C}_r$ to be introduced into the equations of motion. The consistent unbalance vectors for a linear distribution of static and couple unbalance are

$$\mathbf{f} = \mathbf{a}_1 \left\{ \begin{array}{c} \epsilon_{\xi_1} + i\epsilon_{\eta_1} \\ \epsilon_{\xi_2} + i\epsilon_{\eta_2} \end{array} \right\} + \mathbf{a}_2 \left\{ \begin{array}{c} \chi_{\eta_1} - i\chi_{\xi_1} \\ \chi_{\eta_2} + i\chi_{\xi_2} \end{array} \right\}, \qquad (4.86)$$

where

$$\mathbf{a}_1 = \frac{\rho A l}{120(1 - \Phi)} \begin{bmatrix} 42 + 40\Phi & 18 + 20\Phi \\ l(6 + 5\Phi) & l(4 + 5\Phi) \\ 18 + 20\Phi & 42 + 40\Phi \\ -l(4 + 5\Phi) & -l(6 + 5\Phi) \end{bmatrix},$$

$$\mathbf{a}_2 = \frac{\rho I_y}{12(1 - \Phi)} \begin{bmatrix} -6 & -6 \\ l(1 + 4\Phi) & l(-1 + 2\Phi) \\ 6 & 6 \\ l(-1 + 2\Phi) & l(1 + 4\Phi) \end{bmatrix}$$

and the other symbols have the same meaning seen in Example 2-6. The eccentricity of the elements has been defined by the components along the ξ- and the η-axes in correspondence to the two end nodes of the element. In the same way, the couple unbalance has been defined, even if it is difficult to correctly define a distributed couple unbalance.

The matrices of the elements can be assembled and condensed in the same way as for nonrotating structures. Gyroscopic matrices are condensed using the same algorithm as for mass matrices.

The model can also include the stator of the machine as well as various rotors with different rotating speeds, provided that the whole system is axially symmetrical, as in the case of multishaft turbines.

4.5.4 Real versus complex coordinates

The equations of motion of axisymmetrical rotors have been obtained in the preceding sections with reference to a set of complex coordinates of the type defined by equation (4.67). The use of complex coordinates is very convenient in the case of axisymmetrical systems, particularly if damping is neglected, because it allows the study of the system using a model whose size is just half the size of the same problem expressed in real coordinates. Note that even if the coordinates are complex, all relevant matrices are real and the computation of the critical speeds, the undamped Campbell diagram, and the undamped unbalance response does not involve actual working with complex numbers, as will be shown in detail in the following sections.

Another advantage of the use of complex coordinates is that of obtaining the response in terms of rotating vectors that rotate in the physical space, directly giving complete information about the orbits of the various nodes and the direction of the whirling motions. Using real coordinates, however, the rotating vectors are defined with reference to the Argand plane and the whirling motion is obtained as the compositions of harmonic vibrations in the directions of the x- and y-axes. The main limitation of the complex coordinate approach is that it deals with difficulty with elliptical orbits, as those occurring in the case of nonaxisymmetrical machines, as will be seen in Section 4.8.

Equation (4.70) can be written by separating the real part of each equation from the imaginary part, obtaining

$$
\begin{bmatrix} \mathbf{M} & 0 \\ 0 & \mathbf{M} \end{bmatrix} \ddot{x} + \left(\omega \begin{bmatrix} 0 & \mathbf{G} \\ -\mathbf{G} & 0 \end{bmatrix} + \begin{bmatrix} \mathbf{C} & 0 \\ 0 & \mathbf{C} \end{bmatrix} \right) \dot{x} + \tag{4.87}
$$

$$
\left(\begin{bmatrix} \mathbf{K} & 0 \\ 0 & \mathbf{K} \end{bmatrix} + \omega \begin{bmatrix} 0 & \mathbf{C}_r \\ -\mathbf{C}_r & 0 \end{bmatrix} \right) x = \omega^2 \left\{ \begin{array}{c} \Re(\mathbf{f}e^{i\omega t}) \\ \Im(\mathbf{f}e^{i\omega t}) \end{array} \right\},
$$

where $\mathbf{x} = [\Re\mathbf{q}^T, \Im\mathbf{q}^T]^T$ and $\mathbf{C}$ is the total damping matrix, $\mathbf{C} = \mathbf{C}_n + \mathbf{C}_r$. Note that in the case of complex coordinates all matrices were symmetrical, but the use of real coordinates results in skew-symmetrical gyroscopic and rotating damping (circulatory) matrices. The real coordinates $\mathbf{x}$ defined earlier differ from the standard real coordinates used for the study of nonrotating systems for what the sign of the rotational degrees of freedom related to rotation about the x-axis is concerned. In many cases, a generalized form of equation (4.66) is used, in which the sign conventions for rotational coordinates are standard in the FEM. The equation of motion of the damped system, then, is

$$\begin{bmatrix} \mathbf{M}_x & 0 \\ 0 & \mathbf{M}_y \end{bmatrix} \ddot{\mathbf{x}}^* + \left(\omega \begin{bmatrix} 0 & \mathbf{G}_{xy} \\ -\mathbf{G}_{yx} & 0 \end{bmatrix} + \begin{bmatrix} \mathbf{C}_x & 0 \\ 0 & \mathbf{C}_y \end{bmatrix} \right) \dot{\mathbf{x}}^* +$$

$$+ \left(\begin{bmatrix} \mathbf{K}_x & 0 \\ 0 & \mathbf{K}_y \end{bmatrix} + \omega \begin{bmatrix} 0 & \mathbf{C}_{r_{xy}} \\ -\mathbf{C}_{r_{yx}} & 0 \end{bmatrix} \right) \mathbf{x}^* = \qquad (4.88)$$

$$= \omega^2 \left\{ \begin{array}{c} \mathbf{f}_x \cos(\omega t) - \mathbf{f}_y \sin(\omega t) \\ \mathbf{f}_x \sin(\omega t) + \mathbf{f}_y \cos(\omega t) \end{array} \right\} ,$$

where matrices with subscript x are the same as for complex coordinates, matrices with subscript y are similar, except for the sign of elements with subscripts made by two numbers whose sum is an odd number. Matrices with subscripts xy and yx have signs that differ from those of the corresponding matrices in equation (4.87) and are such that the global gyroscopic and rotating damping matrices are skew-symmetrical. Vector $\mathbf{x}^*$ is a vector containing the generalized coordinates and is of the type of the vector of the generalized coordinates of equation (4.66).

The order of the degrees of freedom shown in equations (4.87) and (4.88), with all coordinates related to the xz-plane listed one after the other followed by those related to the other plane, is just an indication because it would lead to matrices with a very large band. Actually, the degrees of freedom are mixed at the element level and the structure of the matrices shown holds for the matrices of the elements and not for those related to the whole structure.

4.5.5 Fixed versus rotating coordinates

Although the flexural behaviour of rotating axisymmetrical systems can be studied using an inertial coordinate frame as seen in the preceding sections, it is possible to use a set of coordinates based on the rotating frame $O\ \Xi H Z$. A set of complex coordinates $\mathbf{r}$, related to the coordinates $\mathbf{q}$, can thus be obtained by the relationship

$$\mathbf{r} = \mathbf{q}e^{-i\omega t} \qquad (4.89)$$

and the equation of motion (4.70) transforms into the following equation:

$$\mathbf{M\ddot{r}} + \left[\mathbf{C}_n + \mathbf{C}_r + i\omega(2\mathbf{M} - \mathbf{G})\right]\mathbf{\dot{r}} +$$
$$+ \left[\mathbf{K} - \omega^2(\mathbf{M} - \mathbf{G}) + i\omega\mathbf{C}_n\right]\mathbf{r} = \omega^2\mathbf{f}. \qquad (4.90)$$

The same considerations seen for equation (4.37) obtained for the Jeffcott model also hold for equation (4.90). Also in the case of rotating coordinates it is possible to chose between the complex coordinates approach seen earlier and an alternative formulation of the problem based on real coordinates. The relevant equation can be easily obtained by separating the real and imaginary parts of equation (4.90). The gyroscopic, Coriolis, and nonrotating damping terms give way to skew-symmetrical matrices. The use of rotating coordinates, which is not very convenient in the case of axisymmetrical systems, becomes advisable when dealing with machines including a nonisotropic rotor (see Section 4.8.3).

4.5.6 State space equations

All the preceding equations can be written with reference to the state space. If complex coordinates are used, the state vector referred to an inertial reference is

$$\mathbf{z} = \left\{ \begin{array}{c} \mathbf{\dot{q}} \\ \mathbf{q} \end{array} \right\}. \qquad (4.91)$$

The corresponding state matrix and product $\boldsymbol{B}\mathbf{u}$ due to unbalance are, respectively,

$$\boldsymbol{A} = \left[\begin{array}{cc} -\mathbf{M}^{-1}\left(\mathbf{C}_n + \Xi_r - i\omega\mathbf{G}\right) & -\mathbf{M}^{-1}\left(\mathbf{K} + \omega^2\mathbf{K}_\omega - i\omega\mathbf{C}_r\right) \\ \mathbf{I} & \mathbf{0} \end{array} \right], \qquad (4.92)$$

$$\mathcal{B}\mathbf{u} = \omega^2 e^{i\omega t} \left\{ \begin{array}{c} \mathbf{f} \\ \mathbf{0} \end{array} \right\}. \tag{4.93}$$

Note that the dynamic matrix is complex, due to the presence of the gyroscopic and rotating damping (circulatory) terms and, as a consequence, its eigenvalues are nonconjugated. When the roots locus, with the spin speed used as a parameter, is plotted, it is not symmetrical with respect to the real axis. This was also the case for the Jeffcott rotor, as shown in Figure 4.8, but in the current case it also occurs for the undamped system owing to the gyroscopic term.

State-space equations can also be written with reference to real coordinates, in which case a real dynamic matrix is always obtained, or rotor-fixed coordinates are used.

4.5.7   Static solution

In this section, the term *static solution* will be referred to as the computation of the deformed shape and of the stress and strain fields of a rotor under the effects of constant forces. The relevant equation yielding the static displacement is easily obtained by adding a vector of static or nonrotating forces $\mathbf{f}_n$ on the right-hand side of equation (4.70) and neglecting all terms containing the time derivatives of the generalized coordinates. There is no difficulty in introducing the hysteretic damping matrix $\mathbf{K}_r''$, i.e., the imaginary part of the complex stiffness matrix related to the rotating part of the system into the equation of motion, following the same procedures as in Section 4.5.5, and remembering that in the case of a static solution the whirl speed λ vanishes. Taking into account centrifugal stiffening, it follows that

$$\left(\mathbf{K} + \omega^2 \mathbf{K}_\omega - i\omega \mathbf{C}_r - i\mathbf{K}_r'' \right) \mathbf{q} = \mathbf{f}_n. \tag{4.94}$$

Neglecting centrifugal stiffening, from equation (4.94) it is clear that, in the case of an undamped rotor the static solution of the rotating system coincides with that of the corresponding nonrotating structure, but when there is damping, the deflected shape is influenced by the spin speed. This effect is easily understood if the force vector is assumed to be real, i.e., if all forces and moments act in the same plane, e.g., in the xz-plane, as in the case of a rotor loaded by self-weight. The presence of imaginary terms in the matrix of the coefficients of equation (4.94) results in a complex displacement vector $\mathbf{q}$. The presence of damping then causes a lateral deviation, which, in the case of viscous damping, depends on the spin speed, of the deformed shape of the rotor from the plane in which all loads are acting. The measurement of such side deviation in a rotating-bending fatigue test has been suggested several times and was actually used to measure the internal damping of materials.

4.5.8 Critical-speed computation

Critical speeds can easily be computed from the homogeneous equation of motion (4.69), yielding free whirling with the added condition $\lambda = \omega$. The following eigenproblem is obtained:

$$\left(-\omega^2(\mathbf{M} - \mathbf{G} - \mathbf{K}_\omega) + \mathbf{K}\right)\mathbf{q} = \mathbf{0}\,. \tag{4.95}$$

Note that the size of the eigenproblem, in which only real quantities are involved, is equal to the number of complex degrees of freedom. Mathematically, the problem is the same as already seen for the computations of the natural frequencies of an undamped vibrating system whose mass matrix is $\mathbf{M} - \mathbf{G} - \mathbf{K}_\omega$. The only difference is that in the current case the mass matrix can be nonpositive definite. All properties of eigenvectors seen for vibrating systems still hold because they were related only to the symmetry of the relevant matrices: If matrix $\mathbf{M} - \mathbf{G} - \mathbf{K}_\omega$ is nonpositive definite, the only consequence is that some of the modal masses are negative and the corresponding eigenvalues are imaginary.[5] The eigenvectors are m-orthogonal only if complex coordinates are used: The real coordinate approach yields a skew-symmetrical gyroscopic matrix and makes it necessary to resort to a set of left and right eigenvectors.

In many cases excitations whose frequency is a multiple of the spin speed can be present: As will be seen later, very often the excitation provided by constant forces on slightly nonsymmetrical rotors is accounted for by intersecting the Campbell diagram with the straight line $\lambda = 2\omega$ while the excitation due to lubricated journal bearings is

---

[5]For a detailed discussion of the meaning of the negative modal masses, see, for example G. Genta and F. De Bona, "Unbalance response of rotors: A modal approach with some extensions to damped natural systems", *J. of Sound and Vibrations*, 140(1), (1990), 129–153.

studied using the straight line $\lambda = \omega/2$. The secondary critical speeds occurring at the intersection of the generic straight line $\lambda = h\omega$ can be computed from the eigenproblem

$$\left(-\omega^2(h^2\mathbf{M} - h\mathbf{G} - \mathbf{K}_\omega) + \mathbf{K}\right)\mathbf{q}_0 = \mathbf{0}. \qquad (4.96)$$

Also, in this case the use of complex coordinates allows the performance of the modal analysis of the system using mode shapes that are k- and m-orthogonal.

4.5.9   Computation of the unbalance response

The response to a generic unbalance distribution can be easily obtained by computing a particular integral of equation (4.69) or (4.70), which has the form $\mathbf{q} = \mathbf{q}_0 e^{i\omega t}$. The algebraic equation yielding the response of the damped system is readily obtained

$$\left(-\omega^2(\mathbf{M} - \mathbf{G} - \mathbf{K}_\omega) + i\omega\mathbf{C}_n + \mathbf{K}\right)\mathbf{q}_0 = \omega^2\mathbf{f}. \qquad (4.97)$$

In the case of undamped systems, equation (4.97) has no solution in case the spin speed coincides with a critical speed and the matrix of the coefficients is singular. If vectors ϵ_i and angles χ_i are not all contained in the same plane, i.e., all phases α_i and β_i are not equal, vector $\mathbf{f}$ is complex and the solution $\mathbf{q}$ is also complex. Physically, this means that the deflected shape is a skew line. However, the matrix of the coefficients of equation (4.97) remains real and no actual working with complex numbers is needed: The real and imaginary parts of the solution can be obtained separately, solving two sets of real linear equations with the same matrix of the coefficients.

The computation of the response to an arbitrary unbalance distribution is formally similar to the computation of the response of a vibrating system excited by a harmonic forcing function, the differences being that the rotational speed ω is used in the equation instead of the frequency λ, the excitation vector $\omega^2\mathbf{f}$ is proportional to ω^2, and the mass matrix of the system is $\mathbf{M} - \mathbf{G} - \mathbf{K}_\omega$, which can be nonpositive definite. Some of its eigenvalues can be negative, in terms of ω^2, and then some of the critical speed can be imaginary, as was seen in detail in Section 4.6.3. The response in terms of displacement is more similar to an inertance than to a dynamic compliance.

When damping is not neglected, the response to unbalance is very similar to the response of a damped vibrating system to harmonic excitation, but only nonrotating damping enters the equation of motion. The matrix of the coefficients never becomes singular, because it has no real eigenvalue. It is, however, complex, and, if software for use with complex numbers is not available, the solution of equation (4.97) involves the solution of a set of equations whose size must be doubled in order to work with real numbers: The number of equations is then equal to the number of real coordinates, and there is little computational advantage to using complex coordinates in this case.

The response to unbalance of an axisymmetrical system could also be easily studied using rotating coordinates. In this case, the response is nothing other than the static response to forces that are constant in both modulus and direction, and the critical speeds are a sort of elastic instability condition, in which the stiffness matrix of the system becomes singular.

4.5.10 Plotting the Campbell diagram and the roots locus

To plot the Campbell diagram it is necessary to compute the whirl frequencies of the system as functions of the spin speed. If the stability of the system has to be studied, damping must also be taken into account. This does not make the problem conceptually more complicated but it makes actual computation take much longer. In the case of viscous damping, the basic equation is the homogeneous equation associated with equation (4.70)

$$\mathbf{M\ddot{q}} + (\mathbf{C}_n + \mathbf{C}_r - i\omega\mathbf{G})\mathbf{\dot{q}} + (\mathbf{K} + \omega^2\mathbf{K}_\omega - i\omega\mathbf{C}_r)\mathbf{q} = 0 \,. \qquad (4.98)$$

The roots locus is directly obtained by computing the eigenvalues of the dynamic matrix (4.92); to take into account structural damping, it is sufficient to add the expression $\pm i\mathbf{K}_n'' \pm i\mathbf{K}_r''$, where the double signs have the same meaning seen in equation (4.31) and in Figure 4.12, to the stiffness matrix $\mathbf{K}$.

However, the Campbell diagram is often referred to a solution of the type $\mathbf{q} = \mathbf{q}_0 e^{i\lambda t}$ instead of $\mathbf{q} = \mathbf{q}_0 e^{st}$. In this case, it is possible to find the complex whirl frequencies λ as eigenvalues of the matrix

$$\begin{bmatrix} \mathbf{M}^{-1}(\omega\mathbf{G} + i\mathbf{C}_n + i\mathbf{C}_r) & \mathbf{M}^{-1}(\mathbf{K} + \omega^2\mathbf{K}_\omega \pm i\mathbf{K}_n'' \pm i\mathbf{K}_r'' - i\omega\mathbf{C}_r) \\ \mathbf{I} & \mathbf{0} \end{bmatrix} .$$

$$(4.99)$$

Operating in this way, a real eigenproblem must be solved in the case of undamped systems.

If the Campbell diagram has to be plotted by scanning the $\lambda\omega$-plane using m values of the spin speed ω, an eigenproblem of order $2n$, where n is the number of complex degrees of freedom, must be solved m times. The computation is then very time-consuming, and large-scale condensation may be necessary to keep computer time within reasonable limits. If the real-coordinate approach were used, the size of the problem would have been doubled and the information on the direction of the whirl motion included in the eigenvalues would have been lost: To distinguish between forward and backward modes it would have been necessary to study the eigenvectors.

In the case of the damped system, the solution of a complex eigenproblem requires a further doubling of the size of the relevant matrices, if software for use with complex algebra is not available. Moreover, if there is structural damping, several eigenproblems with the different signs present in equation (4.99) must be solved for each value of the spin speed. A maximum of four eigenproblems must be solved, in the case where all forms of damping are present. After obtaining the solutions, there is no difficulty checking where the real part of the various solutions is located in the Campbell diagram and then in choosing which eigenvalues are to be discarded, following the scheme of Figure 4.12. The computation of the damped Campbell diagram is then much heavier than that of the undamped diagram. In most cases, however, the presence of damping has little effect on the values of the whirl frequencies, while deeply affecting the decay rate. A good strategy is that of first studying the undamped system in such a way as to locate the critical speeds and the general pattern of the whirl frequencies and then computing the complex whirl frequencies at some selected values of the speed, mainly in the supercritical range, where the occurrence of instability can be suspected.

In the case of undamped systems, a solution of the type $\mathbf{q} = \mathbf{q}_0 e^{i\lambda t}$ allows the obtaining of an eigenproblem with real matrices while assuming that $\mathbf{q} = \mathbf{q}_0 e^{st}$ and the dynamic matrix is complex. On the contrary, if damping is not neglected, the two notations lead to similar complexities and the second solution can thus be used advantageously.

4.5.11 *Acceleration of a torsionally stiff rotor*

If the elastic behaviour of the system is such that axial, torsional, and flexural behaviours are uncoupled, the same considerations seen in Section 4.6.1 for the simple model with six degrees of freedom also hold for more complex models. In particular, within the frame of the linearized theory, the axial degrees of freedom are uncoupled from flexural behaviour. If the rotor is torsionally stiff, i.e., the torsional rotations of all cross sections are equal, and the acceleration is performed with an imposed law $\theta(t)$, the flexural behaviour can also be studied independently using equations of the type of the first two equations (4.59), the second equation (4.60), and equation (4.61). The only equation for the rotational degree of freedom is an equation of the type of the third equation (4.60). By introducing the usual complex coordinates, it follows that

$$\mathbf{M}\ddot{\mathbf{q}} + (\mathbf{C}_n + \mathbf{C}_r - iw\mathbf{G})\dot{\mathbf{q}} + (\mathbf{K} - iw\mathbf{C}_r)\mathbf{q} =$$
$$= (\omega^2 - i\dot{\omega})\mathbf{f}e^{i\theta},$$
$$M_z = J_{p_{tot}}\dot{\omega} + \Im\left(\overline{\mathbf{f}}^T \ddot{\mathbf{q}}e^{-i\theta}\right),$$
(4.100)

where centrifugal stiffening has been neglected and $J_{p_{tot}}$ is the total moment of inertia of the rotor about the z-axis, also taking into account the presence of static and couple unbalance. By introducing $\ddot{\theta} = 0$ and the solution for steady-state whirling into the last equation (4.100), the driving torque needed to maintain a constant angular velocity is readily obtained. To perform the numerical integration in time of equation (4.100) it is easier to resort to rotating coordinates, as seen for the case of the Jeffcott rotor. The relevant equations of motion are then

$$\mathbf{M}\ddot{\mathbf{r}} + \left[\mathbf{C}_n + \mathbf{C}_r - iw(2\mathbf{M} - \mathbf{G})\right]\dot{\mathbf{r}} + \left[\mathbf{K} - \omega^2(\mathbf{M} - \mathbf{G}) + \right.$$
$$\left. + i(\dot{\omega}\mathbf{M} + \omega\mathbf{C}_n)\right]\mathbf{r} = (\omega^2 - i\dot{\omega})\mathbf{f},$$
(4.101)
$$M_z = J_{p_{tot}}\dot{\omega} + \Im\left[\overline{\mathbf{f}}^T\left(\ddot{\mathbf{r}} + i\dot{\omega}\mathbf{r} + 2i\omega\dot{\mathbf{r}} - \omega^2\mathbf{r}\right)\right].$$

Note that the rotating reference frame used for equation (4.101) rotates at a variable spin speed $\omega(t)$, as defined by equations (4.46).

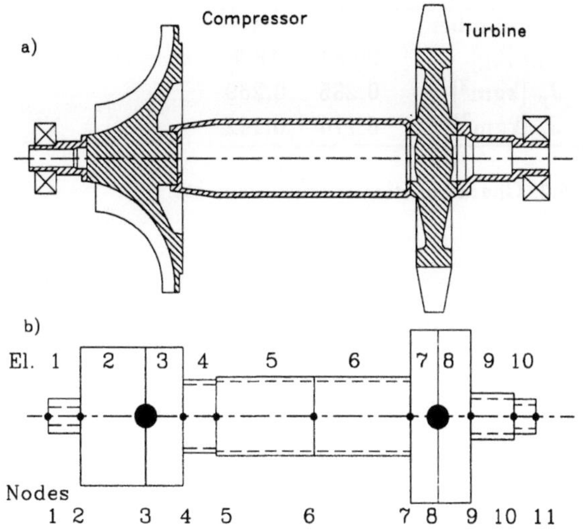

FIGURE 4.20. Sketch of the rotor of a small gas turbine rotor (a) and model for dynamic analysis (b).

El. number #	1	2	3	4	5	6	7	8	9	10
ϕ_i [mm]	15	0	0	55	60	60	0	0	30	20
ϕ_o [mm]	30	120	120	63	68	68	150	150	40	30
l [mm]	30	68.3	34.2	30	88	88	25	30	40	20
ρ [kg/m^2]	7810	0	0	7810	7810	7810	0	0	7810	7810
E [MN/m^2]					210 000					

TABLE 4.1. Characteristics of beam elements.

Example 4-3

Study the dynamic behaviour of the rotor of a small gas turbine sketched in Figure 4.20a using both the FEM and Myklestadt-Prohl method. elements. The finite element model, which can be used for the lumped parameters approach, is sketched in Figure 4.20b. It consists of 10 beam elements and 2 mass The compressor and turbine rotors can be considered rigid bodies; the stiffness of beam elements 2,3,7, and 8 is thus large (large diameter) and their mass is zero (vanishing material density) because the relevant mass is introduced as concentrated mass elements at nodes 3 and 8. The main characteristics of the beam and mass elements are reported in Tables 4.1 and 4.2.

The bearings, located at nodes 1 and 10, are modeled as rigid supports. If, in the construction of the global stiffness matrix, they must be considered elastic bearings, a very high value of the stiffness, $k = 10^{15}$ N/m, can be used.

Element number	1	2
Node number	3	8
m [kg]	20.81	18.2
J_p [kgm^2]	0.285	0.269
J_t [kgm^2]	0.174	0.142

TABLE 4.2. Characteristics of mass elements.

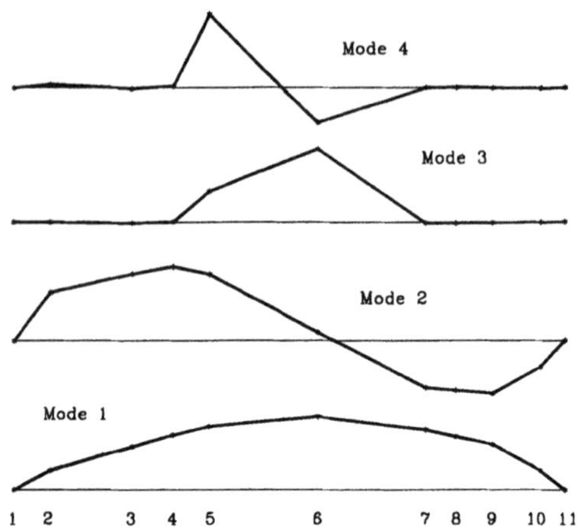

FIGURE 4.21. Mode shapes at the first four critical speeds computed using the model with 20 degrees of freedom.

Solution using the FEM – Consistent mass matrices. If the bearings are modeled as very stiff elastic constraints, the total number of complex degrees of freedom of the model is 22; there are only 20 degrees of freedom if the constraints are considered rigid. The critical speeds are computed using all degrees of freedom and through Guyan reduction. In the latter case two condensation schemes are considered; a very strong reduction in which only two degrees of freedom are retained (translations at nodes 3 and 8) and a more detailed scheme, in which six master degrees of freedom are considered (translations at nodes 3, 4, 6, 7, 8, and 9). The results are reported in the following table:

Master degrees of freedom	2	6	20
ω_{cr_I} [rad/s]	2,080	2,032	2,028
$\omega_{cr_{II}}$ [rad/s]	4,445	4,312	4,304
$\omega_{cr_{III}}$ [rad/s]	–	34,062	32,130
$\omega_{cr_{IV}}$ [rad/s]	–	534,380	795,010

The first four mode shapes obtained from the complete model are reported in Figure 4.21.

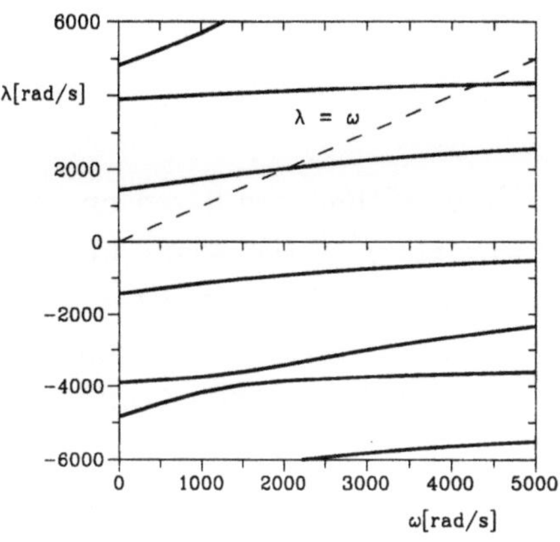

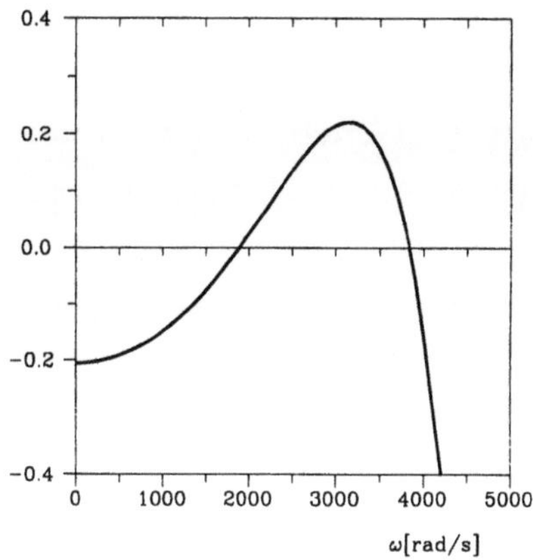

FIGURE 4.22. Campbell diagram of the rotor in Figure 4.20.

FIGURE 4.23. Graph for the computation of the critical speeds of the system in Figure 4.20 using Myklestadt-Prohl method.

From the table, it is clear that the first critical speed can be computed with good precision even with the simplest model, and its precision is still acceptable when searching the second critical speed. From the mode shapes, it is also evident that the third and fourth modes are mainly due to the deformation of the shaft connecting the turbine to the compressor: To compute such critical speeds with greater accuracy, the use of a finer model in the relevant zones is advisable.

The Campbell diagram of the system is reported in Figure 4.22. The intermediate reduction scheme has been used, since the diagram was to be obtained for speeds in excess of the second critical speed.

Solution with Myklestadt-Prohl method. The computation of the first two critical speeds is performed by assuming various values of the speed from 0 to 5,000 rad/s and computing the value of the determinant, which must vanish at the critical speeds. The result is reported in Figure 4.23. The values of the first two critical speeds are $\omega_{cr_I} = 1,888$ rad/s and $\omega_{cr_{II}} = 3,833$ rad/s.

The results so obtained are lower than those previously computed. This is consistent with the consideration that the computation has been performed by lumping the mass of the shaft at the nodes and neglecting the gyroscopic effects of the shafts. Note that higher values would have been obtained if the shear deformation of the shaft were neglected: $\omega_{cr_I} = 2,115$ rad/s and $\omega_{cr_{II}} = 4,919$ rad/s. Shear deformation is important in this case because the shaft is not slender and the results obtainable from a Euler-Bernoulli approach can be affected by large errors.

4.6   Nonisotropic systems

4.6.1   General considerations

All the models studied in the preceding sections are based on the
assumptions of linearity and axial symmetry of the whole model.
These assumptions are usually not verified exactly in real life, and
sometimes a model based on them can be only a very rough approximation of the actual system. In this section, the axial symmetry
assumption will be dropped. If a rotating machine is not isotropic
about the rotation axis, the deviation from symmetry can concern
only the rotor, only the stator, or both. In the first two cases, the
model is not exceedingly complicated; in the third case, the analytical complexity forces only approximate solutions of the dynamic
behaviour.

It must be noted that there are cases in which the model displays
a nonisotropic behaviour, even if all the parts of the machine are
geometrically axisymmetrical. This occurs particularly when the rotor runs on lubricated journal bearings: Under the effect of external
forces the journal takes an eccentric position within the bearing and
reacts in different ways to the forces in the various planes including
the rotation axis.

In the following sections the study of unsymmetrical rotors will
be dealt with in subsequent steps. At first a very simple configuration, based on the Jeffcott rotor, will be studied. After the relevant
phenomena have been qualitatively understood using this simplified
model, a more complete study, which allows quantitative results to
be obtained in the case of very complex systems, will be expounded.

4.6.2   Jeffcott rotor on nonisotropic supports

Consider the case of the rotor in Figure 4.3b but assume now that
the stiffness of the supports is not isotropic in the xy-plane. All other
assumptions made in Section 4.5, in particular the linearity of the
system and the assimilation of the rotor to a point mass, will be
retained. The motion will be studied in the xy-plane. In this plane
the polar diagram of the stiffness of the supports is an ellipse, the
so-called *ellipse of elasticity*.

Without any loss of generality, axes x and y will be assumed to coincide with the axes of the ellipse of elasticity, i.e., to be the principal

axes of elasticity of the supporting structure. Assume that the stiffness along the x-direction is lower than that along the y-direction. The elastic reaction of the shaft in this case is

$$F_x = -k_x x\,, \qquad\qquad F_y = -k_y y\,. \qquad (4.102)$$

By introducing the two different values of the stiffness into the equation of motion (4.10), the latter transforms into

$$\begin{cases} m\ddot{x} + k_x x = m\epsilon[\dot{\theta}^2 \cos(\theta) + \ddot{\theta} \sin(\theta)]\,, \\ m\ddot{y} + k_y y = m\epsilon[\dot{\theta}^2 \sin(\theta) - \ddot{\theta} \cos(\theta)]\,. \end{cases} \qquad (4.103)$$

In the case of constant spin speed $\dot{\theta} = \omega$, equation (4.103) with $\ddot{\theta} = 0$ still holds. From the homogeneous equation, it is clear that there are two natural frequencies, one (the lower) related to the motion in the xz-plane and the other related to the motion in the yz-plane:

$$\lambda_{n_1} = \sqrt{\frac{k_x}{m}}\,, \qquad \lambda_{n_2} = \sqrt{\frac{k_y}{m}}\,. \qquad (4.104)$$

They are not influenced by the spin speed, and then the Campbell diagram is made by two straight lines. Note that in this case the motions in the two planes occur at different frequencies, so the two harmonic motions cannot combine to make circles or ellipses. The fact that the two natural frequencies are independent of the spin speed causes the two critical speeds to coincide with the natural frequencies: $\omega_{cr_1} = \lambda_{n_1}$ and $\omega_{cr_2} = \lambda_{n_2}$. At the first critical speed the motion reduces to a straight vibration along the x-axis, and at the other critical speed it reduces to a straight motion along the y-axis.

The unbalance response can be obtained directly from equation (4.103) with $\ddot{\theta} = 0$. The response is in each plane equal to the response of the Jeffcott rotor, computed using the stiffness related to the plane itself

$$x_0 = \frac{m\epsilon\omega^2}{k_x - m\omega^2}\,, \qquad y_0 = \frac{m\epsilon\omega^2}{k_y - m\omega^2}\,. \qquad (4.105)$$

The unbalance response can be subdivided into three speed ranges (Figure 4.24):

- From standstill to the first critical speed: The responses in the two planes have the same sign and are out of phase from each

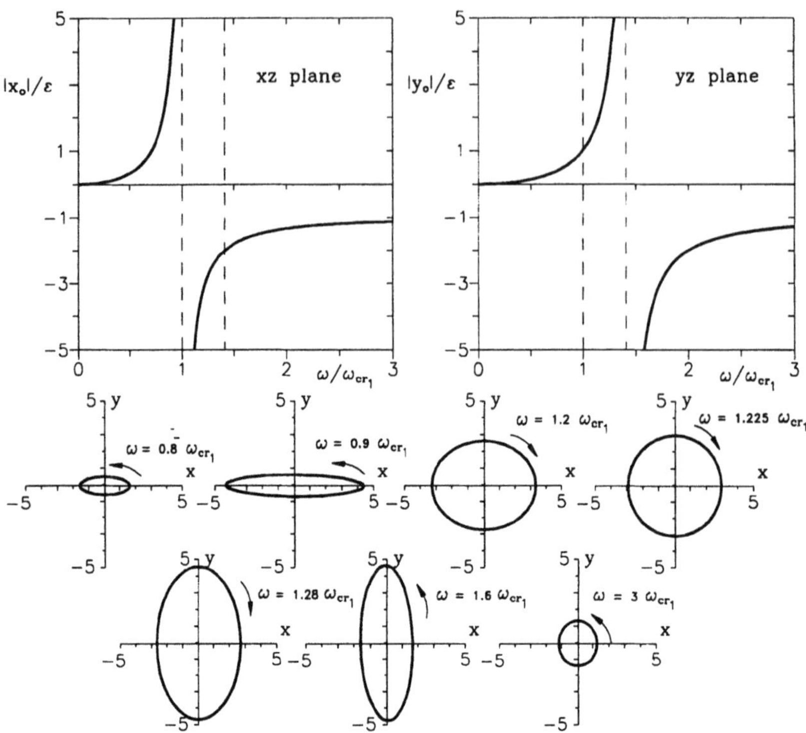

FIGURE 4.24. Unbalance response of a Jeffcott rotor on anisotropic supports. Amplitude of the motion along the x- and y-axes as a function of the speed and orbits at some selected values of ω; ratio between the stiffnesses in the two planes $\alpha = k_y/k_x = 2$.

other by 90°, as clearly seen from equation (4.103), where the excitation is expressed by a sine and a cosine function. The orbit grows mainly along the x-axis and has the shape of an elongated ellipse. Approaching the first critical speed, the axis of the orbit along the x-axis tends to infinity.

- From the first to the second critical speed: The response along the x-axis is negative, having already crossed the critical speed; that along the y-axis is positive. When they combine, they give way to an elliptical motion in the backward direction. Near the first critical speed the ellipse is elongated along the x-axis, and near the second it is elongated in the other direction. There is an intermediate speed at which the amplitudes in the two planes are equal and the orbit is circular. Unbalance, an excitation that by definition is applied in the forward direction, can thus excite a backward synchronous whirling of the rotor.

- Above the second critical speed: The amplitudes in both planes are of the same sign, and they both tend to the same value, namely, to $-\epsilon$. An elliptic forward whirling that tends to become circular with increasing speed is so obtained; self-centering takes place normally. It could easily be expected that in the high supercritical field the elastic anisotropy has little influence on the dynamic behaviour of the rotor: The behaviour of the system is dominated by inertia forces that are clearly isotropic.

The behaviour of the Jeffcott rotor on anisotropic supports has been studied here using real coordinates. Although not very common, it is also possible to use complex coordinates to study non-symmetrical rotors. As usual, it is possible to add the first equation (4.103) to the second equation multiplied by the imaginary unit i. By introducing the mean stiffness $k_m = (k_x + k_y)/2$ and the deviatoric stiffness $k_d = (k_x - k_y)/2$, the equation for the unbalance response in terms of complex coordinates is

$$m\ddot{z} + k_m z + k_d \bar{z} = m\epsilon\omega^2 e^{i\omega t}, \qquad (4.106)$$

where $\bar{z}$ is the conjugate of the complex number z. The solution of the homogeneous equation is of the type

$$z = z_1 e^{i\lambda t} + z_2 e^{-i\bar{\lambda} t},$$

which gives way to elliptical orbits. It can be introduced into the homogeneous equation of motion, yielding

$$\left(-\lambda^2 \begin{bmatrix} m & 0 \\ 0 & m \end{bmatrix} + \begin{bmatrix} k_m & k_d \\ k_d & k_m \end{bmatrix} \right) \left\{ \begin{array}{c} z_1 \\ \bar{z}_2 \end{array} \right\} = \left\{ \begin{array}{c} 0 \\ 0 \end{array} \right\}. \qquad (4.107)$$

Equation (4.107) can easily be solved in λ, yielding the values of the whirl frequencies coinciding with those expressed by equation (4.104).

The particular integral that allows the unbalance response to be computed is

$$z = z_1 e^{i\omega t} + z_2 e^{-i\omega t}.$$

By introducing it into the equation of motion and solving for the amplitudes of the forward and backward components z_1 and z_2, the following unbalance response is obtained:

$$z = \frac{m\epsilon\omega^2}{(k_x - m\omega^2)(k_y - m\omega^2)}\left[(k_m - m\omega^2)e^{i\omega t} - k_d e^{-i\omega t}\right]. \quad (4.108)$$

It is easy to demonstrate that the orbits expressed by equation (4.108) coincide with those expressed by equation (4.105). The possibility of backward whirling to be excited by unbalance is, however, more obvious when complex coordinates are used: At the speed $\omega_b = \sqrt{k_m/m}$ the amplitude of the forward component vanishes, and the orbit is a circular backward whirl with amplitude $z_{0_b} = \epsilon k_d/k_m$.

If the system is damped, the field in which backward whirling occurs is reduced and can, if damping is large enough, disappear altogether. When reaching the first critical speed, the amplitude of the elliptical orbits remains limited while the axis of the ellipse is not exactly aligned with one of the axes of elasticity of the supports. The orbit starts then to rotate in the xy-plane and to become thinner: if at a certain speed it reduces to a line, a reversal of the direction of whirling occurs.

The elliptical backward orbit then continues its rotation, with increasing and then decreasing width, until it again becomes a straight line and a new reversal of the whirling direction takes place. When forward whirling occurs, the direction of the larger axis of the ellipse has a direction close to the other axis of elasticity of the supports and the speed is close to the second critical speed. If damping is large enough, the orbit never reduces to a line and the reversal of its direction does not occur.

The unbalance response of a Jeffcot rotor with rotating and non-rotating damping ratio $\zeta_r = 0.01$ and $\zeta_n = 0.1$ is shown in Figure 4.25. With these values of the damping ratios, backward whirling actually occurs. Due to the complexity of the behaviour, a new representation is used: The orbits are plotted in a tridimensional graph, stacked along the spin-speed axis. It was proposed to designate this representation as an "orbital tube".[6] The tube starts as a point at zero speed and then enlarges, taking an elliptical cross section. At very high speed, due to self-centering, it tends to a circular cylinder with radius equal to the eccentricity.

---

[6]C. Delprete, G. Genta, S. Carabelli, *Control Strategies for Decentralised Control of Active Magnetic Bearings*, 4th Int. Symp. on Magnetic Bearings, Zurich, August 1994.

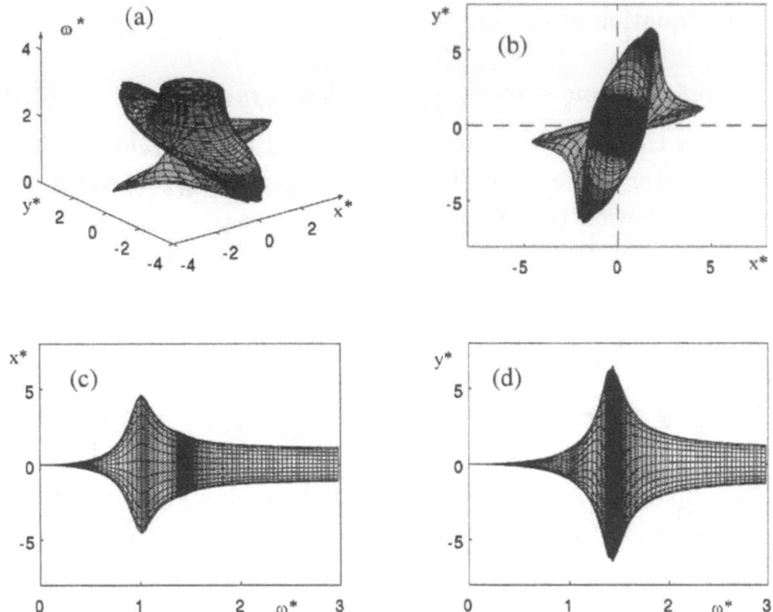

FIGURE 4.25. Nondimensional unbalance response of a damped Jeffcott rotor on anistrotopic supports; (a) orbital tube, i.e., representation of the orbits at different speeds as a tri-dimensional plot; (b) orbital view; (c) and (d) projections on the ωx- and ωy-planes, respectively.

The projection on the xy-plane (orbital view) directly gives the orbits at various speeds, superimposed on each other. The projections on the ωx- and ωy-planes give the peak-to-peak amplitude as a function of the spin speed.

4.6.3 Nonisotropic Jeffcott rotor

Now consider a Jeffcott rotor of the type shown in Figure 4.3a, in which the stiffness of the shaft is not isotropic. The polar diagram of the stiffness is now an ellipse whose axes can be assumed, without loss of generality, to lie along the ξ- and η-axes. It is then possible to write an equation similar to equation (4.102) but referred to the $O\xi\eta z$-frame:

$$F_\xi = -k_\xi\xi, \qquad F_\eta = -k_\eta\eta.$$

When dealing with rotating asymmetry, it is advisable to write the equation of motion with reference to the rotating frame. By introducing the mean and deviatoric stiffness, defined as in the preceding section but with reference to the $O\xi\eta z$-frame and neglecting damping, the equation of motion (4.37) becomes

$$m\ddot{\zeta} + 2i\omega m\dot{\zeta} - m\omega^2\zeta + k_m\zeta + k_d\bar{\zeta} = m\epsilon\omega^2 e^{i\alpha} , \qquad (4.109)$$

where α is the angle between the ξ-axis and the direction of the unbalance vector in the $\xi\eta$-plane (Figure 4.15b). Using real coordinates, the same equation transforms into

$$\begin{bmatrix} m & 0 \\ 0 & m \end{bmatrix} \begin{Bmatrix} \ddot{\xi} \\ \ddot{\eta} \end{Bmatrix} + \begin{bmatrix} 0 & -2m\omega \\ 2m\omega & 0 \end{bmatrix} \begin{Bmatrix} \dot{\xi} \\ \dot{\eta} \end{Bmatrix} + \qquad (4.110)$$

$$+ \begin{bmatrix} k_\xi - m\omega^2 & 0 \\ 0 & k_\eta - m\omega^2 \end{bmatrix} \begin{Bmatrix} \xi \\ \eta \end{Bmatrix} = m\epsilon\omega^2 \begin{Bmatrix} \cos(\alpha) \\ \sin(\alpha) \end{Bmatrix} .$$

The unbalance response is easily obtained as a steady-state solution of equation (4.110)

$$\xi = \frac{m\epsilon\omega^2 \cos(\alpha)}{k_\xi - m\omega^2} , \qquad \eta = \frac{m\epsilon\omega^2 \sin(\alpha)}{k_\eta - m\omega^2} , \qquad (4.111)$$

which represents, in a fixed reference frame, a circular whirling. The denominators of equation (4.111) vanish for two values of the spin speed, which are the critical speeds of the system

$$\omega_{cr\,I} = \sqrt{\frac{k_\xi}{m}} , \qquad \omega_{cr\,II} = \sqrt{\frac{k_\eta}{m}} . \qquad (4.112)$$

The free whirling of the system can easily be obtained from the homogeneous equation (4.109) or (4.110), being immaterial whether real or complex coordinates are used. In the first case the solution of the homogeneous equation (4.110) is

$$\xi = \xi_0 e^{i\lambda' t} , \qquad \eta = \eta_0 e^{i\lambda' t} .$$

λ' is a complex whirl speed in the $\xi\eta$-plane; it does not coincide with the whirl speed λ in the xy-plane but it is linked with the latter by the obvious relationships $\Re(\lambda) = \Re(\lambda') + \omega$, $\Im(\lambda) = \Im(\lambda')$.

The homogeneous equation (4.110) then yields an eigenproblem in λ'

$$\begin{bmatrix} \omega_{cr\,I}^2 - \lambda'^2 - \omega^2 & -2i\omega\lambda' \\ 2i\omega\lambda' & \omega_{cr\,II}^2 - \lambda'^2 - \omega^2 \end{bmatrix} \begin{Bmatrix} \xi_0 \\ \eta_0 \end{Bmatrix} = 0 . \qquad (4.113)$$

By introducing the nondimensional speeds $\lambda'^* = \lambda'/\omega_{cr_I}$, $\lambda^* = \lambda/\omega_{cr_I}$, $\omega^* = \omega/\omega_{cr_I}$ and the stiffness ratio $\alpha^* = k_\eta/k_\xi$, the characteristic equation can be written in nondimensional form

$$\lambda'^{*4} - \lambda'^{*2}(1 + \alpha^* + 2\omega^{*2}) + (1 - \omega^{*2})(\alpha^* - \omega^{*2}) = 0 . \qquad (4.114)$$

The stiffness ratio α^* is greater than unity if $k_\xi < k_\eta$. By first solving equation (4.114) in λ'^{*2} it follows that

$$\lambda'^{*2} = \omega^{*2} + \frac{1 + \alpha^*}{2} \pm \sqrt{2\omega^{*2}(1 + \alpha^*) + \frac{(1 - \alpha^*)^2}{4}} . \qquad (4.115)$$

The expression under the radical sign in equation (4.115) is always positive: The two solutions for λ'^{*2} are then always real. The one with the upper sign $(+)$ is always positive and yields two real solutions in λ'^*, one positive and one negative. The solution with the lower sign $(-)$ is positive only if

$$\omega^{*4} - \omega^{*2}(1 + \alpha^*) + \alpha^* > 0.$$

As α^* was assumed to be greater than unity, the last condition can be written in the form

$$\omega^* < 1; \qquad \alpha^* < \omega^* \text{ , i.e.,} \qquad \omega < \sqrt{\frac{k_\xi}{m}} ; \qquad \sqrt{\frac{k_\eta}{m}} < \omega . \qquad (4.116)$$

If condition (4.116) is satisfied, the characteristic equation (4.114) has four real roots – two positive and two negative. The behaviour of the system is stable. If, however, the value of the spin speed ω lies between the two critical speeds of the system, as shown by condition (4.116), two imaginary roots are found. Because the equation of motion has real coefficients, the solutions are conjugate numbers, or, if they have vanishing real parts, they have opposite signs. A negative imaginary solution then exists, which corresponds to an unstable behaviour of the system, with amplitude growing indefinitely with exponential law. The presence of an elastic anisotropy of the rotating parts of the system causes the occurrence of an instability range that spans from the lowest to the highest critical speed.

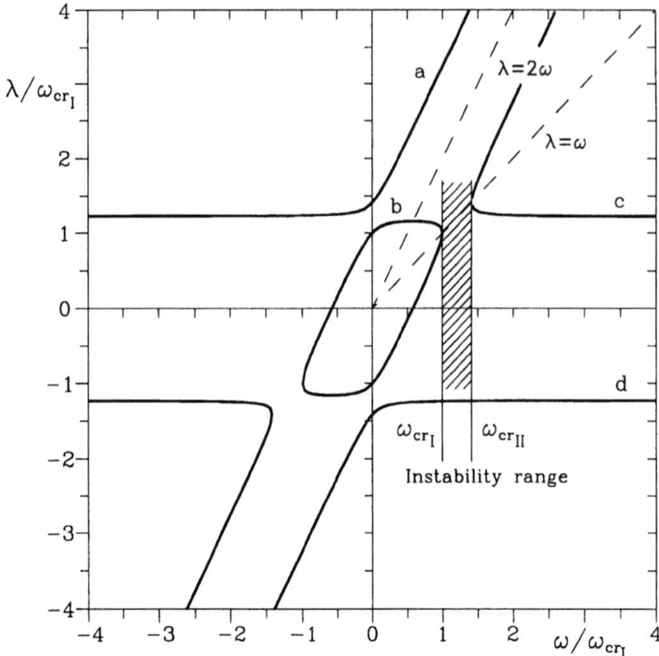

FIGURE 4.26. Campbell diagram for an anisotropic Jeffcott rotor; ratio between the stiffnesses in the two planes $\alpha^* = k_\eta/k_\xi = 2$.

The Campbell diagram $\lambda(\omega)$ for a system with $\alpha^* = k_\eta/k_\xi = 2$ obtained from equation (4.115) is plotted in Figure 4.26. All four quadrants of the λ,ω-plane have been reported, even if just two of them, the first and fourth, for example, give a complete picture of the situation.

The same conclusions already seen can be drawn from the figure. At low speed, up to the first critical speed, there are four solutions. Those on branches a and d of the curve come from the solution (4.115) with the upper sign $(+)$ and are a forward and a backward whirl. Those on branch b come from the solution with the lower sign $(-)$ and are one backward and one forward or both forward, depending on the value of ω. At high speed, above the second critical speed, the situation is similar, the difference being that the solutions coming from the expression with the lower sign $(-)$ are both forward motions and lie on branch c of the curve. If the value of the speed lies in the instability range spanning between the critical speeds, there are two real solutions, on branches a and d of the curve, and two imaginary ones, not reported in the figure. The negative imaginary solution causes the behaviour of the system to become unstable.

Rotating damping, which can cause the system to become unstable at high speed, reduces the instability range between the critical speeds. A similar effect is due to nonrotating damping, which, however, is stabilizing at any speed. If damping is high enough, the instability range gets smaller and can disappear altogether. The homogeneous equation of motion of the damped system in the rotating frame is

$$
\begin{bmatrix} m & 0 \\ 0 & m \end{bmatrix} \begin{Bmatrix} \ddot{\xi} \\ \ddot{\eta} \end{Bmatrix} + \begin{bmatrix} c_n + c_r & -2m\omega \\ 2m\omega & c_n + c_r \end{bmatrix} \begin{Bmatrix} \dot{\xi} \\ \dot{\eta} \end{Bmatrix} +
$$
$$
+ \begin{bmatrix} k_\xi - m\omega^2 & -\omega c_n \\ \omega c_n & k_\eta - m\omega^2 \end{bmatrix} \begin{Bmatrix} \xi \\ \eta \end{Bmatrix} = \{0\} . \tag{4.117}
$$

By introducing the solution for free whirling into equation (4.117), the following characteristic equation is obtained

$$
\det \begin{bmatrix} -m\lambda'^2 + i\lambda'(c_n + c_r) - m\omega^2 + k_\xi & -m\omega(2i\lambda' + c_n) \\ m\omega(2i\lambda' + c_n) & -m\lambda'^2 + i\lambda'(c_n + c_r) - m\omega^2 + k_\eta \end{bmatrix} = 0 . \tag{4.118}
$$

The system of Figure 4.11 can be considered a limiting case of an asymmetrical rotor: The stiffness along the η-axis is infinitely high, causing a second critical speed that tends to infinity. The field of instability then extends for all values of ω that are above the critical speed.

4.6.4 Secondary critical speeds due to rotor weight

All conditions in which there is resonance between one of the natural frequencies of the system and an exciting force different from the rotating force due to unbalance will be referred to as *secondary critical speeds*. It is well known that when constant bending forces, such as the self-weight of a rotor whose axis is horizontal, act on the rotor, the critical speeds and the Campbell diagram are not influenced by the presence of such forces. Whirling takes place about the deflected configuration of the rotor, but, due to linearity, the two effects, namely, static bending and whirling, do not interact. It is, however, well known that the weight of a rotor with a horizontal axis can cause the occurrence of secondary critical speeds, whose values are about half those of primary critical speeds, or, more exactly, are located at the intersections on the Campbell diagram of the curves for free whirling with the straight line $\lambda = 2\omega$. The presence of these secondary critical speeds is linked to the deviations from a perfect axial symmetry of the rotor.

In the case of the Jeffcott rotor studied in the preceding section, these secondary critical speeds are easily deduced from Figure 4.26. In the figure an intersection of branch b of the curve with the ω-axis is clearly visible. The system has, then, at a well-determined speed, a natural frequency that is vanishingly small and then, at that speed, a sort of resonance with a static force, i.e., with a force constant in modulus and direction, is possible.

The phenomenon may be easier to understand with reference to the rotating frame $O\xi\eta z$, in which condition $\lambda = 0$ is seen as $\lambda' = -\omega$. A constant force in the xy-plane, as self-weight of a horizontal rotor, is seen in the $\xi\eta$-plane as a force rotating with speed $-\omega$, which can cause resonance when the natural frequency of the system has the same frequency. The same phenomenon can also be seen in a different way. When the stiffness of the shaft is not isotropic in the $\xi\eta$-plane, its polar diagram is an ellipse, the ellipse of elasticity. The functions of time expressing the stiffnesses $k_x(t)$ and $k_y(t)$ are periodic in time, if the angular speed ω is constant, and their period is equal to half a revolution (frequency equal to 2ω). The conditions for resonance occur when the curves on the Campbell diagram intersect the straight line $\lambda = 2\omega$.

The two ways of seeing the same phenomenon are equivalent: From Figure 4.26 it is clear that at the same value of the speed at which a branch of the curve intersects the ω-axis, another intersects the line $\lambda = 2\omega$. From the characteristic equation, if $\Re(\lambda) = \Re(\lambda') + \omega$ is the real part of a solution for free whirling, $\Re(\lambda) = -\Re(\lambda') + \omega$ is also the real part of another solution. The value of the secondary critical speed for the Jeffcott rotor of Figure 4.26 can be easily computed by introducing the condition $\lambda' = -\omega$ into equation (4.114) and solving it in ω. It follows that

$$\omega^* = \sqrt{\frac{\alpha^*}{2(1+\alpha^*)}} \ , \text{ i.e., } \quad \omega^* = \sqrt{\frac{k_\xi k_\eta}{2(k_\xi + k_\eta)}} . \tag{4.119}$$

If k_ξ and k_η tend to a single value k, $\alpha^* \to 1$, the value of the secondary critical speed tends to

$$\omega_{cr_s} = \frac{1}{2}\sqrt{\frac{k}{m}} = \frac{1}{2}\omega_{cr} \, . \tag{4.120}$$

Equation (4.120) holds only in the case of a Jeffcott rotor; in all other cases the secondary critical speeds can be found at the intersections of the curve $\lambda(\omega)$ with the line $\lambda = 2\omega$. Also inertial anisotropy, i.e., difference in the moments of inertia about transversal baricentrical axes, has effects similar to those seen for elastic anisotropy. All secondary critical speeds characterized by the condition $\lambda > \omega$, as is the case for those excited by self-weight, occur in the subcritical region and usually cannot trigger unstable behaviour. The internal damping of the rotor is in these conditions stabilizing and, generally speaking, no unstable behaviour is expected in the subcritical region.

4.6.5 Equation of motion for an anisotropic machine with many degrees of freedom

Consider a beam element of the type studied in Example 2-6, for which all the assumptions for uncoupling among flexural, torsional, and axial behaviour hold. If the principal axes of inertia and elasticity lie in the xz- and yz-planes (hereafter designated by subscripts x and y), the expressions for the mass and stiffness matrix related to flexural behaviour can be written in the form

$$\mathbf{M} = \begin{bmatrix} \mathbf{M}_x & \mathbf{0} \\ \mathbf{0} & \mathbf{M}_y \end{bmatrix}, \qquad \mathbf{K} = \begin{bmatrix} \mathbf{K}_x & \mathbf{0} \\ \mathbf{0} & \mathbf{K}_y \end{bmatrix}, \tag{4.121}$$

where matrices related to the two bending planes are shown in Section 2.4.1. In the following analytical development, the generalized coordinate for rotation in the yz-plane will be $-\phi_x$ instead of ϕ_x, in such a way that matrices related to the xz- and yz-planes are equal if an element is axially symmetrical. In this way, the introduction of complex coordinates will be straightforward.

When assembling the structure, assume that the global reference frame has the same z-axis as those of each element, but that the x-axes of the elements are rotated of an angle α with respect to the global reference frame. The rotation matrix is

$$\mathbf{R}' = \begin{bmatrix} \cos(\alpha)\,\mathbf{I} & \sin(\alpha)\,\mathbf{I} \\ -\sin(\alpha)\,\mathbf{I} & \cos(\alpha)\,\mathbf{I} \end{bmatrix}, \tag{4.122}$$

and the stiffness matrix in the global reference frame is

$$\mathbf{K}_g = \begin{bmatrix} \cos^2(\alpha)\mathbf{K}_x + \sin^2(\alpha)\mathbf{K}_y & \sin(\alpha)\cos(\alpha)(\mathbf{K}_x - \mathbf{K}_y) \\ \sin(\alpha)\cos(\alpha)(\mathbf{K}_x - \mathbf{K}_y) & \sin^2(\alpha)\mathbf{K}_x + \cos^2(\alpha)\mathbf{K}_y \end{bmatrix}. \tag{4.123}$$

By introducing the mean and deviatoric stiffness matrices of the elements

$$\mathbf{K}_m = \frac{1}{2}(\mathbf{K}_x + \mathbf{K}_y), \qquad \mathbf{K}_d = \frac{1}{2}(\mathbf{K}_x - \mathbf{K}_y),$$

equation (4.123) can be written as

$$\mathbf{K}_g = \begin{bmatrix} \mathbf{K}_m + \mathbf{K}_d\cos(2\alpha) & \mathbf{K}_d\sin(2\alpha) \\ \mathbf{K}_d\sin(2\alpha) & \mathbf{K}_m - \mathbf{K}_d\cos(2\alpha) \end{bmatrix}. \tag{4.124}$$

If the element belongs to the rotor, angle α must be substituted by $\alpha + \theta$, or, in the case of constant spin speed equal to ω, by $\alpha + \omega t$.

All the aforementioned considerations hold for mass and damping matrices and for elements other than beam elements. Note that, in general, all mean and deviatoric matrices of structural elements are symmetrical, with the exception of the stiffness and damping matrices of the elements used to model lubricated journal bearings in linearized theories (see Section 4.10). Once all matrices of the elements have been obtained and expressed in the global reference frame, it is possible to assemble the various elements to obtain the matrices related to the whole structure. Obviously, the rotating elements must be assembled separately from the nonrotating elements.

Due to the presence of the deviatoric matrices (it is sufficient that a single element has a nonvanishing deviatoric matrix), the structure of the assembled matrices is more complex than that of equation (4.121). For the stiffness matrix it follows that

$$\mathbf{K} = \begin{bmatrix} \mathbf{K}_x & \mathbf{K}_{xy} \\ \mathbf{K}_{yx} & \mathbf{K}_y \end{bmatrix}. \tag{4.125}$$

Due to the presence of the coupling terms with xy- and yx-subscripts, a new definition of the mean and deviatoric matrices for the whole structure is needed

$$\begin{cases} \mathbf{K}_m = \dfrac{1}{2}(\mathbf{K}_x + \mathbf{K}_y) + i\dfrac{1}{2}(\mathbf{K}_{yx} - \mathbf{K}_{xy}) \\[2mm] \mathbf{K}_d = \dfrac{1}{2}(\mathbf{K}_x - \mathbf{K}_y) + i\dfrac{1}{2}(\mathbf{K}_{yx} + \mathbf{K}_{xy}). \end{cases} \tag{4.126}$$

Note that, except in the mentioned case of elements used for the linearized modeling of hydrodynamic bearings, the matrices with subscripts xy and yx are equal and the mean matrices are real. On the contrary, deviatoric matrices are, in general, complex. Using the complex-coordinate approach and the definitions of mean and deviatoric matrices given by equation (4.126), the equation of motion describing the flexural behaviour of a general system containing stationary elements and elements rotating at constant spin speed ω can be shown to be[7]

$$\mathbf{M}_m \ddot{\mathbf{q}} + (\mathbf{C}_m - i\omega \mathbf{G})\dot{\mathbf{q}} + (\mathbf{K}_m - i\omega \mathbf{C}_{r_m})\mathbf{q} + \mathbf{M}_{n_d}\ddot{\overline{\mathbf{q}}} +$$

$$+ \mathbf{M}_{r_d} e^{2i\omega t}(\ddot{\overline{\mathbf{q}}} + 2i\omega \dot{\overline{\mathbf{q}}}) + \mathbf{C}_{n_d}\dot{\overline{\mathbf{q}}} + \mathbf{C}_{r_d} e^{2i\omega t}\dot{\overline{\mathbf{q}}} + \tag{4.127}$$

$$+ \mathbf{K}_{n_d}\overline{\mathbf{q}} + (\mathbf{K}_{r_d} - i\omega \mathbf{C}_{r_d})e^{2i\omega t}\overline{\mathbf{q}} = \mathbf{F}_n + \omega^2 \mathbf{F}_r e^{i\omega t}.$$

Matrices and vectors with subscript r are related to the rotating elements, and those with subscript n are related to the stator of the machine. The mean matrices without r or n subscripts are related to the whole system and are obviously the sum of the corresponding nonrotating and rotating mean matrices. The nonrotating force vector is related to static forces, while the rotating vector is related to forces that are stationary in a reference frame rotating at the spin speed ω. Because the latter are usually unbalance forces, their magnitude is proportional to the square of the spin speed. Note that the complex conjugate of the vector of the generalized coordinates is present in the terms related to deviatoric matrices, and that all terms containing the deviatoric matrices of rotating elements have coefficients that are periodic functions of time, with periods equal to half of the period of rotation.

Equation (4.127) is a linear differential equation with periodic coefficients of the type studied by the Floquet theory. If all deviatoric matrices vanish, as with axisymmetrical systems, equation (4.127)

---

[7]G. Genta, "Whirling of unsymmetrical rotors: A finite element approach based on complex coordinates", *J. of Sound and Vibration*, 124(1), (1988), 24–53.

reduces to the constant coefficient equation already studied in Section 4.7 for isotropic rotors.

4.6.6 General system with anisotropic stator

Consider a system whose behaviour is modeled by equation (4.127), but with an isotropic rotor. The simplest of the systems of this type is the Jeffcott rotor on nonisotropic supports studied in Section 4.8.2. Because all deviatoric matrices related to the rotor vanish, equation (4.127) reduces to the following differential equation with constant coefficients:

$$\mathbf{M}_m\ddot{\mathbf{q}} + (\mathbf{C}_m - i\omega\mathbf{G})\dot{\mathbf{q}} + (\mathbf{K}_m - i\omega\mathbf{C}_{rm})\mathbf{q} + \\ + \mathbf{M}_{n_d}\ddot{\overline{\mathbf{q}}} + \mathbf{C}_{n_d}\dot{\overline{\mathbf{q}}} + \mathbf{K}_{n_d}\overline{\mathbf{q}} = \mathbf{F}_n + \omega^2\mathbf{F}_r e^{i\omega t} \tag{4.128}$$

Equation (4.128) involves actually working with complex coordinates, bacause deviatoric matrices are generally complex, and, consequently, the advantage of resorting to complex coordinates depends on the time and cost-effectiveness of the available subroutines for computations involving complex numbers. Alternatively, it is possible to use the same equation of motion written in terms of real coordinates

$$\begin{bmatrix} \mathbf{M}_x & \mathbf{M}_{xy} \\ \mathbf{M}_{yx} & \mathbf{M}_y \end{bmatrix}\ddot{\mathbf{x}} + \left(\omega\begin{bmatrix} 0 & \mathbf{G} \\ -\mathbf{G} & 0 \end{bmatrix} + \begin{bmatrix} \mathbf{C}_x & \mathbf{C}_{xy} \\ \mathbf{C}_{yx} & \mathbf{C}_y \end{bmatrix}\right)\dot{\mathbf{x}} + $$
$$+ \left(\begin{bmatrix} \mathbf{K}_x & \mathbf{K}_{xy} \\ \mathbf{K}_{yx} & \mathbf{K}_y \end{bmatrix} + \omega\begin{bmatrix} 0 & \mathbf{C}_r \\ -\mathbf{C}_r & 0 \end{bmatrix}\right)\mathbf{x} = \tag{4.129}$$
$$= \omega^2\begin{Bmatrix} \Re(\mathbf{f}_r e^{i\omega t}) \\ \Im(\mathbf{f}_r e^{i\omega t}) \end{Bmatrix} + \begin{Bmatrix} \Re(\mathbf{f}_n) \\ \Im(\mathbf{f}_n) \end{Bmatrix},$$

where the real-coordinates vector is defined as in equation (4.87).

The solution for static loading is similar to the corresponding solution for axisymmetrical systems, i.e., a constant vector $\mathbf{q} = \mathbf{q}_0$, leading to the equation

$$(\mathbf{K}_m - i\omega\mathbf{C}_{rm})\mathbf{q}_0 + \mathbf{K}_{n_d}\overline{\mathbf{q}}_0 = \mathbf{f}_n, \tag{4.130}$$

or, if real coordinates are used,

$$\left(\begin{bmatrix} \mathbf{K}_x & \mathbf{K}_{xy} \\ \mathbf{K}_{yx} & \mathbf{K}_y \end{bmatrix} + \begin{bmatrix} \mathbf{0} & \mathbf{C}_r \\ -\mathbf{C}_r & \mathbf{0} \end{bmatrix} \right) \{x_0\} = \left\{ \begin{array}{c} \Re(\mathbf{f}_n) \\ \Im(\mathbf{f}_n) \end{array} \right\}. \tag{4.131}$$

The inflected shape, then, is a line (generally a skew line) that remains fixed in space. Rotating damping couples the behaviour in the xz- and yz-planes, even if the coordinate planes are planes of symmetry for the stator.

The unbalance response is a synchronous elliptical whirling. The solution of the equation of motion can be expressed in the form

$$\mathbf{q} = \mathbf{q}_1 e^{i\omega t} + \mathbf{q}_2 e^{-i\omega t},$$

i.e., as the sum of two circular whirling motions taking place at speed ω in opposite directions. Both $\mathbf{q}_1$ and $\mathbf{q}_2$ are, generally speaking, complex vectors, that physically correspond to elliptical orbits not having axes x and y as axes of symmetry. The unknowns of the problem are then $4n$ in number, i.e., the imaginary and real parts of two vectors of size n. By introducing this solution into the equation of motion (4.131), the latter yields

$$\begin{bmatrix} \mathbf{A}_{11} & \mathbf{A}_{12} \\ \mathbf{A}_{21} & \mathbf{A}_{22} \end{bmatrix} \left\{ \begin{array}{c} \mathbf{q}_1 \\ \overline{\mathbf{q}}_2 \end{array} \right\} = \omega^2 \left\{ \begin{array}{c} \mathbf{f}_r \\ 0 \end{array} \right\}, \tag{4.132}$$

where

$$\begin{aligned} \mathbf{A}_{11} &= -\omega^2(\mathbf{M}_m - \mathbf{G}) + i\omega \mathbf{C}_{n_m} + \mathbf{K}_m, \\ \mathbf{A}_{12} &= -\omega^2 \mathbf{M}_{n_d} + i\omega \mathbf{C}_{n_d} + \mathbf{K}_{n_d}, \\ \mathbf{A}_{21} &= -\omega^2 \mathbf{M}_{n_d} - i\omega \mathbf{C}_{n_d} + \mathbf{K}_{n_d}, \\ \mathbf{A}_{22} &= -\omega^2(\mathbf{M}_m + \mathbf{G}) - i\omega(\mathbf{C}_{n_m} + 2\mathbf{C}_{r_m}) + \mathbf{K}_m. \end{aligned} \tag{4.133}$$

Rotating damping now enters the equation yielding the unbalance response: The shaft no longer rotates in the deformed configuration but actually vibrates, in the sense that each part of it experiences stresses that vary with time.

In the case of undamped systems, equation (4.132) reduces to

$$\left(\begin{bmatrix} \mathbf{K}_m & \mathbf{K}_{n_d} \\ \overline{\mathbf{K}}_{n_d} & \overline{\mathbf{K}}_m \end{bmatrix} - \omega^2 \begin{bmatrix} \overline{\mathbf{M}}_m - \mathbf{G} & \mathbf{M}_{n_d} \\ \overline{\mathbf{M}}_{n_d} & \overline{\mathbf{M}}_m + \mathbf{G} \end{bmatrix} \right) \left\{ \begin{array}{c} \mathbf{q}_1 \\ \overline{\mathbf{q}}_2 \end{array} \right\} =$$

$$= \omega^2 \left\{ \begin{array}{c} \mathbf{f}_r \\ 0 \end{array} \right\}, \tag{4.134}$$

which is real if the stator is symmetrical with respect to the coordinate planes. Note that the mean mass matrix is always real and coincides with its conjugate.

By equating to zero the determinant of the matrix of the coefficients of equation (4.134), an eigenproblem in ω^2 is obtained, which allows the critical speeds to be computed. At certain speeds, vector $\mathbf{q}_1$ vanishes; this physically corresponds to a circular backward whirling motion due to unbalance, as was shown in Section 4.8.2 for the Jeffcott rotor.

The equation corresponding to equation (4.134) but obtained using real coordinates is very similar, and the complexity of the actual computations to be performed is very similar.

In the case of free whirling, the orbits of the system are elliptical. The relevant solution of the homogeneous equation of motion is of the type

$$\mathbf{q} = \mathbf{q}_1 e^{i\lambda t} + \mathbf{q}_2 e^{-i\bar{\lambda} t},$$

which leads to the following algebraic equation

$$\left(-\lambda^2 \begin{bmatrix} \mathbf{M}_m & \mathbf{M}_{n_d} \\ \overline{\mathbf{M}}_{n_d} & \mathbf{M}_m \end{bmatrix} + \lambda\omega \begin{bmatrix} \mathbf{G} & 0 \\ 0 & -\mathbf{G} \end{bmatrix} + i\lambda \begin{bmatrix} \mathbf{C}_m & \mathbf{C}_{n_d} \\ \overline{\mathbf{C}}_{n_d} & \overline{\mathbf{C}}_m \end{bmatrix} + \right.$$

$$\left. + \begin{bmatrix} \mathbf{K}_m & \mathbf{K}_{n_d} \\ \overline{\mathbf{K}}_{n_d} & \overline{\mathbf{K}}_m \end{bmatrix} - i\omega \begin{bmatrix} \mathbf{C}_r & 0 \\ 0 & -\mathbf{C}_r \end{bmatrix} \right) \left\{ \begin{array}{c} \mathbf{q}_1 \\ \overline{\mathbf{q}}_2 \end{array} \right\} = 0. \tag{4.135}$$

The corresponding solution in terms of real coordinates is $\mathbf{x} = \Re(\mathbf{x}_0 e^{st})$, which yields the following algebraic equation

$$\left(s^2 \begin{bmatrix} \mathbf{M}_x & \mathbf{M}_{xy} \\ \mathbf{M}_{yx} & \mathbf{M}_y \end{bmatrix} + s\omega \begin{bmatrix} 0 & \mathbf{G} \\ -\mathbf{G} & 0 \end{bmatrix} + s \begin{bmatrix} \mathbf{C}_{n_x} + \mathbf{C}_r & \mathbf{C}_{n_{xy}} \\ \overline{\mathbf{C}}_{n_{yx}} & \mathbf{C}_{n_y} + \mathbf{C}_r \end{bmatrix} + \right.$$

$$\left. + \begin{bmatrix} \mathbf{K}_x & \mathbf{K}_{xy} \\ \mathbf{K}_{yx} & \mathbf{K}_y \end{bmatrix} + \omega \begin{bmatrix} 0 & \mathbf{C}_r \\ -\mathbf{C}_r & 0 \end{bmatrix} \right) \mathbf{x}_0 = 0. \tag{4.136}$$

In the case of an undamped system whose stator is symmetrical with respect to the coordinate planes, the two approaches are exactly equivalent, as both lead to a set of $2n$ real algebraic equations.

The eigenvalues from equation (4.135) are real, and those from equation (4.136) are imaginary. The eigenvectors of the former are real, and those of the latter are made up of real and imaginary terms, depending on the phasing of the various motions added to give the various orbits. In the most general case, equation (4.135) yields a set of $2n$ complex equations, and equations (4.136) are always real. In the author's opinion, however, the physical interpretation is more straightforward in the case of the former equation, even if it is not sufficient to find out the sign of the eigenvalue λ to assess whether the whirl motion occurs in the forward or backward direction. In fact, if λ is a solution of the eigenproblem, $-\bar{\lambda}$ is also a solution and, consequently, each mode is found twice, with opposite signs of $\Re(\lambda)$. Moreover, there are modes, sometimes referred to as *mixed modes*, in which the whirling occurs in the forward direction at some points of the rotor and in the backward direction at other points. Although only the study of the eigenvectors can make clear which mode occurs in the forward or backward direction, the author feels that the physical interpretation of the solution is somehow more clear when using complex coordinates.

Example 4-4
Consider the rotor of the small gas turbine studied in Example 4-3. Compute the critical speeds and plot the Campbell diagram assuming that the rigid bearings are substituted by nonisotropic supports, whose stiffness is 1.2×10^7 N/m in the vertical plane and 8×10^6 N/m in the horizontal plane.
The dynamic study will be performed using the same FEM model seen in Example 4-4, with the only difference being that when performing matrix condensation, the translational coordinates of nodes 1 and 11 (i.e., the displacements at the supports) are retained. The complex-coordinates approach will be followed, and the mean and deviatoric stiffness of the supports is first computed: $k_m = 1 \times 10^7$ N/m, $k_d = 2 \times 10^6$ N/m.
A first computation is then run, neglecting the deviatoric stiffness, by simply assembling the mean stiffness of the bearings in the stiffness matrix of the rotor, in position 1, 1 and 11, 11.
The values of the critical speeds obtained from a condensation scheme with eight master degrees of freedom (corresponding to the scheme with six master degrees of freedom of Example 4-4) are 652.9 rad/s; 1,352 rad/s; 34,029 rad/s; and 57,920 rad/s.

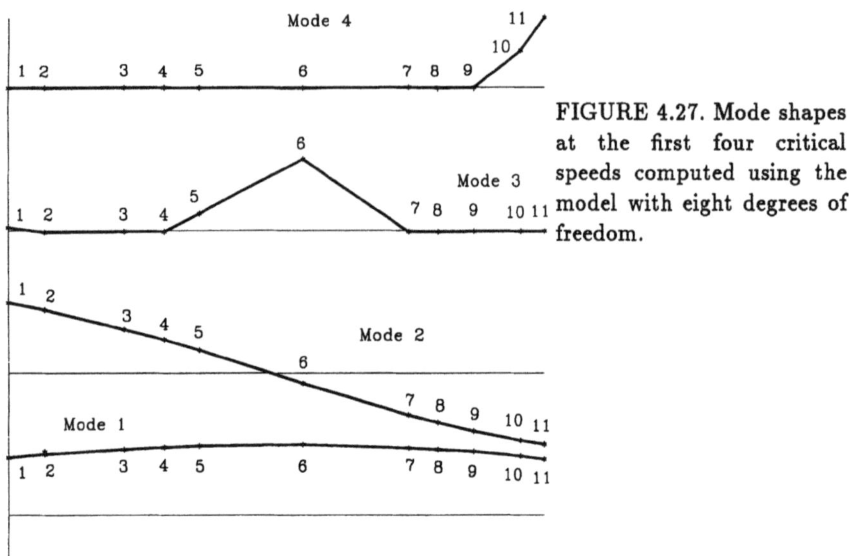

FIGURE 4.27. Mode shapes at the first four critical speeds computed using the model with eight degrees of freedom.

Note that, compared with the case with stiff bearings, the first two critical speeds are reduced, and the third is almost unchanged. The fourth is completely different, as can be explained by plotting the mode shapes (Figure 4.27): In the fourth mode the deformation is mainly localized at the bearing on the turbine side, which in the previous model was assumed to be stiff.

The Campbell diagram is plotted in Figure 4.28.

If the anisotropy of the supports is not neglected, the deviatoric stiffness matrix must be built: It is a matrix with almost all elements equal to zero, except for two on the main diagonal that are equal to k_d. The other deviatoric matrices are equal to zero. Equation 4.134 can be used to compute the following values of the first 10 critical speeds: 588.2 rad/s; 698.0 rad/s; 768.3 rad/s; 1,362 rad/s; 1,463 rad/s; 3,530 rad/s; 33,617 rad/s; 34,029 rad/s; 51,632 rad/s; and 57,938 rad/s. Note that many critical speeds are found; from the undamped analysis it is impossible to state the severity of such critical conditions, which can be obtained only by plotting an unbalance response of the damped system.

The Campbell diagram has been reported in Figure 4.29. Note that only one quadrant of the diagram has been plotted, because it contains all the information. From the diagram, it is impossible to state which modes are related to forward or backward whirling; a clue, however, is that the branches sloping upward are related to forward whirling. On the same plot, the curves computed for the isotropic system are also reported (dotted lines).

Note that the effects of the anisotropic bearings are quite limited, particularly where the higher-order modes are concerned.

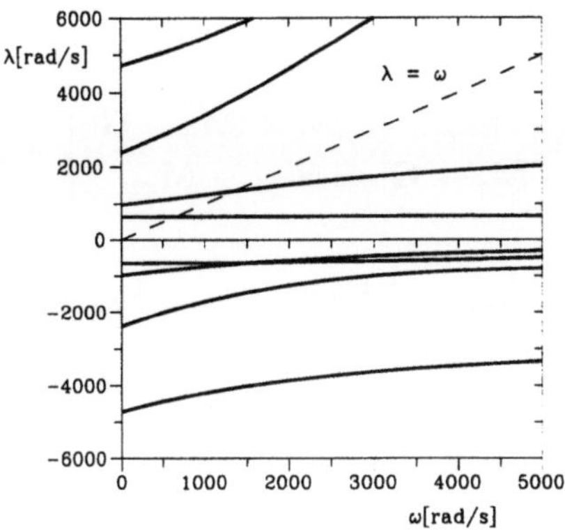
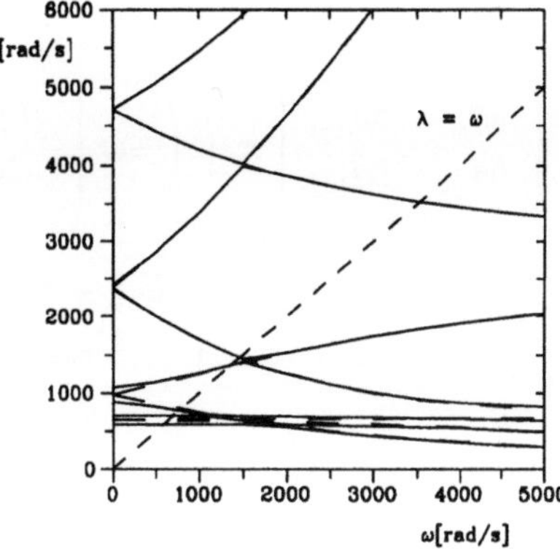

FIGURE 4.28. Campbell diagram of the rotor of Figure 4.20, with nonrigid isotropic bearings.

FIGURE 4.29. Same as Figure 4.28, but with anisotropy of the bearings not neglected. Full lines; anisotropic system; dashed lines; mean isotropic system.

4.6.7   General system with anisotropic rotor

Now consider a system of the type whose behaviour is again modeled by equation (4.127), but whose stator is isotropic. The simplest system of this type is the anisotropic Jeffcott rotor studied in Section 4.8.3. All deviatoric matrices related to the stator vanish, and equation (4.127) reduces to the following differential equation:

$$\mathbf{M}_m\ddot{\mathbf{q}} + (\mathbf{C}_m - i\omega\mathbf{G})\dot{\mathbf{q}} + (\mathbf{K}_m - i\omega\mathbf{C}_{r_m})\mathbf{q}+$$

$$+ \mathbf{M}_{r_d}e^{2i\omega t}(\ddot{\overline{\mathbf{q}}} + 2i\omega\dot{\overline{\mathbf{q}}}) + \mathbf{C}_{r_d}e^{2i\omega t}\dot{\overline{\mathbf{q}}}+ \qquad (4.137)$$

$$+(\mathbf{K}_{r_d} - i\omega\mathbf{C}_{r_d})e^{2i\omega t}\overline{\mathbf{q}} = \mathbf{F}_n + \omega^2\mathbf{F}_r e^{i\omega t}\,.$$

It is easy to verify that equation (4.137) can be transformed into an equation with constant coefficients by resorting to the rotating generalized coordinates defined by equation (4.89)

$$\mathbf{M}_m\ddot{\mathbf{r}} + \left[\mathbf{C}_m + i\omega(2\mathbf{M}_m - \mathbf{G})\right]\dot{\mathbf{r}}+$$

$$+ \left[-\omega^2(\mathbf{M}_m - \mathbf{G}) + \mathbf{K}_m + i\omega\mathbf{C}_n)\right]\mathbf{r}+ \qquad (4.138)$$

$$+\mathbf{M}_{r_d}\ddot{\overline{\mathbf{r}}} + \mathbf{C}_{r_d}\dot{\overline{\mathbf{r}}} + \left(\omega^2\mathbf{M}_{r_d} + \mathbf{K}_{r_d}\right)\overline{\mathbf{r}} = \mathbf{F}_n e^{-i\omega t} + \omega^2\mathbf{F}_r\,.$$

Alternatively, it is possible to use the same equation of motion written in terms of real coordinates

$$
\begin{bmatrix} \mathbf{M}_x & \mathbf{M}_{xy} \\ \mathbf{M}_{yx} & \mathbf{M}_y \end{bmatrix} \ddot{\mathbf{x}} + \left(-\omega \begin{bmatrix} \mathbf{M}_{yx} - \mathbf{M}_{xy} & \mathbf{M}_x + \mathbf{M}_y - \mathbf{G} \\ -\mathbf{M}_x - \mathbf{M}_y + \mathbf{G} & \mathbf{M}_{yx} - \mathbf{M}_{xy} \end{bmatrix} + \right.
$$

$$
\left. + \begin{bmatrix} \mathbf{C}_{r_x} + \mathbf{C}_n & \mathbf{C}_{r_{xy}} \\ \mathbf{C}_{r_{yx}} & \mathbf{C}_{r_y} + \mathbf{C}_n \end{bmatrix} \right) \dot{\mathbf{x}} + \tag{4.139}
$$

$$
+ \left(\begin{bmatrix} \mathbf{K}_x & \mathbf{K}_{xy} \\ \mathbf{K}_{yx} & \mathbf{K}_y \end{bmatrix} - \omega^2 \begin{bmatrix} \mathbf{M}_y - \mathbf{G} & -\mathbf{M}_{yx} \\ -\mathbf{M}_{xy} & \mathbf{M}_x - \mathbf{G} \end{bmatrix} + \right.
$$

$$
\left. + \omega \begin{bmatrix} 0 & -\mathbf{C}_n \\ \mathbf{C}_n & 0 \end{bmatrix} \right) \mathbf{x} = \left\{ \begin{matrix} \Re(\mathbf{f}_n e^{-i\omega t}) \\ \Im(\mathbf{f}_n e^{-i\omega t}) \end{matrix} \right\} + \omega^2 \left\{ \begin{matrix} \Re(\mathbf{f}_r) \\ \Im(\mathbf{f}_r) \end{matrix} \right\},
$$

where the real-coordinates vector is $\mathbf{x} = [\Re(\mathbf{r})^T, \Im(\mathbf{r})^T]^T$.

In the current case, the study of the unbalance response is easier than that of a static loading because the first gives way to a deformed configuration that is stationary with respect to the system of reference. The solution of equation (4.139) for static loading is of the type

$$
\mathbf{r} = \mathbf{r}_1 e^{-i\omega t} + \mathbf{r}_2 e^{i\omega t},
$$

leading to the equation

$$
\begin{bmatrix} \mathbf{A}_{11} & \mathbf{A}_{12} \\ \mathbf{A}_{21} & \mathbf{A}_{22} \end{bmatrix} \left\{ \begin{matrix} \mathbf{q}_1 \\ \overline{\mathbf{q}}_2 \end{matrix} \right\} = \left\{ \begin{matrix} \mathbf{f}_n \\ 0 \end{matrix} \right\}, \tag{4.140}
$$

where

$$
\begin{aligned}
\mathbf{A}_{11} &= -i\omega \mathbf{C}_{r_m} + \mathbf{K}_m, \\
\mathbf{A}_{12} &= -i\omega \mathbf{C}_{r_d} + \mathbf{K}_d, \\
\mathbf{A}_{21} &= -i\omega \overline{\mathbf{C}}_{r_d} + \overline{\mathbf{K}}_d, \\
\mathbf{A}_{22} &= -2\omega^2 (2\mathbf{M}_m - \mathbf{G}) - i\omega(\overline{\mathbf{C}}_{r_m} + 2\overline{\mathbf{C}}_n) + \overline{\mathbf{K}}_m.
\end{aligned}
$$

The same solution can be written with reference to the fixed frame as

$$
\mathbf{q} = \mathbf{r} e^{i\omega t} = \mathbf{r}_1 + \mathbf{r}_2 e^{2i\omega t}.
$$

The obvious meaning of $\mathbf{r}_1$, then, is the mean inflected shape, which is fixed in space, while that of $\mathbf{r}_2$ is a component of the deflected shape, which rotates at a speed equal to 2ω. A static loading then causes the onset of vibrations, which are seen by the stator as occurring with a frequency 2ω and by the rotor with a frequency ω.

By equating to zero the matrix of the coefficients of equation (4.140), an eigenproblem in ω is obtained. It yields the values of the secondary critical speeds due to a constant load distribution, as seen in Section 4.8.4 for the Jeffcott rotor.

The solution of the problem related to a given unbalance distribution is straightforward, leading to a synchronous circular whirling. The solution of the equation of motion is constant, $\mathbf{r} = \mathbf{r}_0$, leading to the equation

$$\left[-\omega^2(\mathbf{M}_m - \mathbf{G}) + \mathbf{K}_m + i\omega\mathbf{C}_n\right]\mathbf{r}_0 + \qquad (4.141)$$

$$+ \left(\omega^2\mathbf{M}_{r_d} + \mathbf{K}_{r_d}\right)\bar{\mathbf{r}}_0 = \omega^2\mathbf{f}_r \, .$$

The constant solution in the rotating frame is equivalent to $\mathbf{q} = \mathbf{r}_0 e^{i\omega t}$ if it is written in the fixed reference frame. The response to unbalance is, consequently, a pure circular synchronous whirling, and rotating damping has no effect on the behaviour of the system because the rotor does not vibrate but merely rotates in the deflected configuration.

The solution for the free whirling of the system is of the type

$$\mathbf{r} = \mathbf{r}_1 e^{i\lambda' t} + \mathbf{r}_2 e^{-i\bar{\lambda}' t} \, ,$$

i.e., an elliptical whirling with reference to the rotating frame $\xi\eta z$. By introducing it into the equation of motion (4.138), the following algebraic equation is obtained:

$$\left(-\lambda'^2\begin{bmatrix}\mathbf{M}_m & \mathbf{M}_d \\ \overline{\mathbf{M}}_d & \overline{\mathbf{M}}_m\end{bmatrix} + \lambda'\omega\begin{bmatrix}2\mathbf{M}_m + \mathbf{G} & 0 \\ 0 & 2\mathbf{M}_m - \mathbf{G}\end{bmatrix} + \right.$$

$$+ i\lambda'\begin{bmatrix}\mathbf{C}_m + \mathbf{C}_{rm} & \mathbf{C}_{r_d} \\ \overline{\mathbf{C}}_{r_d} & \overline{\mathbf{C}}_n + \overline{\mathbf{C}}_{rm}\end{bmatrix} + i\omega\begin{bmatrix}\mathbf{C}_n & 0 \\ 0 & -\overline{\mathbf{C}}_n\end{bmatrix} +$$

$$+ \begin{bmatrix}-\mathbf{M}_m + \mathbf{G} & \mathbf{M}_d \\ \overline{\mathbf{M}}_d & -\mathbf{M}_m + \mathbf{G}\end{bmatrix} + \begin{bmatrix}\mathbf{K}_m & \mathbf{K}_{n_d} \\ \overline{\mathbf{K}}_{n_d} & \overline{\mathbf{K}}_m\end{bmatrix}\right)\left\{\begin{matrix}\mathbf{r}_1 \\ \bar{\mathbf{r}}_2\end{matrix}\right\} = 0.$$

$$(4.142)$$

By expressing the same solution in the fixed frame, it yields

$$\mathbf{q} = \mathbf{r}_1 e^{i\lambda t} + \mathbf{r}_2 e^{i(2\omega - \bar{\lambda})t} \, .$$

The orbits are then elliptical when seen in the rotating frame, but become Lissajous curves when seen in the fixed frame. The equation of motion could be written directly in terms of fixed coordinates, obtaining an equation in which the whirl speed λ is present instead of λ'

$$
\left(-\lambda^2 \begin{bmatrix} \mathbf{M}_m & \mathbf{M}_d \\ \overline{\mathbf{M}}_d & \overline{\mathbf{M}}_m \end{bmatrix} + \lambda\omega \begin{bmatrix} \mathbf{G} & 2\mathbf{M}_d \\ 2\overline{\mathbf{M}}_d & 4\mathbf{M}_m - \mathbf{G} \end{bmatrix} + \right.
$$

$$
+\omega^2 \begin{bmatrix} \mathbf{0} & \mathbf{0} \\ \mathbf{0} & -4\mathbf{M}_m + 2\mathbf{G} \end{bmatrix} + i\lambda \begin{bmatrix} \mathbf{C}_m + \mathbf{C}_{rm} & \mathbf{C}_{rd} \\ \overline{\mathbf{C}}_{rd} & \overline{\mathbf{C}}_n + \overline{\mathbf{C}}_{rm} \end{bmatrix} +
$$

$$
\left. -i\omega \begin{bmatrix} \mathbf{C}_{rm} & \mathbf{C}_{rd} \\ \overline{\mathbf{C}}_{rd} & 2\overline{\mathbf{C}}_n + \overline{\mathbf{C}}_{rm} \end{bmatrix} + \begin{bmatrix} \mathbf{K}_m & \mathbf{K}_{nd} \\ \overline{\mathbf{K}}_{nd} & \overline{\mathbf{K}}_m \end{bmatrix} \right) \begin{Bmatrix} \mathbf{r}_1 \\ \overline{\mathbf{r}}_2 \end{Bmatrix} = \mathbf{0}.
$$

$$
(4.143)
$$

It is then not necessary, even in this case, to resort to the rotating frame. The problem of finding the fields of instability of systems with many degrees of freedom can easily be solved by obtaining the eigenvalues of equation (4.143) and then studying the sign of the imaginary part of the complex frequency.

4.6.8   General system with anisotropic stator and rotor

In the case where neither the stator nor the rotor have axial symmetry, the equation of motion has periodic coefficients in both rotating and nonrotating systems of reference. As already stated in Section 3.7, linear equations with variable coefficients can be solved by adding a particular integral of the complete equation to the general solution of the homogeneous equation. However, no general method exists to reach such a solution, particularly for large sets of equations. Because the coefficients are periodic functions of time with period π/ω, the solution of the homogeneous equation associated with equation (4.127) is of the type

$$
\mathbf{q} = \mathbf{q}_1(t)e^{i\lambda t} + \mathbf{q}_2(t)e^{-i\overline{\lambda} t},
$$

where both vectors $\mathbf{q}_1$ and $\mathbf{q}_2$ are periodic functions of time with the same period, i.e., with fundamental frequency equal to 2ω. The general solution is then the sum of a number of terms of the type mentioned earlier, each with its value of λ, plus a solution of the complete equation. The study of the stability of the system can then be performed in the same way as in the case with constant $\mathbf{q}_1$ and $\mathbf{q}_2$.

As seen in Section 3.4.2 for Hill's equation, unknown functions $\mathbf{q}_1$ and $\mathbf{q}_2$ can be expressed by trigonometric series

$$\mathbf{q} = \sum_{j=-\infty}^{\infty} \left(\mathbf{q}_{1_j} e^{i(\lambda+2j\omega)t} + \mathbf{q}_{2_j} e^{-i(\overline{\lambda}+2j\omega)t} \right). \tag{4.144}$$

By introducing equation (4.144) into the homogeneous equation associated with equation (4.127), the following algebraic equation is readily obtained

$$\sum_{j=-\infty}^{\infty} \left(\mathbf{A}_j \mathbf{q}_{1_j} e^{i(\lambda+2j\omega)t} + \overline{\mathbf{B}}_j \mathbf{q}_{2_j} e^{-i(\overline{\lambda}+2j\omega)t} + \overline{\mathbf{C}}_j \overline{\mathbf{q}}_{1_j} e^{-i(\overline{\lambda}+2j\omega)t} + \right.$$
$$\left. + \mathbf{D}_j \overline{\mathbf{q}}_{2_j} e^{i(\lambda+2j\omega)t} + \overline{\mathbf{E}}_j \overline{\mathbf{q}}_{1_j} e^{-i[\overline{\lambda}+2(j-1)\omega]t} + \mathbf{E}_j \overline{\mathbf{q}}_{2_j} e^{i[\lambda+2(j+1)\omega]t} \right) = 0. \tag{4.145}$$

By separately equating to zero the various terms of equation (4.145), the following infinite set of algebraic equations is readily obtained

$$\begin{bmatrix} \cdots & \cdots & \cdots & & & & & \\ \mathbf{F}_{-2} & \mathbf{A}_{-1} & \mathbf{D}_{-1} & & & & & \\ & \mathbf{C}_{-1} & \mathbf{B}_{-1} & \mathbf{E}_0 & & & & \\ & & \mathbf{F}_{-1} & \mathbf{A}_0 & \mathbf{D}_0 & & & \\ & & & \mathbf{C}_0 & \mathbf{B}_0 & \mathbf{E}_1 & & \\ & & & & \mathbf{F}_0 & \mathbf{A}_1 & \mathbf{D}_1 & \\ & & & & & \mathbf{C}_1 & \mathbf{B}_1 & \mathbf{E}_2 \\ & & & & & & \mathbf{F}_1 & \mathbf{A}_2 & \mathbf{D}_2 \\ & & & & & \cdots & \cdots & \cdots \end{bmatrix} \times \begin{Bmatrix} \cdots \\ \mathbf{q}_{1_{-1}} \\ \overline{\mathbf{q}}_{2_{-1}} \\ \mathbf{q}_{1_0} \\ \overline{\mathbf{q}}_{2_0} \\ \mathbf{q}_{1_1} \\ \overline{\mathbf{q}}_{2_1} \\ \mathbf{q}_{1_2} \\ \cdots \end{Bmatrix} = \{0\}, \tag{4.146}$$

where

$$\mathbf{A}_j = -(\lambda+2j\omega)^2 \mathbf{M}_m + \omega(\lambda+2j\omega)\mathbf{G} + i(\lambda+2j\omega)\mathbf{C}_m + \mathbf{K}_m - i\omega\mathbf{C}_{r_m},$$

$$\mathbf{B}_j = -(\lambda+2j\omega)^2 \mathbf{M}_m - \omega(\lambda+2j\omega)\mathbf{G} + i(\lambda+2j\omega)\overline{\mathbf{C}}_m + \overline{\mathbf{K}}_m + i\omega\overline{\mathbf{C}}_{r_m},$$

$$\mathbf{C}_j = -(\lambda+2j\omega)^2 \overline{\mathbf{M}}_{n_d} + i(\lambda+2j\omega)\overline{\mathbf{C}}_{n_d} + \overline{\mathbf{K}}_{n_d}, \tag{4.147}$$

$$\mathbf{D}_j = -(\lambda+2j\omega)^2 \mathbf{M}_{n_d} + i(\lambda+2j\omega)\mathbf{C}_{n_d} + \mathbf{K}_{n_d},$$

$$\mathbf{E}_j = [-\lambda^2 - 2\lambda\omega(2j-1) - 4j\omega^2(j-1)]\overline{\mathbf{M}}_{r_d} + i(\lambda+2j\omega)\overline{\mathbf{C}}_{r_d} + \overline{\mathbf{K}}_{r_d} - i\omega\overline{\mathbf{C}}_{r_d},$$

$$\mathbf{F}_j = [-\lambda^2 - 2\lambda\omega(2j+1) - 4j\omega^2(j+1)]\mathbf{M}_{r_d} + i(\lambda+2j\omega)\mathbf{C}_{r_d} + \mathbf{K}_{r_d} + i\omega\mathbf{C}_{r_d}.$$

The homogeneous equation (4.146) has solutions different from the trivial one only if the matrix of the coefficients is singular. An eigenproblem, similar to the one linked with Hill's infinite determinant (equation (3.95)) is obtained. The difference is that the bandwidth of the determinant here is $3n$ instead of 3. Approximate solutions for the eigenproblem can then be obtained by considering a limited array of $2m$ rows by $2m$ columns, in terms of matrices (the actual number of rows and columns is $2\ m \times n$), centered on the matrices with subscript 0. Let these limited matrices be denoted by $\mathbf{Z}_m$ and let matrix $\mathbf{A}_0$ be $\mathbf{Z}_0$. It can readily be seen that the solution of the zero-order approximation involving matrix $\mathbf{Z}_0$ coincides with the solution of a symmetrical system having the mean properties of the actual system.

By inspecting equations (4.147), it is then clear that matrices $\mathbf{A}_j$ and $\mathbf{B}_j$ contain the mean properties of the system, matrices $\mathbf{C}_j$ and $\mathbf{D}_j$ are related to the unsymmetrical characteristics of the stator (hence, they vanish when the stator is symmetrical), and matrices $\mathbf{E}_j$ and $\mathbf{F}_j$ are related to the unsymmetrical properties of the rotor.

The first-order approximation obtained by considering matrix

$$\mathbf{Z}_1 = \begin{bmatrix} \mathbf{A}_0 & \mathbf{B}_0 \\ \mathbf{C}_0 & \mathbf{D}_0 \end{bmatrix}$$

coincides with the solution of a system with isotropic rotor (with mean properties) running on an asymmetrical stator (with the actual properties).

The equation that allows the study of a system with a symmetrical stator and an asymmetrical rotor can be written as

$$\begin{bmatrix} \mathbf{B}_0 & \mathbf{E}_1 \\ \mathbf{F}_0 & \mathbf{A}_1 \end{bmatrix} \begin{Bmatrix} \bar{\mathbf{q}}_{2_0} \\ \mathbf{q}_{1_1} \end{Bmatrix} = 0. \qquad (4.148)$$

The infinite set of equations (4.146) and the corresponding nonhomogeneous set that includes the forcing functions, represent a general model for rotating machinery. The solution of the eigenproblem, however, is not easy. The equations must be rearranged to obtain an eigenproblem in standard form, which results in a further doubling of the size of the problem, becoming $4mn$ and then a problem of a very large order must be faced. It must be noted that if $\Re(\lambda)$ is the real part of the solution of the eigenproblem, then $\Re(\lambda) + 2j\omega$ and $\Re(\lambda) - 2j\omega$ (for any value of j) is the real part of a solution. This explains the apparent inconsistency of the number of the eigenvalues, which tend to infinity when increasing the size of the matrix $\mathbf{Z}_m$: While obtaining a better precision, solutions whose real parts are equal to those already obtained plus a multiple of 2ω are obtained. Vectors $\mathbf{q}_j$ with j greater than 1 usually add only a small ripple on the basic solution, which is given by the vectors with $j \leq 1$. Third-order approximations should, consequently, give results that accurately simulate the behaviour of the actual system.

Solutions for nonhomogeneous problems, like those related to the response to static loading and unbalance, can be obtained in a similar way. In the former case the solution can be expressed by the series

$$\mathbf{q} = \mathbf{q}_0 + \sum_{j=1}^{\infty} \mathbf{q}_{1_j} e^{2ij\omega t} + \sum_{j=1}^{\infty} \mathbf{q}_{2_j} e^{-2ij\omega t}. \qquad (4.149)$$

In equation (4.149) the mean response, the forward components, and the backward components are written separately. By using this expression of the deflected shape, the following infinite set of equations can be obtained from the equation of motion (4.127)

$$\begin{cases} \mathbf{A}_0\mathbf{q}_0 + \mathbf{C}_0\overline{\mathbf{q}}_0 + \mathbf{E}_1\overline{\mathbf{q}}_{1_1} = \mathbf{f}_n \\ \mathbf{F}_0\overline{\mathbf{q}}_0 + \mathbf{A}_1\mathbf{q}_{1_1} + \mathbf{D}_1\overline{\mathbf{q}}_{2_1} = \{0\} \\ \mathbf{C}_1\mathbf{q}_{1_1} + \mathbf{B}_1\overline{\mathbf{q}}_{2_1} + \mathbf{E}_2\mathbf{q}_{1_2} = \{0\} \\ \mathbf{F}_1\overline{\mathbf{q}}_{2_1} + \mathbf{A}_2\mathbf{q}_{1_2} + \mathbf{D}_1\overline{\mathbf{q}}_{2_2} = \{0\} \\ \cdots\cdots\cdots\cdots\cdots\cdots\cdots\cdots\cdots\cdots\cdots\cdots, \end{cases} \qquad (4.150)$$

where the relevant matrices can be obtained from equation (4.146) by setting $\lambda = 0$. Here, again, a good approximation can be obtained by considering only the components $\mathbf{q}_0$, $\mathbf{q}_{11}$, and $\mathbf{q}_{21}$, i.e., by resorting to a reduced set formed by the first three equations (4.150) in which the term in $\mathbf{E}_2$ has been neglected. The size of the problem is thus reduced to $3n$ equations.

The response to an arbitrary unbalance distribution is similarly obtained by assuming a solution of the type

$$\mathbf{q} = \sum_{j=1}^{\infty} \mathbf{q}_{1_j} e^{i(2j+1)\omega t} + \sum_{j=1}^{\infty} \mathbf{q}_{2_j} e^{-i(2j-1)\omega t}. \qquad (4.151)$$

By using the expressions (4.151) for the deflected shape, the following infinite set of equations can be obtained from the equation of motion (4.127)

$$
\begin{bmatrix}
\mathbf{A}_0 & \mathbf{D}_0 & & & \\
\mathbf{C}_0 & \mathbf{B}_0 & \mathbf{E}_1 & & \\
& \mathbf{F}_0 & \mathbf{A}_1 & \mathbf{D}_1 & \\
& & \mathbf{C}_1 & \mathbf{B}_1 & \mathbf{E}_2 \\
& & \cdots & \cdots & \cdots
\end{bmatrix}
\begin{Bmatrix}
\mathbf{q}_{1_0} \\
\overline{\mathbf{q}}_{2_0} \\
\mathbf{q}_{1_1} \\
\overline{\mathbf{q}}_{2_1} \\
\cdots
\end{Bmatrix}
= \omega^2
\begin{Bmatrix}
\mathbf{f}_r \\
\{0\} \\
\{0\} \\
\{0\} \\
\cdots
\end{Bmatrix} . \qquad (4.152)
$$

Here the relevant equations can be obtained from equation (4.146) by setting $\lambda = \omega$. In this case a good approximation can be obtained by considering only the first four equations in which the term in $\mathbf{E}_2$ has been neglected. The size of the problem thus reduces to $4n$ equations.

4.7 Introduction to nonlinear rotor dynamics

4.7.1 General considerations

It is well known that the behaviour of both lubricated journal bearings and rolling element bearings is strongly nonlinear and can cause rotors to behave in a nonlinear way. However, many other mechanisms can have similar effects, including nonlinear elasticity or dry friction. Often such nonlinearities can be neglected; nonlinear parts can be modeled as linear, or even rigid, elements. This is usually the case, for example, of bearings when they are much stiffer than other parts of the machine. There are, however, cases in which this simple approach is not suitable and, consequently, the nonlinear problem must be faced. In such a case, the determination of the critical speed, at least in the usual terms, and the plotting of the Campbell diagram become impossible, and the study is generally limited to the computation of the response of the system to given forcing causes, usually unbalance distributions. It is still possible to define critical speeds as those speeds at which the response of the system becomes very large, but the fact that these speeds depend on the particular unbalance distribution that has been assumed makes their definition of limited use.

The nonlinearity of the problem makes it impossible to study separately the effects of static and dynamic loads and the free whirling of the system because the dynamic behaviour can be strongly influenced by the presence of static loads.

In the case of nonlinear rotors there is a deep difference between the behaviour of axisymmetrical and nonisotropic systems. If the system is axially symmetrical, circular whirling is an exact solution of the problem, although the nonlinearity of the system makes possible the existence of other solutions. Contrary to what was seen for general nonlinear vibrating systems, in this case it is possible to obtain an exact solution of the equation of motion in closed form. This is generally impossible in the case of anisotropic systems, for which the usual approximation techniques seen in Chapter 3 must be used.

4.7.2 Nonlinear Jeffcott rotor: Equation of motion

Consider a Jeffcott rotor of the type seen in Section 4.5, but with restoring and damping forces that are nonlinear functions of the displacement and velocity, respectively. The axial symmetry of the system, however, causes the restoring force to have the same direction as the displacement and the damping force to have the same direction as the velocity. By resorting to the complex coordinate z, defined in equation (4.5), the restoring force and the forces due to nonrotating and rotating damping can be expressed in the form

$$\begin{cases} \vec{F}_e = -kz\left[1 + f(|z|)\right], \\ \vec{F}_{sn} = -c_n\dot{z}\left[1 + \beta_n(|\dot{z}|)\right], \\ \vec{F}_{sr} = -c_r(\dot{z} - i\omega z)\left[1 + \beta_r(|\dot{z} - i\omega z|)\right]. \end{cases} \qquad (4.153)$$

Note that no interaction between restoring and damping force has been assumed: The current model is not the most general nonlinear Jeffcott rotor that can be built. By introducing the expressions of the nonlinear forces obtained earlier into the equation of motion of the Jeffcott rotor (equation (4.18)) written for constant spin speed, it follows

$$m\ddot{z} + (c_n + c_r)\dot{z} + (k - i\omega c_r)z + \left[c_n\beta_n(|\dot{z}|) + c_r\beta_r(|\dot{z} - i\omega z|)\right]\dot{z} + \\ + kf(|z|)z - i\omega c_r\beta_r(|\dot{z} - i\omega z|)z = m\epsilon\omega^2 e^{i\omega t} + F_n.$$

$$(4.154)$$

Equation (4.154) can be used to study the free behaviour of the system (free circular whirling), the effect of a static load F_n (of weight, for example), or that of an eccentricity of mass m. The presence of nonlinear terms, however, makes it impossible to perform a general study by superimposing the various solutions and to obtain noncircular whirling by adding forward and backward whirling of different amplitudes.

4.7.3   Unbalance response

Consider the undamped system whose model can be obtained from equation (4.154) by neglecting all terms related to damping, linear and nonlinear, and the nonrotating force. A possible solution to the equation is $z = z_0 e^{i\omega t}$, which allows the response to the static unbalance $m\epsilon$ to be computed. Note that it represents an exact solution of the nonlinear equation of motion and not just an approximation of the fundamental harmonic of the response. As a consequence, there are no higher-order harmonics in the response. This does not mean that thre cannot be higher-order harmonics, because the circular whirling solution is only one of the possible solutions; however, it can be demonstrated to be stable and has been found both experimentally and by numerical experimentation. Introducing this solution into the equation of motion, the following algebraic equation, which allows the amplitude of the response to be computed, is readily obtained:

$$\left[k - m\omega^2 + kf(|z_0|) \right] z_0 = m\epsilon\omega^2 . \tag{4.155}$$

In this case it is also possible to obtain a backbone of the response, by assuming that the unbalance is vanishingly small, i.e., by solving the homogeneous equation associated with equation (4.155). Because the backbone defines, in the case of vibrating systems, the conditions for a sort of nonlinear resonance, here the situation on the backbone is similar to that occurring at a critical speed.

Example 4-5
Compute the unbalance response of an undamped rotor characterized by a function $f(|z|)$ of the same type as that seen for Duffing's equation:
$f(|z_0|) = \mu|z_0|^2$.
The algebraic equations allowing computation of the amplitude of the circular whirling and the backbone are

$$\left(1 - \frac{m}{k}\omega^2 + \mu|z_0|^2 \right) z_0 = \frac{m}{k}\epsilon\omega^2 ,$$

$$1 - \frac{m}{k}\omega^2 + \mu|z_0|^2 = 0 \,.$$

The equation yielding the backbone is

$$z_0 = \sqrt{\frac{m\omega^2 - k}{k\mu}} \,.$$

A result that is quite similar to that obtained for Duffing's equation is obtained. Some important differences, however, must be mentioned: ω here is the spin speed and not a circular frequency, and the excitation is proportional to ω^2. As already stated, in the case of Duffing's equation, the solution obtained was only a first approximation of the fundamental harmonic of the response, while here the solution gives the whole response. This can be related to the fact that in this case the nonlinear element does not oscillate along the force-displacement characteristic but actually rotates, maintaining a given deformation. Most considerations regarding multiple solutions and the jump phenomenon, however, hold. These considerations hold in general, regardless of the particular law $f(|z|)$.

The same circular whirling solution also holds in the case of damped systems. The algebraic equation that can be obtained in the usual way is

$$\left[1 - \frac{m}{k}\omega^2 + f(|z_0|) + i\omega\frac{c_n}{k} + i\omega\frac{c_n}{k}\beta_n(|\omega z_0|) \right] z_0 = \frac{m}{k}\epsilon\omega^2 \,. \quad (4.156)$$

As easily predictable, rotating damping plays no role in determining the conditions for synchronous whirling. The amplitude z_0 is, in this case, expressed by a complex number, as already seen in the linear case. By separating the real and imaginary parts of equation (4.156) and introducing the phase Φ of the response with respect to the unbalance vector, which is, as usual, assumed to be along the ξ-axis, it follows that

$$\begin{cases} z_0 \left[1 - \frac{m}{k}\omega^2 + f(|z_0|) \right] = \frac{m}{k}\epsilon\omega^2 \cos(\Phi) \,, \\ z_0 c_n \left[1 + \beta_n(|\omega z_0|) \right] = -m\epsilon\omega \sin(\Phi) \,, \end{cases} \quad (4.157)$$

which yields the amplitude and phase of the response

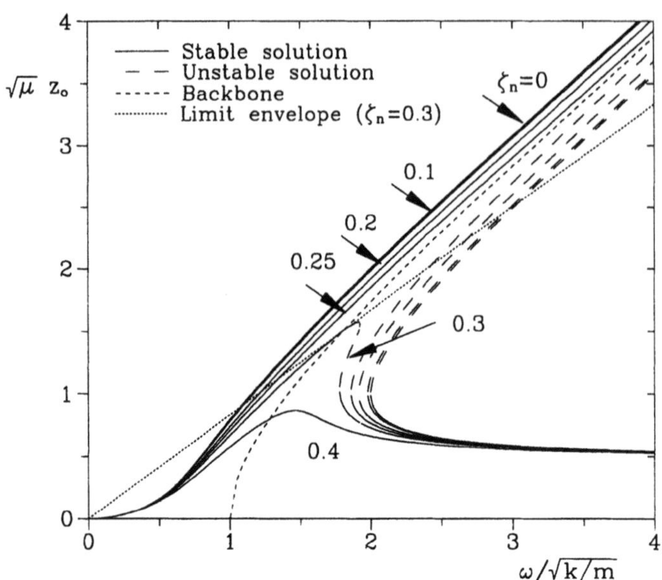

FIGURE 4.30. Unbalance response for a nonlinear Jeffcott rotor with linear damping. The characteristic of the nonlinear element is assumed to be of the hardening Duffing type.

$$\begin{cases} |z_0|^2 \left\{ \left[1 - \dfrac{m}{k}\omega^2 + f(|z_0|) \right]^2 + \left(\dfrac{\omega c_n}{k} \right)^2 [1 + \beta_n(|\omega z_0|)]^2 \right\} = \left(\dfrac{m}{k}\epsilon \right)^2 \omega^4, \\[4mm] \Phi = -\arctan\left[\omega c_n \dfrac{1 + \beta_n(|\omega z_0|)}{k - m\omega^2 + k f(|z_0|)} \right]. \end{cases}$$

$$(4.158)$$

Also, in the case of the nonlinear Jeffcott rotor, it is possible to obtain a limit envelope of the response. By stating that the phase Φ is equal to $-90°$, the second equation (4.157) directly yields the equation of the limit envelope

$$z_0 c_n [1 + \beta_n(|\omega z_0|)] = m\epsilon\omega. \qquad (4.159)$$

If damping is linear, the limit envelope is just a straight line whose equation is $z_0 = \omega m\epsilon/c_n$. Some results obtained for a restoring force of the Duffing type with hardening characteristic and linear damping are shown in nondimensional form in Figure 4.30. The plot has been obtained for a nondimensional nonlinear parameter $\mu^* = \mu\epsilon^2 = 0.25$ and various values of the nondimensional damping $\zeta_n = c_n/2\sqrt{km}$.

The differences with the usual results obtained for damped Duffing's equation are that the peak amplitude occurs now at the right of the backbone instead of on the left (the same thing happens in linear systems), and for low values of the damping, the curves do not close. This suggests a greater difficulty in experiencing the jump phenomenon and, consequently, self-centering.

It is easily predictable that self-centering is much more difficult in the case of nonlinear rotors than in the linear case, and that higher nonrotating damping is needed to work in the supercritical range, because the jump, which is needed to obtain the self-centered configuration, takes place when the energy supplied by the forcing function is not sufficient to sustain the motion with higher amplitude. In rotating systems the forcing function is due to unbalance, and the amount of energy supplied by the centrifugal field is large. This can be seen from the shape of the limit envelope, which may not cross the backbone.

4.7.4 Free circular whirling

A possible solution of the homogeneous equation associated with equation (4.154), i.e., of the model for a perfectly balanced Jeffcott rotor, is $z = z_0 e^{i\lambda t}$. Note again that this solution, representing a circular free whirling of the system, is an exact solution of the equation of motion. To avoid greater analytical complexities without losing generality of the results, damping will be assumed to be linear. By introducing the solution for circular free whirling into the homogeneous equation of motion, it follows that

$$\left\{ 1 - \frac{m}{k}\lambda^2 + f(|z_0|) + i\left[\lambda\omega\frac{c_n}{k} + (\lambda - \omega)\frac{c_r}{k} \right] \right\} z_0 = 0 . \qquad (4.160)$$

Equation (4.160) has a solution different from the trivial solution $z_0 = 0$ only if the expression in braces is equal to zero. An equation in λ is so obtained, which yields

$$\lambda_R^* = \pm\sqrt{ \Gamma^* + \sqrt{ \Gamma^{*2} + \omega^{*2}\zeta_r^2 } } ,$$

$$\lambda_I^* = \zeta_n + \zeta_r \mp \sqrt{ -\Gamma^* + \sqrt{ \Gamma^{*2} + \omega^{*2}\zeta_r^2 } } , \qquad (4.161)$$

where the nondimensional speeds ω^* and λ^* and the damping ratios ζ_n and ζ_r are defined in the usual way, with reference to the linearized system, and

$$\Gamma^* = [1 + f(|z_0|) - (\zeta_n + \zeta_r)^2]/2 .$$

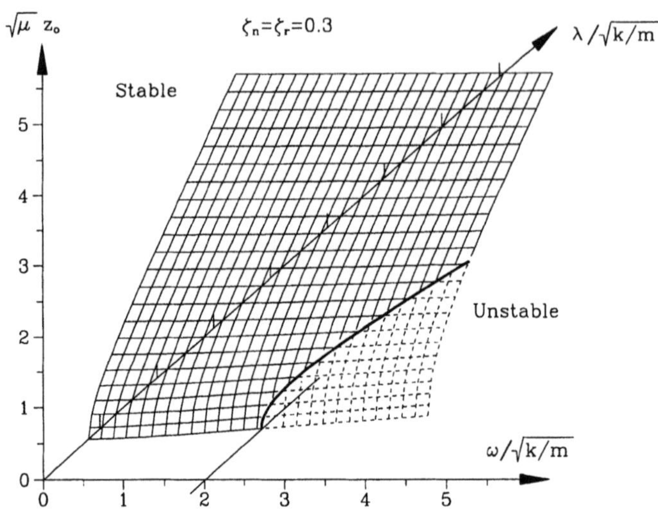

FIGURE 4.31. Tridimensional plot for the study of the free whirling of the non-linear Jeffcott rotor studied in Figure 4.30.

Equation (4.161) is very similar to the corresponding expression for the linear case, with the difference that in the current case the whirl speed is a function of the amplitude of the motion. As is obvious for a nonlinear system, the Campbell diagram loses any meaning. A three-dimensional plot in which the amplitude $|z_0|$ is reported as a function of ω and λ_R can, however, be introduced. The ω,λ_R–plane of the tridimensional plot coincides with the Campbell diagram of the linearized system. The first equation (4.161) defines a surface expressing all the possible conditions of free whirling; it can be considered the backbone of the system at varying spin speed ω. The intersection of the surface with the $|z_0|,\lambda_R$–plane expresses the relationship linking the natural frequency with the amplitude at standstill: It is then the backbone of the nonrotating system for a whirling mode in the xy-plane.

One of the mentioned plots has been reported in nondimensional form in Figure 4.31. The figure has been plotted for a system with $\zeta_n = \zeta_r = 0.3$. The Campbell diagram of the linearized system coincides with that plotted in Figure 4.7, (curve for the corresponding value of the damping ratio). The stability threshold of the linearized system is at $\omega^* = 2$.

The intersection of the surface with the plane of equation $\lambda = \omega$ gives the conditions for free synchronous whirling: It then coincides with the backbone of the unbalance response shown in Figure 4.30.

From the second equation (4.161) it is possible to obtain the condition for stability

$$\omega < \sqrt{1 + f(|z_0|)}\sqrt{\frac{k}{m}\left(1 + \frac{c_n}{c_r}\right)} \, . \qquad (4.162)$$

The threshold of stability depends on the amplitude, as shown in Figure 4.31, where the unstable part of the surface is dashed. If the spin speed ω is lower than the threshold of instability of the linearized system, the motion is always stable and the amplitude of free whirling decays to zero. On the contrary, when the linearized analysis shows an unstable behaviour in the small, the nonlinear effects reduce the instability, and the result is a motion with growing amplitude until the border separating the full lines from the dotted lines is reached. The amplitude corresponding to these conditions is that of a sort of limit cycle that constitutes an attractor for all free whirl motions. The amplitude of the limit cycle is a function of the speed and grows with increasing ω.

The situation described earlier holds for hardening systems and qualitatively can be extended to systems with many degrees of freedom.

4.7.5 Stability of the equilibrium position

The stability of the equilibrium position for the unbalance response of the system can be obtained in the same way as for linear systems. Consider a damped nonlinear Jeffcott rotor whose restoring and damping forces have a cubic (Duffing type) and linear characteristic, respectively, and neglect nonrotating forces. The equation of motion in the rotating $\xi\eta$ reference frame can be easily obtained:

$$\begin{cases} m\ddot{\xi} + (c_n + c_r)\dot{\xi} - 2m\omega\dot{\eta} + \left[k - m\omega^2 + \mu(\xi^2 + \eta^2)\right]\xi - c_n\omega\eta = m\epsilon\omega^2 \,, \\ m\ddot{\eta} + (c_n + c_r)\dot{\eta} + 2m\omega\dot{\xi} + \left[k - m\omega^2 + \mu(\xi^2 + \eta^2)\right]\eta + c_n\omega\xi = 0 \,. \end{cases}$$
$$(4.163)$$

The behaviour of the system depends on three nondimensional parameters; the damping ratios ζ_n and ζ_r and the nonlinearity parameter $\epsilon^2\mu$. Equation (4.163) can also be used to study the motion of the system in nonstationary conditions with reference to the rotating frame. The unbalance response can be immediately computed

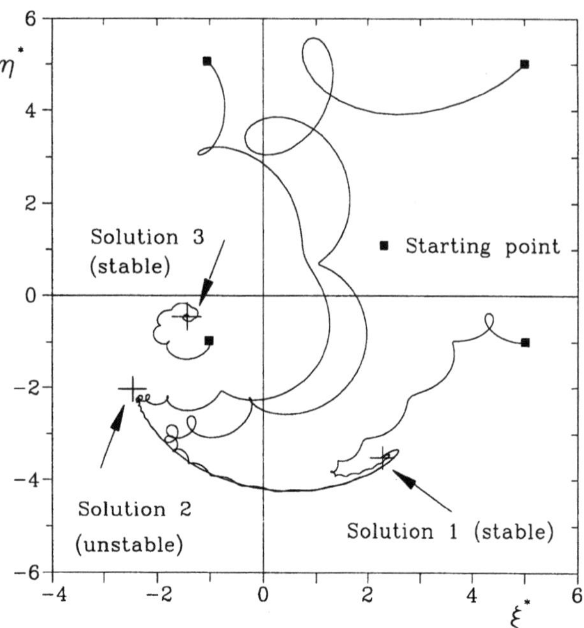

FIGURE 4.32. Trajectories in the rotating plane starting from four different points, one belonging to the domain of attraction of the lowest solution and three to the other domain. Values of parameters: $\omega^* = 2$, $\mu^* = 0.2$, $\zeta_n = 0.2$, $\zeta_r = 0.03$.

by assuming a stationary solution in the $\xi\eta$-plane. The result so obtained coincides with the solution already obtained in Section 4.9.3. The study of the stability in the small about the equilibrium position can be performed in the usual way. If $\xi_1(t)$ and $\eta_1(t)$ are small displacements from the equilibrium position ξ_0, η_0, the equation of motion can be linearized, obtaining, in nondimensional terms,

$$\begin{bmatrix} 1 & 0 \\ 0 & 1 \end{bmatrix} \left\{ \begin{matrix} \xi_1^{*''} \\ \eta_1^{*''} \end{matrix} \right\} + \begin{bmatrix} 2(\zeta_n + \zeta_r) & -2\omega^* \\ 2\omega^* & 2(\zeta_n + \zeta_r) \end{bmatrix} \left\{ \begin{matrix} \xi_1^{*'} \\ \eta_1^{*'} \end{matrix} \right\} +$$

$$+ \begin{bmatrix} 1 - \omega^{*2} + \mu^* \left(3\xi_0^{*2} + \eta_0^{*2}\right) & 2\left(-\omega^*\zeta_n + \mu^*\xi_0^*\eta_0^*\right) \\ 2\left(\omega^*\zeta_n + \mu^*\xi_0^*\eta_0^*\right) & 1 - \omega^{*2} + \mu^*\left(\xi_0^{*2} + 3\eta_0^{*2}\right) \end{bmatrix} \left\{ \begin{matrix} \xi_1^* \\ \eta_1^* \end{matrix} \right\} = 0$$

$$(4.164)$$

where, apart from the damping ratios and the usual nondimensional speed, the other nondimensional quantities are defined as $\xi^* = \xi/\epsilon$, $\eta^* = \eta/\epsilon$, $\tau = t\sqrt{k/m}$, $\mu^* = \mu\epsilon^2$, and prime denotes differentiation with respect to τ.

The stability of the motion in the vicinity of an equilibrium po-

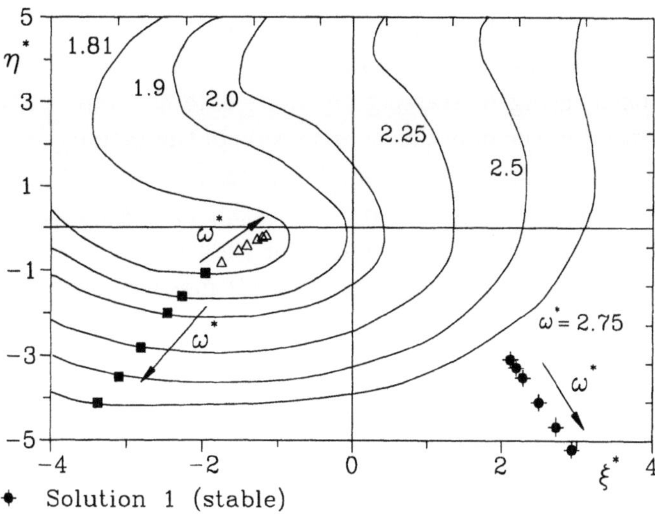

♦ Solution 1 (stable)

■ Solution 2 (unstable)   △ Solution 3 (stable)

FIGURE 4.33. Basins of attraction of the equilibrium positions at different values of the speed computed by numerically integrating the equations of motion; same values of the parameters as in Figure 4.32.

sition can be assessed by studying the sign of the decay rate of the small oscillations. Because the solution of a simple fourth-degree algebraic equation is required, the study can be carried on in closed form. Proceeding in this way it follows that, in the case of hardening systems, when three equilibrium positions are found, the solution with the largest amplitude is always stable. The intermediate solution is found to be unstable (saddle point), while the smallest solution (the self-centered one) is stable only up to a certain speed, which depends essentially on the ratio between the nonrotating and rotating damping. Note that the nonlinearity of the system ensures that the stability condition for the unbalance response will not coincide with the stability condition of the free whirling, i.e., of the perfectly balanced system.

Equation (4.163) can be used to study numerically the behaviour of the system in nonstationary conditions with the aim of assessing the stability in the large and the possibility of the existence of steady-state solutions different from those already obtained. An example of trajectories is shown in Figure 4.32. The figure has been obtained in conditions yielding three stationary solutions, two of which are stable.

The trajectories tend to the stable solutions, even if a certain attraction is also felt toward the saddle point, which is unstable. By integrating the equations of motion of the system for different values of the speed, using as starting positions different pairs of values of coordinates ξ,η, the domains of attraction of the various solutions at different speeds can be obtained. The results obtained for some values of the speed are reported in Figure 4.33. The figure was obtained using the same parameter values as for Figure 4.32, numerically integrating the equations of motion 81,000 times for each value of the speed until a stable equilibrium position was found. In some cases, a few hundred steps were sufficient; in others, the motion had to be followed for thousands of steps.

The two domains of attraction are clearly well defined, and the unstable solution lies on the separatrix of the two attraction domains. At low speed, when the self-centered solution just starts existing, its domain of attraction is very small. By increasing the speed, the domain of attraction of the self-centered solution grows, and at the speed at which the jump occurs it extends for the whole plane. The physics of the phenomenon consequently does not show any critical dependence on the initial conditions, and chaotic behaviour was never encountered.

Some solutions found in the literature in which chaotic behaviour of an axisymmetrical rotor has been found are available. However, a key factor that can trigger chaotic behaviour of rotating systems seems to be the lack of axial symmetry, either due to geometric or material anisotropy or to the presence of mechanisms such as bearing clearance or static loading, which make an isotropic system operate in an offset position. At any rate, it must be stressed that the domains of attraction should be plotted in phase space, which in the current case has four dimensions (two positions and two velocities) and not in the space of the configurations as in Figure 4.33.

4.7.6 Systems with many degrees of freedom

A very general mathematical model of a nonlinear rotor can be obtained from equation (4.163) by adding a generic vector function $\mathbf{f}(q_i,\dot{q}_i,\theta,t)$ to take into account the behaviour of the nonlinear part of the system and by introducing the term $(\omega^2 - ia)$ instead of ω^2 to allow us to take into account angular accelerations

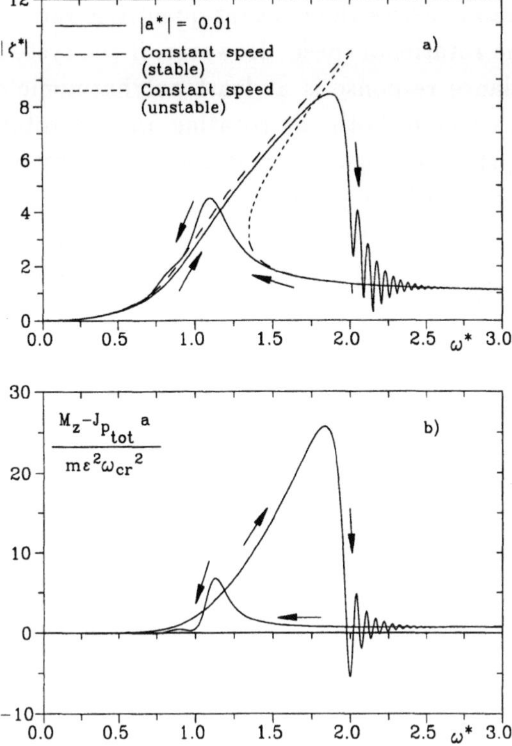

FIGURE 4.34. Acceleration and deceleration of a nonlinear Jeffcott rotor through the critical speed; (a) amplitude of the motion, compared with the amplitude in steady-state operation; (b) driving torque. (a^* = 0.01, ζ_r = 0.01, ζ_n = 0.10 and $\mu\epsilon^2$ = 0.03; ω_{cr} is the critical speed of the linearized system).

$$+\mathbf{M}_m\ddot{\mathbf{q}} + \left(\mathbf{C}_m - i\omega\mathbf{G}\right)\dot{\mathbf{q}} + \left(\mathbf{K}_m - i\omega\mathbf{C}_{r_m}\right)\mathbf{q} + \mathbf{M}_{n_d}\ddot{\overline{\mathbf{q}}} +$$

$$+\mathbf{M}_{r_d}e^{2i\theta}\ddot{\overline{\mathbf{q}}} + \mathbf{C}_{n_d}\dot{\overline{\mathbf{q}}} + \left(\mathbf{C}_{r_d} + 2i\omega\mathbf{M}_{r_d}\right)e^{2i\theta}\dot{\overline{\mathbf{q}}} + \mathbf{K}_{n_d}\overline{\mathbf{q}} +$$

$$+\left(\mathbf{K}_{r_d} + i\omega\mathbf{C}_{r_d}\right)e^{2i\theta}\overline{\mathbf{q}} + \mathbf{f}(q_i,\dot{q}_i,\theta,t) = (\omega^2 - ia)\mathbf{f}_r e^{i\theta} + \mathbf{f}_n .$$

$$(4.165)$$

Equation (4.165) has been obtained with only the assumption of uncoupling among flexural, axial, and torsional behaviour and neglecting centrifugal stiffening. The latter can be accounted for simply by introducing appropriate stiffness matrices proportional to ω^2.

The equation that allows us to describe the rotational degree of freedom of the system, which is just one due to the assumption of a torsionally rigid rotor, is again equation (4.100), which is not affected by either nonlinearities or deviations from axial symmetry.

In general, it is reasonable to expect that the time history of the accelerating system has a simpler expression in the rotating frame, where a slow variation of generalized coordinates in time should occur, than in the fixed frame, where the relevant quantities vary with a frequency equal to the rotational speed. However, if the system is nonisotropic, the unbalance response is at least polyharmonic and then fast variation can occur in both the rotating and nonrotating frames. By introducing the rotating coordinates expressed by equation (4.89) into equation (4.165), it yields

$$
\begin{aligned}
\mathbf{M}_m\ddot{\mathbf{r}} + &\left[\mathbf{C}_m + i\omega\left(2\mathbf{M}_m - \mathbf{G}\right)\right]\dot{\mathbf{r}} + \left[\mathbf{K}_m - \omega^2\left(\mathbf{M}_m - \mathbf{G}\right) + \right. \\
&+ i\left(a\mathbf{M}_m + \omega\mathbf{C}_{r_m}\right)\bigg]\mathbf{r} + \mathbf{M}_{n_d}e^{-2i\theta}\ddot{\bar{\mathbf{r}}} + \left(\mathbf{C}_{n_d} + 2i\omega\mathbf{M}_{n_d}\right)e^{-2i\theta}\dot{\bar{\mathbf{r}}} + \\
&+ \left[\mathbf{K}_{n_d} - \omega^2\mathbf{M}_{n_d} - i\left(a\mathbf{M}_{n_d} + \omega\mathbf{C}_{n_d}\right)\right]e^{-2i\theta}\bar{\mathbf{r}} + \mathbf{M}_{r_d}\ddot{\bar{\mathbf{r}}} + \mathbf{C}_{r_d}\dot{\bar{\mathbf{r}}} + \\
&+ \left(\mathbf{K}_{r_d} + \omega^2\mathbf{M}_{r_d} - ia\mathbf{M}_{r_d}\right)\bar{\mathbf{r}} + \mathbf{f}'(r_i,\dot{r}_i,\theta,t)\}e^{-i\theta} = \\
&= \left(\omega^2 - ia\right)\mathbf{f}_r + \mathbf{f}_n e^{-i\theta} .
\end{aligned}
$$

$$(4.166)$$

The equation yielding the driving torque is still the second equation (4.101).

If the angular acceleration is not vanishingly small, equation (4.166) must be integrated numerically in time, and as a consequence there is no conceptual difficulty taking into account both nonlinearities and asymmetry. If software allowing us to deal with complex quantities is available, there is no need to split the equation into its real and imaginary parts before integrating: Once laws $a(t)$, $\omega(t)$, and $\theta(t)$ are stated, the numerical integration is straightforward. In general, equation (4.166) is a powerful tool to study by numerical integation in time many difficult rotordynamics problems, like constant speed whirling of unsymmetrical rotors or chaotic behaviour due to the simultaneous presence of nonlinearities and asymmetry.

A simple example of the behaviour of a nonlinear rotor performing a constant rate acceleration-deceleration cycle is shown in Figure 4.34. The system is a nonlinear Jeffcott rotor of the same type already studied in Figure 4.30, and the results are shown in nondimensional form, as in Figure 4.13. The values of the nondimensional parameters used for the simulation are $a^* = 0.01$, $\zeta_r = 0.01$, $\zeta_n = 0.10$, and $\mu\epsilon^2 = 0.03$. In the same figure, the solutions obtained from the usual steady-state approach are reported (dashed lines). Note that the values of the parameters are such that the system can experience a downward jump and achieve a self-centered condition. As in the linear case, the presence of the angular acceleration reduces the peak amplitude; however, in the linear case the displacement peak occurs at a higher speed than that characterizing the maximum steady-state amplitude, but the jump here occurs at a lower speed. The curve related to spin down is different from that related to the acceleration phase. This occurs also for linear rotors, when angular acceleration is accounted for, but here this effect is far greater, because both acceleration and nonlinearity contribute to it.

If the system is axially symmetrical and the spin speed is constant, equation (4.165) can be simplified as

$$\mathbf{M}\ddot{\mathbf{q}} + \left(\mathbf{C}_n + \mathbf{C}_r - i\omega\mathbf{G}\right)\dot{\mathbf{q}} + \left(\mathbf{K} - i\omega\mathbf{C}_r\right)\mathbf{q} + \\ + \mathbf{f}(q_{0_i},\dot{q}_{0_i}) = \omega^2\mathbf{f}_r e^{i\omega t} + \mathbf{f}_n . \tag{4.167}$$

If no nonrotating force acts on the structure, the unbalance response is a circular whirling that can be expressed by the solution $\mathbf{z} = \mathbf{z}_0 e^{i\omega t}$. It must be stressed again that circular whirling is an exact solution of the equation of motion, not just an approximation of the fundamental harmonic of the response, as is customary in nonlinear vibrating systems. Introducing this solution into equation (4.167) and considering the presence of a nonrotating structural damping matrix $\mathbf{K}_n''$, the unbalance response of a nonlinear rotor with many degrees of freedom is readily obtained

$$\left(\mathbf{K} - \omega^2(\mathbf{M} - \mathbf{G}) + i(\omega\mathbf{C}_n + \mathbf{K}_n'')\right)\mathbf{z}_0 + \mathbf{f}_i(z_{0_i}) = \omega^2\mathbf{f}_r. \tag{4.168}$$

To reduce the size of the problem, the condensation technique described in Section 3.6.7 can be used, particularly when fewer generalized coordinates are directly linked with nonlinearities, as when dealing with a rotor that is linear in itself but runs on nonlinear bearings. A first reduction of the number of degrees of freedom is performed by eliminating those degrees of freedom not directly involved in nonlinearities with which a small fraction of the total mass is associated. This procedure, basically a Guyan reduction, introduces approximations, which are usually very small if the degrees of freedom to be dropped are chosen with care. The equations of motion are then divided into two groups: a linear and a nonlinear set of equations. A second dynamic condensation procedure, which must be applied at each value of the spin speed at which the unbalance response is to be obtained, can then follow to eliminate all the linear degrees of freedom.

In some particular cases the solution of the nonlinear problem can easily be performed, because it reduces to that of a single nonlinear equation. This includes the case of an undamped linear rotor supported by two nonlinear bearings, one of which has a behaviour that can be approximated by a cubic characteristic. For the solution of the general problem, in which the nonlinear functions can have complicated expressions and there are many degrees of freedom, the Newton-Raphson method seems to be the most appropriate choice. In this case it is advisable to write the equation of motion using real coordinates instead of complex ones.

The iterative algorithm allowing the computation of the solution of equation (4.168), after referring to a set of real coordinates, at the $(i+1)$-th iteration from that at the ith iteration is

$$\mathbf{x}_{i+1} = \mathbf{x}_i - h\mathbf{S}(x_i)^{-1}\mathbf{p}(x_i), \qquad (4.169)$$

where h is a relaxation constant and the vector of the unknown $\mathbf{x}$, the elements of Jacobian matrix $\mathbf{S}(x)$, and the functions $\mathbf{p}(x)$ are defined as

$$\mathbf{x} = \left\{ \begin{array}{c} \Re(\mathbf{z})_0 \\ \Im(\mathbf{z})_0 \end{array} \right\} = \left\{ \begin{array}{c} \mathbf{x}_\xi \\ \mathbf{x}_\eta \end{array} \right\}, \qquad S_{ij} = \frac{\partial p_i(x)}{\partial x_j},$$

$$\mathbf{p}(\mathbf{x}) = \left[\begin{array}{cc} \mathbf{K} - \omega^2(\mathbf{M} - \mathbf{G}) & -\omega(\mathbf{C}_n + \mathbf{K}_n'') \\ \omega(\mathbf{C}_n + \mathbf{K}_n'') & \mathbf{K} - \omega^2(\mathbf{M} - \mathbf{G}) \end{array} \right] \mathbf{x} +$$

$$+ \{f(\mathbf{x})\} - \omega^2 \left\{ \begin{array}{c} \Re(\mathbf{f})_r \\ \Im(\mathbf{f})_r \end{array} \right\}. \qquad (4.170)$$

If there is no damping in the system and the unbalance distribution is contained in a plane (i.e., vector $\mathbf{f}_r$ is real), the computation of the response is much simpler because vector $\mathbf{z}_0$ is also real. All the aforementioned equations still hold, but they can be written directly using the unknowns $\mathbf{z}_0$. The number of equations is, consequently, halved, and, in some particular cases, the solution of the nonlinear part can be reduced to the solution of a single nonlinear equation.

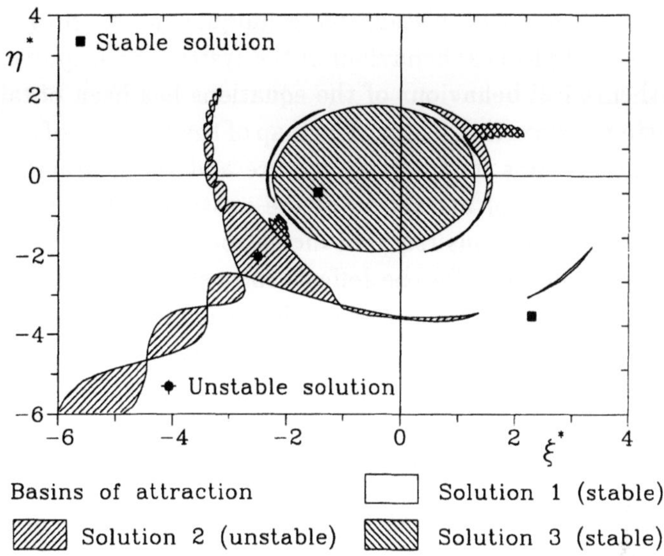

FIGURE 4.35. Basins of attraction of the solutions for the equilibrium positions of the system of Figure 4.30 at a speed $\omega^* = 2$, computed through the Newton-Raphson iterative technique.

Because the backbone curves are related to the undamped system, they can be computed using the same approach as for the latter, simply by neglecting the forcing vector $\mathbf{f}_r$ in the definition of functions $\mathbf{p}(\mathbf{z}_0)$.

The convergence of the generalized Newton method can be, in some cases, a source of potential problems. In many cases, the basins of attraction of the solutions take very complicated shapes, or the computation can lock itself into a cycle without reaching any actual solution of the basic equation. For example, the domains of attraction of the various solutions plotted for the Jeffcott rotor already studied in Figure 4.33 are reported in Figure 4.35. The map has been plotted for a value of the nondimensional speed $\omega^* = 2$, at which three solutions (two stable and an unstable one) exist.

The structure of the map is fractal, as can be seen by enlarging selected zones, and the unstable solution also has a nonvanishing attraction domain. A consequence of the fractal structure of the map is the fact that there are zones in which very small changes of the initial values assumed for the computation cause large differences on the solution obtained. When comparing the maps of Figures 4.32 and 4.34 it is clear that the fractal nature of the second is strictly linked with the mathematical procedure used for the solution of the equation (i.e., the Newton-Raphson technique) and has nothing to do with the actual physical behaviour of the system. No improvement of the mathematical behaviour of the equations has been obtained by introducing a relaxation factor. The map of the domains of attraction changes but retains its fractal structure and the characteristics of showing a domain of attraction of the unstable solution; however, the number of iterations required increases.

The results obtained for the Jeffcott rotor show fairly good convergence characteristics of the Newton-Raphson technique. Obviously this does not guarantee an equally well-behaved nature of the general mathematical model for systems with many degrees of freedom. The following considerations, however, can be extrapolated and used as guidelines in the solution of more complex problems:

- It seems that numerical damping is not of great use in avoiding nonconvergence.

- Attractive stable cycles can be found, particularly in the fields just above a critical speed of the linearized system. After each iteration, it is necessary to check not only whether convergence has been obtained, but also whether the computation has been locked in a stable cycle.

- If this circumstance occurs, the computation can be started again using a different set of initial values. When operating in a self-centered branch, a good guess could be trying to start from a vector $\mathbf{x}$ obtained by multiplying the one of the preceding attempt by a constant smaller than one; in a high branch of the response a constant greater than 1 can be used.

- The fractal nature of the domains of attraction can cause large differences in the results to be produced by small changes in the trial vector $\mathbf{x}$. The solution can also converge on an unstable branch of the response.

- The stable branches should be obtained using as a trial vector the result obtained in the previous computation.

- The aforementioned considerations also hold for the computation of the various branches of the backbone. Here, however, a trivial solution with all the elements of $\mathbf{x}$ vanishingly small exists and, therefore, the procedure must be started outside the basin of attraction of the trivial solution. A suggestion is to start at a speed just above a critical speed of the linearized

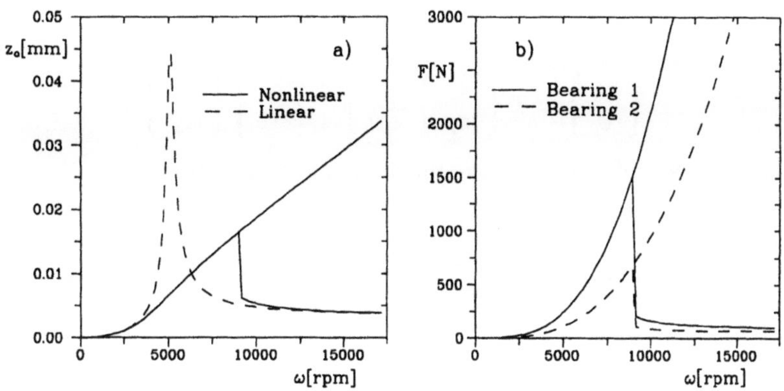

FIGURE 4.36. Amplitude of the orbit of the center of gravity of the (a) nonlinear and linearized rotors and (b) bearing forces as functions of the speed.

system (for hardening systems) with a trial vector proportional to the eigenvector of the linearized system corresponding to the mentioned critical speed.

Example 4-6
A flywheel, with a working range between 8,500 and 17,000 rpm, has a mass $m = 125$ kg and moments of inertia $J_p = 2.272$ kg m^2 and $J_t = 1.477$ kg m^2.
It runs on a pair of ball bearings whose nonlinear characteristic can be approximated by the expression $F = -2 \times 10^7(1 + 10^{10}(|z|)^2)z$ and whose damping properties can be modeled as hysteretic damping applied only to the linear part of the stiffness with loss factor $\eta = 0.08$. Knowing that the span between the bearings is 400 mm and that the center of gravity is at 30% of the span, study the response to a static unbalance $m\epsilon = 438 \times 10^{-6}$ kg m and state whether self-centering is possible. Compute the forces exerted on the stator.
Because the distances of the center of gravity of the rotor from the bearings are $a = 120$ mm and $b = 280$ mm, respectively, the matrices entering into the mathematical model of the linearized system are

$$\mathbf{M} = \begin{bmatrix} 125 & 0 \\ 0 & 1.477 \end{bmatrix}, \quad \mathbf{G} = \begin{bmatrix} 0 & 0 \\ 0 & 2.272 \end{bmatrix},$$

$$\mathbf{K} = k \begin{bmatrix} 2 & a - b \\ a - b & a^2 + b^2 \end{bmatrix} = \begin{bmatrix} 40 & -3.2 \\ -3.2 & 1.856 \end{bmatrix} \times 10^6.$$

In this case it is expedient to use as generalized coordinates the displacements at the bearing locations. The complex coordinates z and ϕ can be obtained from the complex coordinates z_1 and z_2 using a transformation matrix $\mathbf{T}$

$$\left\{ \begin{array}{c} z \\ \phi \end{array} \right\} = \mathbf{T} \left\{ \begin{array}{c} z_1 \\ z_2 \end{array} \right\} = \frac{1}{l} \left[\begin{array}{cc} b & a \\ 1 & -1 \end{array} \right] \left\{ \begin{array}{c} z_1 \\ z_2 \end{array} \right\} .$$

The equation of motion of the undamped system is then

$$\mathbf{M}^* \left\{ \begin{array}{c} \ddot{z}_1 \\ \ddot{z}_2 \end{array} \right\} - i\omega \mathbf{G}^* \left\{ \begin{array}{c} \dot{z}_1 \\ \dot{z}_2 \end{array} \right\} + \mathbf{K}^* \left\{ \begin{array}{c} z_1 \\ z_2 \end{array} \right\} + k\mu \left\{ \begin{array}{c} |z_1|^2 z_1 \\ |z_2|^2 z_2 \end{array} \right\} = \omega^2 \mathbf{f}^* e^{i\omega t},$$

where

$$\mathbf{M}^* = \mathbf{T}^T \mathbf{M} \mathbf{T}, \qquad \mathbf{G}^* = \mathbf{T}^T \mathbf{G} \mathbf{T}$$

$$\mathbf{K}^* = \mathbf{T}^T \mathbf{K} \mathbf{T} = \left[\begin{array}{cc} k & 0 \\ 0 & k \end{array} \right], \qquad \mathbf{f}^* = \mathbf{T}^T \mathbf{f} = \frac{m\epsilon}{l} \left\{ \begin{array}{c} b \\ a \end{array} \right\} .$$

The solution for circular whirling can be obtained from equation (4.168):

$$\left(-\omega^2 (\mathbf{M}^* - \mathbf{G}^*) + \mathbf{K}^* (1 + i\eta) \right) \left\{ \begin{array}{c} z_{1_0} \\ z_{2_0} \end{array} \right\} + k\mu \left\{ \begin{array}{c} |z_{1_0}|^2 z_{1_0} \\ |z_{2_0}|^2 z_{2_0} \end{array} \right\} = \omega^2 \mathbf{f}^* .$$

The solution can be performed using the Newton-Raphson algorithm. The expression of the Jacobian matrix is simply

$$\mathbf{S} = \left(-\omega^2 (\mathbf{M}^* - \mathbf{G}^*) + \mathbf{K}^* (1 + i\eta) \right) + 3k\mu \left[\begin{array}{cc} |z_{1_0}|^2 & 0 \\ 0 & |z_{2_0}|^2 \end{array} \right] .$$

Note that all matrices and vectors are complex and, if software allowing the use of complex numbers is available, there is no need to double the number of equations to work with real numbers.

The solution is performed twice. By increasing the speed from zero, the higher branch is obtained, and by reducing the speed from the maximum value the lower, self-centered branch is computed. At each speed, the solution obtained in the previous computation is chosen as the starting solution. The results are plotted in Figure 4.36. From the figure it is clear that self-centering is not possible, due to the low value of nonrotating damping. Correspondingly, the forces on the bearings are very high.

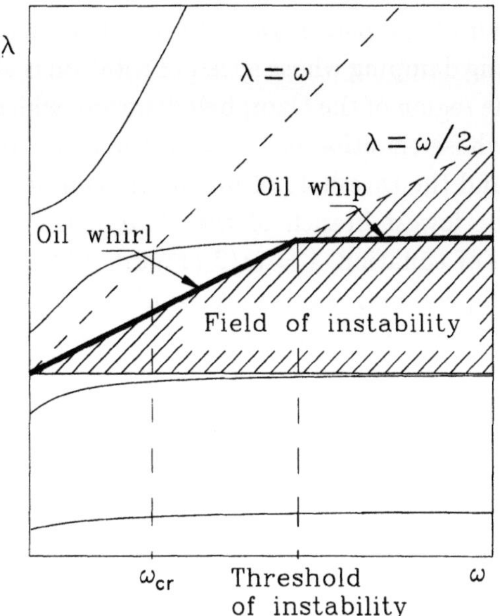

FIGURE 4.37. Oil whip and oil whirl on the Campbell diagram.

4.8 Rotors on hydrodynamic bearings (oil whirl and oil whip)

4.8.1 General considerations

It is very well known that rotors running on journal bearings show particular dynamic problems linked intrinsically with the behaviour of the fluid film within the bearings. These problems are usually referred to as *oil whirl* and *oil whip*.

The first phenomenon is a whirling of the rotor, which takes place at a frequency of about half the speed of rotation (sometimes referred to as *half-frequency whirl*) and is superimposed on the other whirling motions, particularly the synchronous whirling due to unbalance. Its amplitude is usually not large and does not constitute an actual problem. At a speed that is usually not far from twice the first critical speed, the motion becomes more severe and can rapidly degenerate in a very violent, sometimes destructive, whirling that takes place at a frequency almost independent of the speed and coincident with the first natural whirl frequency of the rotor. This motion is usually referred to as *oil whip*.

A very simple heuristic explanation of the phenomenon accounts for the fact that the whirling takes place at about half the rotational frequency by noting that the oil in the bearing moves around at a speed that is about half of the peripheral velocity of the journal, providing a sort of rotating damping whose speed of rotation is about half the spin speed. In the region of the Campbell diagram, which lies below the straight line whose equation is $\lambda = \omega/2$, the behaviour of the system is unstable, and the threshold of instability can be found easily by intersecting the lowest branch of the Campbell diagram with the mentioned line of equation $\lambda = \omega/2$ (Figure 4.37). If the lowest branch of the Campbell diagram is a horizontal straight line, as is the case of the Jeffcott rotor, the threshold of instability occurs at twice the critical speed of the rotor.

The mentioned heuristic explanation also accounts for the fact that the frequency of the oil whirl is slightly lower than half the rotational speed, usually in the range of 0.45 to 0.48 ω. The average velocity of the oil film is slightly lower than half the peripheral speed of the journal, depending on the clearance and the exact velocity profile.

However, the phenomenon is much more complex, and the actual behaviour of the lubricant film must be modeled in some detail. In particular, the intrinsic nonlinear nature of the bearing cannot be neglected, which makes the study of the dynamic behaviour of the system much more complex than the simple linear rotor dynamics study that is often sufficient when no allowance is taken for the compliance of journal bearings. In particular, the whirl-whip transition is not as abrupt as shown in Figure 4.37.

4.8.2   Forces exerted by the oil film on the journal

Consider the journal bearing sketched in Figure 4.38a. Assume that the bearing is perfectly aligned, i.e., that the axes of the bearing and the journal are parallel. The nonrotating reference frame $Oxyz$ is centered in the center of the bearing and has the directions of its axes fixed in space, while the directions of the axes of reference frame $Ox'y'z$, whose axis x' contains the center of the journal C, are not fixed in space if point C moves about point O. In this section, the way of working of the bearing will be defined as *stationary* if the coordinates x and y of the center of the journal are independent from time t and, consequently, reference frame $Ox'y'z$ is also fixed in space and angle β is constant.

Assume that the pressure is linked to the thickness of the fluid film by the well-known Reynolds equation[8]

$$\frac{1}{6}\left[\frac{1}{R_j^2}\frac{\partial}{\partial\theta}\left(\frac{h^3}{\mu}\frac{\partial p}{\partial\theta}\right)+\frac{\partial}{\partial z}\left(\frac{h^3}{\mu}\frac{\partial p}{\partial z}\right)\right]=\omega\frac{\partial h}{\partial\theta}+2\frac{\partial h}{\partial t} \qquad (4.171)$$

where μ is the viscosity of the lubricant and $\omega = \omega_j - \omega_b$, subscripts b and j referring to the bearing and the journal, respectively. Usually, the bearing does not rotate; hereafter it will be assumed that $\omega_b = 0$ and $\omega_j = \omega$.

The film thickness h is easily expressed as a function of the coordinates of the center of the journal with respect to the center of the bearing

$$h = c[1 - x^* \cos(\theta) - y^* \sin(\theta)] = c[1 - \epsilon^* \cos(\theta')], \qquad (4.172)$$

where the clearance c is simply given by the difference of the radii $c = R_b - R_j$ and the nondimensional eccentricity and coordinates are $\epsilon^* = \epsilon/c$, $x^* = x/c$ and $y^* = y/c$.

Equation (4.171) can be used to obtain the pressure distribution in the fluid film. However, if a numerical solution is considered as not general enough, some simplifications must be introduced to allow the pressure to be computed in closed form. If the bearing is assumed to be very long, it is possible to neglect the fluid flow and pressure gradient in axial direction, obtaining the so-called long-bearing approximation, often associated with the name of Sommerfeld. Equation (4.171) reduces to

$$\frac{1}{6R_j^2}\frac{\partial}{\partial\theta}\left(\frac{h^3}{\mu}\frac{\partial p}{\partial\theta}\right)=(\omega x - 2\dot{y})\sin(\theta)-(\omega y + 2\dot{x})\cos(\theta). \quad (4.173)$$

If the center of the journal does not move, i.e., in stationary conditions, the pressure distribution around the journal can be expressed by the relationships

$$p - p_0 = -6\mu\omega\left(\frac{R_j}{c}\right)^2\frac{\epsilon^*}{2+\epsilon^{*2}}\frac{2-\epsilon^*\cos(\theta')}{[1-\epsilon^*\cos(\theta')]^3}\sin(\theta'). \qquad (4.174)$$

---

[8]O. Reynolds, "On the theory of lubrication and its applications to Mr. Towers' experiments", *Phil. Trans. Soc., London*, Vol. 177, (1886), 154–234.

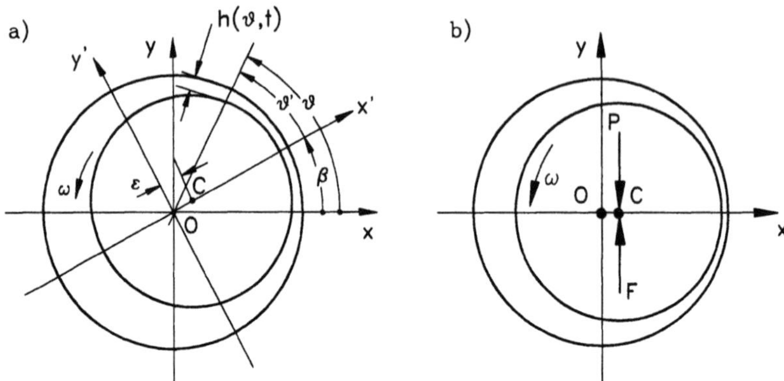

FIGURE 4.38. Lubricated journal bearing; (a) geometrical definitions, (b) position of the journal when a load P acts along the y-axis; solution obtained for stationary conditions from equation (4.175).

Pressure p_0 is the pressure at $\theta' = 0$, i.e., in the point at which the oil film is at its minimum thickness. It can be easily computed by assuming that the pressure attains a known value at the locations in which the oil supply is located. The components of the force the journal receives from the oil film in the x'- and y'-directions can be obtained simply by integrating the pressure profile on the journal surface

$$
\begin{cases}
F_{x'} = \displaystyle\int_0^{2\pi} (p - p_0)\cos(\theta')d\theta \,, \\[2mm]
F_{y'} = \displaystyle\int_0^{2\pi} (p - p_0)\sin(\theta')d\theta = 12\pi\mu\omega R_j l \left(\dfrac{R_j}{c}\right)^2 \dfrac{\epsilon^*}{(2+\epsilon^{*2})\sqrt{1-\epsilon^{*2}}} \,,
\end{cases}
$$
$$(4.175)$$

where l is the length of the bearing in the axial direction.

The force exerted by the oil film on the journal is then directed along the y'- axis and the displacement is directed along the x'-axis. The journal is displaced in a direction perpendicular to the direction of the load, as shown in Figure 4.38b, in which the load P is assumed to act in the vertical direction. By linearizing the expression of the forces about the central position, the following expression is obtained:

$$
\left\{ \begin{array}{c} F_x \\ F_y \end{array} \right\} = -\mathbf{K} \left\{ \begin{array}{c} x \\ y \end{array} \right\} = -6\pi\mu\omega l \left(\frac{R_j}{c}\right)^3 \left[\begin{array}{cc} 0 & 1 \\ -1 & 0 \end{array} \right] \left\{ \begin{array}{c} x \\ y \end{array} \right\} .
$$
$$(4.176)$$

Note that the stiffness matrix obtained from the linearization of the bearing has vanishing elements on the main diagonal and is not symmetrical. Altough the first feature is linked to the particular over-simplified formulation used, the second one is general. In a similar way, if the center of the journal C is also allowed to move, i.e., if its x- and y-coordinates are not assumed to be constant, it is possible to compute the forces due to this motion and then, by linearizing about the central position of the journal, to obtain a damping matrix

$$\left\{ \begin{array}{c} F_x \\ F_y \end{array} \right\} = -\mathbf{C} \left\{ \begin{array}{c} \dot{x} \\ \dot{y} \end{array} \right\} = -12\pi\mu l \left(\frac{R_j}{c} \right)^3 \left[\begin{array}{cc} 1 & 0 \\ 0 & 1 \end{array} \right] \left\{ \begin{array}{c} \dot{x} \\ \dot{y} \end{array} \right\} . \quad (4.177)$$

Consider a circular whirling with very small amplitude ϵ and whirl speed λ. The position and velocity of the center of the journal are

$$\left\{ \begin{array}{c} x \\ y \end{array} \right\} = \epsilon \left\{ \begin{array}{c} \cos(\lambda t) \\ \sin(\lambda t) \end{array} \right\} , \qquad \left\{ \begin{array}{c} \dot{x} \\ \dot{y} \end{array} \right\} = \epsilon \left\{ \begin{array}{c} -\sin(\lambda t) \\ \cos(\lambda t) \end{array} \right\} , \quad (4.178)$$

and the force that the journal receives from the oil film is

$$\left\{ \begin{array}{c} F_x \\ F_y \end{array} \right\} = -12\pi\mu l\epsilon \left(\frac{R_j}{c} \right)^3 \left(\frac{\omega}{2} - \lambda \right) \left\{ \begin{array}{c} \sin(\lambda t) \\ -\cos(\lambda t) \end{array} \right\} . \quad (4.179)$$

It is easy to verify that this force is directed tangentially to the orbit of point C and, if $\lambda < \omega/2$, its direction is the same as the velocity of point C. The force drives the shaft along its motion with obvious unstabilizing effects. If, however, $\lambda > \omega/2$, the direction of the force is opposite that of the velocity and it opposes the whirling motion with stabilizing effects. The same results already seen and sketched in Figure 4.37 are then obtained. They are, however, affected by the assumptions used, mainly by the linearization about the central position, which is acceptable only in the case of very lightly loaded bearings, and by the pressure distribution expressed by equation (4.174) and plotted as a function of angle θ' in Figure 4.39 for some selected values of the eccentricity ϵ^*.

From the figure, it is clear that when the eccentricity is high and the inlet pressure low, very low values of the absolute pressure can be reached in some parts of the oil film, or even negative absolute pressures, which is physically without meaning. When the pressure

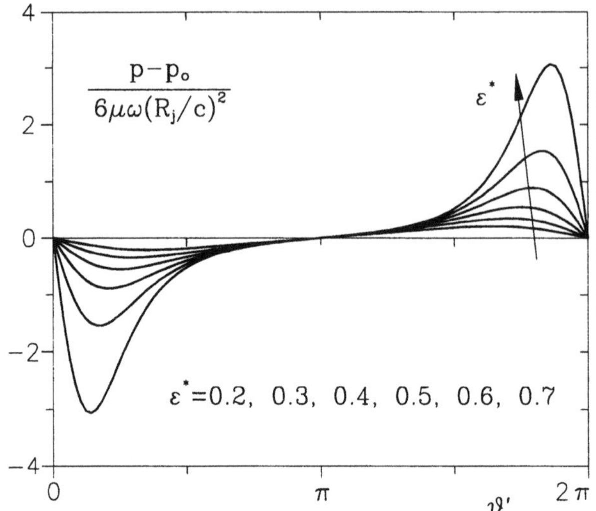

FIGURE 4.39. Pressure distribution on the journal along angle θ' for various values of the nondimensional eccentricity ϵ^*. Long-bearing assumption, equation (4.174).

becomes lower than the vapour pressure of the lubricant at the relevant temperature, cavitation occurs and the oil film ruptures. A usual approach for the study of bearings with partially cavitated oil film is assuming that the pressure in the noncavitated part of the bearing is equal to that which can be computed by assuming a complete oil film. Because the pressure in the cavitated part of the bearing can be neglected, the same formulas already seen for the computation of the forces can be used, provided that the integration is performed between angles θ_1 and θ_2, which define the region on which the oil film extends.

A simple approach, usually referred to as *fully cavitated bearing*, is assuming that the oil film extends between $\theta' = \pi$ and $\theta' = 2\pi$, i.e., in the region in which the value of the pressure is higher than p_0. Once angles θ_1 and θ_2 have been defined, there is little difficulty computing the static forces corresponding to a given displacement of the journal, i.e., to a pair of values x_{st} and y_{st} of the coordinates of point C. For small motions about the static equilibrium position it is possible to linearize the expressions of the forces and to compute the stiffness and damping matrices of the bearing. In the case of the long-bearing assumption, the relevant equations are

$$\mathbf{f} = \frac{6\mu R_j^3 l\omega}{c^2} \left\{ \begin{array}{c} I_7 \\ I_8 \end{array} \right\},$$

$$\mathbf{C} = -\left[\left(\frac{\partial F_i}{\partial \dot{x}_j}\right)_{\substack{x_1 = x_{st} \\ x_2 = y_{st}}}\right] = \frac{6\mu R_j^3 l}{n^2 c^3}\left[\begin{array}{cc} I_9 & I_{10} \\ I_{11} & I_{12} \end{array}\right], \qquad (4.180)$$

$$\mathbf{K} = -\left[\left(\frac{\partial F_i}{\partial x_j}\right)_{\substack{x_1 = x_{st} \\ x_2 = y_{st}}}\right] = -\frac{6\mu R_j^3 l\omega}{(2+n^2)c^3}\left[\begin{array}{cc} I_1 + I_2 & -I_4 + I_5 \\ I_6 + I_5 & -I_1 + I_3 \end{array}\right],$$

where

$$I_1 = \int_{\theta_1}^{\theta_2} \frac{1 + h_{st}^*}{h_{st}^{*3}} \sin(\theta)\cos(\theta)d\theta, \qquad I_2 = \int_{\theta_1}^{\theta_2} (2 + h_{st}^*)\frac{\Lambda}{h_{st}^*}\cos^2(\theta)d\theta,$$

$$I_3 = \int_{\theta_1}^{\theta_2} (2 + h_{st}^*)\frac{\Lambda}{h_{st}^*}\sin^2(\theta)d\theta, \qquad I_4 = \int_{\theta_1}^{\theta_2} \frac{1 + h_{st}^*}{h_{st}^{*3}}\cos^2(\theta)d\theta,$$

$$I_5 = \int_{\theta_1}^{\theta_2} (2 + h_{st}^*)\frac{\Lambda}{h_{st}^*}\sin(\theta)\cos(\theta)d\theta, \quad I_6 = \int_{\theta_1}^{\theta_2} \frac{1 + h_{st}^*}{h_{st}^{*3}}\sin^2(\theta)d\theta,$$

$$I_7 = \int_{\theta_1}^{\theta_2} \Lambda\cos(\theta)d\theta, \qquad I_8 = \int_{\theta_1}^{\theta_2} \Lambda\sin(\theta)d\theta,$$

$$I_9 = \int_{\theta_1}^{\theta_2} (Ry_{st}^* + Sx_{st}^*)\cos(\theta)d\theta, \qquad I_{10} = \int_{\theta_1}^{\theta_2} (-Rx_{st}^* + Sy_{st}^*)\cos(\theta)d\theta,$$

$$I_{11} = \int_{\theta_1}^{\theta_2} (Ry_{st}^* + Sx_{st}^*)\sin(\theta)d\theta, \quad I_{12} = \int_{\theta_1}^{\theta_2} (-Rx_{st}^* + Sy_{st}^*)\sin(\theta)d\theta,$$

$$h_{st}^* = 1 - x_{st}^*\cos(\theta) + y_{st}^*\sin(\theta), \qquad \Lambda = \frac{x_{st}^*\sin(\theta) - y_{st}^*\cos(\theta)}{h_{st}^{*3}},$$

$$R = 2\Lambda h_{st}^*\frac{1 + h_{st}^*}{2 + \epsilon^{*2}}, \qquad S = \frac{1}{(1+\epsilon^*)^2} - \frac{1}{h_{st}^{*2}}.$$

The integrals in equations (4.180) must be solved numerically, but this does not imply long and costly computations.

If the bearing is relatively short and the long-bearing approach does not seem applicable, the flow in the circumferential direction can be neglected obtaining the so-called short-bearing approximation. By neglecting the term linked with the circumferential pressure gradients in Reynolds equation (4.171) and introducing into the latter the expression of the film thickness, it follows that

$$\frac{1}{6}\frac{\partial}{\partial z}\left(\frac{h^3}{\mu}\frac{\partial p}{\partial z}\right) = (\omega x - 2\dot{y})\sin(\theta) - (\omega y + 2\dot{x})\cos(\theta). \qquad (4.181)$$

By operating in the same way as for the long-bearing Sommerfeld approximation, the following expressions for the forces in static conditions and the stiffness and damping matrices are obtained

$$\mathbf{f} = \frac{\mu R_j l^3 \omega}{2c^2} \left\{ \begin{array}{c} I_7 \\ I_8 \end{array} \right\} ,$$

$$\mathbf{C} = - \left[\left(\frac{\partial F_i}{\partial \dot{x}_j} \right)_{\substack{x_1 = x_{st} \\ x_2 = y_{st}}} \right] = \frac{\mu R_j l^3}{c^3} \left[\begin{array}{cc} I_4 & I_1 \\ I_1 & I_6 \end{array} \right] , \qquad (4.182)$$

$$\mathbf{K} = - \left[\left(\frac{\partial F_i}{\partial x_j} \right)_{\substack{x_1 = x_{st} \\ x_2 = y_{st}}} \right] = - \frac{\mu R_j l^3 \omega}{2c^3} \left[\begin{array}{cc} I_1 + I_2 & -I_4 + I_5 \\ I_6 + I_5 & -I_1 + I_3 \end{array} \right] ,$$

where

$$I_1 = \int_{\theta_1}^{\theta_2} \frac{\sin(\theta)\cos(\theta)}{h_{st}^{*3}} d\theta , \qquad I_2 = 3 \int_{\theta_1}^{\theta_2} \frac{\Lambda}{h_{st}^*} \cos^2(\theta) d\theta ,$$

$$I_3 = 3 \int_{\theta_1}^{\theta_2} \frac{\Lambda}{h_{st}^*} \sin^2(\theta) d\theta , \qquad I_4 = \int_{\theta_1}^{\theta_2} \frac{\cos^2(\theta)}{h_{st}^{*3}} d\theta ,$$

$$I_5 = 3 \int_{\theta_1}^{\theta_2} \frac{\Lambda}{h_{st}^*} \sin(\theta)\cos(\theta) d\theta , \qquad I_6 = \int_{\theta_1}^{\theta_2} \frac{\sin^2(\theta)}{h_{st}^{*3}} d\theta ,$$

$$I_7 = \int_{\theta_1}^{\theta_2} \Lambda \cos(\theta) d\theta , \qquad I_8 = \int_{\theta_1}^{\theta_2} \Lambda \sin(\theta) d\theta ,$$

$$h_{st}^* = 1 - x_{st}^* \cos(\theta) - y_{st}^* \sin(\theta) , \qquad \Lambda = \frac{x_{st}^* \sin(\theta) - y_{st}^* \cos(\theta)}{h_{st}^{*3}} .$$

For the intermediate case of a bearing with nonnegligible length, several authors have suggested approximate solutions, usually obtained from the Sommerfeld solution by multiplying by a function of the axial coordinate. Such approximate pressure distributions are usually more satisfactory than either of the two solutions mentioned earlier, because they take into account both axial and circumferential flow.

A solution of this type was proposed by Warner.[9] The pressure distribution is obtained by multiplying the pressure computed for an infinitely long bearing p_∞ (Sommerfeld solution) by a coefficient

---

[9]P.C. Warner, "Static and dynamic properties of partial journal bearings", *J. of Basic Engineering*, Trans. ASME, Series D, 85, (1963), 244.

$$C_W = 1 - \frac{\cosh\left[\left(\frac{2z}{l}\right)\left(\frac{\gamma l}{2R_j}\right)\right]}{\cosh\left(\frac{\gamma l}{2R_j}\right)}, \qquad (4.183)$$

where

$$\gamma^2 = \frac{\displaystyle\int_{\theta_1}^{\theta_2} [1 + \epsilon^* \cos(\theta)]^3 \left(\frac{dp_\infty}{d\theta}\right)^2 d\theta}{\displaystyle\int_{\theta_1}^{\theta_2} [1 + \epsilon^* \cos(\theta)]^3 p_\infty^2 \, d\theta}.$$

In the case of the Warner solution, if the derivatives of coefficient C_W with respect to the position and velocity are neglected, the expressions of the forces and of the stiffness and damping matrices can be directly obtained by multiplying those obtained for the long-bearing case by the factor

$$C_W' = 1 - \frac{2R_j}{\gamma l} \tanh\left(\frac{\gamma l}{2R_j}\right) \qquad (4.184)$$

obtained by integrating coefficient C_W in the axial direction.

Apart from the plain lubricated bearing that was studied earlier, there are many other bearing types, mostly evolved with the aim of reducing the inherent instability of plain bearings. In the case of complex geometries there is no chance of obtaining a closed-form solution, or at least a solution computed with simple numerical integrations. The possible approaches then either use experimental results or perform a more complex numerical modeling of the fluid film. When the results obtained experimentally or by numerical modeling are presented in the form of general charts applicable to a given family of bearings, nondimensional parameters, related to the average pressure $p_m = F/2R_j l$ are commonly used. One of them is the Sommerfeld number, defined as

$$S = \frac{\mu\omega}{2\pi p_m}\left(\frac{R_j}{c}\right)^2 = \frac{\mu\omega R_j l}{\pi F}\left(\frac{R_j}{c}\right)^2. \qquad (4.185)$$

The Sommerfeld number is well suited to the study of long bearings. For the short-bearing model, the load factor

$$O = \frac{2p_m}{\mu\omega}\left(\frac{c}{R_j}\right)^2\left(\frac{2R_j}{l}\right)^2 = \frac{F}{\mu\omega l R_j}\left(\frac{c}{R_j}\right)^2\left(\frac{2R_j}{l}\right)^2 \qquad (4.186)$$

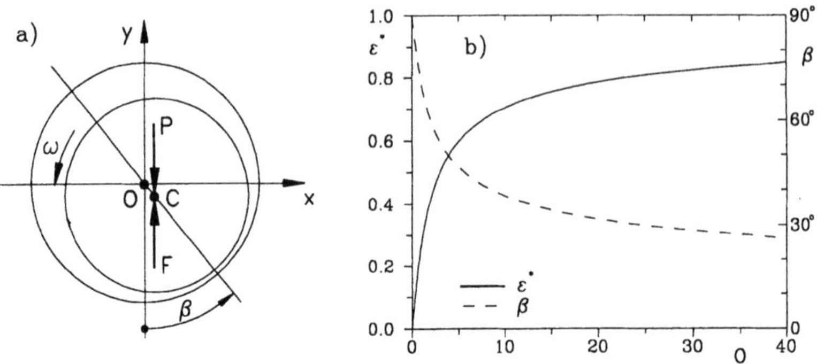

FIGURE 4.40. Lubricated journal bearing; (a) definition of the attitude angle β (load P acting along y-axis); (b) eccentricity and attitude angle as functions of the load factor for a fully cavitated short bearing.

is commonly used. Barwell in 1956 proposed calling the load factor the Ocvirk number, and will be referred to here with the symbol O. The two nondimensional parameters are linked by the relationship

$$O = \frac{1}{\pi S} \left(\frac{2R_j}{l} \right)^2 . \tag{4.187}$$

The position of the center of the journal in stationary conditions is univocally determined, given a certain type of bearing, once the Sommerfeld number (or any other relevant nondimensional parameter) is stated. Charts giving the nondimensional coordinates of the center of the bearing x^* and y^*, or better, the eccentricity ϵ^* and the attitude angle β, as functions of the Sommerfeld number, then summarize the static behaviour of the bearing. The attitude angle defines the direction of the displacement of the center of the journal with respect to the direction of the force F, as shown in Figure 4.40a. In the case of an uncavitated long bearing, the attitude angle is then equal to 90° for any value of the Sommerfeld number.

In many cases, the minimum thickness of the oil film $h^*_{min} = 1 - \epsilon^*$ is given instead of the eccentricity. The graph giving the eccentricity and the attitude angle as functions of the load factor for a fully cavitated short bearing is reported in Figure 4.40b.

In Figure 4.41 the elements of the stiffness and damping matrices for the same bearing in Figure 4.40b are reported in nondimensional form as functions of the load factor. Similar graphs for a grooved bearing with ratio $l/2R_j = 0.5$ are reported in Figure 4.42. Graphs of the type shown in Figures 4.39, 4.40, and 4.41 are reported in the

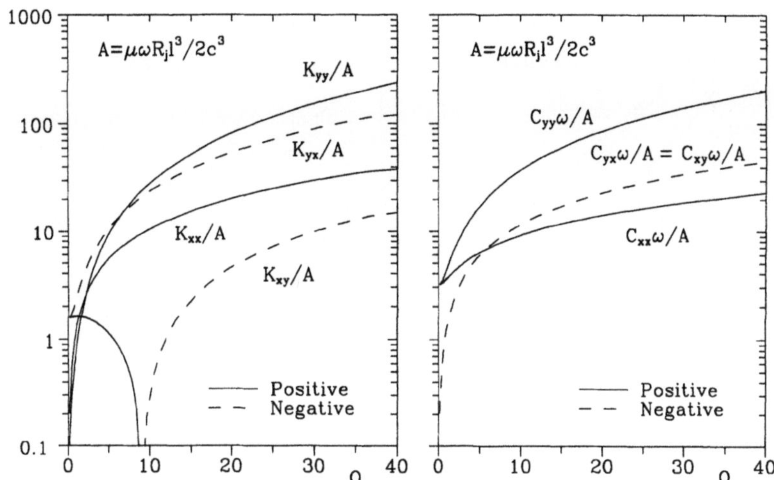

FIGURE 4.41. Elements of the stiffness and damping matrices for the same bearing as Figure 4.40b as functions of the load factor. The static force is assumed to act in the y-direction.

literature for different types of lubricated bearings.[10]

4.8.3 Interaction between the behaviour of the oil film and that of the structure

The behaviour of lubricated journal bearings was studied first by Robertson in 1933,[11] who investigated the stability of the ideal 360° infinitely long journal bearing. Using the expressions for the film forces obtained by Harrison in 1913,[12] he concluded that the rotor will be unstable at all speeds rather than at speeds above twice the first critical speed. This is easily ascribed to the fact that with an uncavitated oil film the attitude angle, i.e., the angle between the direction of the load and that of the displacement of the journal, is 90° and the radial stiffness of the bearing vanishes.

Once a model of the bearing is introduced into the model of the rotor, a nonlinear problem results. The simplest way to solve it is to first obtain the static equilibrium position, i.e., the position the

---

[10]Many charts and tables for bearings of different types are reported in T. Someya, *Journal-Bearing Databook*, Springer, Berlin, 1989.

[11]D. Robertson, "Whirling of a journal in a sleeve bearing", *Phil. Mag.*, Series 7, Vol. 15, (1933), 113-130.

[12]W.J. Harrison, "The hydrodynamical theory of lubrication with special reference to air as lubricant", *Trans. of Cambridge Phil. Soc.*, Vol. 22, (1913), 39.

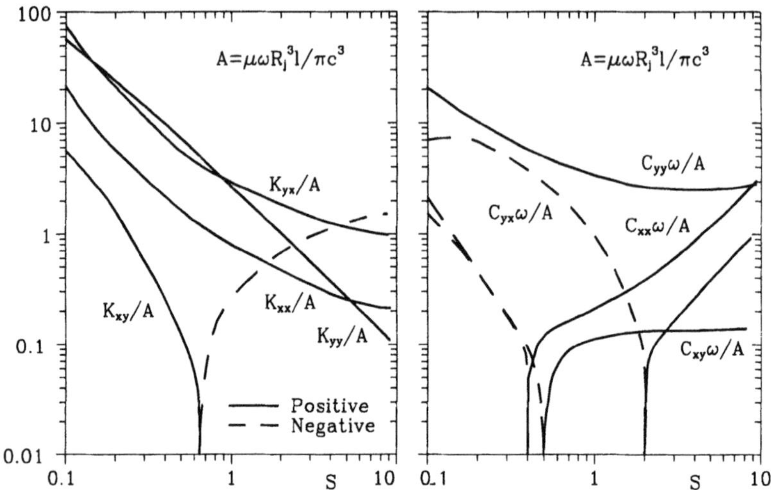

FIGURE 4.42. Elements of the stiffness and damping matrices for a grooved bearing as functions of the Sommerfeld number; ratio $l/2R_j = 0.5$. (J.S. Rao, *Rotor dynamics*, Wiley Eastern, New Delhi, 1983, 110–111). The plots have been transformed to report the curves in a way that is more consistent with Figure 4.41.

rotating shaft takes under the effects of static loads, and then to linearize the motion about this equilibrium position. This approach is very well suited to having a general picture of the behaviour of the rotor where the small oscillations are concerned and to study its stability in the small.

If a small unbalance is present, provided it gives way to displacements small enough not to go beyond the field in which the linearization holds, the unbalance response can be studied and the relevant results can be superimposed on those related to the free whirling. The well-known statement that the whirling (and eventually whipping) due to the lubricant does not interact with unbalance response holds exactly in this condition.

In the following section the rotor will be assumed to be axially symmetrical; no similar assumption is made with regard to the stator. The intrinsic asymmetry of a loaded journal bearing makes it impossible to exploit the symmetrical characteristics of the stator, and on the contrary, an asymmetrical rotor loaded by static forces will produce an elliptical whirling, which is incompatible with the assumption of the existence of a stationary equilibrium position. If the rotor is anisotropic, it must be substituted with a symmetrical rotor with average properties in the computation of the static equilibrium position. When the amplitude grows larger the nonlinear behaviour cannot be neglected where the dynamic behaviour is concerned. The usual approach is numerically integrating the equations of motion to study the stability in the large. In this case the behaviour of the fluid film can be studied in greater detail, and the hydrodynamic equations can be integrated numerically to study such effects as those of misalignments between the bearing and the journal, e.g., due to bending deformations of the shaft, on cavitation.

For the study in the small, the journal bearings are modeled using the conventional eight-coefficient linearized model, i.e., by computing a stiffness and a damping matrix as shown in Section 4.10.2. Due to the influence of the static loads on the dynamic properties of the bearings, a nonlinear static problem is first solved for each value of the spin speed at which the dynamic computation is to be performed. The definition of the parameters of the bearing then follows and the eigenproblem yielding the eigenfrequencies of the linearized system is solved. Their analysis allows information about the stability of the free whirling of the system to be immediately obtained. The response to a small unbalance can also be obtained without any difficulty and the applicability of the result obtained to the actual system can be checked by verifying whether the computed motion of the journal can actually be considered small.

It is also well known that in the case of rotors supported on more than two bearings, their misalignment can have a large effect on the dynamic behaviour of the machine by inducing preloads that affect the stiffness of the bearings. This fact can also be used by inducing by purpose such preloads through a control system to tailor the dynamic characteristics of the machine. The displacements of the bearing center with respect to the nominal position can be introduced into the model to also allow the study of this effect.

If the system is statically determined, there is no difficulty computing the static loads on the bearings. Once the forces are known, it is possible to compute the two components of the relative displacement of the journals (with respect to the bearings) and then the eight coefficients that characterize their dynamic behaviour. If graphs of the type of Figure 4.41 are not available, it is possible, at least when the long- or the short-bearing assumptions can be made, to resort to an iterative technique such as the Newton-Raphson method. Note that the solution of the static problem is usually unique, and severe convergence problems are not expected. If the system is not statically determinate, however, the loads on the bearings depend on the deformation of both the stator and the rotor and on possible misalignments. A coupled problem that is far more complex must be solved. The equation allowing the static deflected configuration of the rotor to be studied is equation (4.94). By also introducing structural damping and writing explicitly the forces due to weight, it yields

$$\left[\mathbf{K} - i(\omega\mathbf{C} + \mathbf{K}'')\right]\mathbf{q} = \mathbf{f}_n + g\mathbf{M}(\delta_x + i\delta_y)\,, \qquad (4.188)$$

where vector $\mathbf{f}_n$ contains static forces, not including weight, while the elements of vector $\delta_x + i\delta_y$ vanish in correspondence with rotational degrees of freedom and are equal to the cosines of the angle between the vertical direction and the x- and y-axes for translational coordinates. Note that the deformed equilibrium position depends on the angular velocity only if there is viscous damping. The internal damping of the rotor will be assumed to be of the structural (hysteretic) type, because it allows the computations referred to the rotor to be performed only once. However, there is no difficulty modifying the equations to also take into account viscous damping.

As when the Newton-Raphson technique is used for the solution of the nonlinear set of equations there is some advantage to resorting to real coordinates, equation (4.188) can be rewritten in the form

$$\mathbf{K}^* \left\{ \begin{array}{c} \mathbf{x}_r \\ \mathbf{y}_r \end{array} \right\} = \mathbf{f}_r^*, \qquad (4.189)$$

where

$$\mathbf{K}^* = \left[\begin{array}{cc} \mathbf{K} & \mathbf{K}'' \\ -\mathbf{K}'' & \mathbf{K} \end{array} \right], \qquad \mathbf{f}_r^* = \left\{ \begin{array}{c} \mathbf{f}_{\mathrm{xn}} \\ \mathbf{f}_{\mathrm{yn}} \end{array} \right\} + g \left\{ \begin{array}{c} \mathbf{M}\delta_x \\ \mathbf{M}\delta_y \end{array} \right\}$$

and vectors $\mathbf{x}_r$ and $\mathbf{y}_r$ contain the real and imaginary parts, respectively, of the complex coordinates of the rotor. The stiffness matrix is singular because the rotor has been considered unsupported and this part of the system is underconstrained.

It is possible to separate the vectors of the generalized coordinates into two subsets: the first, labeled with subscript 1, containing the displacements at the supporting points (i.e., the centers of the journals), and the second, with subscript 2, containing all other generalized displacements. The hydrodynamic bearings will be assumed to react only to translations of the journal, so the first set contains a number of elements equal to twice the number of the bearings. If the model of the bearing is modified to also include the moment due to angular displacements, the inclusion of the rotational degrees of freedom in this set of displacements is straightforward. By partitioning accordingly the relevant matrices and vectors, it follows that

$$
\begin{bmatrix} \mathbf{K}_{11}^* & \mathbf{K}_{12}^* \\ \mathbf{K}_{21}^* & \mathbf{K}_{22}^* \end{bmatrix}
\left\{ \begin{Bmatrix} \mathbf{x} \\ \mathbf{y} \end{Bmatrix}_1 \atop \begin{Bmatrix} \mathbf{x} \\ \mathbf{y} \end{Bmatrix}_2 \right\}
= \left\{ \begin{matrix} \mathbf{f}_1^* \\ \mathbf{f}_2^* \end{matrix} \right\} .
\tag{4.190}
$$

By applying the usual techniques of static reduction and assuming the generalized coordinates of the first group as master degrees of freedom, equation (4.188) reduces to

$$
\mathbf{K}_{cond_r} \begin{Bmatrix} \mathbf{x} \\ \mathbf{y} \end{Bmatrix}_1 = \mathbf{f}_{cond_r} ,
\tag{4.191}
$$

where the expressions of the condensed matrices are the usual ones and the internal generalized coordinates of the rotor are expressed by equation (2.139).

The model of the stator can be built in a way similar to that seen for the rotor, with two important differences. The stator does not need to be axially symmetrical and, because it is stationary, its static deformation is not affected by its damping. The equilibrium equation for the stator is equation (4.189) where subscript r has been substituted with s. Matrices $\mathbf{M}^*$ and $\mathbf{K}^*$ are, in general, symmetrical and contain the coupling terms between the behaviour in the xz- and yz-planes. Also, in this case it is possible to separate the vectors of the generalized coordinates into two subsets: The first, labeled with subscript 3, containing the displacements at the supporting points (i.e., the center of the bearings) and the second, with subscript 4, containing all other generalized displacements, and then to resort to static reduction techniques.

The equilibrium equation referred to the displacements of the bearings is then equation (4.191) where subscripts r and 1 have been substituted with s and 3, respectively. It can be reduced with the usual algorithm.

Note that because the interface between stator and rotor is represented by the bearings, which react only to translations, the sets of generalized coordinates with subscripts 1 and 3 contain only displacements and no rotations. This allows the use of conventions for the rotations of the stator and the rotor that are not consistent and the use of any standard FEM code to build the model of the stator even if the conventions for rotations about the x-axis are different from those used in the model of the rotor. Any type of element can be used in both models, provided that the displacements at the interface are measured with reference to the same axes used for the rotor.

The relative displacements x_i and y_i of the center of the ith journal with respect to the center of the ith bearing can be easily obtained from the displacements x_{i_1} and y_{i_1} of the former and the displacements x_{i_3} and y_{i_3} of the latter, where subscripts 1 and 3 refer to the partitioning of the generalized coordinates. Taking into account the fact that the center of the ith bearing can be displaced by the quantities Δ_{i_x} and Δ_{i_y}, with respect to the nominal position when no force acts on the bearing, the relative displacement of the journal with respect to the bearing can be expressed as

$$\left\{ \begin{array}{c} x_i \\ y_i \end{array} \right\} = \left\{ \begin{array}{c} x_{i_1} - x_{i_3} - \Delta_{i_x} \\ y_{i_1} - y_{i_3} - \Delta_{i_y} \end{array} \right\}. \tag{4.192}$$

Equation (4.192) allows the computation of the relative displacement of the bearing and then of the forces F_x and F_y that the journal receives from the oil film through numerical integration of the bearing model or the use of experimental graphs. The interaction between stator and rotor can then be expressed by adding the forces due to the oil film to equations (4.191) for the rotor and the corresponding one for the stator

$$\left\{ \begin{array}{c} \mathbf{K}_{cond_r} \left\{ \begin{array}{c} x \\ y \end{array} \right\}_1 + \left\{ \begin{array}{c} \mathbf{f}_x \\ \mathbf{f}_y \end{array} \right\} = \mathbf{f}_{cond_r} \, , \\[3ex] \mathbf{K}_{cond_s} \left\{ \begin{array}{c} x \\ y \end{array} \right\}_3 - \left\{ \begin{array}{c} \mathbf{f}_x \\ \mathbf{f}_y \end{array} \right\} = \mathbf{f}_{cond_s} \, . \end{array} \right. \tag{4.193}$$

If the number of bearings is m, equation (4.193) is a set of $4m$ nonlinear equations with the $4m$ unknowns representing the displacements of the journals and the bearings in the x- and y-directions. The number of nonlinear equations can, however, be reduced because the actual unknowns of the nonlinear part of the equation are the differences between the displacements of the rotor and the stator, which are only $2m$ in number. Because the reduced stiffness matrix of the rotor is generally singular and cannot be inverted, the second set of equations (4.193) can be multiplied by $\mathbf{K}_{cond_r}\mathbf{K}_{cond_s}^{-1}$ and added to the first, obtaining the following nonlinear equation, which can be solved by resorting to the Newton-Raphson iterative technique

$$\mathbf{K}_{cond_r}\mathbf{x} + \mathbf{A}\left\{\begin{array}{c} \mathbf{f}_x(\{X - \Delta\}) \\ \mathbf{f}_y(\{X - \Delta\}) \end{array}\right\} = \mathbf{f}^{**}, \qquad (4.194)$$

where

$$\mathbf{x} = \left\{\begin{array}{c} \mathbf{x}_1 - \mathbf{x}_3 \\ \mathbf{y}_1 - \mathbf{y}_3 \end{array}\right\}, \qquad \mathbf{A} = \mathbf{I} + \mathbf{K}_{cond_r}\mathbf{K}_{cond_s}^{-1},$$

$$\mathbf{f}^{**} = \mathbf{f}_{cond_r} - \mathbf{K}_{cond_r}\mathbf{K}_{cond_s}^{-1}\mathbf{f}_{cond_s}.$$

Once the values of the relative displacements $\mathbf{x}$ of the bearings in the static equilibrium position have been computed, the eight coefficients that characterize the dynamic behaviour of the bearing can be easily obtained. Note that the solution of the static problem must be repeated for each value of the speed at which the natural frequencies are to be computed. Even if no viscous damping is considered and the relevant matrices can be computed only once, the forces in the bearing are strongly influenced by the spin speed.

The dynamic study in the small can be performed using the model described in Section 4.8.6, because the presence of the lubricated journal bearing makes the nonrotating parts of the machine behave in an anisotropic way. Both real or complex coordinates can be used. Free whirling and unbalance response can be studied, the latter only if the amplitude of the response is small enough not to exceed the linearity limits. As a last consideration, it must be noted that if the static forces acting on the bearings are vanishingly small, as in the case of a vertical bearing without static loading, the linearized model can supply only a very rough approximation of the actual behaviour of the system. The system is generally unstable in the small and only a true nonlinear analysis allows the limit cycle to be found.

To study the motion in the large and to find the limit cycles that can occur in the cases in which the motion in the small is unstable, it is necessary to resort to numerical integration of the equations of motion. This can be performed using the same model described earlier, even if more detailed models of the oil film, which also include angular misalignments of the bearings, are described in the literature.

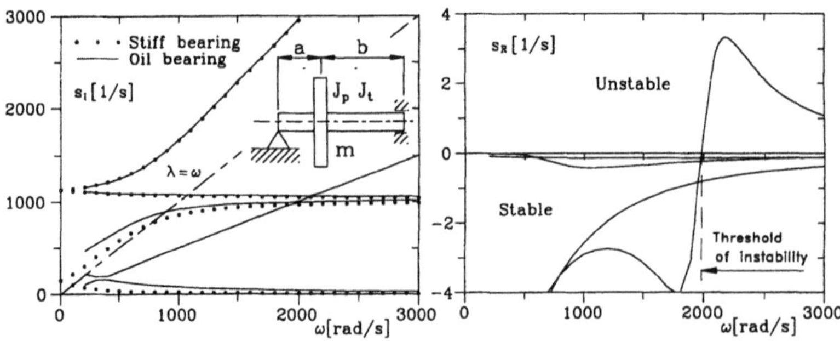

FIGURE 4.43. Sketch of the system used as example, together with the plot of the real and imaginary parts of the eigenvalues as functions of the spin speed.

It must be noted that the inherent instability of plain journal bearings makes them unsuitable for high-speed supercritical machinery and many other bearing configurations have been developed with the aim of overcoming this difficulty. In particular, tilting pad bearings allow the instability problem to be solved completely at the cost of a reduction of damping at low speed and of added overall complexity. Among the many papers and books existing on this subject, those by Tondl and Muszynska[13] are worth mentioning.

Problems similar to those linked with lubricated journal bearings are also encountered in other cases in which a fluid is interposed between the stator and the rotor, as, for example, in labyrinth and liquid seals. In all cases, one of the most effective measures aimed at reducing the instability problems is the decrease of the peripheral velocity of the fluid around the shaft. This can be done using anti-swirl vanes, by roughening the stator walls, or by injecting the fluid in the tangential, backward direction.

Example 4-7
Consider as an example the rotor sketched in Figure 4.43. It is made of a massless flexible shaft on which a disk with mass m and moments of inertia J_p and J_t is fitted. The shaft is supported on one side on a bearing that is assumed to be rigid and on the other side on a hydrodynamic journal bearing.

---

[13]See, for example, A. Tondl, *Some problems in rotor dynamics*, Czechoslovak Academy of Sciences, Prague, Czechoslovakia, 1965, and A. Muszynska, *Rotor instability*, Senior Mechanical Engineering Seminar, Carson City, June, 1984.

Due to the nonlinearity of the system, it is impossible to perform a general study and actual data must be stated. Let rotor $a = 85$ mm, $b = 255$ mm, diameter of shaft $= 25.4$ mm, $m = 20$ kg, $J_p = 1$ kgm^2, $J_t = 0.7$ kgm^2; bearings $l = 16$ mm, $c = 35.2$ μm, and $\mu = 0.02$ Ns/m^2. Let the bearing be modeled using the short-bearing assumption, with a fully cavitated film, i.e., with the oil film extending only for $180°$.

The imaginary and real parts of the eigenvalues, i.e., the actual whirl frequency and the decay rate, are plotted in Figure 4.43. The whirl frequency is compared to the values obtained assuming that both bearings are rigid. Due to the use of real coordinates (the size of the relevant matrices is 6, after reduction), it is not possible to distinguish immediately between forward and backward whirl motions without analyzing the eigenvectors. However, a clue is that the modes related to branches that slope downward on the Campbell diagram are backward modes.

From the plot it is clear that the presence of the hydrodynamic bearing has little influence on the whirl frequencies of the system but causes an added frequency to be present. This new mode, which is clearly an oil whirl, follows almost exactly the line $\lambda = \omega/2$. The decay rate is always negative, i.e., the motion is stable, except for the oil whirl mode, which becomes unstable at about 2,000 rad/s. The transition between oil whirl and oil whip occurs exactly when the line $\lambda = \omega/2$ crosses the first forward whirl mode, as predicted by the usual approximate criteria. As a conclusion, the behaviour of the system is practically coincident with what could be expected using the approximate criteria, and the picture of the phenomenon shown in Figure 4.37 holds. This is, however, not a general result, because in more complicated cases the presence of the bearings can have a more important effect on the dynamic behaviour of the system.

4.9   Flexural vibration dampers

As seen in the previous sections, in many instances it is important to increase nonrotating damping to achieve the required stability. The easiest way is by introducing one or more dampers, which must be nonrotating i.e., located within the stator or interposed between the stator and the rotor, being careful in the second case that the element that dissipates energy is restrained from rotating. In machines containing rotors that work at different speeds, like multishaft turbines, it is also possible to use intershaft dampers, i.e., dampers located between two shafts rotating at different speeds; their use, however, must be managed with great care because they can have a destabilizing effect in particular working conditions.

When a damper is nonrotating, the only constraint in its place-

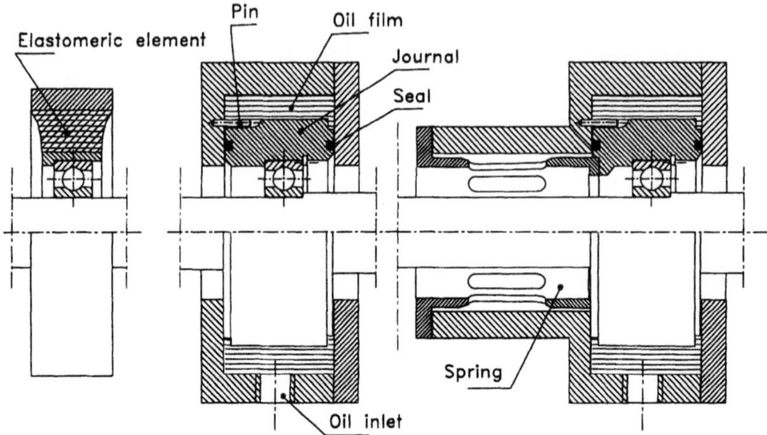

FIGURE 4.44. Flexural vibrations dampers: (a) elastomeric type; (b) squeeze-film damper; (c) damped support of the squeeze-film type.

ment is that it must be between two points of the machine with relative displacement in the modes of vibrations that must be controlled as large as possible. It is possible to use pure dampers, namely, elements that react to a relative velocity between the two attachment points but not to a relative displacement or elements that can also supply a restoring force apart from the damping force.

Generally speaking, flexural vibration dampers are of the dissipative type and use the viscosity of a liquid, usually oil, or the internal damping of some material, usually an elastomer. Active or passive electromagnetic dampers are also entering the application stage. An example of an elastomeric damper is shown in Figure 4.44a. It is a damped support, made by a cylindrical elastomeric element into which a rolling element bearing is inserted. Devices of the type shown are simple and low cost, but their drawback is that they are subject to overheating. The energy dissipated by the damper is totally converted into heat within the elastomeric mass and, due to the low thermal conductivity of most elastomeric materials, can cause its temperature to increase, even to the point of causing a deterioration of the characteristics of the material. If the damper starts overheating, the internal damping of the material usually decreases, causing an increase of the amplitude of the vibrations, which, in turn, increases the heat generation in the damper. An unstable increase of temperature that leads quickly to the complete destruction of the element can result. Dampers of this type can be used only after an accurate thermal analysis has demonstrated their applicability.

An oil damper, of the type generally referred to as a squeeze-film damper, is shown in Figure 4.44b. A journal, which is restrained from rotating by a pin, is connected to the rotor through a rolling element bearing. The journal is located in a bearing and the oil between them is prevented from moving axially by seals at the ends. Radial movements of the journal cause the oil to move circumferentially, and this movement provides the required damping. Using the models seen in the previous section it is easy to state that the stiffness of the oil film is vanishingly small, because no relative rotation between journal and bearing occurs, and that there is no unstabilizing effect, linked to the circumferential motion of the oil. If the behaviour of the device about the centered position can be linearized and the long-bearing model is used because of the axial seals, the damping matrix can be expressed by equation (4.177). In the damper of Figure 4.44c, the journal is connected to the bearing by a spring and the device can provide both a damping and a restoring force. More than a damper, it can then be defined as a damped support.

Squeeze-film dampers, which are very common, introduce nonlinearities into the system because their damping depends on the amplitude of the motion. The aforementioned linearization holds only for motions with very small amplitude.

4.10 Signature of rotating machinery

The experimental analysis of the vibrations caused by rotating machinery yields much useful information on their working conditions and allows the discovery of possible problems that have been incurred before consequences become too severe and, in some cases, even predicts their occurrence. The ultimate aim of the analysis is to diagnose the state of the machine to be able to perform preventive maintenance. To monitor the dynamic behaviour of the machine, it is often sufficient to attach transducers that can measure the acceleration, velocity, or displacement in selected locations. In some machines, the transducers can be mounted permanently, and the signal they supply can be monitored continuously or at regular intervals or even only when working anomalies become apparent. Different kinds of transducers are available, but now they are almost always connected to electronic data-acquisition systems that can perform various types of analyses and supply the relevant information in the

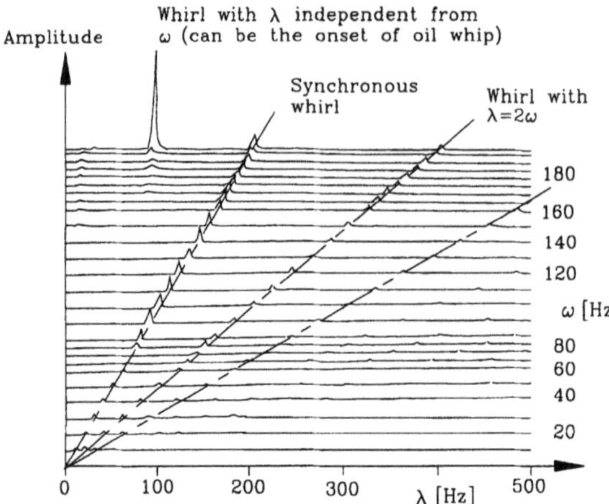

FIGURE 4.45. Cascade plot.

form the user feels is more useful. In particular, it is very useful to perform a harmonic analysis of the output of the transducers to obtain the acceleration or displacement spectrum. A very common and useful way of representing these spectra is the so-called cascade plot: a sort of tridimensional plot in which the spectra obtained at different spin speeds are reported in the planes with y constant in a tridimensional space xyz (Figure 4.45). Often, the frequency is reported in Hz, and the spin speed is in revolutions per minute or revolutions per second.

Usually, the terms *cascade plot* and *waterfall plot* are used for two different types of diagrams. In the first, the spectra are plotted at different spin speeds; in the second, they are obtained at the same speed but at different times. The waterfall plot is not used to study the behaviour of the machine in different working conditions but to follow the evolution in time of its dynamic behaviour. The cascade plot has some similarity with the tridimensional plot of Figure 4.30, except that λ and ω are exchanged. There is, however, a major difference: In the plot of Figure 4.30, the amplitude is the total amplitude of a monoharmonic motion, or better, of circular whirling, while in the cascade plot the amplitude of the various harmonic components is reported as a function of the frequency after performing the harmonic analysis of the nonharmonic waveform. In the plot, it is possible to identify, at each speed, the frequencies of the various motions of the rotor. If different transducers are placed in different radial planes, then by comparing the phases of the relevant signals, it is possible to assess the direction of the various whirl components.

	Forced or resonant vibrations	Self-excited vibrations
Relationship between frequency and speed	Frequency is equal to (i.e., synchronous with) the spin speed or a whole number or rational fraction of spin speed.	Frequency is nearly constant and essentially independent of spin speed or any external excitation or/and is at or near one of the shaft natural frequencies.
Relationship between amplitude and speed	Amplitude will peak in a narrow band of spin speed wherein the rotor's natural frequency is equal to the spin speed or to a whole number multiple or a rational fraction of the spin speed on external excitation.	Amplitude will suddenly increase at a threshold speed and continues at high or increasing levels as spin speed is increased.
Whirl direction	Almost always forward whirling.	Generally forward whirling, but backward whirling has been reported.
Rotor stressing	Static stressing in case of synchronous whirling.	Oscillatory stressing at frequency equal to $\lambda - \omega$.
Correcting actions	1. Introduce damping to limit peak amplitudes at critical speeds. 2. Tune the system's critical speeds to be outside the working range. 3. Eliminate all deviations from axial symmetry in the system as built or as induced during operation (e.g., balancing).	1. Increase damping to increase the threshold of instability above the operating speed range. 2. Raise the rotor natural frequencies as high as possible. 3. Identify and eliminate the instability mechanism.
Influence of damping	Addition of damping may reduce peak amplitude but does not affect the spin speed at which it occurs.	Addition of damping may raise the speed at which instability occurs but usually does not affect the amplitude after onset.
Influence of system geometry	Excitation level and hence amplitude are dependent on some lack of axial symmetry in the rotor mass distribution or geometry or external forces applied to the rotor. Amplitudes may be reduced by refining the system to make it more axisymmetric or balanced.	Amplitudes are independent of system axial symmetry. Given an infinitesimal deflection to an otherwise axisymmetric system, the amplitude will self-propagate for whipping speeds above the threshold of instability.

TABLE 4.3. Characterization of forced and self-excited rotor vibrations.

Mechanism	Ratio λ/ω	Direction
Internal rotor damping	$0,2 < \lambda/\omega < 1$ ($\lambda/\omega = 0,5$)	Forward
Hydrodynamic bearings, labyrinth, or liquid seals	$\lambda/\omega < 0,5$ ($0,45 < \lambda/\omega < 0,48$)	Forward
Blade-tip clearance excitation	Dependent on fluid force levels	Forward
Centrifugal pump and compressor whirl	Dependent on fluid force levels	Forward
Propeller and turbomachinery whirl	Dependent on fluid force levels	Backward, if the vertex of the cone described in the whirl motion is after the rotor (referring to the direction of the fluid flow). Forward in the opposite case
Excitation due to fluid trapped in rotors	$0,5 < \lambda/\omega < 1,0$ ($0,7 < \lambda/\omega < 0,9$)	Forward

TABLE 4.4. Diagnostic table of self-excited vibrations of rotating machinery.

From the frequency and phase information, it is possible to study the causes that produce the vibration and to decide the proper correcting actions.

For example, the synchronous component is usually linked with unbalance and can be corrected by performing a more accurate balancing, while a component with frequency equal to twice the rotational speed is generally linked with rotor anisotropy and can be corrected by making it more symmetrical. As a general rule, balancing the rotor has very little effect on all the nonsynchronous components.

Each machine produces a characteristic vibration spectrum, which is often referred to as the mechanical signature of the machine. Any alteration in time of the signature, as evidenced by a waterfall plot, is the symptom of an anomaly of the working conditions and must be considered very carefully. It can actually be linked with a problem that has occurred or is developing. The importance of correctly diagnosing problems before they actually occur in reducing costs linked with maintenance and to the unavailability of the machine, is clear.

Some tables that can help in the identification of the causes of anomalies in the dynamic behaviour of rotating machinery are here reported from a paper by F. Ehrich and D. Childs[14] (Tables 4.3, 4.4

---

[14]F. Ehrich, D. Childs, "Self-excited vibration in high performance turbomachinery", *Mech. Eng.*, May, (1984).

Mechanism	Correcting action
Internal rotor damping	Minimize number of separate elements in rotor; restrict span of rabbets and shrink-fitted parts; provide secure lock up on assembled elements.
Hydrodynamic bearings	Install tilting pad or rolling elements bearings.
Labyrinth seals	Add swirl brakes at seal inlets to reduce the inlet tangential velocity. Replace rotor-mounted labyrinth vanes with stator-mounted vanes. Replace labyrinth seals with honeycomb seals.
Liquid seals for pumps	Introduce swirl webs or brakes at seal inlet; roughen stator elements.
Blade-tip clearance	No ready measures that do not affect the unit excitation operating efficiency.
Centrifugal pump and compressor whirl	Not well understood.
Propeller and turbomachinery whirl	Modify mode shapes to minimize angular motion of the plane of turbomachinery.
Excitation due to fluid trapped in rotors	Introduce drain holes to eliminate fluid accumulation or axial webs to inhibit rotation of fluid.

TABLE 4.5. Design correcting actions to reduce instability.

and 4.5).

4.11 Rotor balancing

4.11.1 General considerations

All rotors, particularly those intended to operate at high rotational speed, must be balanced before they start their service life. Sometimes balancing procedures must be repeated during the life of a machine. The designer must then take into account this requirement to provide the possibility of removing or adding masses in proper locations from the early design stages. Balancing must be regarded as one of the construction stages, to be performed after assembling the whole rotor or before, on its component, if they must be balanced separately (which usually does not avoid a further balancing process on the assembly), and balancing tolerances must be stated in a way that is not conceptually different from what is done for other types of tolerances, dimensional or geometrical. The balance conditions of a rotor can change in time, and periodic rebalancing may be needed. In some cases this phenomenon can be quite severe and is usually referred to as *wandering unbalance*. It can be caused by thermal deformations of the rotor, material inhomogeneity, cracks, loose tolerances in built-up rotors, and the like. Some rotors must be balanced several times during the first runs at subsequent higher speeds, in order to reach good balancing conditions at operating speed and running temperature. Poor balance conditions can be encountered during start-up, until steady-state conditions are reached.

Rotor balancing has been the object of unification and designers must refer to the standards in stating balancing tolerance at the design stage. Standards are stated for the various types of machines, but it is the duty of the designer to verify that the stresses and deformations caused by the maximum residual unbalance prescribed are not beyond allowable limits. He must also be sure that the prescribed balancing tolerances are strict enough to prevent the rotor from being a source of unwanted vibrations and noise for the surrounding environment. As with all tolerances, it must be remembered that it is impossible to reach a perfect balancing and that it is not necessary, and, generally, not advisable (at least from the economical point of view) to impose too-strict balancing requirements.

From the point of view of balancing, rotors are usually divided into two categories: rigid and deformable rotors. This subdivision, which is accepted by ISO standards, is in a certain sense arbitrary, because no rigid body exists in the real word. A rotor can belong to either class, depending on the speed at which it is supposed to operate and, in particular, a speed at which any rigid rotor ceases to behave as such always exists. The balancing of rigid and deformable rotors will be only briefly summarized in the following sections: The reader can find all the required details in specialized monographs, in particular, those published by firms that build balancing machines.[15]

4.11.2 Rigid rotors

Following the ISO 1925 standard, a rotor can be considered rigid if it can be balanced by adding or removing mass in two arbitrarily chosen planes perpendicular to the rotation axis and if its balance conditions are practically independent from speed up to the maximum allowable speed. If this condition is satisfied, the rotor can be assimilated to a rigid body. Its dynamic behaviour can be studied using the approach in Section 4.6. Actually, if the inertia of the stator is

---

[15] As an example, see H. Schneider, *Balancing technology*, Schenck, Darmstadt, 1974.

neglected and its elastic properties, referred to the center of gravity of the rotor, are summarized in the stiffness matrix $\mathbf{K}$, a model with four real degrees of freedom (two complex ones) is adequate to study its flexural dynamic behaviour. The balance conditions of the machine can then be summarized by two parameters: the eccentricity ϵ (or the static unbalance $m\epsilon$) and angle χ between the principal axis of inertia of the rotor and the rotation axis (or the couple unbalance $(J_p - J_t)\chi$).

Standards use the peripheral velocity of the center of mass of the rotor as a measure of the static unbalance

$$V_G = \omega_{max}\epsilon_{max} \tag{4.195}$$

and define a quality grade for balancing, usually referred to as G, as the maximum allowable peripheral velocity of the center of mass, expressed in mm/s. A rotor that has to be balanced in the class G=2.5 at 10,000 rpm, for example, must have an eccentricity smaller than $\epsilon_{max} = V_G/\omega_{max} = 2.38\mu$m, which results in a peripheral velocity of the mass center of 2.5 mm/s.

The maximum values of the residual eccentricity are plotted as functions of the maximum operating speed for different values of the quality grade in Figure 4.45. The quality grades suggested by ISO standard 1940 for different types of rotors are reported in Table 4.6.

The mentioned standards consider only eccentricity and then static unbalance. In the case of a rigid rotor running on two bearings, a couple unbalance due to two static unbalances, each equal to half the maximum allowable static unbalance placed at the bearing locations and phased at $180°$, is considered a limit. Let d be the distance between the bearings; it follows that

$$|\chi(J_p - J_t)|_{max} = m\epsilon_{max}\frac{d}{2} . \tag{4.196}$$

Rigid rotors are normally balanced using balancing machines. They can be either the high-stiffness or the low-stiffness type. Rigid (high-stiffness) balancing machines are machines in which the rotor can be spun on two very stiff supports provided with force transducers. The unbalance condition of the rotor is obtained from the measurement of the forces it exerts on its supports. Low-stiffness machines are similar to the former but the supports are more or less free to move and the transducers measure a quantity linked with

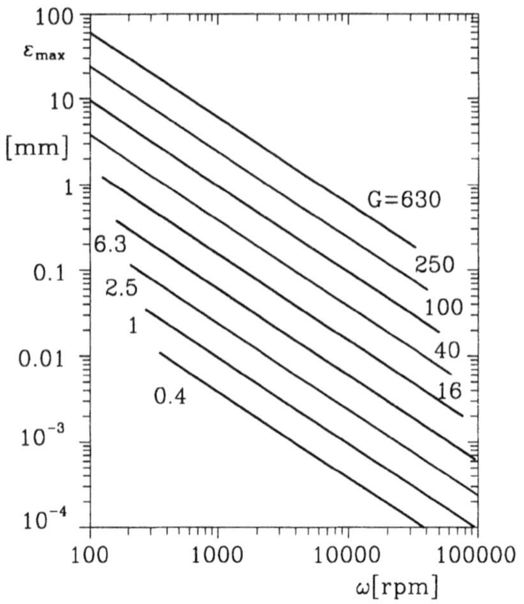

FIGURE 4.46. Maximum residual eccentricity as a function of the maximum operating speed for quality grades between $G = 0.4$ and $G = 630$ mm/s.

Grade	Examples
G 4000	Crankshaft drives of rigidly mounted slow marine diesel engines with an uneven number of cylinders.
G 1600	Crankshaft drives of rigidly mounted large two-cycle engines.
G 630	Crankshaft drives of rigidly mounted large four-cycle engines, crankshaft drives of elastically mounted marine diesel engines.
G 250	Crankshaft drives of rigidly mounted fast four-cylinder diesel engines.
G 100	Crankshaft drives of fast diesel engines with six or more cylinders, complete engines (gasoline or diesel) for cars, trucks, and locomotives.
G 40	Car wheels, wheel rims, wheel sets, drive shafts, crankshaft drives of elastically mounted fast four-cycle engines (gasoline or diesel) with six or more cylinders, crankshaft drives for engines of cars, trucks, and locomotives.
G 16	Drive shafts (propeller shafts, cardan shafts) with special requirements, parts of crushing machinery, parts of agricultural machinery, individual components of engines (gasoline or diesel) for cars, trucks, and locomotives, crankshaft drives of engines with six or more cylinders under special requirements.
G 6.3	Parts of process plant machinery, marine main turbine gears (merchant service), centrifuge drums, fans, assembled aircraft gas turbine rotors, flywheels, pump impellers, machine tools and general machinery parts, normal electrical armatures, individual components of engines under special requirements.
G 2.5	Gas and steam turbines, including marine main turbines (merchant service), rigid turbogenerator rotors, rotors, turbocompressors, machine-tool drives, medium and large electrical armatures with special requirements, small electrical armatures, turbine-driven pumps.
G 1	Tape recorder and phonograph (gramophone) drives, grinding machine drives, small electrical armatures with special requirements.
G 0.4	Spindles, discs, and armatures of precision grinders, gyroscopes.

TABLE 4.6. Quality grades G suggested for different types of rotors (ISO 1940).

their motion, usually the acceleration but sometimes the displacement or velocity. The unbalance of the rotor is obtained from the displacement measurements. Two planes perpendicular to the rotation axis are chosen, as far from each other as possible, on which masses can be either added or removed. The maximum values of the residual unbalance on the two balancing planes are computed from the allowable residual unbalance corresponding to the required quality grade. The actual unbalance state of the rotor is then determined, spinning the rotor on a balancing machine. Modern balancing machines are provided for data-acquisition systems that perform all needed computations and directly supply the values of the unbalance on the two correction planes and all information on the masses to be added or removed in them (amount of mass, radius, and angular position). Once the correction has been performed, a further measurement aimed at checking whether the required tolerance has been achieved usually follows.

The need to add or remove mass in the two planes at the same angular position is a clear symptom of a static unbalance. If the corrections must be phased 180° from each other, the unbalance is purely a couple unbalance. Generally, the phasing is neither at 0° nor at 180°, corresponding to a general state of dynamic unbalance, defined as the sum of static plus couple unbalance. In some cases, the various parts that constitute a rotor must be balanced separately. In this case, the balancing tolerances of the various parts must be stated, remembering that unbalance is a vector quantity and that in the assembly process they usually add to each other in a random way. The absolute value of the unbalance is, in the most unfavourable case, equal to the sum of the absolute values of the unbalances of the parts. Dimensional tolerances of the various parts and their effects on the relative position of the various parts must also be considered. If possible, the rotor must be balanced after assembly, possibly on its bearings in such a way that the tolerances of the bearings and their seats are also accounted for. In some cases (rotors whose size exceeds the possibilities of available balancing machines, high-precision machinery), the rotor is balanced directly on-site. The machine is instrumented and the synchronous component of the vibration of the machine is monitored at different speeds. The amplitude and phase of the synchronous component give all the information needed to identify and correct the unbalance.

4.11.3 Flexible rotors

Balancing flexible rotors is much more difficult than balancing rigid rotors and, strictly speaking, it cannot be performed on balancing machines. The stiffness of the supports of the machine has, in fact, great influence on the deflected shape of the rotor, and then balancing should be performed directly on the whole machine. The process of balancing the whole machine is generally referred to as *field balancing*. It is, however, often still possible to resort to a balancing machine, provided that its stiffness is not too different from that of the stator of the actual machine and proper allowance is taken for the unavoidable differences. Strictly speaking, a flexible rotor is balanced only if each of its cross sections is statically and dynamically balanced. In practice, it is not necessary that this condition be met to achieve the required aim, which is that of maintaining the effects of unbalance (vibrations, stresses, noise, etc.) within tolerable limits in the whole working range of the machine.

The flexible nature of the rotor and the difficulties that can be encountered in the balancing process are directly linked with the ratio between the maximum operating speed and the first flexural critical speed due to a bending mode of the rotor. If the rotor has to operate at speeds far lower than the first critical speed (below about half of it), it can be assumed to be rigid. When the operating speed is near the first critical speed, the possibility that the rotor inflects, assuming a shape not far from the first mode shape, must be taken into account. Correspondingly, near the nth critical speed, the inflected shape is not dissimilar from the nth mode shape (nth eigenvector or eigenfunction, depending on the model used for the dynamic analysis). If damping is neglected, the inflected shape can always be considered a linear combination of the mode shapes[16] and if the rotor operates between two critical speeds, the dominant modes are those related to them. If the rotor has to operate above the nth critical speed, in order to be balanced in the whole working range it must be balanced with reference to the first $n + 1$ mode shapes.

From a practical point of view, ISO standards subdivide rotors into five classes. Rigid rotors, as described in the preceding section,

---

[16]This property holds also if the gyroscopic effect is taken into account, see G. Genta and F. De Bona, "Unbalance response of rotors: a modal approach with some extensions to damped natural systems", *J. of Sound and Vibration*, 140 (1), 1990, 129–153.

belong to the first class. Rotors of the second class are defined as semirigid, i.e., rotors that cannot be considered rigid but can be balanced in a low-speed balancing machine. This class is subdivided into eight subclasses, from 2a to 2h, as shown in Table 4.7. The third class contains the true flexible rotors, which cannot be balanced using a low-speed balancing machine, but require balancing at high speed. From this class, the rotors belonging to classes 4 and 5 are excluded. Rotors of the fourth class are said to be special flexible rotors, which are defined as rotors that would be rigid or semirigid but to which one or more flexible components have been added. The fifth class contains flexible rotors, which would belong to class 3 or 4, but for which balancing at a single speed (generally at the operating speed) is required.

Two procedures are usually considered for field balancing of flexible rotors: modal balancing and the influence coefficients method. The first practice is based on balancing the rotor at the various critical speeds, starting from the lowest. The correction masses must be located in the planes in which the relevant mode shape has large displacements. Generally speaking, the number of the correction planes must be equal to the order of the critical speed. It is possible to demonstrate that it is possible to correct the unbalance at the general ith critical speed without disturbing the balance conditions of the previous $i - 1$ modes, which have already been balanced.

Consider a rotor modeled as a discretized, multi-degree-of-freedom system. The unbalance response can be computed using equation (4.97). To balance the system at the first critical speed means to add to the unbalance distribution $\mathbf{f}$ another unbalance distribution $\mathbf{f}_{b1}$, which causes the modal force due to unbalance related to the first mode to vanish

$$\overline{f}_1 = \mathbf{q}_1^T(\mathbf{f} + \mathbf{f}_{b1}) = 0 \,. \tag{4.197}$$

If balancing is performed by adding the unbalance $m_1\epsilon_1$ at the jth generalized coordinate (vector $\mathbf{f}_{b1}$ has all elements equal to zero except the jth, whose value is $m_1\epsilon_1\omega^2$), the modal unbalance that has been added is $\mathbf{q}_1^T\mathbf{f}_{b1} = m_1\epsilon_1 q_{j1}\omega^2$. From equation (4.197) the value of the unbalance to be added to balance the first mode shape is immediately obtained

$$m_1\epsilon_1 q_{j1}\omega^2 = -\mathbf{q}_1^T\mathbf{f} \,. \tag{4.198}$$

Class	Description	Example
2a	A rotor with a single transverse plane of unbalance, e.g., a single mass on a shaft with negligible unbalance.	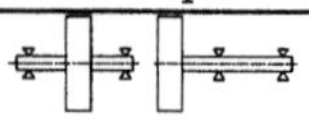
2b	A rotor with two transverse planes of unbalance, e.g., two masses on a shaft with negligible unbalance.	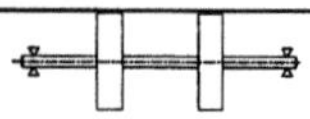
2c	A rotor with more than two transverse planes of unbalance and flexible shaft.	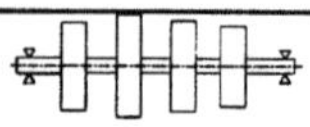
2d	A rotor with uniformly distributed unbalance.	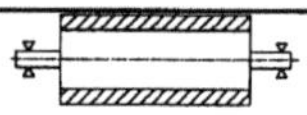
2e	A rotor consisting of a rigid mass of significant axial length supported by a flexible shaft with negligible unbalance.	
2f	A symmetrical rotor with two end correction planes, whose maximum speed does not significantly approach second critical speed, whose service speed range does not contain first critical speed and with component balanced separately.	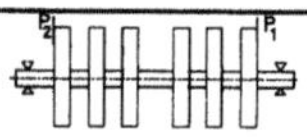
2g	As 2f, but with a third intermediate correction plane.	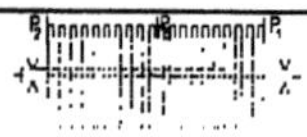
2h	As 2f, but without a symmetry plane at midspan.	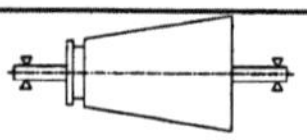

TABLE 4.7. Rotors of class 2 (semirigid rotors).

To balance the second mode, the two unbalances $m_{2r}\epsilon_{2r}$ and $m_{2s}\epsilon_{2s}$ are added corresponding to the rth and sth generalized coordinates. The second mode is balanced if

$$(m_{2r}\epsilon_{2r}q_{r2} + m_{2s}\epsilon_{2s}q_{s2})\omega^2 = -\mathbf{q}_2^T\mathbf{f} + \mathbf{f}_{b1} \,. \tag{4.199}$$

The second mode can be balanced without disturbing the balancing, already achieved, of the first mode if the modal force corresponding to the first mode remains equal to zero after the addition of the new balancing masses at the generalized coordinates r and s

$$m_{2r}\epsilon_{2r}q_{r1} + m_{2s}\epsilon_{2s}q_{s1} = 0 \,. \tag{4.200}$$

Equations (4.199) and (4.200) allow computation of the two correction unbalances $m_{2r}\epsilon_{2r}$ and $m_{2s}\epsilon_{2s}$.

The correction of the third mode can be performed in the same way: The three correction unbalances can be computed using an equation of the type of equation (4.199), stating that the modal force corresponding to the third mode must vanish, and two equations of the type of equation (4.200) stating that the balancing of the third mode does not affect the balancing conditions at the previously balanced modes. In a similar way, all other modes can be balanced. What has been shown is actually a demonstration that modal balancing is possible, not the procedure to practically implement the balancing process. The knowledge of the unbalance distribution $\mathbf{f}$, which is generally unknown, and of the mode shapes is not required. It is, however, clear that the computation of the mode shapes allows an easier way to identify the planes in which the correcting action is most effective, because the balancing masses must be located at the loops of the mode shape, or not too near to the nodes, where they would be ineffective.

Before starting modal balancing, the rotor can be balanced as a rigid body on a balancing machine. There is no agreement on the advisability of this practice, but, generally speaking, rigid-body balancing can be omitted if the starting unbalance is not severe enough to prevent operating near the first critical speed, as needed to start the modal balancing process.

It can happen that a rotor, which has been balanced at the first n critical speeds, causes strong vibrations at the maximum operating speed, which is between the nth and $(n+1)$-th critical speeds, without reaching the latter to complete the balancing procedure.

The strong vibrations are caused, in this case, by the unbalanced higher modes and, in particular, the $(n+1)$-th mode. In this case the last corrections are implemented, taking into account the mode shape that is predicted at the $(n+1)$-th critical speed, even if it is not materially possible to reach it.

The influence coefficients method is based on the observation that the vibrations detected in a number of measuring points can be considered the effect of a number of concentrated unbalances in a number of arbitrarily located planes. This addition process implies that the behaviour of the system is linear. Let the number of measuring points be n, the number of speeds at which the balancing process is performed be m, and the number of planes in which the correction masses are to be located be q. Due to the linearity of the system, the $m \times n$ responses z_i obtained in the n measuring points at the m test speeds are linked to the q unknown unbalances $m_i \epsilon_i$ in the q correction planes by the general linear relationship

$$\{z\}_{(m \times n)} = [A]_{(m \times n) \times q} \{m\epsilon\}_q. \qquad (4.201)$$

Note that both responses z_i, which can be displacements, as implicitly assumed in equation (4.201), but also accelerations or velocities, and the unbalances $m_i \epsilon_i$ are vector quantities and coincide with the complex quantities z and $m\epsilon$ in the preceding sections. If the influence coefficients in matrix $\mathbf{A}$, whose size is $(m \times n) \times q$, were known, equation (4.201) could be used directly to compute the unknown unbalances from the vibration measurements, provided that matrix $\mathbf{A}$ is square, i.e., the number of correction planes q is equal to the product of the number of test speeds m by the number of measuring points n, and is nonsingular.

The coefficients of influence are easily determined. A known unbalance $m_j r_j$ is introduced in a generic correction plane (the jth), in a known angular position. The tests are repeated, and a set of new $m \times n$ responses z_i^* are measured. They are linked to vector $\{m\epsilon\}^*$ through equation (4.201): $\mathbf{z}^* = \mathbf{A}\{m\epsilon\}^*$, where vector $\{m\epsilon\}^*$ has all elements that coincide with those of the vector in equation (4.201) except the jth element, which has been incremented by the known unbalance $m_j r_j$.

By subtracting equation (4-200) written for unbalances $\{m\epsilon\}$ and $\{m\epsilon\}^*$, it follows that

$$\mathbf{z}^* - \mathbf{z} = \mathbf{A} \begin{bmatrix} 0 & 0 & \dots & m_j r_j & \dots & 0 \end{bmatrix}^T. \qquad (4.202)$$

The jth column of matrix $\mathbf{A}$ is then simply obtained by dividing the difference between vectors $\mathbf{z}^*$ and $\mathbf{z}$ by $m_j r_j$. The procedure can be repeated by removing the previously added unbalance from the jth plane and adding a new known unbalance in another plane. The matrix of the coefficients of influence is thus obtained, column after column, and the unbalance distribution $\{m\epsilon\}$ is obtained:

$$\{m\epsilon\} = \mathbf{A}^{-1}\mathbf{z}. \qquad (4.203)$$

The rotor is then balanced by introducing in the q correction planes suitable unbalances equal and opposite to the ones computed. The balanced conditions of the rotor are thus achieved, at least at the speeds at which the tests have been performed.

The procedure described here is the extension to flexible rotors of the usual procedure for the calibration of the balancing machine for rigid rotors. In the latter case, only one test speed is chosen ($m = 1$), because the balance conditions are independent of the speed. By using two measuring points ($n = 2$), the number of the required correction planes reduces to 2 ($q = 2$).

The quality of the balancing of a flexible rotor can only be stated from the amplitude of the synchronous component of the vibrations measured on the running machine. Table 4.8, from an ISO recommendation, can supply detailed indications.

Example 4-8
Consider the turbomolecular pump studied in Example 4-2. Compute the unbalance response at operating speed, knowing that the balancing quality grade required is $G = 2.5$.

	Range of effective pedestal synchronous vibration velocity (mm/s), r.m.s.			Correction factor		
	.28 .45 .71 1.12 1.8 2.8 4.5 7.1 11.2 18 28 45 71			C_1	C_2	C_3
	A B C D					
I	Small electric motors (up to 15 kW)			.63		
	Superchargers			.63	2	
	Gyroscopes			.63	2	
	A B C D					
II	Paper making machines			.63		
	Med. size elec. motors and gener. (17-75 kW), norm. foundations			.63	4	
	Electric motors and gener. up to 3000 kW, special foundations			.63	4	20
	Pumps and compressors			.63	8	15
	Small turbines			.63	4	8
	A B C D					
III	Large electric motors			.63	5	
	Turbines and generators, rigid and heavy foundations			.63	5	20
	A B C D					
IV	Large electric motors, turbines and generators, light foundations			.63	3	10
	Small jet engines			.63		
	A B C D					
V	Jet engines larger than cathegory IV			.63	2	10

Balance quality
A = Precision quality
B = Commercially acceptable
C = In need of attention at next overhaul
D = In need of imediate attention

Correction factors
C_1 = Measurement in high speed balancing machine at service speed where bearing conditions are different from service conditions
C_2 = Shaft vibrations measured in or adjacent to the bearings
C_3 = Shaft vibrations measured at location of maximum shaft lateral deflection.

TABLE 4.8. Balance quality criteria for flexible rotors.

The peripheral velocity of the center of mass corresponding to a quality grade $G = 2.5$ is of 2.5 mm/s. The maximum value of the eccentricity is then $\epsilon_{max} = V_G/\omega_{max} = 0.00714$ mm. Because the distance between the two balancing planes is 260 mm, the maximum value of the couple unbalance is $|\chi(J_p - J_t)|_{max} = m\epsilon_{max}d/2 == 5.57 \times 10^{-6}$ kgm^2.
The responses to static and couple unbalance at 30,000 rpm can be computed from the equations

$$\begin{bmatrix} 5.715 & -5.591 \\ -5.591 & 129 \end{bmatrix} \times 10^7 \mathbf{q} = \left\{ \begin{array}{c} 422.9 \\ 0 \end{array} \right\},$$

$$\begin{bmatrix} 5.715 & -5.591 \\ -5.591 & 129 \end{bmatrix} \times 10^7 \mathbf{q} = \left\{ \begin{array}{c} 0 \\ 54.99 \end{array} \right\},$$

which yield the following values of the displacements and rotations:

$$\mathbf{q} = \left\{ \begin{array}{c} -4.0990 \\ -0.3088 \end{array} \right\} \times 10^{-6} \qquad \text{(static unbalance)},$$

$$\mathbf{q} = \left\{ \begin{array}{c} -4.001 \\ 4.089 \end{array} \right\} \times 10^{-8} \qquad \text{(couple unbalance)}.$$

From the results obtained, it is clear that at 30,000 rpm the rotor is almost completely self-centered, where static unbalance is concerned, and couple unbalance causes very small deformations of the shaft.

4.12   Exercises

Exercise 4-1

Consider the transmission shaft of Exercise 2-12. Plot the Campbell diagram and compute the critical speeds.

Exercise 4-2

Consider the same system studied in the preceding exercise. Assuming that the central joint is balanced to a quality $G = 16$ for a maximum speed of 10,000 rpm and that the central support has a hysteretic damping with loss factor $\eta = 0.1$, compute the unbalance response up to a speed equal to 1.5 times the first critical speed.

Exercise 4-3

Compute the driving torque and power needed to overcome the drag due to nonrotating damping in the case of the shaft of Exercise 4-2. Compute the minimum torque and power needed to avoid stalling when crossing the critical speed.

Exercise 4-4

A flywheel must store 1 kWh at 17,000 rpm, with a working range between 8,500 and 17,000 rpm. Knowing that the energy density at the maximum speed of the rotor is 8 Wh/kg and that the transversal moment of inertia is equal to 65% of the polar moment of inertia, compute its mass and moments of inertia.

The flywheel runs on a pair of ball bearings. Knowing that the span between the bearings is 400 mm, that the center of gravity is at 30% of the span, and the stiffness of each bearing is 10^8 N/m, compute the Campbell diagram and the critical speeds. If a critical speed falls within a range spanning between 60% of the minimum operating speed and 130% of the maximum operating speed, change the stiffness of the bearings in such a way that it is moved either below or above the stated range.

Exercise 4-5

Compute the unbalance response of the system of the previous exercise. If two solutions were obtained, compute both responses. Compute the balance quality grade needed to keep the total force due to static unbalance on the stator of the machine within 200 N in the whole working range (compute the forces from the amplitude of the orbits at the bearing locations). Compute the forces due to a couple unbalance corresponding to the same quality grade.

Exercise 4-6

Assume that the bearings of the flywheel of the preceding exercise are characterized by a loss factor $\eta = 0.08$. Compute the unbalance response from standstill to the maximum operating speed with the balance quality grade computed and, if a critical speed must be passed, the maximum forces exerted on the stator. Compute the driving torque and power needed to overcome the drag due to nonrotating damping and the minimum torque and power needed to avoid stalling when crossing the critical speed.

Exercise 4-7

The flywheel of the preceding exercise is operated by an electric motor whose maximum power is 20 kW. Assuming that the controller maintains constant the driving torque during acceleration from standstill, compute the time needed to reach the maximum operating speed, neglecting the flexural behaviour of the rotor. By integrating numerically in time the equations of motion, compute the unbalance response during an acceleration with the constant rate computed earlier. Compute the driving torque actually needed to follow the stated velocity time history and compare it with the torque supplied by the motor.

Exercise 4-8

Compute the unbalance response of the flywheel of Exercise 4-6 assuming that the bearings have the hardening characteristic $F = 2 \times 10^7(1+10^{10}z^2)z$ and that the hysteretic damping, with $\eta = 0.2$, applies only on the linear part of the stiffness.

Study the response obtained and state whether self-centering is possible. Compute the forces exerted on the stator.

Exercise 4-9

Consider the same rigid rotor used for the preceding examples, but substitute the ball bearings with a pair of plain journal bearings. Modeling the bearing using the short-bearing assumption, with a fully cavitated film, i.e., with the oil film extending only for 180°, compute the position of the center of the journals at all speeds up to the maximum operating speed, perform the dynamic analysis in the small, and plot the Campbell diagram, studying the stability of the system. Bearing data: diameter of shaft = 60 mm, l = 40 mm, $c = 40$ μm, $\mu = 0.02$ Ns/m^2. Let the bearing be modeled using the short-bearing assumption, with a fully cavitated film, i.e., with the oil film extending only for 180°.

Exercise 4-10

Consider the gas turbine of Example 4-3. Repeat the dynamic analysis substituting the rigid bearings with two journal bearings of the type seen in Exercise 4-9, but with a journal diameter of 35 mm and a length of 30 mm.

5
Dynamic Problems of Reciprocating Machines

5.1   Specific problems of reciprocating machines

Machines that contain elements provided of reciprocating motion have some characteristic dynamic problems. In many cases, their solution is difficult and designers often cannot ensure working conditions that are as smooth as those that can be reached with machines containing only rotating elements. In many cases it is already a difficult task to avoid strong vibrations, which are uncomfortable for users and dangerous to structural safety. It is often necessary to resort to extensive experimentation, even to reach this limited goal.

Most reciprocating machines are based on the crank mechanism, often in the form of a crankshaft with several connecting rods and reciprocating elements. Such devices cannot, in general, be exactly balanced: The inertial forces they exert on the structure constitute a system of forces whose resultant is not vanishingly small and is variable in time. The very geometric configuration of the system made by the crankshaft, the connecting rods, and the reciprocating elements can be quite complex, and crankshafts not only do not possess axial symmetry but often do not have symmetry planes. In these conditions, the uncoupling among axial, torsional, and flexural behaviours, seen in the preceding sections, is no longer possible, if not as a very rough approximation, and vibration modes become very complicated. The external forces acting on the elements of reciprocating machines are usually variable in time, often with periodic law, like, for example, the forces exerted by hot gases on the pistons of reciprocating internal-combustion engines. Their periodic time histories are not harmonic but, once harmonic analysis has been performed, they can be considered the sum of many harmonic components whose frequencies are usually multiples, by a whole number or a rational fraction, of the rotational speed of the machine. There may be many possibilities of resonance between these forcing functions and the natural frequencies of the system.

In most cases, the most dangerous vibrations are linked to modes that are essentially torsional. In this section the attention will be concentrated on the torsional vibrations of reciprocating machines, dealing with axial and composite vibrations only marginally. It must, however, be kept in mind that all vibrations of crankshafts are composite vibrations, and no uncoupling is strictly possible.

5.2   Equivalent system for the study of torsional vibrations

5.2.1   Equivalent inertia

The traditional approach to the study of torsional vibrations of reciprocating machines is based on the reduction of the actual system made of crankshafts, connecting rods, and reciprocating elements to an equivalent system, usually modeled as a lumped-parameters system, whose torsional behaviour can be studied separately (Figure 5.1). The model of the reciprocating machine so obtained is then coupled to a model, usually based on the lumped-parameters approach, including all the rotating elements connected to the crankshaft, e.g., the driven machine in the case of an engine or the motor that operates the compressor.

The inertia of the cranks and the reciprocating elements is lumped into a number (one for each crank) of moments of inertia (flywheels). The moments of inertia are then connected to each other by straight shafts, whose diameter is equal to the diameter of the relevant part of the actual shaft, or, more often, has a standard conventional value, and whose length is computed in such a way that the torsional stiffness of the equivalent shaft is as close as possible to the actual stiff-

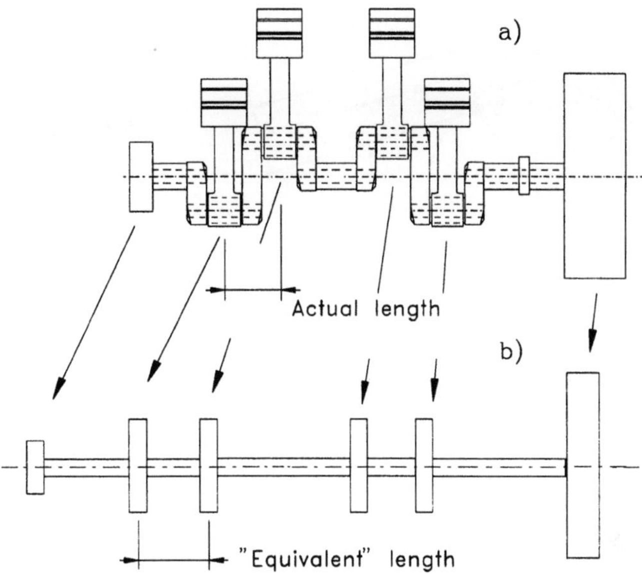

FIGURE 5.1. Sketch of the crankshaft: (a) actual system; (b) equivalent system, lumped-parameters model.

ness of the shaft. Note that the equivalent system is made by a straight shaft on which moments of inertia are located and then the uncoupling among torsional, axial, and flexural behaviours is possible and the dynamic study is straightforward. The construction of the equivalent system then reduces to the computation of the moment of inertia of the flywheels simulating the inertia of the cranks and of the equivalent lengths of the shafts.

Consider the crank mechanism sketched in Figure 5.2. It is made of a disc, with a crankpin in B on which the connecting rod PB, whose center of mass is G, is articulated. The reciprocating parts of the machine are articulated to the connecting rod in P. The actual position of the center of mass of the reciprocating elements, which can include the piston as well as the crosshead and other parts, is not important in the analysis; in the following study this point will be assumed to be located directly in P. Note that the axis of cylinder, i.e., the line of motion of point P, does not necessarily pass through the axis of the shaft; the offset d will, however, be assumed to be small. Let J_d, J_b, m_b, and m_p be the moments of inertia of the disc that constitutes the crank and of the connecting rod (about its center of gravity G) and the masses of the connecting rod and of the reciprocating parts, respectively.

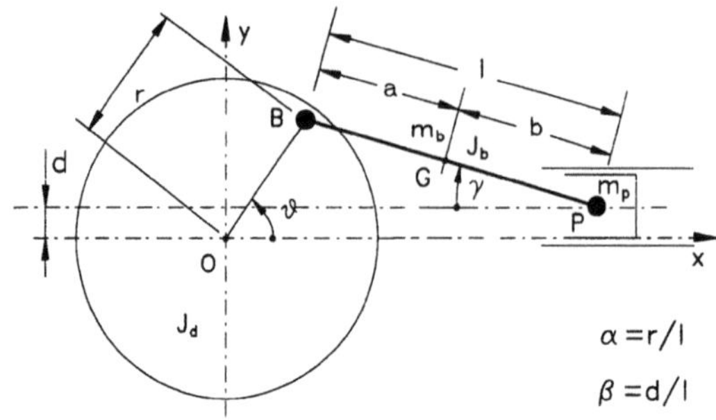

FIGURE 5.2. Sketch of the crank mechanism.

The coordinates of points B, G, and P can be expressed in the reference frame Oxy shown in Figure 5.2 as functions of the crank angle θ as

$$(\overline{B-O}) = \left\{ \begin{array}{c} r\cos(\theta) \\ r\sin(\theta) \end{array} \right\}, \qquad (\overline{G-O}) = \left\{ \begin{array}{c} r\cos(\theta) + a\cos(\gamma) \\ r\sin(\theta) - a\sin(\gamma) \end{array} \right\},$$

$$(5.1)$$

$$(\overline{P-O}) = \left\{ \begin{array}{c} r\cos(\theta) + l\cos(\gamma) \\ d \end{array} \right\}.$$

Angle γ is linked to angle θ by the equation

$$r\sin(\theta) = d + l\sin(\gamma), \qquad (5.2)$$

i.e.,

$$\sin(\gamma) = \alpha\sin(\theta) - \beta,$$

where $\alpha = r/l$ and $\beta = d/l$. Ratios α and β are expressed by numbers smaller than 1, and in practice they are quite small; usually $\alpha \leq 0.3$ and $\beta = 0$. In the case of an ideal crank mechanism, in which the connecting rod is infinitely long ($\alpha = 0$) and the axis of the cylinder passes through the axis of the crank ($\beta = 0$), the motion of the reciprocating masses is harmonic when the crank speed is constant.

Because $\dot{\theta}$ is the angular velocity of the crank, its kinetic energy is simply

$$\mathcal{T}_d = \frac{1}{2}J_d\dot{\theta}^2. \qquad (5.3)$$

The speed of the reciprocating masses can be easily obtained by differentiating the third equation (5.1) with respect to time and obtaining the expression for $\dot{\gamma}$ from equation (5.2):

$$V_p = -r\dot{\theta}\sin(\theta) - l\dot{\gamma}\sin(\gamma) = -r\dot{\theta}\left[\left(1 + \alpha\frac{\cos(\theta)}{\cos(\gamma)}\right)\sin(\theta) - \beta\frac{\cos(\theta)}{\cos(\gamma)}\right].$$
$$(5.4)$$

The kinetic energy of the reciprocating masses is

$$\mathcal{T}_p = \frac{1}{2}m_p r^2 \dot{\theta}^2 f_1(\theta),\tag{5.5}$$

where

$$f_1(\theta) = \left[\sin(\theta) + \alpha\frac{\sin(2\theta)}{2\cos(\gamma)} - \beta\frac{\cos(\theta)}{\cos(\gamma)}\right]^2.$$

Instead of computing the kinetic energy of the connecting rod by writing the velocity of its center of gravity G, it is customary to substitute the rod with a system made of two masses m_1 and m_2, located at the crankpin B and the wrist pin P, respectively, and a moment of inertia J_0. To correctly simulate the connecting rod, such a system must have the same total mass, moment of inertia, and position of the center of gravity. These three conditions allow for three equations yielding the following values for m_1, m_2, and J_0:

$$m_1 = m_b\frac{b}{l}, \qquad m_2 = m_b\frac{a}{l},$$
$$J_0 = J_b - (m_1 a^2 + m_2 b^2) = J_b - m_b ab.\tag{5.6}$$

Generally speaking, the moment of inertia of masses m_1 and m_2 is greater than the actual moment of inertia of the connecting rod and, consequently, the term J_0 is negative. The kinetic energy of mass m_1 can be computed simply by adding a moment of inertia $m_1 r^2$ to that of the crank. Similarly, the kinetic energy of mass m_2 can be accounted for by adding m_2 to the reciprocating masses. The effect of the moment of inertia J_0 can be easily computed

$$\mathcal{T}_{J_0} = \frac{1}{2}J_0\dot{\gamma}^2 = \frac{1}{2}J_0\dot{\theta}^2 f_2(\theta),\tag{5.7}$$

where

$$f_2(\theta) = \alpha^2\left[\frac{\cos(\theta)}{\cos(\gamma)}\right]^2.$$

The total kinetic energy of the system shown in Figure 5.2 is, consequently,

$$T = \frac{1}{2}\dot{\theta}^2 \left[J_d + m_1 r^2 + (m_2 + m_p)r^2 f_1(\theta) + J_0 f_2(\theta) \right] = \frac{1}{2} J_{eq}(\theta)\dot{\theta}^2 \, .$$

$$(5.8)$$

It is then clear that the whole system can be modeled, from the viewpoint of the kinetic energy, by a single moment of inertia variable with the crank angle $J_{eq}(\theta)$, rotating at the angular velocity $\dot{\theta}$.

In the study of the torsional vibrations, the motion of each section of the crankshaft can be expressed as the superimposition of a rigid-body rotation with angular velocity ω and a vibrational motion expressed by the torsional rotation $\phi_z(t)$. Angle θ can be expressed as $\theta(t) = \omega t + \phi_z(t)$, and the kinetic energy (equation (5.8)) is

$$T = \frac{1}{2} J_{eq}(\theta)(\omega + \dot{\phi}_z)^2 \, . \tag{5.9}$$

Equation (5.9) holds in general, even if the average velocity ω of the shaft varies in time; in all subsequent computations, however, only constant-rate rotation will be assumed and derivatives of ω with respect to time will be neglected. To study the vibrations in a shaft rotating at variable speed, a fixed law $\omega(t)$ can be assumed, and the equations that follow can be accordingly modified.

The equation of motion of the crank in terms of the generalized coordinate $\phi_z(t)$ can be obtained from the Lagrange equation. By performing all the relevant derivatives, it follows that

$$J_{eq}\ddot{\phi}_z + \frac{1}{2}(\omega + \dot{\phi})^2 \frac{\partial J_{eq}(\theta)}{\partial \theta} = M \, , \tag{5.10}$$

where M is the moment due to all external actions on the system (elastic and damping reaction of the shaft, forces applied on the piston, etc.).

Equation (5.10) is a second-order differential equation in ϕ_z. It is nonlinear even if the moment M is a linear function of the generalized coordinate ϕ_z (as when the shaft has a linear elastic behaviour), or of its derivative (as when linear damping is present), due to the fact that its coefficients, apart from being variable in time, are functions of the coordinate ϕ_z. The conventional approach to the study of reciprocating machines is based on a number of simplifications of the equation of motion (5.10) of the crank mechanism aimed at obtaining a linear differential equation with constant coefficients.

The first simplification is to substitute a constant moment of inertia $\overline{J}_{eq}$ in the term $J_{eq}\ddot{\phi}_z$ to the actual value that is a function of θ. A second simplification is to assume that angle ϕ_z is small enough to be neglected in the expression of θ in the second term of equation (5.10), which consequently reduces to

$$\overline{J}_{eq}\ddot{\phi}_z = M - \frac{1}{2}\omega^2\left[\frac{\partial J_{eq}(\omega t)}{\partial(\omega t)}\right] . \tag{5.11}$$

The last term of equation (5.11) is independent from angle ϕ_z and its derivatives. Consequently, it is a known function of angle ωt and hence of time and can be dealt with as an external excitation applied to the system. The aforementioned approach clarifies why and under which assumptions it is possible to substitute each crank with a constant moment of inertia on which suitable inertia forces (inertia torques), variable in time with known time history, act together with the actual external forces due to the working fluid and the other parts of the system. The study of the free oscillations of the system, performed by resorting to the homogeneous equation associated with equation (5.11), does not take into account such inertia forces and considers only the mean moment of inertia $\overline{J}_{eq}$. For the computation of both $\overline{J}_{eq}$ and the inertia forces, it is possible to resort to a series expansion of functions $f_1(\theta)$ and $f_2(\theta)$, which can be expressed as a trigonometric series of angle θ, whose coefficients are functions of nondimensional parameters α and β.

In the limiting case of $\alpha = \beta = 0$, corresponding to an infinitely long connecting rod (piston moving with harmonic time history), the expressions for $f_1(\theta)$ and $f_2(\theta)$ are particularly simple

$$f_1(\theta) = \sin^2(\theta) = \frac{1 - \cos(2\theta)}{2}, \qquad f_2(\theta) = 0 . \tag{5.12}$$

In practice, it is impossible to neglect the fact that the length of the connecting rod is finite, even if α is usually not greater than 0.3. By expressing $\cos(\gamma)$ as $\sqrt{1 - [\alpha\sin(\theta) - \beta]^2}$, computing the Taylor series in α and β for functions $f_1(\theta)$ and $f_2(\theta)$, and truncating it retaining all terms containing powers up to the fifth, the following trigonometric polynomials can be obtained:

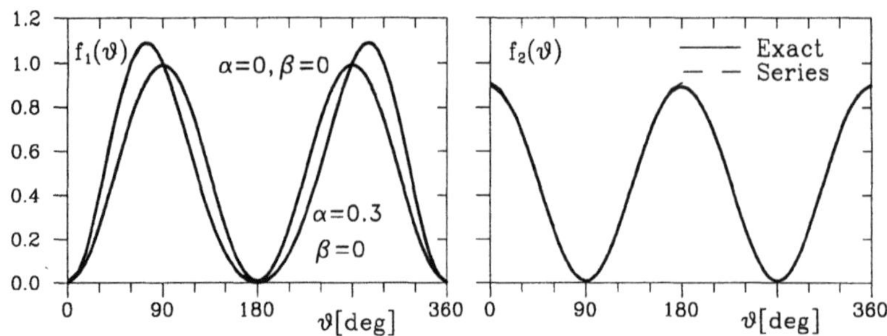

FIGURE 5.3. Functions $f_1(\theta)$ and $f_2(\theta)$.

$$f_1(\theta) = a_0 + \sum_{i=1}^{7} a_i \cos(i\theta) + \sum_{i=1}^{6} b_i \sin(i\theta)$$
$$f_2(\theta) = c_0 + \sum_{i=1}^{4} c_i \cos(i\theta) + \sum_{i=1}^{3} d_i \sin(i\theta)$$

$$(5.13)$$

The coefficients of the trigonometric series are

$$a_0 = \frac{8+2\alpha^2(1+6\beta^2)+8\beta^2(1+\beta^2)+\alpha^4}{16},$$

$$a_1 = \alpha\frac{128+16\alpha^2(2+15\beta^2)+48\beta^2(4+5\beta^2)+15\alpha^4}{256},$$

$$a_2 = -\frac{16+\alpha^4-16\beta^2(1+\beta^2)}{32},$$

$$a_3 = -\alpha\frac{128+24\alpha^2(2+15\beta^2)+48\beta^2(4+5\beta^2)+27\alpha^4}{256},$$

$$a_4 = -\alpha^2\frac{2+\alpha^2+12\beta^2}{16},$$

$$a_5 = \alpha^3\frac{16+15\alpha^2+120\beta^2}{256},$$

$$a_6 = \frac{\alpha^4}{32},$$

$$a_7 = -\frac{3\alpha^5}{256},$$

$$b_1 = -\beta\alpha\frac{1+\alpha^2+2\beta^2}{2},$$

$$b_2 = -\beta\frac{128+48\alpha^2(2+5\beta^2)+16\beta^2(4+3\beta^2)+75\alpha^4}{128},$$

$$b_3 = -\alpha\beta\frac{2+\alpha^2+4\beta^2}{2},$$

$$b_4 = \alpha^2\beta\frac{12+15\alpha^2+30\beta^2}{32},$$

$$b_5 = \beta\frac{\alpha^3}{4},$$

$$b_6 = -\beta\frac{15\alpha^4}{128},$$

$$c_0 = \alpha^2\frac{4+\alpha^2(1+6\beta^2)+4\beta^2}{8},$$

$$c_2 = \alpha^2\frac{1+\beta^2}{2},$$

$$c_4 = -\frac{\alpha^4}{8},$$

$$d_1 = d_3 = -\beta\frac{\alpha^3}{2},$$

$$c_1 = c_3 = d_2 = 0,$$

If the axis of the cylinder passes through the center of the crank ($\beta = 0$), as is usually the case, all coefficients b_i and c_i vanish, as was easily predictable, because for symmetry reasons $f_1(\theta)$ and $f_2(\theta)$ are even functions of θ. Functions $f_1(\theta)$ and $f_2(\theta)$ are reported in Figure 5.3 for two different values of α with $\beta = 0$. The results obtained using the series (5.13) with simplified expressions of the coefficients (considering only the first three nonvanishing harmonic terms for $f_1(\theta)$ and the first two for $f_2(\theta)$ and only the powers of α up to the third) are compared with the exact solutions. The lines are completely superimposed, and the approximation obtainable in this way is clearly very good. By retaining the first seven harmonic terms in $\cos(\theta)$ and the first six terms in $\sin(\theta)$, the equivalent moment of inertia, consequently, can be expressed as

$$J_{eq} = \overline{J}_{eq} + \sum_{i=1}^{7} J_{c_i} \cos(i\theta) + \sum_{i=1}^{6} J_{s_i} \sin(i\theta), \qquad (5.14)$$

where

$$\overline{J}_{eq} = J_d + m_1 r^2 + a_0(m_2 + m_p)r^2 + J_0 c_0,$$
$$J_{c_i} = a_i(m_2 + m_p)r^2 + J_0 c_i,$$
$$J_{s_i} = b_i(m_2 + m_p)r^2 + J_0 d_i.$$

Also in this case, all terms in sine vanish if the axis of the cylinder passes through the center of the crank $(\beta = 0)$.

If $\alpha = 0$, the expression for the mean equivalent moment of inertia reduces to

$$\overline{J}_{eq} = J_d + m_1 r^2 + r^2 \frac{m_2 + m_p}{2}. \qquad (5.15)$$

Some authors[1] suggest using for $\overline{J}_{eq}$ the reciprocal of the average value of function $1/\overline{J}_{eq}$ on a complete revolution of the shaft. This procedure, usually implemented by plotting function $1/\overline{J}_{eq}$ and then graphically computing the mean, is based on the observation that the square of the torsional natural frequency is proportional to $1/\overline{J}_{eq}$. The equivalent system so computed has a natural frequency equal to the mean square value of the natural frequencies of the actual system, computed in the various positions. By introducing the series computed earlier for functions $f_1(\theta)$ and $f_2(\theta)$ in the last term of the equation of motion (5.10), approximate expressions for the inertia forces are obtained. This part of the analysis of the system will be included in the study of the forced oscillations of the system, because the inertia forces are introduced into the computation as external excitations.

Instead of resorting to expressions of the type of equation (5.13), it is simpler to use the expressions of $f_1(\theta)$ and $f_2(\theta)$ in which $\cos(\gamma)$ has been substituted by its expression as a function of angle θ

---

[1]See, for example, C.B. Biezeno and R. Grammel, *Technische Dynamik*, Springer, Berlin, 1953.

$$f_1(\theta) = \left[\sin(\theta) + \frac{\alpha \sin(\theta)\cos(\theta) - \beta \cos(\theta)}{\sqrt{1 - [\alpha \sin(\theta) - \beta]^2}} \right]^2 ,$$

$$f_2(\theta) = \frac{\alpha^2 \cos^2(\theta)}{1 - [\alpha \sin(\theta) - \beta]^2}$$

(5.16)

and to numerically compute the values of $\overline{J}_{eq}$, J_{c_i}, and J_{s_i} through a fast Fourier transform of function $J_{eq}(\theta)$ computed using equation (5.14).

The function usually needs to be computed in a small number of points (32 or 64) but if many harmonics are required, more points can be needed. However, even if 1,024 or 2,048 points are needed, the computation is very fast on any computer.

The approximate procedure outlined here leads to results that can be very rough, particularly in those cases in which the inertia forces due to reciprocating masses are high. Note that one of the aims of designers of reciprocating machinery is reduction of the mass of reciprocating elements, particularly in the case of fast machines. Even with the mentioned limitations, the approach described here is the only one that allows the behaviour of a complex machine to be predicted. In Section 5.5, an approach aimed at performing a more detailed analysis for very simple geometries, such as a single-cylinder engine, will be described.

Example 5-1
Compute values of $\overline{J}_{eq}$, J_{c_i}, and J_{s_i} up to the seventh harmonic for a crank and piston unit with the following data: stroke: $2r = 85$ mm; mass of complete piston: $m_p = 0.540$ kg; mass of connecting rod: $m_b = 0.420$ kg; length of connecting rod: $l_b = 181$ mm; distance a (Figure 5.2): $a = 78$ mm; moment of inertia of connecting rod: $J_b = 0.0015$ kg m^2; moment of inertia of crank: $J_d = 0.00447$ kg m^2.
Compare the results obtained from equation (5.13) with those obtained through fast Fourier transform of function $J_{eq}(\theta)$ computed using equation (5.14).

The value of ratio α is $\alpha = r/l = 0.235$ and $\beta = 0$. From equation (5.6) it follows that $m_1 = 0.239$ kg, $m_2 = 0.181$ kg, and $J_0 = -0.0019$ kg m^2. Because $\beta = 0$, all coefficients b_i and d_i vanish; coefficients a_i and c_i are: $a_0 = 0.5071$, $a_1 = 0.1191$, $a_2 = -0.5001$, $a_3 = -0.1199$, $a_4 = -007082$, $a_5 = 0.0008509$, $a_6 = 0.00009499$, $a_7 = -0.000008364$, $c_0 = 0.0279$, $c_1 = 0.0001551$, $c_2 = -0.0007029$, $c_3 = 0.0001562$, $c_4 = -0.000008510$, $c_5 = 0.000001108$, $c_6 = 0.0000001237$, and $c_7 = -0.00000001089$. All J_{s_i} vanish, and the results for $\overline{J}_{eq}$ and J_{c_i} are reported in the following table:

	Eq. (5.13)	FFT 32 points	FFT 128 or more points
J_{eq}	5.50969×10^{-3}	5.50968×10^{-3}	5.50968×10^{-3}
J_{c_1}	1.55056×10^{-4}	1.55056×10^{-4}	1.55056×10^{-4}
J_{c_2}	-7.02940×10^{-4}	-7.02937×10^{-4}	-7.02937×10^{-4}
J_{c_3}	-1.56153×10^{-4}	-1.56157×10^{-4}	-1.56157×10^{-4}
J_{c_4}	-8.51037×10^{-6}	-8.49719×10^{-6}	-8.49719×10^{-6}
J_{c_5}	1.10817×10^{-6}	1.11079×10^{-6}	1.11079×10^{-6}
J_{c_6}	1.23709×10^{-7}	1.20490×10^{-7}	1.20466×10^{-7}
J_{c_7}	-1.08929×10^{-8}	-1.16848×10^{-8}	-1.18246×10^{-8}

Note that the results obtained from the FFT are to be considered more correct than those computed through equation (5.13) because the latter are obtained by truncating the series in the powers of α and β.

5.2.2 Equivalent length

To completely define the equivalent system, the stiffness of the portions of shaft that connect the flywheels must be computed. The equivalent system is modeled as a lumped-parameters system and, consequently, the various flywheels have a vanishing axial length: The portions of the actual shaft that must be modeled by the equivalent stiffnesses must be contiguous, each starting in the same position in which the previous one ends. It is not possible to compute the stiffness by modeling each part or the crank as a simple body (beams loaded in torsion for the journals, beams loaded in bending and torsion for the crankpins, beams loaded in bending for the crank webs, etc.). The geometric complexity, the presence of radii, and the very small slenderness of the beams make it impossible to resort to such modeling.

There are three ways to evaluate the equivalent stiffness: experimental evaluation, use of semiempirical methods, and numerical modeling, mainly using the FEM. The experimental evaluation is clearly the way that gives the most reliable results, but it obviously cannot be performed at the design stage without additional costs and increases in the time required for dynamic analysis, to build models or prototypes. Empirical and semiempirical formulas, which allow at least approximate evaluations to be obtained, have been suggested by many authors and can be found in several handbooks[2].

They are mainly based on simple mathematical models and modified on the base of experimental results. They yield the equivalent stiffness or the equivalent length of a shaft with a given cross section, which is the same. They allow good results to be obtained, provided they are used with care, without attempting to use them outside the field for which they are suggested. One of the mentioned formulas, usually referred to as Carter's formula, is

$$l_{eq} = (2c + 0{,}8b) + \frac{3}{4}a\frac{D^4 - d^4}{D'^4 - d'^4} + \frac{3}{2}r\frac{D^4 - d^4}{bs^3}, \qquad (5.17)$$

where the notation refers to Figure 5.4. It is particularly suited for fast internal-combustion engines. Another common formula is the so-called Tuplin's formula, particularly suited for large diesel engines (same notation as Figure 5.4)

$$l_{eq} = \frac{2c + 0{,}15D}{1 - \left(\frac{d}{D}\right)^4} + \frac{(a + 0{,}15D)(D^4 - d^4)}{\left[1 - \left(\frac{d'}{D'}\right)^4\right](D'^4 - d'^4)} + \left[2b - 0{,}15(D + D')\right] \times$$
$$\times \frac{D^4 - d^4}{s^4 - d^4} + r\left(0{,}065\frac{D}{b} + 0{,}58\right)\frac{D^4 - d^4}{bs^3} + 0{,}016\frac{D^4 - d^4}{b^2 s}.$$
$$(5.18)$$

It is possible to resort to a third approach based on the construction of numerical models of a single crank and to evaluate their static stiffness by numerical methods, mainly the FEM. This is much simpler than the complete numerical simulation of the crankshaft using the same numerical approach. Only one crank (or half, for symmetry) needs to be modeled, if all cranks are equal, and the computation reduces to a static evaluation.

Nevertheless, the geometric complexity and the uncertainties on how constraining the mathematical model, can make this computation more difficult than it appears.

---

[2]See, for example, E.J. Nestorides, *A Handbook on Torsional Vibration*, Cambridge Univ. Press, 1958; or W. Ker Wilson, *Torsional Vibration Problems*, Chapman & Hall, 1963.

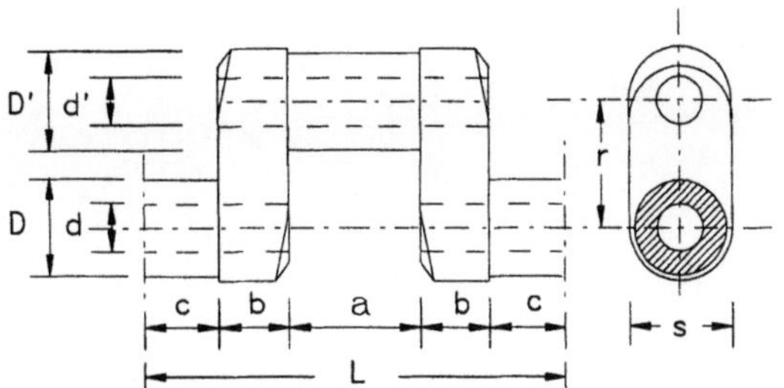

FIGURE 5.4. Sketch of the crankshaft for the computation of the equivalent length.

Strictly speaking, the lack of symmetry couples torsional and flexural deformations, and the stiffness of the basement of the engine and the presence of the oil films in the bearings can affect the results. The equivalent stiffness and the equivalent length, computed through any of the mentioned approaches, are linked to each other through the obvious formula

$$k = G\frac{I_p}{l_{eq}}\,.$$

(5.19)

5.2.3 Geared systems

Consider the system sketched in Figure 5.5a in which the two shafts are linked by a pair of gear wheels, with transmission ratio τ. For the study of the torsional vibrations of the system, it is possible to substitute the system with a suitable equivalent system, in which one of the two shafts is substituted by an expansion of the other (Figure 5.5b). This substitution can be performed only if no allowance is taken for backlash, which would introduce nonlinearities. Assuming also that the deformation of gear wheels is negligible, the equivalent rotations ϕ_i^* can be obtained from the actual rotations ϕ_i simply by dividing the latter by the transmission ratio $\tau = \omega_2/\omega_1$: $\phi_i^* = \phi_i/\tau$.

The kinetic energy of the ith flywheel, whose moment of inertia is J_i, and the elastic potential energy of the ith span of the shaft are, respectively,

$$\begin{aligned}
\mathcal{T} &= \tfrac{1}{2}J_i\dot{\phi}_i^2 = \tfrac{1}{2}J_i^*\dot{\phi}_i^{*2}\,, \\
\mathcal{U} &= \tfrac{1}{2}k_i\left(\phi_{i+1}^2 - \phi_i^2\right) = \tfrac{1}{2}k_i^*\left(\phi_{i+1}^{*2} - \phi_i^{*2}\right),
\end{aligned}$$

(5.20)

where the equivalent moment of inertia and stiffness are, respectively, $J_i^* = \tau^2 J_i$ and $k_i^* = \tau^2 k_i$.

If the system includes a planetary gear train, the computation

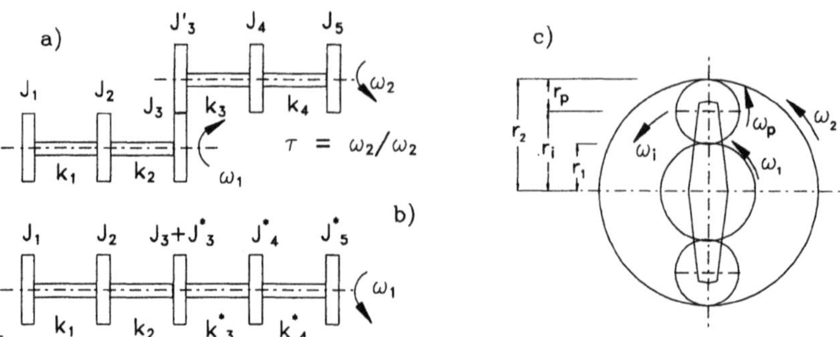

FIGURE 5.5. Geared system: Sketch of the (a) actual system and (b) equivalent system; (c) planetary gear train. Sketch of the system and notation.

can be performed without difficulties. The equivalent stiffness can be computed simply from the overall transmission ratio, and in the computation of the equivalent inertia the total kinetic energy of the rotating parts must be taken into account. The angular velocities of the central gear ω_1, of the ring gear ω_2, of the revolving carrier ω_i, and of the intermediate pinions ω_p of the planetary gear shown in Figure 5.5c are linked by the equation

$$\frac{\omega_1 - \omega_i}{\omega_2 - \omega_i} = -\frac{r_2}{r_1}, \qquad \omega_p = (\omega_1 - \omega_i)\frac{r_1}{r_p} - \omega_i. \qquad (5.21)$$

The equivalent moment of inertia of the system made of the internal gear, with moment of inertia J_1, the ring gear, with moment of inertia J_2, the revolving carrier, with moment of inertia J_i, and n intermediate pinions, each with mass m_p and moment of inertia J_p, referred to the shaft of the internal gear is

$$J_{eq} = J_1 + J_2 \left(\frac{\omega_2}{\omega_1}\right)^2 + (J_i + nm_p r_i^2)\left(\frac{\omega_i}{\omega_1}\right)^2 + nJ_p\left(\frac{\omega_p}{\omega_1}\right)^2. \qquad (5.22)$$

If the deformation of the meshing teeth must be accounted for, it is possible to introduce into the model two separate degrees of freedom for the two meshing gear wheels, modeled as two different inertias, and to introduce between them a shaft whose compliance simulates the compliance of the transmission. This is particularly important when a belt or flexible transmission of some kind is used instead of the stiffer gear wheels. In a machine there may be several shafts connected to each other, in series or in parallel, by gear wheels with different transmission ratios. The equivalent system is referred to one of the shafts and the equivalent inertias and stiffnesses of the elements of the others are all computed using the ratios between the speeds of the relevant element and the reference one. The equivalent system will then be made of a set of elements, in series or in parallel, following the scheme of the actual system, but with rotations that are all consistent. If the compliance of the gears is to be taken into account in detail, the nonlinearities due to the contacts between the meshing teeth and backlash must be considered. The study becomes far more complex and the methods seen in Chapter 3 for nonlinear systems must be used.

5.3   Computation of the natural frequencies

In many cases, the equivalent system reduces to a single shaft on which a number of moments of inertia are located. The system then has the aspect of an in-line system, and it is possible to use the transfer-matrices method. The Holzer method, described in detail in Section 2.6.3, has for decades been the most common tool for dealing with this problem. It is, however, possible, and it is more computationally efficient to resort to a stiffness approach. There is no difficulty writing the stiffness matrix of each span of the shaft, modeled as a beam element and assembling them into a global stiffness matrix. In the case of in-line systems, a tridiagonal stiffness matrix is obtained. The stiffness matrix is singular because there is no constraint to prevent the system from performing rigid rotations: It cannot be inverted, and the compliance matrix is not available. The mass matrix is even simpler: Because the inertia properties are lumped, the mass matrix is diagonal.

The equation of motion for the study of the free behaviour of the system of Figure 5.5b is, for example,

$$
\begin{bmatrix}
J_1 & 0 & 0 & 0 & 0 \\
0 & J_2 & 0 & 0 & 0 \\
0 & 0 & J_3 + J_3^* & 0 & 0 \\
0 & 0 & 0 & J_4^* & 0 \\
0 & 0 & 0 & 0 & J_5^*
\end{bmatrix}
\begin{Bmatrix}
\ddot{\phi}_1 \\
\ddot{\phi}_2 \\
\ddot{\phi}_3 \\
\ddot{\phi}_4^* \\
\ddot{\phi}_5^*
\end{Bmatrix}
+ \tag{5.23}
$$

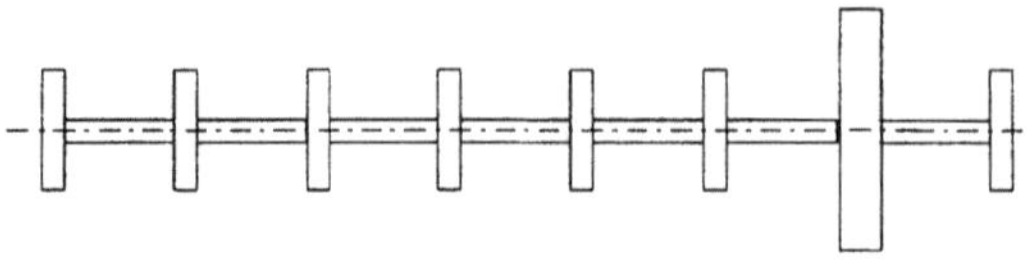

FIGURE 5.6. Equivalent system.

$$+ \begin{bmatrix} k_1 & -k_1 & 0 & 0 & 0 \\ -k_1 & k_1 + k_2 & -k_2 & 0 & 0 \\ 0 & -k_2 & k_2 + k_3^* & -k_3^* & 0 \\ 0 & 0 & -k_3^* & k_3^* + k_4^* & -k_4^* \\ 0 & 0 & 0 & -k_4^* & k_4^* \end{bmatrix} \begin{Bmatrix} \phi_1 \\ \phi_2 \\ \phi_3 \\ \phi_4^* \\ \phi_5^* \end{Bmatrix} = \begin{Bmatrix} 0 \\ 0 \\ 0 \\ 0 \\ 0 \end{Bmatrix}.$$

The system is underconstrained, the first natural frequency is equal to zero and the corresponding mode is a rigid-body rotation about the axis of the shaft. Usually only the nonvanishing natural frequencies are considered: They are $n - 1$ if the system has n degrees of freedom. If the system is damped the study can be performed without difficulty using the FEM, but the presence of damping can introduce some complexities when using transfer matrices.

In the case of in-line systems, it is possible to resort to the transfer-matrices approach, but when the geometry is more complicated, it is necessary to resort to the stiffness approach. However, in the case of branched systems, when the secondary branches stemming from the main system have little influence on the dynamic behaviour of the latter, it is still possible to use the Holzer method by approximating each branch with a single inertia located at the connection point. The simplifications that can be introduced in this way are, however, very small, if modern computing facilities are used, and they do not justify the approximations so introduced. The stiffness matrix of a branched or multiply connected system can easily be built using the assembly procedures in Section 4.7.2. It is no longer tridiagonal, even if it usually still has a band structure, and it is possible to resort to the usual algorithms to reorder the list of the generalized coordinates to reduce the bandwidth to a minimum. The mass matrix, on the contrary, is always a diagonal matrix if the model is based on a lumped-parameters approach. The damping matrix usually has the same type of structure of the stiffness matrix, except that a number of elements can give vanishingly small contributions to damping.

Example 5-2

Compute the torsional natural frequencies of a diesel six-cylinder, four-stroke-cycle engine driving an electric generator unit.

The main characteristics of the system are: bore: $\phi = 300$ mm; area of the pistons: $A = 706.8$ cm^2; stroke: $2r = 450$ mm; firing order: 1-5-3-6-2-4; speed: $\omega = 500$ rpm (52.33 rad/s); mass of complete piston: $m_p = 70.5$ kg; mass of connecting rod: $m_b = 100.6$ kg; length of connecting rod: $l_b = 950$ mm; distance a (Figure 5.2): $a = 270$ mm; moment of inertia of connecting rod: $J_b = 16.383$ kg m^2. Geometric data of crankshaft (Figure 5.4): $a = 136$ mm, $b = 92$ mm, $c = 73$ mm, $D = 240$ mm, $d = 80$ mm, $D' = 230$ mm, $d' = 80$ mm, $s = 400$ mm. Moment of inertia of crank web: $J_d = 14.880$ kg m^2; moment of inertia of flywheel: $J_6 = 98$ kg m^2; stiffness engine-flywheel shaft: $k_6 = 66 \times 10^6$ Nm/rad; moment of inertia of generator: $J_7 = 49$ kg m^2; stiffness flywheel-generator shaft: $k_7 = 15 \times 10^6$ Nm/rad.

Equivalent system: Moments of inertia. The value of ratio α is $\alpha = r/l = 0.2368$. From equation (5.6) it follows that $m_1 = 72.008$ kg, $m_2 = 28.591$ kg, $J_0 = -2.087$ kg m^2. The mean equivalent moment of inertia is then $\overline{J}_{eq} = 21$ kg m^2.

Equivalent system: Stiffness of cranks. From equation (5.18) (Tuplin's formula), it follows that $l_{eq} = 513.79$. The stiffness of the shaft that separates each pair of contiguous equivalent inertias is then $k = GJ_p/l_{eq} = 51 \times 10^6$ Nm/rad. The scheme of the equivalent system is reported in Figure 5.6.

Natural frequencies and mode shapes. The mass matrix of the system is the following diagonal matrix:

$$\mathbf{M} = \text{diag}\left\{\; 21 \quad 21 \quad 21 \quad 21 \quad 21 \quad 21 \quad 98 \quad 49 \;\right\}.$$

The stiffness matrix is a tridiagonal matrix:

$$\mathbf{K} = \begin{bmatrix}
51 & -51 & 0 & 0 & 0 & 0 & 0 & 0 \\
-51 & 102 & -51 & 0 & 0 & 0 & 0 & 0 \\
0 & -51 & 102 & -51 & 0 & 0 & 0 & 0 \\
0 & 0 & -51 & 102 & -51 & 0 & 0 & 0 \\
0 & 0 & 0 & -51 & 102 & -51 & 0 & 0 \\
0 & 0 & 0 & 0 & -51 & 117 & -66 & 0 \\
0 & 0 & 0 & 0 & 0 & -66 & 81 & -15 \\
0 & 0 & 0 & 0 & 0 & 0 & -15 & 15
\end{bmatrix} \times 10^6.$$

The natural frequencies and mode shapes can easily be obtained by solving the eigenproblem, $\det(\mathbf{K} - \lambda^2 \mathbf{M}) = 0$, which yields the following matrix of the eigenvalues:

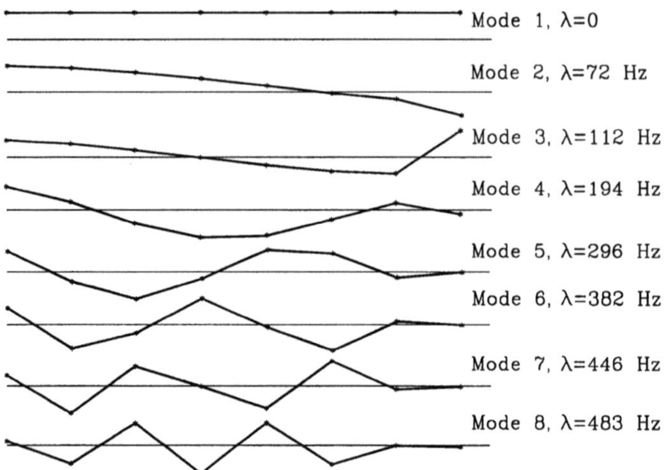

FIGURE 5.7. Torsional modes of the system in Figure 5.6.

$$[\lambda^2] = \mathrm{diag}\left\{\ 0\quad 0{,}210\quad 0{,}499\quad 1{,}486\quad 3{,}468\quad 5{,}789\quad 7{,}867\quad 9{,}245\ \right\}\times 10^6.$$

The natural frequencies of the system are then $\lambda_0 = 0$, $\lambda_1 = 458$ rad/s $=$ 72 Hz, $\lambda_2 = 704$ rad/s $= 112$ Hz, $\lambda_3 = 1{,}219$ rad/s $= 194$ Hz, $\lambda_4 = 1{,}862$ rad/s $= 296$ Hz, $\lambda_5 = 2{,}406$ rad/s $= 382$ Hz, $\lambda_6 = 2{,}804$ rad/s $= 446$ Hz, $\lambda_7 = 3{,}040$ rad/s $= 483$ Hz.
The mode shapes are reported in Figure 5.7.

5.4   Forced vibrations

5.4.1   Driving torque

The elements of reciprocating machines are loaded by forces variable in time, which in many instances can excite torsional vibrations of the shaft. In this section, only the forces originating the driving torque in reciprocating engines will be studied in detail but the same considerations can be referred to reciprocating compressors and similar machines. The fact that a term of the equation of motion coming from the derivatives of the kinetic energy, i.e., from the inertia of the system, can, under very widely used assumptions, be considered due to inertia forces applied to the system with known time history was explained in detail in Section 5.2.3. The corresponding inertia forces will then be included in the external forcing functions. On each flywheel of the equivalent system, a moment, varying in time with known time history, equal to the sum of a driving torque and an inertia torque, will be considered.

The pressure of the gases contained in the cylinder $p(t)$ varies in time during the working cycle of the engine. The virtual displacement δs of the piston corresponding to a virtual displacement $\delta\theta$ of the crank and the corresponding virtual work $\delta\mathcal{L}$ performed by the pressure can be expressed, respectively, as

$$\delta s = \frac{V_p}{\theta}\delta\theta, \qquad \delta\mathcal{L} = p(t)A\delta s = p(t)rA\sqrt{f_1(\theta)}\delta\theta, \qquad (5.24)$$

where function $f_1(\theta)$ is given by equation (5.5) and A is the area of the piston. The generalized force M_m due to the pressure $p(t)$ is, consequently,

$$M_m = \frac{d(\delta\mathcal{L})}{d(\delta\theta)} = p(t)rA\sqrt{f_1(\theta)}. \qquad (5.25)$$

It must be noted that $M_m(t)$ is a function of the generalized coordinate $\phi_z(t)$ through function $f_1(\theta)$. Usually this dependence is neglected, introducing into equation (5.25) a simplified expression of the crank angle $\theta = \omega t$. This assumption introduces negligible errors, at least in the case of vibrations with small amplitude, and is usually considered acceptable; at any rate, it is consistent with all the other simplifications seen earlier. In the case of gas compressors, steam engines, or two-stroke-cycle internal-combustion engines, function $p(t)$ is periodical with fundamental frequency equal to the rotational velocity ω of the shaft. In the case of four-stroke-cycle internal-combustion engines, the period of function $p(t)$ is doubled, i.e., its fundamental frequency is equal to $\omega/2$. Because the generalized force (moment) $M_m(t)$ is periodical, with the same frequency of law $p(t)$, it can be expressed by a trigonometric polynomial, truncated after m harmonic terms

$$M_m = A_0 + \sum_{k=1}^{m} A_k \cos(k\omega't) + \sum_{k=1}^{m} B_k \sin(k\omega't), \qquad (5.26)$$

where the frequency ω' of the fundamental harmonic is equal to ω, except in the case of four-stroke-cycle internal-combustion engines in which $\omega' = \omega/2$, and

$$\begin{cases} A_0 = \dfrac{1}{2\pi} \displaystyle\int_0^{2\pi} M_m \, d(\omega' t)\,, \\[2ex] A_k = \dfrac{1}{\pi} \displaystyle\int_0^{2\pi} M_m \cos(k\omega' t)\,d(\omega' t)\,, & k = 1, 2, \ldots, \\[2ex] B_k = \dfrac{1}{\pi} \displaystyle\int_0^{2\pi} M_m \sin(k\omega' t)\,d(\omega' t)\,, & k = 1, 2, \ldots\,. \end{cases} \qquad (5.27)$$

The coefficients of the polynomial can be computed from the theoretical or experimental law $p(t)$, and empirical expressions can be found in the literature. The driving torque depends at any rate on the working conditions of the machine; the coefficients A_i and B_i can be assumed to be proportional to the average driving torque A_0 or to the product of half the displacement (the area of the piston times crank radius) by the mean indicated pressure.

5.4.2 Inertia torque

The inertia torque, i.e., the moment acting on the flywheel simulating the crank as a result of the variability of the equivalent moment of inertia, can be obtained by introducing equation (5.8) into equation (5.11)

$$M_i = -\frac{1}{2}\omega^2 \left[r^2(m_2 + m_p)\frac{df_1(\theta)}{d\theta} + J_0\frac{df_2(\theta)}{d\theta} \right]\,, \qquad (5.28)$$

where functions $f_1(\theta)$ and $f_2(\theta)$ are expressed by equations (5.5) and (5.7). By introducing the series (5.13) and remembering that $\theta \approx \omega t$, it follows that

$$M_i = \omega^2 \left[\sum_{k=1}^{\infty} a_k^* \cos(k\omega t) + \sum_{k=1}^{\infty} b_k^* \sin(k\omega t) \right]\,. \qquad (5.29)$$

Coefficients a_k^* and b_k^* can be computed from those of series (5.13)

$$\begin{cases} a_k^* = \dfrac{k}{2}\left[r^2(m_2 + m_p)a_k + J_0 c_k \right]\,, \\[2ex] b_k^* = -\dfrac{k}{2}\left[r^2(m_2 + m_p)b_k + J_0 d_k \right]\,. \end{cases} \qquad (5.30)$$

Note that the fundamental harmonic of the inertia torque M_i is always ω and that, if the axis of the cylinder passes through the center of the crank ($\beta = 0$), there are only harmonics in sine (all $b_k^* =$

0). The harmonics in the series for M_i are the same that are in the series for the driving torque M_m; there is no difficulty obtaining the series for the total forcing torque acting on the equivalent flywheels simulating the cranks by adding the relevant terms.

5.4.3 Torsional critical speeds

The frequency of each harmonic of the forcing function acting on the cranks is proportional to the rotational speed of the machine. In the case of two-stroke-cycle internal-combustion engines or other reciprocating machines in which the duration of the working cycle corresponds to a single revolution of the crankshaft, the frequency of the various harmonics is equal to whole multiples of the rotational speed $\lambda_i = i\omega$. In the case of four-stroke-cycle internal-combustion engines the frequency of the fundamental harmonic is equal to half of the rotational speed ω and the frequency of the ith harmonic is $\lambda_i = i\omega/2$.

The torsional natural frequencies of the equivalent system are independent of the rotational speed. The resonance conditions, defining the torsional critical speeds, can then be studied using a Campbell diagram of the type shown in Figure 5.8. There are many resonance conditions, and it is very difficult, usually impossible, to avoid some of them being located within the working range of the machine. Not all resonance conditions are equally dangerous, and the dynamic stressing of the shaft at the various critical speeds must be evaluated to understand their severity.

Example 5-3

Compute the torsional critical speeds of the diesel engine-electric generator unit studied in Example 5-2.

The Campbell diagram of the system is plotted in Figure 5.8. In the speed range from 0 to 600 rpm (overspeed of 20%) there are six critical speeds linked with the first 20 harmonics of the forcing function. The ones nearest to the maximum speed are those related to the 14th and 15th harmonic ($\lambda = 7\omega$ and $\lambda = 15\omega\ /2$). They are: $\omega_{cr_{14}} = 624.8$ rpm and $\omega_{cr_15} = 583.1$ rpm.

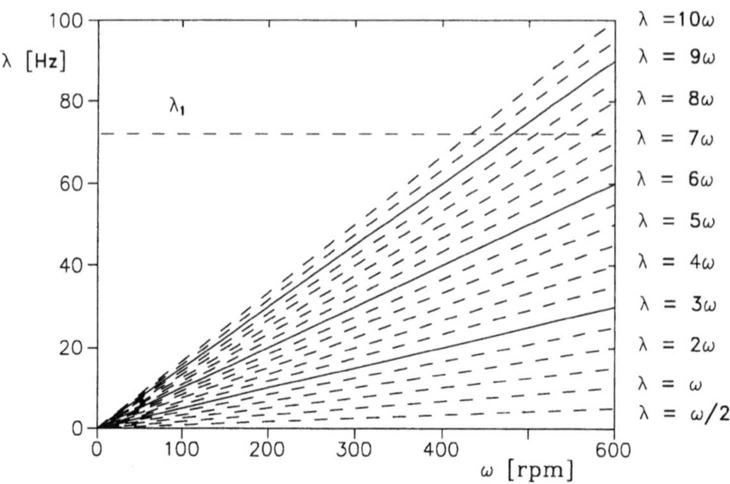

FIGURE 5.8. Campbell diagram for a six-cylinder four-stroke in-line internal-combustion engine for the computation of the resonance conditions: the first 20 harmonics of the forcing function and the first torsional natural frequency. The continuous lines are related to the major harmonics, those in which the excitation of all cylinders are in phase ($\lambda = 3\omega$, $\lambda = 6\omega$, $\lambda = 9\omega$, ...).

5.4.4 Forcing functions on the cranks of multicylinder machines

The total torque acting on the ith crank, obtainable as a sum of moments M_m and M_i, can be expressed as a trigonometric polynomial, whose terms are characterized by different amplitudes and phases. In the case of machines with a number of cranks, if the various cranks, reciprocating parts, and working cycles are all equal, the time histories of the moments acting on the various nodes of the equivalent system are all equal but are timed in a different way. Because each harmonic component of the moment acting on the cranks can be represented as the projection on the real axis of a vector rotating in the Argand plane with constant angular velocity, it is possible to draw, for each harmonic, a plot in which the various vectors acting on the different cranks of the machine are represented. If the amplitudes of these vectors are all equal, as in the case when all cranks are equal, the diagram is useful only to compare the phases of the vectors, and they are traditionally plotted with unit amplitude. The phasing of the vectors depends on the geometric characteristics of the machine and, in the case of four-stroke-cycle engines, on the firing order. These diagrams are usually referred to as *phase angle diagrams*.

Consider, for example, an in-line four-stroke-cycle internal-combustion engine. If the working cycles of the various cylinders are evenly spaced in time, the cranks that fire subsequently must make an angle of $4\pi/n$ rad, where n is the number of cylinders. In the case of a four-in-line engine, this angle is $180°$, and the most common geometric configuration of the crankshaft is that shown in Figure 5.9a, chosen because it allows the best balancing of inertia forces. In the same figure, the configuration of the crankshaft of a six-in-line engine is also shown.

In the case of the four-cylinder engine, the possible firing orders are two: 1-2-3-4 and 1-3-4-2. In both cases, it is impossible to prevent two contiguous cylinders from firing immediately one after the other. The phase-angle diagrams for the first four harmonics are plotted in Figure 5.9b, for the second of the two firing orders.

If the order of the harmonic is a whole multiple of the number of cylinders, all rotating vectors are superimposed, i.e., the forcing functions acting on all cranks are all in phase. These harmonics are usually the most dangerous and are often referred to as major harmonics. The phase-angle diagrams for the $(n + i)$-th harmonic coincides with that related to the ith harmonic and, consequently, only the first n phase-angle diagrams are usually plotted.

Note that the phase-angle diagrams have been plotted in such a way that they directly supply the phasing of the excitation on the various cranks with respect to that acting on a crank chosen as reference, usually the first one. Each harmonic then has a phasing with respect to the fundamental harmonic, and this must be taken into account when the effects of the various harmonics are added. The forcing function acting on the jth crank can be expressed by the following series, which has been truncated at the mth harmonic

$$M_{tot_j} = M_{m_j} + M_{i_j} = \qquad\qquad (5.31)$$

$$= \sum_{k=0}^{m} M_{m_k} e^{i\left(k\omega'i + \Phi_{m_k} + \delta_{j_k}\right)} + \sum_{i=0}^{m} M_{i_k} e^{i\left(k\omega'i + \Phi_{i_k} + \delta_{j_k}\right)},$$

where

- M_{m_k} and Φ_{m_k} are the amplitude and phase of the kth harmonic of the driving torque, respectively. With reference to the series (5.26) approximating the driving torque, their values are $M_{m_k} = \sqrt{A_k^2 + B_k^2}$ and $\Phi_{m_k} = \arctan(A_k/B_k)$, respectively.

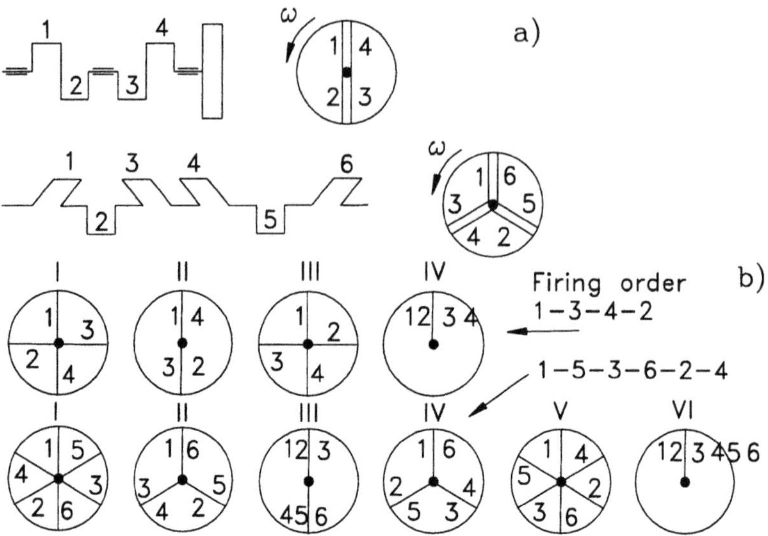

FIGURE 5.9. (a) Configuration of the crankshaft and crank angle diagrams for in-line four-stroke-cycle four- and six-cylinder internal-combustion engines. In the latter case the configuration shown is just one of the possible choices; (b) phase-angle diagrams for the same engines.

- δ_{j_k} is the phase of the kth harmonic acting on the jth crank, as obtained from the phase-angle diagram. If the diagram is referred to the first crank, $\delta_{j_k} = 0$ for $j = 1$.

- M_{i_k} e Φ_{i_k} are the amplitude and phase, respectively, of the kth harmonic (referred to the fundamental harmonic ω' and not to the rotational speed ω) of the inertia torque. If $\omega' = \omega$, the order of the harmonic to be introduced into equation (5.31) coincides with that in equation (5.29); if, on the contrary, $\omega' = \omega/2$, the amplitude of the odd harmonics vanishes while even harmonics correspond with those whose order is halved in equation (5.29). If the axis of the cylinder passes through the center of the shaft, the phases ϕ_{i_k} are all equal to $\pi/2$, as all terms in $\cos(k\omega t)$ vanish.

Phases δ_{j_k} of the harmonics of the driving torque M_{m_k} coincide with those of the harmonics of the inertia torque M_{i_k}.

5.4.5 Evaluation of the damping of the system

Although the response of the system in conditions far from resonance can be computed from an undamped model, the response at resonance can be obtained only after the damping of the system has been evaluated, and the precision of the results is strictly dependent on the precision with which damping is known. Because the resonant conditions are usually the most dangerous, this part of the dynamic analysis is very important, and the difficulty in achieving a good estimate of the damping is one of the factors limiting the usefulness of the dynamic analysis in the current case and compel one to resort to extensive experimentation. If damping were mostly due to the internal damping of the material constituting the crankshaft, there would be no difficulty introducing a proportional damping with modal damping ratio equal for all modes: $\zeta_j = \eta/2$, where η is the loss factor of the material of the crankshaft.

Actually, damping is due to many causes, among which friction between moving parts, including that between the piston and the cylinder wall, electromagnetic forces, and the presence of fluid in which some rotating parts move, can be important. Neglecting them would lead to a large underestimate of damping, and it is usually necessary to resort to experimental results, obtained from machines similar to the one under study and to empirical or semiempirical formulas and numerical values reported in the literature.

Usually the damping due to the crank mechanism is evaluated introducing a damping force acting on the crankpin proportional to the area of the piston and the velocity of the crankpin. The damping moment acting on the jth crank is

$$M_d(t) = k'Ar^2 \, , \tag{5.32}$$

where k' is a coefficient whose dimension is a force multiplied by time and divided by the third power of a length. In S.I. units, it is expressed in Ns/m^3. Values of k' included in the range between 3,500 and 10,000 Ns/m^3 for in-line aircraft engines and between 15,000 and 1.5×10^6 Ns/m^3 for large internal-combustion engines can be found in the literature. The lower and upper values of these ranges are very different, and they must be regarded only as indicative values; only experimental results on machines similar to the one under study can be reliable. The use of equation (5.32) leads to the assumption that in each crank there is a viscous damper with damping coefficient equal to $c_j = k'Ar^2$.

The damping of the other elements connected to the shaft (propellers, brakes, electrical machines) can be evaluated by assuming that they provide a braking torque on the relevant node that is proportional to the instantaneous angular velocity at the power p through coefficient k''

$$M_{d_j}(t) = k'' \left[\omega + \sum_{k=1}^{m} \dot{\phi}_{z_{j_k}}(t) \right]^p , \qquad (5.33)$$

where subscript j refers to the jth node of the model and k denotes the kth harmonic. Note that all harmonics are assumed to act in phase, which is clearly an approximation but does not affect the results obtained later. This type of damping is clearly nonlinear, but can be linearized by introducing in the computation a viscous damping for each harmonic, which is equivalent from the viewpoint of energy dissipation. By introducing the harmonic time history of each component of the velocity $\dot{\phi}_{z_{j_k}}(t)$, the energy dissipated in a cycle is

$$\mathcal{L} = \int_0^T k'' \left[\omega + \sum_{k=1}^{m} k\omega' \phi_{z_{j_{k_0}}} \cos(k\omega' t) \right]^{p+1} dt . \qquad (5.34)$$

By computing the $(p+1)$-th power of the binomial within the integral sign, it follows that

$$\mathcal{L} = \int_0^T k'' \omega^{p+1} dt + \binom{p+1}{1} \int_0^T k'' \omega^p \omega' \sum_{k=1}^{m} k\phi_{z_{j_{k_0}}} \cos(k\omega' t) dt +$$

$$+ \binom{p+1}{2} \int_0^T k'' \omega^{p-1} \omega'^2 \left[\sum_{k=1}^{m} k\phi_{z_{j_{k_0}}} \cos(k\omega' t) \right]^2 dt + \ldots$$

$$(5.35)$$

The first integral of equation (5.35) is constant and corresponds to the average braking power $P = k'' \omega^{p+1}$ applied in the relevant node. It can be used to compute coefficient k'', as the average power is generally known.

The second integral vanishes, while the third contains terms in the squares of cosine function and terms in which there are products of cosines with different arguments; the latter vanish once they are integrated over a whole period. The fourth term also vanishes. If the series (5.35) is truncated after the fourth term, the braking torque, excluding the contribution due to the constant term, is

$$\mathcal{L} = \left(\begin{array}{c} p+1 \\ 2 \end{array}\right) \pi k'' \omega^{p-1} \omega' \sum_{k=1}^{m} k\phi_{z_{j_{k_0}}}^2 \ , \qquad (5.36)$$

corresponding to that due to a viscous damper with equivalent damping coefficient

$$c_{eq} = \left(\begin{array}{c} p+1 \\ 2 \end{array}\right) k'' \omega^{p-1} \ . \qquad (5.37)$$

If $p = 1$, there is a true viscous damper and c_{eq} coincides with k''. If $p = 2$, as is often assumed in the case of propellers, k'' is expressed in S.I. units in as Nms^2 and $c_{eq} = 3k''\omega$. If $p = 3$, k'' is expressed in S.I. units in as Nms^3 and $c_{eq} = 6k''\omega^2$.

The damping matrix of the system can be obtained by applying in the various nodes the viscous dampers whose damping coefficients have been computed above and is diagonal. This way of evaluating the damping of the system is clearly an approximation that can, in many cases, be very rough. However, it is often used in practice because there are no simple alternatives.

5.4.6 Forced response of the system

Once the forcing functions have been obtained, there is no difficulty, at least from a theoretical viewpoint, to compute the response of the system at various speeds. Usually the problem of computing the forced response of the system is subdivided into two parts: the evaluation of the response to the harmonics of the forcing function that are not in resonance and that related to the resonant harmonics. The first problem can be solved by resorting to the undamped system; for the second it is necessary to resort to the damped model.

A vector of the driving functions, whose terms are expressed by the series (5.31), can be added on the right-hand side of the equation of motion (5.23), to which the damping matrix obtained in Section 5.4.5 has also been added

$$\mathbf{J}\{\ddot{\phi}_z\} + \mathbf{C}\{\dot{\phi}_z\} + \mathbf{K}\{\phi_z\} = \sum_{(k=1)}^{m} \mathbf{m}_{s_k} \sin(k\omega't) + \sum_{(k=1)}^{m} \mathbf{m}_{c_k} \cos(k\omega't),$$

$$(5.38)$$

where, to avoid confusion with the moments M, the symbol $\mathbf{J}$ (not to be confused with the identity matrix $\mathbf{I}$) has been used for the mass matrix because all its elements are mass moments of inertia; **m** indicates the vectors in which the moments are listed. The coefficients of the terms in cosine and sine, respectively, expressed by equation (5.31) have been collected in vectors:

$$\mathbf{m}_{c_k} = \{m_{m_k}\cos(\varPhi_{m_k} + \delta_k)\} + \omega^2\{m_{i_k}\cos(\varPhi_{i_k} + \delta_k)\},$$
$$\mathbf{m}_{s_k} = \{m_{m_k}\sin(\varPhi_{m_k} + \delta_k)\} + \omega^2\{m_{i_k}\sin(\varPhi_{i_k} + \delta_k)\}. \tag{5.39}$$

They are clearly functions of ω^2, due to the presence of inertia torques. The series have been truncated at the mth harmonic, and the static component of the torque has been neglected. In the case of major harmonics of a machine in which all elements are equal and distance d is equal to zero, all δ_{j_k} vanish and $\varPhi_{i_k}$ are all equal to $\pi/2$.

The solution of equation (5.38) can be obtained directly by adding the responses to the various harmonic components:

$$\boldsymbol{\phi} = \sum_{k=1}^{m} \boldsymbol{\phi}_{z_k} = \sum_{k=1}^{m} \boldsymbol{\phi}_{z_{sk}} \sin(k\omega't) + \sum_{k=1}^{m} \boldsymbol{\phi}_{z_{ck}} \cos(k\omega't). \tag{5.40}$$

The amplitudes of the components in sine and cosine of the response can be obtained by solving the following linear sets of equations:

$$\begin{bmatrix} \mathbf{K} - k^2\omega'^2\mathbf{J} & -k\omega'\mathbf{C} \\ k\omega'\mathbf{C} & \mathbf{K} - k^2\omega'^2\mathbf{J} \end{bmatrix} \begin{Bmatrix} \boldsymbol{\phi}_{z_{sk}} \\ \boldsymbol{\phi}_{z_{ck}} \end{Bmatrix} = \begin{Bmatrix} \mathbf{m}_{s_k} \\ \mathbf{m}_{c_k} \end{Bmatrix}. \tag{5.41}$$

If damping is neglected, as is customary when the response in conditions that are far from resonance is to be obtained, equation (5.41) uncouples into two separate linear sets of equations. Equation (5.41) must be solved at each rotational speed and for each harmonic of the forcing function; however, the computation cannot be exceedingly heavy, because the order of the matrices, equal to the number of the degrees of freedom of the equivalent system, is usually small.

The amplitude of the oscillations at each node of the system, due to each harmonic of the forcing function, can be easily computed. Once the amplitudes are known, there is no difficulty obtaining the dynamic stressing of each span of the shaft. The maximum value of the shear stress in the shaft spanning from the jth to the $(j+1)$-th node is

$$\tau_{max_j}(t) = \frac{k_j(\phi_{z_{j+1}} - \phi_{z_j})}{W_j}, \qquad (5.42)$$

where W_j is the torsional section modulus of the relevant shaft element. The torsional displacements at the nodes are not in phase, and then it is impossible to obtain the twist angle of each span as difference of amplitude between the end sections. The components in phase and in quadrature must be accounted for separately, and the amplitude of the twist angle, needed to compute the stress, must be obtained from the two components. The shear stress so computed is variable in time with polyharmonic time history; the time histories of the stresses due to the various harmonics, each with its amplitude, phase, and frequency, should then be added to each other and to the static stressing, and the fact that their consequences on the overall fatigue of the shaft are different should be considered. In practice, a much simpler approach is followed, not because of the long computations involved, which with a modern computer could be dealt with without problem, but because the phasing of the harmonics can be difficult to evaluate (near the resonance the phase is quickly variable) and predicting the fatigue life of a machine element subject to polyharmonic stressing is still difficult. The amplitudes of the stress cycles due to the various harmonics are computed and added together, often limiting the sum to the few most important harmonics. This procedure leads to overestimating the amplitude of the variable component of the stress and then is conservative.

The shear stresses so computed must be added to those due to other causes and, using a suitable failure criterion, to stresses due to bending, axial forces, shrink fitting, surface forces, etc. The stress concentrations due to the complex geometric shape of crankshafts can be quite high, and the analyst cannot neglect them.

5.4.7 Modal computation of the response

The computation of the response of the system can also be performed following a modal approach, which is essentially equivalent to uncoupling the equations of motion by neglecting the elements of the modal damping matrix outside the main diagonal. The equation yielding the jth modal amplitude is then

$$\ddot{\eta}_j + 2\zeta_j\lambda_j\dot{\eta}_j + \lambda_j^2\eta_j = \sum_{(k=1)}^{m} \left[\mathbf{q}_j^T \mathbf{m}_{s_k} \sin(k\omega't) + \mathbf{q}_j^T \mathbf{m}_{c_k} \cos(k\omega't) \right] ,$$

(5.43)

where $\mathbf{q}_j$ is the jth eigenvector of the undamped system, normalized in such a way that the corresponding modal mass has a unit value, and the modal damping ratio ζ_j is

$$\zeta_j = \frac{\mathbf{q}_j^T \mathbf{C} \mathbf{q}_j}{2\lambda_j} .$$

(5.44)

The number of modal equations (5.43) is equal to the number of modes with nonvanishing eigenfrequency; however, only a few modes need to be considered, and in most cases only the response corresponding to the first mode is computed.

Usually further approximations are introduced and the computation of the response follows the approach seen in Figure 1.9. On the base of the considerations seen in Section 1.7, a harmonic of the forcing function with frequency $k\omega'$ is resonant with the sth mode if $\lambda_s\sqrt{1 - 2\zeta_s} \leq k\omega' \leq \lambda_s\sqrt{1 + 2\zeta_s}$. Outside the mentioned frequency range, the undamped system can be used to computate the response of the relevant (jth) mode:

$$\eta_j = \sum_{k=1}^{m} \left[\frac{\mathbf{q}_j^T \mathbf{m}_{s_k} \sin(k\omega't) + \mathbf{q}_j^T \mathbf{m}_{c_k} \cos(k\omega't)}{\lambda_j^2 + k^2\omega'^2} \right] .$$

(5.45)

Once the modal responses have been recombined, the amplitudes $\boldsymbol{\phi} = \boldsymbol{\Phi}\boldsymbol{\eta}$ and the shear stresses are obtained.

If one of the harmonics of the forcing function is near one of the natural frequency, the relevant amplitude must be obtained from the damped model. The deformed shape in resonance is assumed to be equal to the corresponding mode shape, which amounts to the assumption that the modal responses are uncoupled even if the system is damped, i.e., to neglect the out-of-diagonal elements of the modal damping matrix, and to neglect the contribution of the other modes. Note that the displacements are then in phase with each other because the eigenvector is a real eigenvector of an undamped system. Note that the oscillations in the various points of the shaft are in phase with each other, but they are neither in phase nor in quadrature with the forcing functions, which are not in phase with each other. Only in the case of the major harmonics are the forcing functions in phase with each other and, in resonance, in quadrature with the response.

Consider the case in which the kth harmonic is in resonance with the sth mode, i.e., $k\omega' = \lambda_s$. Because the deformed shape is assumed to be proportional to the sth mode shape, it follows that

$$\boldsymbol{\phi}(t) = \alpha \mathbf{q}_s \sin(k\omega't), \tag{5.46}$$

where α is a proportionality coefficient that depends on the criterion used for normalizing the eigenvectors and the phasing of the response has been assumed to be equal to 0, i.e., the shaft is assumed to be in the undeformed configuration at time $t = 0$. If the largest value of the eigenvector is assumed to be equal to unity, α is the maximum amplitude of the deflected shape: In modal terms, α is nothing other than the amplitude η_{s_0} of the sth modal coordinate. It can be computed through equation (1.75), written using the modal quantities

$$\alpha = \frac{\overline{F}_s}{\overline{C}_s \lambda_s} = \frac{\overline{F}_s}{\overline{C}_s k\omega'}, \tag{5.47}$$

where $\overline{C}_s = \mathbf{q}_s^T \mathbf{C} \mathbf{q}_s = \sum_{j=1}^n c_{eq_j} q_{s_j}^2$. The computation of the modal force is more complex. If M_{o_j} is the amplitude of the kth harmonic of the forcing function applied to the jth node, and δ_j is the phase lag between the moment applied to the jth node and that applied to the first node obtainable from the phase-angle diagram ($\delta_1 = 0$), it follows that[3]

$$M_j(t) = M_{0_j} \sin(k\omega't + \delta_j) = \tag{5.48}$$

$$= \left[M_{0_j} \cos(\delta_j)\right] \sin(k\omega't) + \left[M_{0_j} \sin(\delta_j)\right] \cos(k\omega't).$$

The in-phase and in-quadrature (with reference to the first node) components of the modal force must be computed separately. They are

---

[3]In the remainder of this section modal forces are written in phase with the forcing function on the first node, while rotations are written in phase with the rotation of the same node. Because only the amplitude is required, the choice of the instant in which $t = 0$ is immaterial.

$$\overline{F}_{s_{phase}} = \sum_{j=1}^{n} q_{s_j} M_{0_j} \cos(\delta_j), \qquad \overline{F}_{s_{quad}} = \sum_{j=1}^{n} q_{s_j} M_{0_j} \sin(\delta_j).$$

$$(5.49)$$

The amplitude of the modal response and the phase delay between the response and the kth harmonic of the forcing function applied at the first node are, respectively,

$$\alpha = \frac{\sqrt{\left[\sum_{j=1}^{n} q_{s_j} M_{0_j} \cos(\delta_j)\right]^2 + \left[\sum_{j=1}^{n} q_{s_j} M_{0_j} \sin(\delta_j)\right]^2}}{k\omega' \sum_{j=1}^{n} c_{eq_j} q_{s_j}^2}, \qquad (5.50)$$

$$\beta = \arctan\left[\frac{\overline{F}_{s_{fase}}}{\overline{F}_{s_{quad}}}\right] = \arctan\left[\frac{\sum_{j=1}^{n} q_{s_j} M_{0_j} \cos(\delta_j)}{\sum_{j=1}^{n} q_{s_j} M_{0_j} \sin(\delta_j)}\right].$$

In the literature, the same result is often obtained using energetic reasoning: The energy dissipated by damping is equated to the work of the forcing functions. The amplitude is often obtained using the geometric construction shown in Figure 5.10, where the modal force $\overline{F}_s$ for the resonance of the fifth harmonic of the forcing function with the first natural frequency of a four-stroke-cycle in-line six-cylinder engine is computed graphically.

Because the rotations at the various nodes are assumed to be in phase at resonance, the computation of the stresses is particularly simple; the amplitude of the twist angle of each span is just the difference between the amplitudes of the rotations at the ends. Note that in many cases the approximations linked with the whole modeling of the system, in particular where the equivalent inertia and damping are concerned, justify the use of the approximations seen earlier, particularly those linked with modal uncoupling, and make it useless to compute the response with greater precision. However, there are cases where the contribution of the nonresonating harmonics to the total stressing is not small, even for harmonics that are very far from resonance.

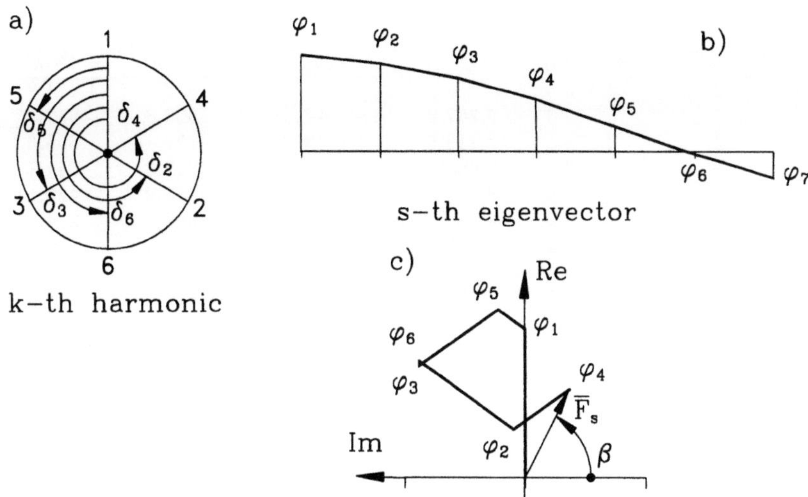

FIGURE 5.10. Graphical computation of the modal force for the resonance between the fifth harmonic and the first natural frequency ($k = 5$, $s = 1$) in an in-line four-stroke-cycle six-cylinder internal-combustion engine: (a) phase-angle diagrams for the fifth harmonic; (b) mode shape of the first natural frequency; (c) computation of the modal force.

The modal approximation becomes very rough in this instance because the overall shape of the response can be very far from the first mode shape, and the response is far better computed using a nonmodal approach (see Example 5-4).

Example 5-4

Consider the diesel engine–generator unit of Example 5-2. Compute the stresses due to torsional vibrations in the most critical part of the crankshaft using both modal and nonmodal approaches.

When damping is required, assume a coefficient $k' = 2 \times 10^5$ Ns/m^3 for the cranks and use equation (5.37) with $p = 3$ for the generator (compute coefficient k'' from the steady-state power).

Assume that the diameters of shafts number 6 and 7 are 200 and 130 mm, respectively. The harmonic components of the driving torque at full power are:

Harmonic	A_k [Nm]	B_k [Nm]	Harmonic	A_k [Nm]	B_k [Nm]
0	3,448.4	–	11	−839.7	1,114.9
1	5,763.3	5,267.8	12	−808.8	853.3
2	3,176.8	8,119.6	13	−727.1	655.6
3	779.8	8,050.4	14	−662.8	541.7
4	−446.4	6,435.8	15	−635.9	420.1
5	−743.0	4,853.6	16	−592.2	274.1
6	−820.8	3,795.8	17	−518.9	170.2
7	−963.5	2,962.4	18	−453.4	111.1
8	−1,024.0	2,193.0	19	−396.9	46.6
9	−939.4	1,645.6	20	−324.0	−13.8
10	−855.0	1,350.3			

The model of the system and the mass and stiffness matrices are the same as in Example 5-2. The damping matrix is a diagonal matrix; the terms corresponding to the cranks are $c_{ii} = k'Ar^2 = 2 \times 10^5 \times 706.8 \times 10^{-4} \times 0.225^2$ Nms/rad.

For the damping in node 8 (generator), coefficient k'' can be easily computed from the steady-state component of the torque. At 500 rpm = 52.4 rad/s it leads to a power of 180.6 kW per cylinder, i.e., to a total power of 1,084 kW; as $p = 3$, the values of k'' and c_{88} are $k'' = P/\omega^{p+1} = 0.144$ Nms^3 and $c_{88} = 6k''\omega^2 = 2.364$ Nms/rad.

However, this value of the damping is linked with the load on the generator. Because in some cases the most dangerous conditions can be those in which the engine works with no load (only inertia torques act on the cranks), it is conservative to neglect the damping on the generator, which leads in most cases to an overestimate of the dynamic stresses.

The static stresses can be easily computed, obtaining the following values for the shafts from the first to the last: 1.46, 2.93, 4.39, 5.86, 7.32, 13.17, and 47.96 MN/m^2. The most stressed shaft is that linking the engine to the generator. The dynamic stresses are computed using two different procedures, modal and nonmodal. For the modal approach, the modal mass, stiffness, and damping for the first mode are computed, obtaining $\overline{M}_1 = 103.35$, $\overline{K}_1 = 2.171 \times 10^7$, $\overline{C}_1 = 1,949$ ($\overline{\zeta}_1 = 0.021$); the first modal system is very underdamped. The maximum dynamic stresses in the first mode occur in element number 7. The dynamic stresses in this element corresponding to the various modes and the total dynamic stress are plotted in Figure 5.11a as functions of the speed.

When performing the nonmodal computation, the element in which the maximum stress takes place depends on the harmonic considered: It is then advisable to plot the total stress due to all harmonics in each element (Figure 5.11b). From the comparison of the results obtained through the two procedures, it is clear that in this case the approximations due to the modal approach are not acceptable, because of the large contribution of the nonresonating harmonics to the overall deformation, resulting in a response shape that is very different from the first mode shape. For example, at a speed of 500 rad/s, the stresses in MN/m^2 due to the 18th harmonic (very near resonance) in the various elements are:

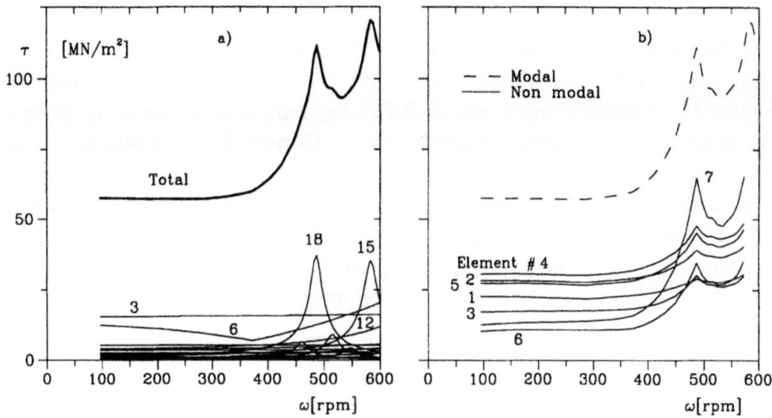

FIGURE 5.11. Dynamic stresses computed through modal approximation with (a) a single mode and (b) nonmodal approach.

Element	1	2	3	4	5	6	7
Modal	2.509	4.800	6.677	7.976	8.585	12.666	28.336
Nonmodal	2.512	4.815	6.716	8.058	8.728	12.998	28.223

The two approaches lead to very close results. The same results, computed for the first harmonic, which is far from resonance, are

Element	1	2	3	4	5	6	7
Modal	0.4822	0.9226	1.283	1.533	1.650	2.434	5.446
Nonmodal	3.3200	3.3194	0.007	3.321	3.325	0.010	0.013

The results in this case are very different, and the modal computation gives unreliable results.

The maximum value of the dynamic stress within the working range occurs in element number 7 at 487 rpm (resonance of the 15th harmonic), and takes a value of 65.04 MN/m^2 (the modal approach would lead to a value of 111.075 MN/m^2). The total stresses are then 47.09 $\pm$ 65.04 MN/m^2.

Example 5-5

Consider a four-stroke-cycle six-cylinder in-line engine with a total capacity of 3,000 cm^3. Assume that the engine is coupled with the transmission through a very soft coupling, which causes a complete dynamic uncoupling of the engine from the transmission in the whole frequency field of interest. The equivalent system is then made of seven equivalent inertias (the six cranks and the flywheel) and six equivalent shafts (Figure 5.12a).

The main characteristics of the system are: number of cylinders: 6; bore: $\phi = 86$ mm; area of the pistons: $A = 58.09$ cm^2; stroke: $2r = 85$ mm; firing order: 1-5-3-6-2-4; speed: $\omega = 6,000$ rpm (638.3 rad/s); reciprocating masses (1 cylinder): $m_p + m_2 = 0.720$ kg; length of connecting rod: $l_p = 181$ mm; crankpin: outer diameter $d_0 = 45$ mm, inner diameter $d_i = 10$ mm; ratio $\alpha = 0.235$.

Equivalent moments of inertia of the cranks and the flywheel [kg m^2]: $\overline{J}_{eq_1} = 0.00551$; $\overline{J}_{eq_2} = \overline{J}_{eq_5} = 0.00489$; $\overline{J}_{eq_3} = \overline{J}_{eq_4} = 0.00602$; $\overline{J}_{eq_6} = 0.00625$; $\overline{J}_{eq_7} = 0.06753$. Equivalent stiffness of the shafts [MNm/rad]: $k_1 = k_2 = k_3 = k_4 = k_4 = 0.392$; $k_6 = 0.455$.

Computation of the natural frequencies. The mass and stiffness matrices of the system are

$$\mathbf{M} = \mathrm{diag}\left\{\; \{551 \quad 489 \quad 602 \quad 602 \quad 489 \quad 625 \quad 6753\;\} \times 10^{-5}\right.,$$

$$\mathbf{K} = \begin{bmatrix}
392 & -392 & 0 & 0 & 0 & 0 & 0 \\
-392 & 784 & -392 & 0 & 0 & 0 & 0 \\
0 & -392 & 784 & -392 & 0 & 0 & 0 \\
0 & 0 & -392 & 784 & -392 & 0 & 0 \\
0 & 0 & 0 & -392 & 784 & -392 & 0 \\
0 & 0 & 0 & 0 & -392 & 847 & -455 \\
0 & 0 & 0 & 0 & 0 & -455 & 455
\end{bmatrix} \times 10^3.$$

The natural frequencies and mode shapes can be easily obtained by solving the usual eigenproblem, which yields the following natural frequencies and eigenvector for the first mode with natural frequency different from zero, normalized in such a way that the largest element has a unit value

$$
\begin{aligned}
\lambda_0 &= 0, \\
\lambda_1 &= 2{,}445 \text{ rad/s} = 391 \text{ Hz}, \\
\lambda_2 &= 6{,}242 \text{ rad/s} = 993 \text{ Hz}, \\
\lambda_3 &= 9{,}607 \text{ rad/s} = 1{,}529 \text{ Hz}, \\
\lambda_4 &= 12{,}485 \text{ rad/s} = 1{,}987 \text{ Hz}, \\
\lambda_5 &= 15{,}300 \text{ rad/s} = 2{,}435 \text{ Hz}, \\
\lambda_6 &= 16{,}349 \text{ rad/s} = 2{,}602 \text{ Hz},
\end{aligned}
\qquad
\{q_1\} = \left\{
\begin{array}{c}
1.000000 \\
0.915283 \\
0.761752 \\
0.537715 \\
0.263907 \\
-0.029741 \\
-0.280269
\end{array}
\right\}.
$$

The corresponding mode shape is plotted in Figure 5.12b. The modal mass and stiffness for the first mode are, respectively,

$$\overline{M}_1 = \mathbf{q}_1^T \mathbf{M} \mathbf{q}_1 = 0.0205, \qquad \overline{K}_1 = \mathbf{q}_1^T \mathbf{K} \mathbf{q}_1 = 123{,}450.$$

Relationship between shaft twisting and shear stresses. If one of the shafts is twisted of angle θ_i, the corresponding torsional moment is $M_t = \theta_i k_i$. If the inner and outer diameters of the most loaded part, usually the crankpin, are d_i and d_0, respectively, for all cranks, the maximum value of the shear stress is $\tau_{max} = M_t/W_t = 2.191 \times 10^{10}\theta_i$ N/m^2.

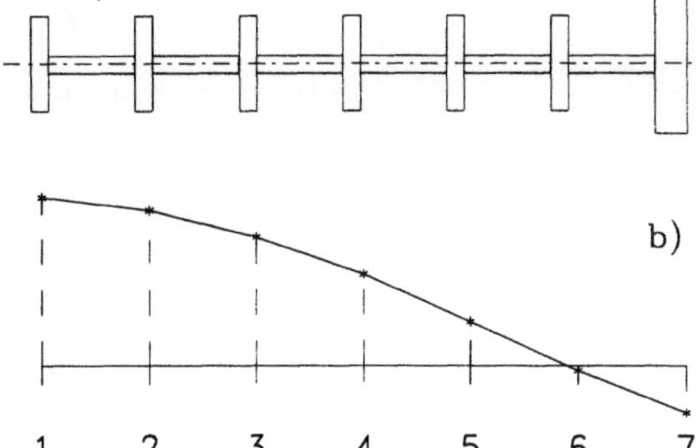

FIGURE 5.12. (a) Equivalent system; (b) first torsional mode.

When the shaft vibrates with the first mode shape, the most stressed crank is the fifth. The corresponding twist angle is $\theta_5 = 0.294q_1$. The proportionality relationship between the maximum value of the shear stress and the amplitude of the first eigenvector is then $\tau_{max}/q_1 = 0.6434 \times 10^{10}$ N/m^{2}rad $= 6{,}434$ MN/m^{2}rad.

Computation of the harmonics of the forcing function and of the modal forces. The curves of the power P and the mean indicated pressure p_{mi} as functions of the engine speed are plotted in Figure 5.13. The curve expressing the mean indicated pressure can be approximated by the expression $p_{mi} = -0.960 \times 10^{-5}\omega^2 + 8.26 \times 10^{-3}\omega - 0.475$, where the angular velocity is in rad/s, and the pressure is in MPa.

To evaluate the harmonics of the driving torque, the empirical formula for the four-stroke-cycle engines reported on the already mentioned book by Ker Wilson (*Practical Solution of Torsional Vibration Problems*, Vol.2, Chapman & Hall, London, 1963, 218) will be used. Using the notation in this book, and particularly using the symbol k for the order of the harmonic, the formula can be written in the form

$$M_{m_0} = p_{mi}rA\frac{1}{2\pi}, \qquad M_{m_k} = p_{mi}rA\frac{25}{50\sqrt{\frac{2}{k} + 5k^2}}.$$

Only the amplitude of each harmonic can be computed from the formula; the phase remains unknown. In practice, the phase is important only to add the inertia torque to the driving torque, and this is important only for harmonics 2, 4, 6, and 8. In the mentioned book, these harmonics are reported separately for the components in phase and in quadrature. The nondimensional amplitudes $M_{m_k}^* = M_{m_k}/p_{mi}rA$, are reported in the following table:

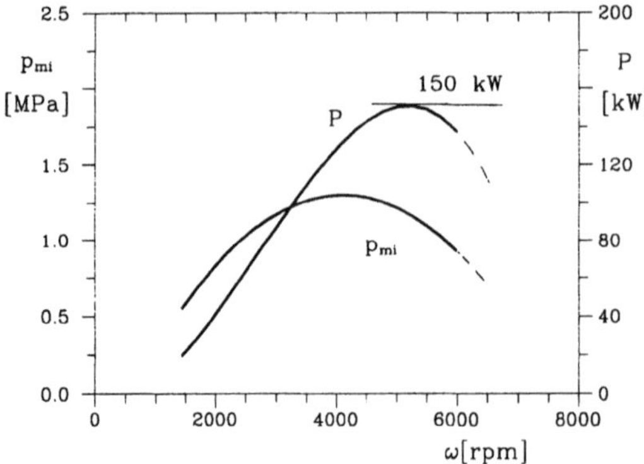

FIGURE 5.13. Power and the mean indicated pressure as functions of the engine speed.

k	A_k^*	B_k^*	$M_{m_k}^*$	k	$M_{m_k}^*$
0	–	–	0.160	11	0.040
1	–	–	0.330	12	0.034
2	0.330	0.140	0.360	13	0.029
3	–	–	0.290	14	0.025
4	0.220	−0.040	0.220	15	0.022
5	–	–	0.160	16	0.019
6	0.110	−0.048	0.120	17	0.017
7	–	–	0.092	18	0.015
8	0.051	−0.051	0.072	19	0.014
9	–	–	0.058	20	0.012
10	–	–	0.048		

Note that the fundamental frequency is equal to $\omega/2$. If the moment of inertia J_0 and all terms in α with powers higher than 2 are neglected, then the four nonvanishing harmonics of the inertia torque can be expressed as $M_{i_k} = m_a r^2 \omega^2 b_k$, where m_a is the total mass of the reciprocating elements and nondimensional coefficients b_k are $b_2 = 0.0588$; $b_4 = -0.5$; $b_6 = -0.176$; $b_8 = -0.0138$.

The modal forces can be computed by considering the six moments acting on the cranks, which are equal in amplitude but out of phase from each other, following the phase-angle diagram. The amplitude of the force due to the kth harmonic is

$$\overline{F}_{1_k} = \{q_1\}^T \{M\}_k \,, \text{ i.e., } |F_{1_k}| = \left| \sum_{j=1}^{6} q_{1_j} M_{k_j} \right| \,.$$

The sum must be considered a vector sum. The phase-angle diagrams are those shown in Figure 5.9 for the six-cylinder engine. From the figure, it is clear that the diagram for harmonics 2 and 4 are the mirror image of each other, and then the relevant modal forces are equal. The same holds for harmonics 1 and 5. Remembering that the diagrams for harmonics beyond the sixth are equal to those of the first six, only four types of phase-angle diagrams exist. Considering the first 24 harmonics, they are as follows.

Group 1: Harmonics 1, 7, 13, and 19 and their mirror images 5, 11, 17, and 23. The phase angles are those shown in the first diagram. The in-phase and in-quadrature components of the modal force are, from equation (5.49),

$$\overline{F}_{1_{k_{phase}}} = \mathbf{q}_1^T\{M_j\cos(\delta_j)\} = M_k\big[q_{11}\cos(0)+$$

$$+q_{12}\cos(120°) + q_{13}\cos(240°) + q_{14}\cos(60°)+$$

$$+q_{15}\cos(300°) + q_{16}\cos(180°)\big] = 0.592M_k,$$

$$\overline{F}_{1_{k_{quad}}} = \mathbf{q}_1^T\{M_j\sin(\delta_j)\} = M_k = \big[q_{11}\sin(0)+$$

$$+q_{12}\sin(120°) + q_{13}\sin(240°) + q_{14}\sin(60°)+$$

$$+q_{15}\sin(300°) + q_{16}\sin(180°)\big] = 0.370M_k,$$

The amplitude of the modal force is then $|\overline{F}_{1_k}| = 0.698M_k$.

Group 2: Harmonics 2, 8, 14, and 20 and their mirror images 4, 10, 16, and 22. By operating as for the first group, it follows that $|\overline{F}_{1_k}| = 0.290M_k$.

Group 3: Harmonics 3, 9, 15, and 21. $|\overline{F}_{1_k}| = 1.90M_k$.

Group 4: Major harmonics 6, 12, 18, and 24. $|\overline{F}_{1_k}| = 3.45M_k$.

Response of nonresonant harmonics.
Computing of the response in conditions far from resonance is straightforward. The modal amplitude of the response is

$$|\eta_{1_k}| = \frac{|\overline{F}_{1_k}|}{\overline{K}_1 - \left(k\frac{\omega}{2}\right)^2\overline{M}_1}.$$

Compute, for example, the response to the sixth harmonic at 6,000 rpm. The mean indicated pressure at that speed is 0.924 MPa. The sixth harmonic belongs to group four (major harmonics). The forcing moment due to the driving torque has components A_k and B_k, which can be computed using the data reported in the previous table: $A_6 = 0.11p_{mi}Ar = 25.11$, $B_6 = -0.048p_{mi}Ar = -10.96$. The moment due to inertia torques is $M_{i_6} = -0.176m_a\omega^2r^2 = -90.36$, and the total moment is $M_6 = \sqrt{(A_6 + M_{i_6})^2 + B_6^2} = 66.16$.

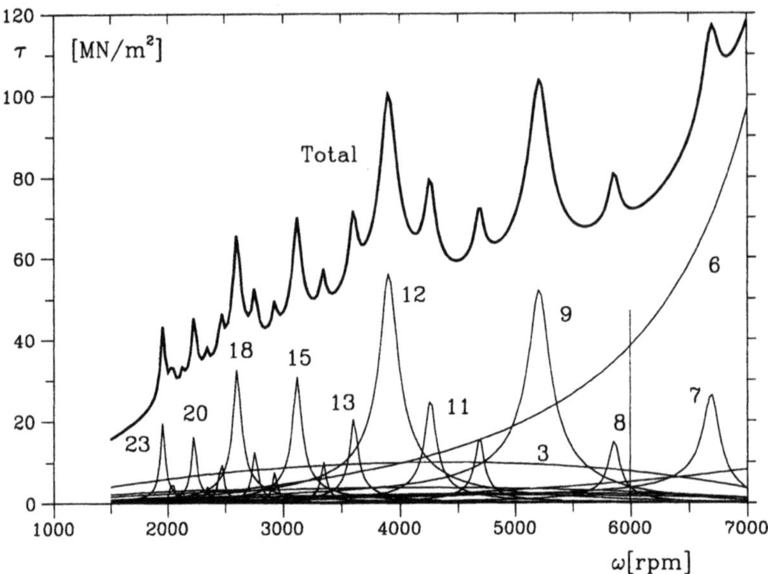

FIGURE 5.14. Dynamic shear stresses in the most loaded part as functions of the speed: the first 24 harmonics of the forcing function and first torsional natural frequency.

The total moment, due to gas pressure and inertia forces, is smaller than the moment due to inertia forces alone. The most critical condition for the harmonic considered at the speed of 6,000 rpm, occurs when the engine is driven from the outside, without any working fluid in it. The stress analysis will then be performed in the corresponding condition. The value of the modal force is $|\overline{F}_{1_6}| = 3.45 M_{i_6} = 311.74$.

The amplitude of the modal response at 6,000 rpm is then

$$\eta_{1_6} = \frac{|\overline{F}_{1_k}|}{\overline{K}_1 - \left(k\frac{\omega}{2}\right)^2 \overline{M}_1} = 6.159 \times 10^{-3},$$

and the corresponding maximum value of the stress is $\tau_{max} = 6{,}434\, q_1 = 39.63$ MN/m^2.

Response of resonant harmonics. To compute the response in resonant conditions, it is necessary to assume a value of the damping of the system. An empirical procedure reported in the already mentioned book by Ker Wilson (page 706 and following) will be followed: The amplitude at resonance is obtained as the product of the static deformation of the modal system by a suitable amplification factor $\eta_{i_{max}} = H_{i_{max}} \overline{F}_i / \overline{K}_i$.

The amplification factor is obtained through the empirical formula

$$H_{i_{max}} = \frac{500}{\sqrt{16 + 145\tau^*}}.$$

Shear stress τ^* is the maximum stress, in MN/m^2, that would be in the shaft if it were deformed following the first mode, in static conditions, by the relevant modal force.

Consider the resonance between the seventh harmonic and the first natural frequency. The critical speed is $\omega_{cr_7} = 2\lambda_1/7 = 701.43$ rad/ s $= 6,698$ rpm; the mean indicated pressure at that speed is $p_{mi} = 0.597$ MPa. The amplitude of the driving torque and the corresponding modal force are

$$M_{m_7} = 0.092 P_{mi} Ar = 13.56\,, \qquad |\overline{F}_{1_7}| = 0.70\,, \qquad M_7 = 9.46\,.$$

The amplitude of the static deformation and the static maximum value of the shear stress corresponding to the modal force are, respectively,

$$\eta_{1_{st}} = \frac{|\overline{F}_{1_7}|}{\overline{K}_1} = 7.669 \times 10^{-5}\,, \qquad \tau^* = 6{,}434 q_{1_{st}} = 0.494 \text{ MN/m}^2\,.$$

The amplification factor, computed using the aforementioned formula, and the damping ratio are, respectively,

$$H_{7_{max}} = \frac{500}{\sqrt{16 + 145\tau^*}} = 53.43\,, \qquad \zeta_7 = \frac{1}{2H_{7_{max}}} = 0.0094.$$

The amplitude of the modal response and the corresponding value of the shear stress are

$$\eta_{1_7} = H_{max}\eta_{1_{st}} = 4.09 \times 10^{-3}\,, \qquad \tau_{max} = 6434 q_1 = 26.3 \text{ MN/m}^2.$$

The responses in terms of maximum shear stress for the first 24 harmonics are shown in Figure 5.14. They have been computed considering the modal system as a damped system with a single degree of freedom whose damping is obtained, for each harmonic, following the same procedure for the seventh. The shear stresses due to the various harmonics are then combined by simply adding the amplitude. This procedure, which neglects the phase, is clearly conservative.

Total stresses. The maximum value of the total stress due to torsional vibration in the working speed range of the machine occurs at a speed of 5,200 rpm in correspondence to the resonance peak of the ninth harmonic. The value of the total shear stress is $\tau_{max} = 104$ MN/m^2. At the speed of 5,200 rpm (544.5 rad/s), the engine can produce a power of 150 kW. The more loaded crank is the fifth, where the static stress is equal to 5/6 of the stress corresponding to the maximum power

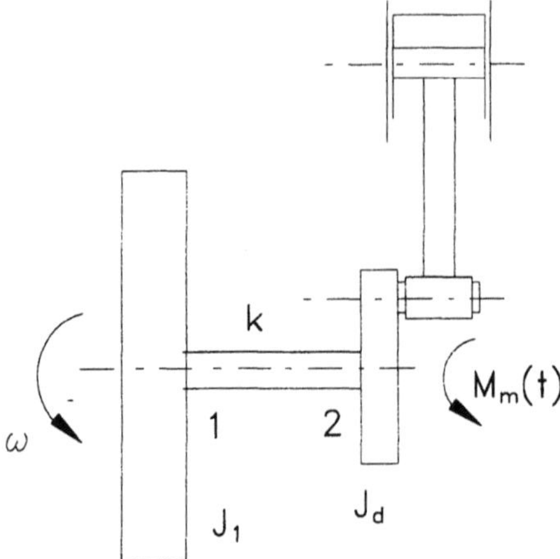

FIGURE 5.15. Crank mechanism: model used for the study of the torsional instability.

$$\tau_{max_{st}} = \frac{5}{6}\frac{P_{max}}{\omega}\frac{16d_e}{\pi(d_e^4 - d_i^4)} = 12.83 \text{ MN/m}^2.$$

The shaft undergoes fatigue loading with cycles included between the limits $\tau = 12.83 \pm 104$ MN/m^2. Because the stressing is mostly due to the resonant ninth harmonic, the frequency of the fatigue loading can be assumed to be equal to the first natural frequency of the shaft, 391 Hz. Note that in the current case the nonmodal computation would have yielded very similar results because most of the stressing is due to resonating harmonics (Exercise 5-7).

5.5   Torsional instability of crank mechanisms

A second-approximation formulation of the equation of motion of the system consisting of the crank, the connecting rod, and the reciprocating parts, can be obtained by introducing the trigonometric polynomial (5.14) expressing the equivalent moment of inertia into the equation of motion (5.10)

$$\overline{J}_{eq}\ddot{\phi}_z + \left\{\sum_{j=1}^{n} J_{c_j}\cos[j(\phi_z + \omega t)] + \sum_{j=1}^{n} J_{s_j}\sin[j(\phi_z + \omega t)]\right\}\ddot{\phi}_z +$$

$$+\frac{1}{2}(\omega + \dot{\phi}_z)\left\{-\sum_{j=1}^{n} jJ_{c_j}\sin[j(\phi_z + \omega t)] + \sum_{j=1}^{n} jJ_{s_j}\cos[j(\phi_z + \omega t)]\right\} = M.$$

$$(5.51)$$

Equation (5.51) would be the exact equation of motion of the system if the number of harmonic terms considered was infinitely large. Actually a very good approximation is obtained by resorting to a small value of n. If the amplitude of the oscillations is small enough, the trigonometric functions of angle $j\phi_z$ can be linearized. By remembering a few trigonometric identities and linearizing the resulting equation, neglecting the products of small quantities, equation (5.51) reduces to

$$\overline{J}_{eq}\ddot{\phi}_z + \ddot{\phi}_z \sum_{j=1}^{n}\left[J_{c_j}\cos(j\omega_t) + J_{s_j}\sin(j\omega_t)\right] - \dot{\phi}_z\omega\sum_{j=1}^{n}j\left[J_{c_j}\sin(j\omega_t)+\right.$$

$$\left. -J_{s_j}\cos(j\omega_t)\right] - \frac{1}{2}\phi_z\omega^2\sum_{j=1}^{n}j^2\left[J_{c_j}\cos(j\omega_t) + J_{s_j}\sin(j\omega_t)\right] +$$

$$-\frac{1}{2}\omega^2\sum_{j=1}^{n}j\left[J_{c_j}\sin(j\omega_t) + J_{s_j}\cos(j\omega_t)\right] = M\ .$$

$$(5.52)$$

Equation (5.52) is a linear differential equation with coefficients that are functions of time. It can be used as a starting point to build a second-approximation model of the reciprocating machine. The complexity of this approach is, however, considerable and only a very simplified case is reported here.[4] Consider a single-cylinder engine with a simplified geometry such that $\alpha = 0$ and $\beta = 0$. Assume that the crank is connected to a large flywheel, so large that its motion is not affected by torsional vibrations (Figure 5.15). Assuming that the flywheel rotates at constant speed ω, the system has a single degree of freedom, namely, the angle of torsion ϕ_z of the shaft. The behaviour of the shaft is linear, with stiffness k, and a viscous damper acts on the crank with damping coefficient c, producing a moment proportional to the speed $\dot{\phi}_z + \omega$, and the driving torque $M_m(t)$. The equation of motion (5.52) reduces to

$$\ddot{\phi}_z[1 - \epsilon\cos(2\omega t)] + 2\epsilon\omega^2\phi_z\cos(2\omega t) + 2\epsilon\omega\dot{\phi}_z\sin(2\omega t)+$$

$$+2\zeta\lambda_0(\omega + \dot{\phi}_z) + \lambda_0\phi_z = -\epsilon\omega^2\sin(2\omega t) + \lambda_0^2\frac{M_m(t)}{k}, \qquad (5.53)$$

---

[4]For a detailed study, see E. Brusa, C. Delprete, G.Genta, "Torsional Vibration of Crankshafts; Effects of Nonconstant Moments of Inertia", *Journal of Sound and Vibration*, 205, 2, (1997), 135–150.

where $\lambda_0 = \sqrt{k/\overline{J}_{eq}}$, $\zeta = c/2\sqrt{k\overline{J}_{eq}}$, and $\epsilon = (m_2 + m_p)r^2/2\overline{J}_{eq}$.

The inertial term on the right-hand side of equation (5.53) coincides with the inertia torque acting on the equivalent system of the simplified model. If the terms containing the products of the small quantity ϵ by the displacement ϕ_z and its derivatives are neglected, equation (5.53) reduces to the equation of motion of the equivalent system studied in the preceding section, with the added conditions that the connecting rod is infinitely long and damping is included in the form studied in Section 5.4.5.

Equation (5.53) cannot be assimilated into either a Mathieu or a Hill equation but could be solved by using the same methods in Section 3.7 and particularly by resorting to a series expansion of the solution and finding the coefficients using an infinite determinant. Due to the complexity of the study, only some results obtained by M.S.Parisha and W.D. Carnegie[5] through numerical experimentation will be reported here. The equation for the free oscillations of the system obtained from equation (5.53), by neglecting both the driving torque and the drag torque linked with the constant angular velocity ω, can be written in the nondimensional form

$$\frac{d^2\phi_z}{d\tau^2}[1 - \epsilon\cos(2\tau)] + 2\frac{d\phi_z}{d\tau}[\epsilon\sin(2\tau) + \zeta\frac{\lambda_0}{\omega}] + \\ +\phi_z\left[2\epsilon\cos(2\tau) + \left(\frac{\lambda_0}{\omega}\right)^2\right] = 2\zeta\frac{\lambda_0}{\omega} - \epsilon\sin(2\tau),\qquad(5.54)$$

where $\tau = \omega t$ is the nondimensional time. Because the behaviour of the system depends only on three nondimensional parameters – ϵ, ζ and λ_0/ω – the study of the stability of the system can be easily performed by numerically integrating the equation of motion with different values of the parameters and checking whether it develops an unstable behaviour. Some of the plots that define the zones in which the system is unstable are summarized in Figure 5.15. The undamped system then has two instability ranges, one at a rotational speed equal to half the natural frequency of the equivalent system

[5]M.S. Parisha e W.D. Carnegie, "Effect of the variable inertia on the damped torsional vibrations of diesel engine systems", *J. of Sound and Vibration*, 46, 3, (1976). See also E. Brusa, C. Delprete, G.Genta, "Torsional vibration of crankshafts; effects of nonconstant moments of inertia", *Journal of Sound and Vibration*, 205, 2, (1997), 135–150.

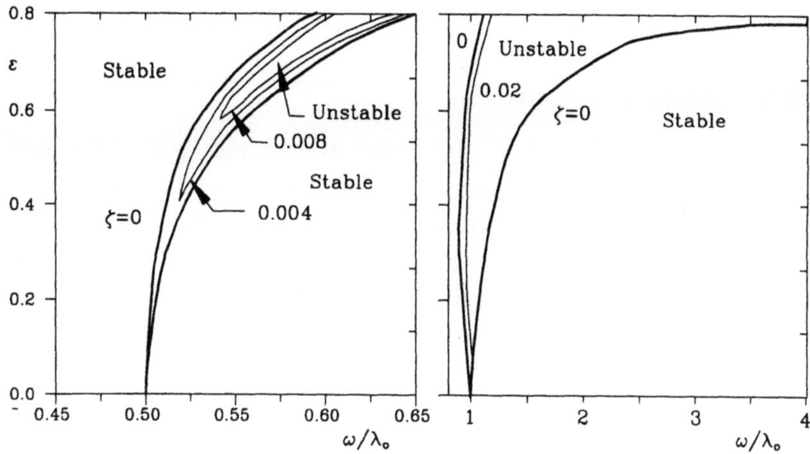

FIGURE 5.16. Instability ranges of the system sketched in Figure 5.14 with different values of damping.

and the other at a speed equal to λ_0. If ϵ tends to zero, the same results seen in the previous sections are obtained: An infinitely small instability range is found in correspondence with the torsional critical speeds found on the Campbell diagram at the intersections of the horizontal straight line $\lambda = \lambda_0$ with the lines $\lambda = 2\,\omega$ and $\lambda = \omega$. The instability range reduces to the torsional critical speeds due to the first two harmonics of moment M_i. With increasing ϵ, the two instability ranges grow larger and, with values of ϵ large enough, it is impossible to operate at values of the speed in excess of the natural frequency. The presence of damping reduces the fields of instability and makes it possible to operate at any speed, provided that ϵ, i.e., the mass of the reciprocating elements, is not too large.

The instability range located near the critical speed due to the harmonic with $\lambda = \omega$ is less affected by damping than that corresponding to condition $\lambda = 2\omega$ and then is the most dangerous. The effect of damping reduces with increasing values of ϵ. The mentioned results confirm that there is no instability range if ϵ is lower than 4ζ in the zone near $\lambda/\omega = 1$ and lower than $4\sqrt{2}\zeta$ in the zone near $\lambda/\omega = 1/2$. These results are applicable, at least from a qualitative viewpoint, well beyond the simplified model used in obtain them. They also hold for the cranks of multi-cylinder engines for the evaluation of the effects of the mass of the reciprocating elements on the overall stability of the system.

5.6 Dampers for torsional vibrations

5.6.1 Dissipative dampers

In many cases, when it is impossible to prevent one of the critical speeds yielding severe dynamic stressing from falling within the working range or, more generally, when the amplitude of the torsional vibrations is incompatible with the safe operation of the machine, suitable damping devices must be introduced into the system. Many different types of such elements have been developed, each with its field of application due to the different mechanical, thermomechanical, and cost characteristics.

Often torsional vibration dampers are applied at one end of the crankshaft and are made of a flywheel (usually referred to as *seismic mass*), whose geometric configuration can be of a wide variety of types, connected to the shaft by suitable elastic and damping elements. In many applications there may be only one of these elements, and sometimes the restoring force can be supplied by the centrifugal field due to rotation and the seismic mass can have the shape of a counterbalance of the crankshaft. The conceptual scheme of almost all torsional vibration dampers can be reduced to that of the damped vibration absorber of Figure 5.17, with the obvious difference that in the current case it is applied to a multi-degree-of-freedom system and that instead of masses, stiffnesses, and dampers, there are moments of inertia, torsional stiffnesses, and torsional dampers. Without including all the possible types of torsional vibration dampers, they can be subdivided into three types: dissipative dampers, damped vibration absorbers, and rotating pendulum vibration absorbers.

The conceptual scheme of a dissipative torsional vibration damper is that of the Lanchester damper described in Example 1-9. It is applied to one of the ends of the crankshaft and consists of a flywheel, generally shaped as a ring free to rotate within a casing filled with a fluid with high viscosity, for example a silicon-based oil (Figure 5.17a).

Damping in this case is of the viscous type, i.e., the drag torque is proportional to the angular velocity, and the damping coefficient depends on the clearance between the ring and the housing and on the characteristics of the fluid. Note that it can be greatly influenced by the temperature of the fluid. The model for the dynamic study of the system must be modified by adding the moment of inertia of

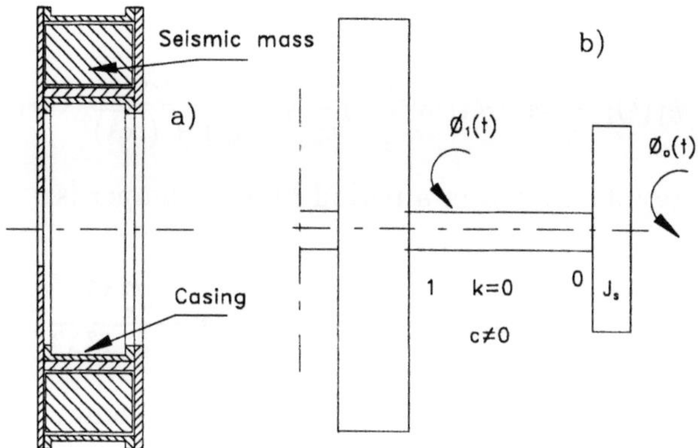

FIGURE 5.17. Dissipative torsional vibration damper; (a) sketch of the system; (b) model for the dynamic study.

the casing, in the node in which the damper is applied, and adding a new node in which the inertia of the ring is lumped. The two nodes are connected by a viscous damper and a spring with zero stiffness.

For a first-approximation evaluation of the optimum damping of the system, it is possible to assume that the presence of the damper does not significantly affect the natural frequencies of the system. Under this assumption, it is possible to study the behaviour of the damper separately, assuming the time history of the motion of the node in which it is applied. With reference to Figure 5.17b, assume that the time history at node 1 of the system, in which the damper is applied, (the node in which the seismic mass is lumped is labeled node 0), is harmonic, $\phi_1(t) = \phi_{1_0} \sin(\lambda t)$. Also, the time history at node 0 is harmonic but characterized by a phase lag ϕ: $\phi_0(t) = \phi_{0_0} \sin(\lambda t + \phi)$. The model so obtained is formally identical to that of a system with a single degree of freedom excited by the displacement of the supporting point. The response can be computed using equations (1.78), in which m is substituted by J_s and $k = 0$. Remembering that $2\zeta\lambda_n = c/m$ and introducing the ratio $\alpha = J_s/c$, it follows that

$$\Re(H) = \frac{1}{1 + (\alpha\lambda)^2}, \qquad \Im(H) = \frac{(\alpha\lambda)}{1 + (\alpha\lambda)^2}. \qquad (5.55)$$

If time $t = 0$ is chosen as the time when the angular displacement between the seismic mass and its housing reaches its maximum, the relative displacement can be expressed as

$$[\phi_0(t) - \phi_1(t)] = |\phi_0 - \phi_1| \cos(\lambda t) = \phi_{1_0} \frac{\alpha\lambda}{\sqrt{1 + (\alpha\lambda)^2}} \cos(\lambda t). \quad (5.56)$$

The energy dissipated in a period by the damper is

$$E_d = \int_0^T c \left[\dot{\phi}_0(t) - \dot{\phi}_1(t) \right]^2 dt = c\phi_{1_0}^2 \pi \frac{J_s^2 \lambda^3}{c^2 + J_s^2 \lambda^2}. \quad (5.57)$$

It is easy to verify that both conditions $c = 0$ and $c \to \infty$ lead to a vanishingly small energy dissipation: in the first case because the seismic mass does not interact with the system and in the second case because nodes 1 and 0 are rigidly connected. The value of the damping coefficient that leads to a maximum energy dissipation can be simply obtained by differentiating equation (5.57) and equating the derivative to zero. The value of the optimum damping so obtained is $c_{opt} = J_s\lambda$. Even if the value of the optimum damping depends on the frequency, dampers of the type described here allow a substantial reduction of the amplitude of vibration in a large range of frequency.

The type shown in Figure 5.17 is quite common, but other types are also used. The type shown in Figure 5.18a, for example, although giving way to viscous damping, is quite different. The seismic mass is made by a hollow flywheel inside which a sort of lever or balance wheel is located. Some cavities inside the flywheel are filled with a high-viscosity fluid that is pumped by the balance wheel from one cavity to the other through calibrated passages during torsional vibrations. Tuning screws allow the passages to be restricted in order to tune the value of the damping coefficient.

Another type of damper is shown in Figure 5.18b. In this case the seismic mass is made of two disc flywheels pressed against a disc rigidly connected to the shaft by spring-loaded set screws. The mating surfaces of the flywheels and the disc are lined by friction material, which constitutes the damping element. In this case the characteristics of the damping are not those of viscous damping but of dry friction, and the behaviour of the system is nonlinear. The term *Lanchester damper* should be used, strictly speaking, only for dampers of this type, not for all dissipative dampers with zero stiffness.

Because all dissipative dampers convert mechanical energy into heat, they are subject to heating, which can be very strong. It is

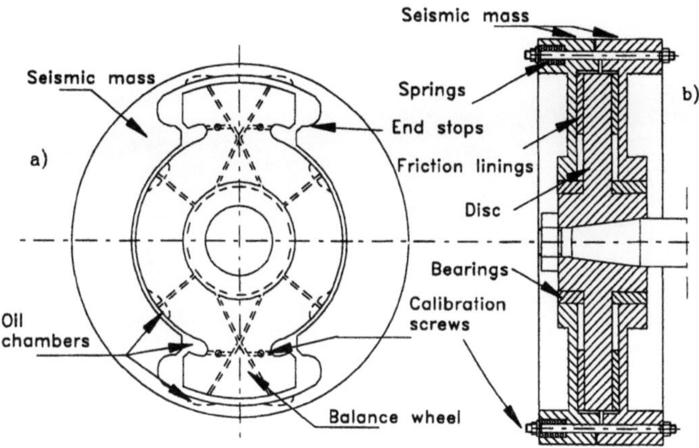

FIGURE 5.18. Dissipative torsional vibration dampers: (a) viscous damper; (b) dry friction damper (Lanchester damper).

then necessary to verify that they can dissipate all the thermal energy they produce, which can be computed using formulas of the type of equation (5.57), at least in terms of average power in a given period of time. Usually a limit to the ratio between the thermal power and the external surface of the damper is assumed. For the type in Figure 5.17a, for example, it is suggested not to exceed 1.9×10^4 kW/m^2 in continuous working and 5×10^4 kW/m^2 for short periods of time.

5.6.2 Damped vibration absorbers

Vibration absorbers were dealt with in Example 1-8. Also in the case of torsional vibration absorbers, they introduce a new natural frequency of the system and change the natural frequency on which they are tuned. The distance between the two resonance peaks increases with the increasing moment of inertia of the seismic mass. Because an undamped vibration absorber is effective in a very narrow frequency range, outside which it is not only ineffective but it can also cause new resonances, the seismic mass is connected to the shaft through a system that has a certain amount of damping. In this case it is possible to obtain a response that is fairly flat in an ample range of frequencies, as clearly shown in Figure 1.31d.

From a practical viewpoint, all dampers shown in the previous section can be converted into damped vibration absorbers simply by adding an elastic element between the shaft and the seismic mass, which allows tuning the damper on the required frequency. The seismic mass can be shaped as a disc or a balance wheel, as in the case

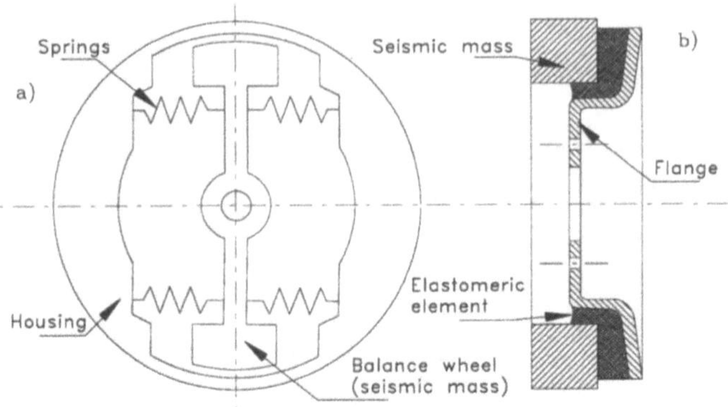

FIGURE 5.19. Damped torsional vibration absorbers: (a) viscous damper with balance wheel; (b) elastomeric damper.

of the damper of Figure 5.19a, which is very similar to that shown in Figure 5.18a, except that in the former the balance wheel is the seismic mass and the casing is attached to the shaft.

The elastomeric dampers widely used on small automotive diesel engines (Figure 5.19b) can be considered dissipative vibration absorbers. The elastomeric elements provide both elasticity and damping and can be shaped in such a way as to obtain the required values for both. Also in this case it is necessary to verify that the heat generated within the damper is not too much, particularly considering the low thermal conductivity and poor high-temperature characteristics of elastomeric materials. Overheating is very dangerous because it usually causes the damping capacity of the material to decrease and, consequently, the amplitude of the vibration to grow, with further increase of heating. This easily results in a complete destruction of the damper and severe fatigue problems of the whole system.

Example 5-6
Add a damped vibration absorber to the front end of the diesel engine of Example 5-4. Assume a moment of inertia of 6 kg m^2 for the casing, connected to the first crank through a shaft whose stiffness is equal to the shaft connecting the last crank to the flywheel, and a moment of inertia of 26 kg m^2 for the seismic mass. Compute the stiffness needed to tune the damper on the first natural frequency of the system and assume for the damping the optimum value computed in Section 5.6.1 for the springless damper. Repeat the computation of the stresses and compare the results.

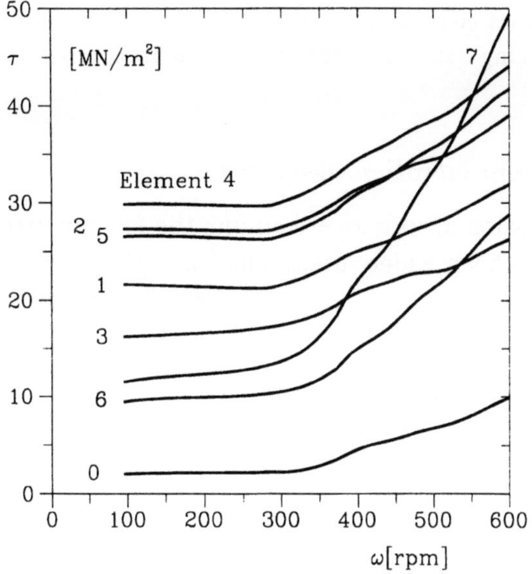

FIGURE 5.20. Dynamic stresses computed through a nonmodal approach.

The equivalent system now includes 10 nodes, with node 1 located at the seismic mass and node 2 at the damper case. The diameter of element 2 is assumed to be 200 mm. A first computation is run with a zero value of the stiffness of the damper, obtaining the first natural frequency of the system without seismic mass of 445.2 rad/s = 70.86 Hz; the casing of the damper has little effect on the first natural frequency.

To tune the damper on the first natural frequency, its stiffness must be $k_{damp} = J_{damp}\lambda_1^2 = 5.53 \times 10^6$ Nm/rad. The nonvanishing natural frequencies of the undamped system are then computed: 55, 67, 114, 192, 290, 376, 442, 482, and 627 Hz.

The presence of the damper introduces two new values in place of the old value for the first mode and a high natural frequency linked mostly to the presence of the damper casing. The remaining values are little affected. The value of the damping can be assumed to be $c_{damp} = J_{damp}\lambda_1 = 1.16 \times 10^4$ Nms/rad.

The computation of the dynamic response is only performed using a non-modal approach, because the presence of the damper makes the assumption of modal uncoupling even less realistic than in Example 5-4 (Figure 5.20). The maximum value of the dynamic stress within the working range occurs in element number 4 (not taking into account the shaft between damper and first crank, which will be indicated as element number 0) at 500 rpm and takes a value of 38.70 MN/m^2. The total stresses are then 47.09 ± 38.70 MN/m^2.

The presence of the damper is quite effective, and reduces dynamic stressing by about 40%. Note that element number 7 is the most stressed only at speeds above the working range and that no well-defined resonance peak occurs.

5.6.3 Rotating-pendulum vibration absorbers

Rotating pendulum vibration absorbers are among the few nondissipative dampers that have been widely used. The conceptual scheme of such a device was shown in Figure 4.1, the only difference being that in the current case the pendulum is constrained to move in the xy-plane. The natural frequency of the free oscillations of the pendulum is proportional to the rotational speed: $\lambda = \omega\sqrt{r/l}$. Because the various harmonics of the forcing function are characterized by a frequency proportional to the rotational speed, once the vibration absorber is tuned at a certain speed on a given harmonic, it remains locked on that in the whole working range of the machine.

To analyze the behaviour of rotating-pendulum vibration absorbers, a more complex model than that shown in Figure 4-1 is needed, as the torsional vibration of the disc at which the pendulum is attached cannot be neglected. Consider the scheme of Figure 5.21, in which a pendulum of mass m and length l is hinged to a disc whose moment of inertia is J. Let the disc be in node 1 of the discretized mathematical model and the node at which the system is connected, through a shaft with stiffness k, be node 0. The aim of the analysis is to build a rotating-pendulum vibration absorber element, to be assembled, using the already-seen procedure, into the finite element model of the system. The velocity of point P can be computed as in Example 4-1, where ϕ_2 and $\omega + \dot{\phi}_1$ are substituted for θ and ω, respectively, and angle ϕ is assumed to be zero, because the pendulum is constrained to remain in the xy-plane. The kinetic and potential energies of the system sketched in Figure 5.19 are, respectively,

$$2T = J(\omega + \dot{\phi}_1)^2 + m|V_P|^2 = (J + mr^2)(\omega + \dot{\phi}_1)^2 +$$
$$+ ml^2(\omega + \dot{\phi}_1 + \dot{\phi}_2)^2 + mrl(\omega + \dot{\phi}_1)(\omega + \dot{\phi}_1 + \dot{\phi}_2)\cos(\phi_2),$$
$$2U = k\phi_1^2 .$$

$$(5.58)$$

By performing the required derivatives, linearizing the trigonometric functions of angle ϕ_2, and neglecting the terms containing products of generalized coordinates and their derivatives, the equa-

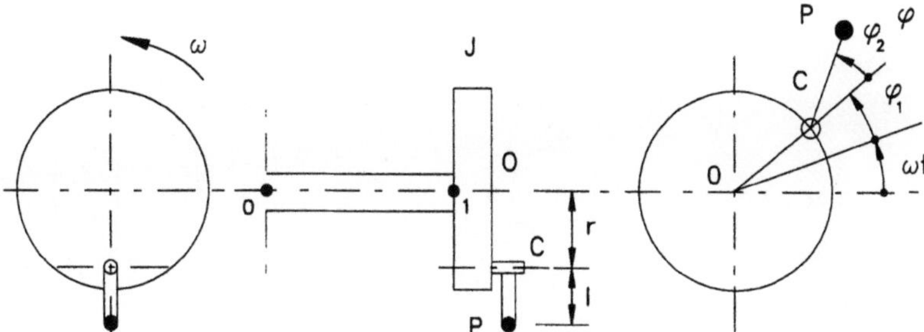

FIGURE 5.21. Rotating-pendulum vibration absorber; model with two degrees of freedom.

tion of motion of the free oscillations of the system sketched in Figure 5.21 is

$$
\begin{bmatrix} J + m(r+l)^2 & ml(r+l) \\ ml(r+l) & ml^2 \end{bmatrix} \left\{ \begin{array}{c} \ddot{\phi}_1 \\ \ddot{\phi}_2 \end{array} \right\} \begin{bmatrix} k & 0 \\ 0 & mrl\omega^2 \end{bmatrix} \left\{ \begin{array}{c} \phi_1 \\ \phi_2 \end{array} \right\} \left\{ \begin{array}{c} 0 \\ 0 \end{array} \right\}.
$$

(5.59)

It is easy to verify that if a moment with harmonic time history with frequency λ is applied to node 1 or the same node is excited by a harmonic motion of the constraint, the ratio between the amplitude of the oscillation of the disc and that of the pendulum is

$$
\frac{\phi_{1_0}}{\phi_{2_0}} = \frac{\frac{l}{r} - \left(\frac{\omega}{\lambda}\right)^2}{1 + \frac{l}{r}}.
$$

(5.60)

The amplitude at node 1 then vanishes if the pendulum is tuned on the frequency λ, i.e., if $r/l = (\lambda/\omega)^2$.

The presence of a rotating-pendulum vibration absorber in the nth node of the system can be accounted for in the mathematical model simply by adding a further degree of freedom, the oscillation angle of the pendulum, and assembling the following mass and stiffness matrices between the nth node and the added one:

$$
\mathbf{M} = m \begin{bmatrix} (r+l)^2 & l(r+l) \\ l(r+l) & l^2 \end{bmatrix} , \qquad \mathbf{K} = \omega^2 \begin{bmatrix} 0 & 0 \\ 0 & mrl \end{bmatrix} .
$$

(5.61)

The length of the pendulum must, in general, be very small, particularly when the vibration absorber is to be tuned on a high-order harmonic. If tuning has to be performed on the sixth harmonic of

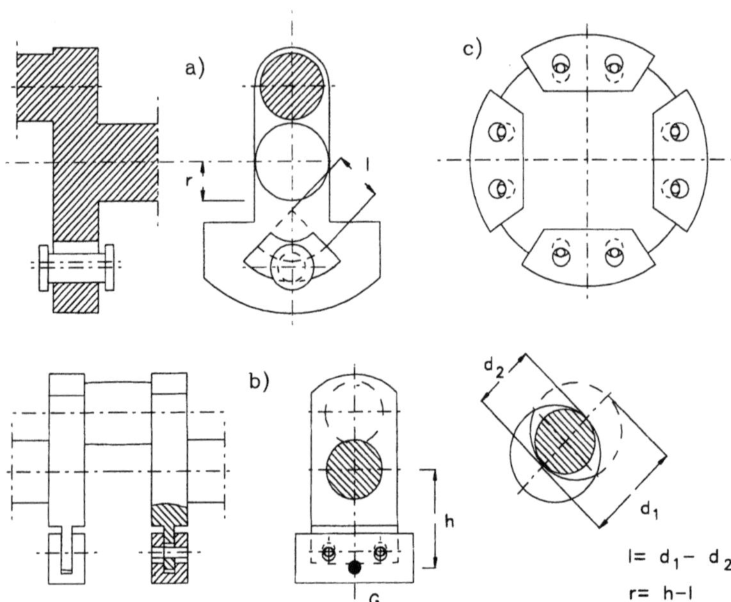

FIGURE 5.22. Rotating pendulum vibration absorbers: (a) roller type, dimensions refer to the case in which the rotational inertia of the roller is neglected, (b) and (c) bifilar pendulum type.

a four-stroke-cycle internal-combustion engine, for example, characterized by ratio $\lambda/\omega = 3$, the length of the pendulum must be equal to 1/9 of the radius at which it is hinged. Very short pendulums result from the need of tuning at higher-order frequencies. Different solutions to the problem of building very short pendulums with sufficient mass have been forwarded. Two of them are sketched in Figure 5.22. The rotating pendulum of Figure 5.22a is made by a roller, free to move in a circular slot in one of the crank webs. The solution of Figure 5.22b is equivalent to a bifilar pendulum and allows very small values of length l to be achieved. The system of Figure 5.22c is made by four rotating pendulums of the type shown in Figure 5.22b attached to a single disc. They can have different values of the length l, in order to be tuned on different harmonics of the forcing function.

5.7 Experimental measurement of torsional vibrations

The torsional vibrations of reciprocating machines are particularly dangerous because they have little effect on the overall motion of the machine and, hence, are very difficult to detect without suitable instrumentation. This is because the shaft is not constrained against rotation and the inertia forces due to torsional vibration are completely balanced, at least in the case in which the relevant modes are completely uncoupled from other modes. Their effects on the nonrotating parts of the machine, which would be completely nil in the case of the equivalent system, are, at any rate, very small. Those warning signs, such as noise and vibrations transferred to other parts of the machine that usually allow detection of the occurrence of dynamic problems, are absent or at least very small. An engine or a compressor can thus work with high levels of dynamic stresses in the shaft without the operator realizing it until failure occurs. Only the use of suitable instruments allows verification that the level of torsional vibrations is low enough not to cause problems.

The preceding statement is not completely true for modern car engines, which drive a number of accessories (air conditioner, a large alternator, etc.) through belts. When there are torsional vibrations, the latter exert on the supports of the accessories forces that are variable in time, and the resulting vibration can be detrimental to the acoustical and vibrational comfort of the vehicle and to the structural safety. Recently, torsional vibration dampers, which were limited to diesel engines, started to be used on many spark-ignition engines and complicated types aimed to reduce the transmission of vibrations to the accessories have appeared.

Torsional vibrations can be detected from cyclic variations of the instant rotational velocity or directly from measurements of the dynamic stressing of the rotating parts of the machine. Speed variations can be detected using the devices usually included in all rotating machinery to measure the rotational speed, provided they are sensitive enough. A common device is a magnetic pick-up that produces a pulse each time a tooth of a gear wheel made of magnetic material passes in front of it. A usual arrangement is to use a wheel with 60 teeth in such a way that the frequency in Hz of the pulses from the pick-up is equal to the rotational speed in revolutions per minute. If the speed varies periodically in time, the frequencies of the torsional oscillations can be obtained from a Fourier analysis of the signal. However, this method is limited where the maximum vibration frequency detectable is concerned, because the period of time between the passages of two subsequent teeth must be much smaller than the period of the frequency to be detected, particularly if the amplitude also has to be measured. Laser velocity transducers can detect the peripheral velocity of the rotor, and from this measurement very accurate assessments of the torsional vibrations of the machine can be made. An advantage of these methods is the absence of the need to transfer signals from the rotor to the stationary parts of the machine.

Alternatively, it is possible to use electric strain gauges on the shaft to directly measure the variable components of the stresses. The strain gauges must be located near the nodes of the mode shape of the harmonic under study, because the stresses reach their maximum value in these locations. When using strain gauges it is necessary to transfer the signal from rotating to stationary parts of the machine. This can be done using slip rings and brushes, which are similar to those used in electrical machines, or by more modern magnetic, radio, or optical contactless systems. Instead of locating the strain gauges directly on the shaft, it is possible to add a seismic mass connected to the shaft through a low-stiffness system and to locate the strain gauges on these springs. If the natural frequency of the seismic mass is lower than the first torsional natural frequency of the system, all the harmonics of the oscillation can be detected. Also in this case, a device that can transfer the signal to the stationary parts of the machine must be provided. Many types of instruments, mostly of the mechanical type, which were used in the past, are now obsolete except for particular applications.

5.8 Axial vibrations of crankshafts

Crankshafts of reciprocating machines can be subject to axial vibrations, or, more correctly because the modes of the various types are all coupled, to vibration modes that are prevalently axial. They can be dangerous because of the possibility of excitation by gas pressures and inertia forces of reciprocating parts due to the coupling of the various modes. In the past, axial vibrations were an actual danger only in a few cases, mainly linked with slow large internal-combustion engines, but the modern tendency toward higher material stress levels, speeds, and power/mass ratios caused them to become an important factor in the design of wider classes of reciprocating machines.

The simplified models used for the study of axial vibrations are very similar to those seen in the context of torsional vibrations.

An equivalent system is obtained by lumping the masses in some nodes and evaluating the axial stiffness of the various parts of the crankshaft. Usually, the largest compliance is due to the cranks, because the crankwebs and crankpins bend during axial vibrations, and the straight parts of the shaft are much stiffer and can often be considered rigid in the study of the lower modes. The stiffness can be computed using the semiempirical or empirical formulas reported in the literature, numerical methods, or experimental tests. The equivalent masses are, in this case, constant and no inertia forces must be taken into account as in the case of torsional vibrations. The masses are only those of the shaft since, because, due to the axial clearance in the bearings, the mass of the connecting rods does not take part to axial vibrations.

The equivalent system for the study of torsional vibrations is underconstrained, but that for axial vibrations is axially kept in position by the thrust bearing, usually modeled as an elastic constraint. All other bearings are usually not considered; they allow axial displacements of the order of magnitude of those encountered in axial vibrations. Once the equivalent system has been obtained, there is no difficulty studying the free vibrations. Usually, an in-line system is obtained and transfer-matrices procedures, like the Holzer method, can be used even if more modern approaches based on the FEM are more common now. Because the equivalent system has been obtained as a lumped-parameters system, there is no difficulty obtaining the modal forces due to the connecting rods by measuring or computing, using empirical formulas or numerical methods, the axial displacements due to unit forces applied by the connecting rods. The study of forced vibrations can thus be performed.

Also in this case there can be dangerous resonance conditions, even if the excitation is much smaller for axial vibrations than for torsional vibrations, due to the lower damping of the system. The same procedures seen for torsional vibrations allow the amplitudes of vibration and then the stresses to be obtained. If the amplitude exceeds the allowable limits, it is possible to resort to suitable dampers. They can be made of a seismic mass, free to move in the axial direction, or a piston that can move in axial direction (and to rotate, because it is connected to the shaft) together with the shaft in a cylinder, rigidly or elastically mounted to the basement of the machine, full of viscous fluid.

Note that the elastomeric damper shown in Figure 5.22b also acts as an axial vibration damper. It is possible to tune the damper on two different frequencies, for torsional and axial vibrations, by tailoring the shape of the elastomeric element and then the ratio between the axial and torsional stiffness. The ratio between the mass and the moment of inertia can also be changed, at least within some limits, to achieve the required tuning.

5.9 Short outline on balancing of reciprocating machines

Reciprocating machines are a source of vibrations, not only as a result of the variation in time of the driving torque, which causes torsional vibrations of the crankshaft and a reaction torque on the basement that varies in time, but also for the inertia forces due to reciprocating parts. The crank mechanism itself is a source of vibrations even when it produces no driving or braking torque. Crank mechanisms are included in a wide variety of very common machines, and the problem of balancing the inertia forces due to them has been widely studied; whole books have been devoted to this subject.[6] Many devices, sometimes quite complicated, have been suggested, and, in many cases, used with the aim of improving the smoothness of reciprocating engines and compressors. Due to the complexity of the problem, only a short outline of the subject is presented here.

The inertia forces due to the crank mechanism can be studied using the same scheme seen in Chapter 4 for the study of the effects of the static unbalance of a rotor. Consider a single-cylinder machine supported on elastic mountings that allow the whole system to be displaced in the x- and y-directions (Figure 5.23a). The crank is assumed to be made of a statically balanced disc D, the crankpin B, whose mass is included in the mass m_1 of the connecting rod, and a counterweight C, that is considered a point mass. If the rotational speed ω of the disc is assumed to be constant, remembering the definition of function $f_1(\theta)$ (equation 5.5) the velocities of points O, C, B, and P are

---

[6]See, for example, W.Thomson, *Fundamentals of Automobile Engines Balancing*, Mech. Eng. Publ. Ltd., 1978).

$$V_O = \left\{ \begin{array}{c} \dot{x} \\ \dot{y} \end{array} \right\}, \qquad V_B = \left\{ \begin{array}{c} \dot{x} - r_B\omega\sin(\omega t) \\ \dot{y} + r_B\omega\cos(\omega t) \end{array} \right\},$$

$$V_C = \left\{ \begin{array}{c} \dot{x} + r_C\omega\sin(\omega t) \\ \dot{y} - r_B\omega\cos(\omega t) \end{array} \right\}, \qquad V_P = \left\{ \begin{array}{c} \dot{x} + r_B\omega\sqrt{f_1(\omega t)} \\ \dot{y} \end{array} \right\}.$$

$$(5.62)$$

If the connecting rod is modeled in the usual way (two point masses m_1 in B and m_2 in P and a massless moment of inertia J_0), the kinetic energy of the whole system made by the crank, the connecting rod, and the reciprocating masses is

$$T = \frac{1}{2}m_t(\dot{x}^2 + \dot{y}^2) + \frac{1}{2}J_{eq}\omega^2 + \omega(m_C r_C - m_1 r_B)\times \qquad (5.63)$$

$$\times\,[\dot{x}\sin(\omega t) - \dot{y}\cos(\omega t)] - (m_2 + m_p)r_B\dot{x}\omega\sqrt{f_1(\omega t)},$$

where

$$m_t = m_d + m_C + m_1 + m_2 + m_P,$$

$$J_{eq} = J_d + m_C r_c^2 + m_1 r_B^2 + (m_2 + m_P)r_B^2 f_1(\omega t) + J_0 f_2(\omega t).$$

In computing the derivatives of the kinetic energy that enter the equation of motion, the rotational velocity ω will be assumed as a constant. If this assumption is dropped, a further generalized coordinate, for example, the rotation angle θ, must be introduced, and the torsional behaviour of the shaft can be studied with the effects of unbalance. By introducing the series for $1/\cos(\gamma)$ truncated at the second term into the expression for $\sqrt{f_1(\theta)}$, it follows that

$$\sqrt{f_1(\theta)} = \sin(\theta) + \frac{\alpha}{4}(4 + \alpha^2 + 2\beta^2)\sin(2\theta) + \qquad (5.64)$$

$$-\frac{\alpha^3}{8}\sin(4\theta) - \frac{\alpha^2\beta}{2}\cos(\theta) + \frac{\alpha^2\beta}{2}\cos(3\theta).$$

By performing the relevant derivatives, using equation (5.64) for $\sqrt{f_1(\theta)}$ and assuming that the axis of the cylinder passes through the center of the crank ($\beta = 0$), the following equations of motion are obtained:

$$m_t\ddot{x} = Q_x - \omega^2(m_C r_C - m_1 r_B)\cos(\omega t) + \omega^2(m_2 + m_P)r_B^2 \times$$
$$\times\left[\cos(\omega t) + \frac{\alpha}{2}(4 + \alpha^2)\cos(2\omega t) - \frac{\alpha^3}{2}\cos(4\omega t)\right],$$
$$m_t\ddot{y} = Q_y - \omega^2(m_C r_C - m_1 r_B)\sin(\omega t)\,.$$

$$(5.65)$$

The last terms on the right-hand side of equation (5.65) are the unbalance forces acting on the crank. If the first-order moment $m_C r_C$ is equal to that of mass m_1, the inertia forces due to rotating masses are balanced and no force acts in the y-direction. Nevertheless, reciprocating masses m_2 and m_P cause inertia forces in the x-direction to be produced. They are periodic in time with fundamental frequency equal to ω and cannot be balanced by simple means such as adding counterweights. The unbalance forces in the x-direction consist of a first-order component with amplitude F_1 and frequency ω, a second-order force with amplitude F_2 and frequency 2ω, and a third-order component with amplitude F_3 and frequency 4ω. Their amplitudes are

$$F_1 = \omega^2 r_B(m_2 + m_P)\,,$$
$$F_2 = \omega^2 \frac{\alpha(4+\alpha^2)}{2} r_B(m_2 + m_P)\,, \qquad (5.66)$$
$$F_3 = \omega^2 \frac{\alpha^3}{2} r_B(m_2 + m_P)\,.$$

If more terms in the series for $1/\cos(\gamma)$ were considered, slightly different expressions of the forces would have been obtained together with higher-order terms. From a practical viewpoint, however, all forces of third and high order are considered negligible.

A first action aimed at reducing unbalance forces is that of over-balancing the shaft, i.e., of using a counterbalance greater than that needed to balance the rotating masses alone. If half of the reciprocating masses are balanced, i.e., if

$$m_C r_C = m_1 r_B + \frac{1}{2} r_B(m_2 + m_p)\,, \qquad (5.67)$$

the unbalance forces acting on the system in the x- and y-directions are, respectively,

$$\begin{cases} F_x = \tfrac{1}{2}F_1\cos(\omega t) + F_2\cos(2\omega t) + F_3\cos(4\omega t)\,, \\ F_y = -\tfrac{1}{2}F_1\sin(\omega t)\,. \end{cases} \qquad (5.68)$$

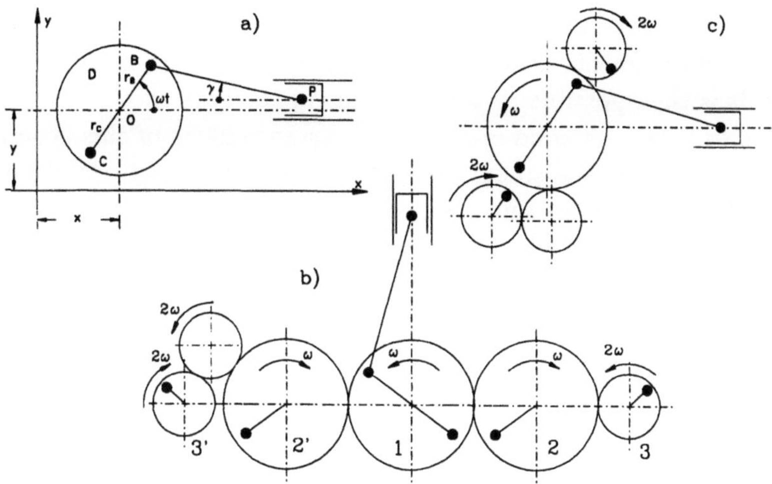

FIGURE 5.23. (a) Model for the study of the inertia forces acting in the x- and y-directions in a crank system, (b) scheme of the arrangement for balancing first- and second-order forces in a single-cylinder machine. First-order moments of the counterweights: shaft 1: $m_1 r_B + F_1/2\omega^2$; shafts 2 and 2' : $F_1/4\omega^2$; shafts 3 and 3': $F_2/8\omega^2$; (c) balancing second-order forces.

The resultant of the unbalanced first-order forces is a force with amplitude $F_1/2$, rotating with angular velocity ω in a direction opposite the direction of rotation. The first-order force is so halved and transformed from a force pulsating in the x-direction to a force rotating in the backward direction. It can be completely balanced by a counterweight with first-order moment $(m_2 + m_P)r_B/2$ rotating in the backward direction.

To avoid inertia torques, such a counterweight should rotate in the same plane of disc D and have the same axis of rotation. In practice, two counterweights located at the sides of the crank, each with a first-order moment equal to half that computed, must be used. In a similar way, second-order forces can be balanced by using two counterbalances, each having a first-order moment equal to $F_2/8\omega^2$, rotating at a speed that is double the rotational speed in opposite directions. The scheme of Figure 5.23b shows a possible arrangement in which first- and second-order forces are completely balanced. However, this scheme is very complicated, and in practice single-cylinder machines are seldom balanced.

In the case of multicylinder machines, first-order forces can be balanced simply by choosing a suitable phasing of the cranks. Because the cranks are located in different planes, care must be taken to balance both the forces and the moment due to unbalance. All first-order forces and moments are balanced in the case of the schemes of Figure 5.9. It is then possible to use a scheme of the type shown in Figure 5.23c to balance second-order forces. Arrangements of this type are used to improve the smoothness of the engines in high-class motor cars, for example. In this case, often the axes of the two balancing shafts are not located in a plane perpendicular to the axis of the cylinder to produce a torque with a frequency equal to 2ω, which can balance the harmonic with the same frequency as the driving torque of the engine (Figure 5.23c). This can be done only approximately, because the inertia torque is proportional to the square of the rotational speed, while the driving torque varies in time in an arbitrary way, being controlled by the driver.

5.10   Exercises

Exercise 5-1

A large marine drive system consists of a four-cylinder diesel engine with a detuner and a long drive shaft with variable diameter.

The geometry of the system can be summarized by subdividing it into 17 elements connecting 18 stations. In station 1 is the seismic mass of the detuner, in station 2 the part of the detuner connected to the crankshaft, in stations 3, 4, 5, and 6 the cranks, in station 7 the flywheel, and in station 18 the propeller.

The inertial and geometric properties of the fields from 2 to 17, which can be either cylindrical or conical, are reported in Table 5.1.

The stiffness of the detuner (field 1) is equal to 7.8×10^6 Nm/rad and the Young's modulus of the material is $E = 2.1 \times 10^{11}$ N/m^2. Draw a detailed sketch of the system. Compute the natural torsional frequencies of the system, with and without the detuner, by using a finite element model with more than 18 nodes to model the conical parts of the propeller shaft with prismatic elements. For the eigensolution, the model can be condensed using only eight master degrees of freedom. Plot the first four mode shapes, computing the rotations in all nodes, so as to obtain a detailed deformed shape of the propeller shaft.

Exercise 5-2

Consider the marine drive of the previous exercise. In node 8 the engine shaft ends with a gear wheel, which meshes with two gear wheels located on propeller shafts identical to the one in the previous exercise. Neglect the compliance of the meshing gears.

Compute the torsional eigenfrequencies and the mode shapes of the system, with the following data: moment of inertia of the propellers, the gear

Node	J [kg m²]	Node	J [kg m²]
1	280	5	2,580
2	111	6	2,218
3	2,260	7	450
4	2,620	18	4,180

Field	Type	D [mm]	l [mm]	Field	Type	D [mm]	l [mm]
2	cyl.	500	1,584	10	con.	585–495	145
3	cyl.	500	2,592	11	cyl.	495	276
4	cyl.	500	3,264	12	con.	495–585	145
5	cyl.	500	2,592	13	cyl.	585	5,532
6	con.	500–450	2,520	14	con.	585–495	145
7	cyl.	450	1,200	15	cyl.	495	276
8	con.	450–585	145	16	con.	495–585	145
9	cyl.	585	3,276	17	cyl.	585	5,040

TABLE 5.1. Exercise 5-1; inertial and geometrical properties.

wheel on the engine shaft and the propeller shafts are 3,200 kg m², 150 kg m², and 220 kg m², respectively; gear ratio is 0.86.

Exercise 5-3

Repeat the computations of the previous exercise, assuming that the stiffness of the meshing gears, relative to the driven systems, is 120×10^6 Nm/rad.

Exercise 5-4

Repeat the study of Example 5-6 using a springless dissipative damper.

Exercise 5-5

Consider the engine of Example 5-5 and repeat the dynamic analysis using a nonmodal approach. Because the way of taking into account damping used in the example is strictly linked with the modal approach, use the procedure described in Section 5.4.5 with a coefficient $k' = 56,000$. Assume that the diameter of element 6 is 45 mm.

Exercise 5-6

Repeat the computation of Exercise 5-5 using the modal approach, with the damping defined earlier and compare the results with those obtained in Example 5-5 and in Exercise 5-5.

Exercise 5-7

The engine of Exercise 5-5 is connected to a blower with a moment of inertia of 0.6 kg m² through a shaft of length 3 m and a joint whose stiffness is 20 kNm/rad. The moment of inertia of the part of the joint on the blower shaft is 0.05 kg m², and the inertia of the other part of the joint is already included in the moment of inertia of the flywheel. Compute the diameter to the drive shaft in order not to exceed a static stressing of 40 MN/m² in steady-state working and repeat the eigenanalysis of the system. Assume that the damping of the blower can be computed through equation (5.37) with $p = 2$ and compute the dynamic stressing of the system. Does the presence of the blower significantly affect the behaviour of the engine?

Exercise 5-8

Just after the elastic joint of the system of Exercise 5-7, connect a water pump, whose moment of inertia is 0.1 kg m^2, through a couple of gear wheels with transmission ratio 1.3. The pinion gear has a diameter of 200 mm, and both gears are 45 mm thick. Assume that the shaft leading to the pump is 4 m long and has a diameter of 50 mm, and estimate the moments of inertia of the gear wheels, assuming that they are made of steel. Neglect the compliance of the meshing gears. Repeat the dynamic computations, with the main aim of evaluating the dynamic torque on the gear wheels and the dynamic stressing on the shaft of the pump.

Exercise 5-9

A marine unit is made by four engines of the type described in Example 5-4 driving two propellers through gear wheels. The engines work in pairs, the two engines being assembled with the flywheels one near the other, connected through a shaft 600 mm long, whose diameter is computed in order not to exceed a shear stress of 30 MN/m^2 in static conditions. The two pairs are located side by side, driving a common shaft through gear wheels. The transmission ratio is 0.9, the diameter of the pinion gear is 208 mm, and the thickness of the gear wheels is 30 mm. The common shaft is 4 m long, and its diameter is chosen for the same value of the static shear stress as the other shafts connecting the engines. Two propeller shafts, 6 m long, are then driven with a transmission ratio of 0.4. The diameter of the pinion gear is 200 mm, and the thickness of the wheels is 30 mm.

Draw a detailed sketch of the system. Knowing that the moments of inertia of the propellers are equal to 12 kg m^2, estimating the moments of inertia and the geometric properties of all elements, and neglecting damping due to the propellers, compute the natural frequencies and the dynamic stressing of the system.

Exercise 5-10

Study a dissipative damper to be added at the free end of each crankshaft of the engines of Exercise 5-9 to reduce dynamic stressing.

6

Short Outline on Controlled and Active Systems

6.1 General considerations

Consider a structure[1] provided with a number of actuators and modeled as a discrete system (Figure 6.1). Its equation of motion in the space of the configuration can be expressed in the form

$$\mathbf{M}\ddot{\mathbf{x}} + \mathbf{C}\dot{\mathbf{x}} + \mathbf{K}\mathbf{x} = \mathbf{f}_c + \mathbf{f}_e \,, \qquad (6.1)$$

where the control forces exerted by the actuators and the disturbances or external forces can be expressed, respectively, as $\mathbf{f}_c = \mathbf{T}_c\mathbf{u}_c$ and $\mathbf{f}_e = \mathbf{T}_e\mathbf{u}_e$, where $\mathbf{u}_c$ and $\mathbf{u}_e$ are the control and external inputs. Matrix $\mathbf{T}_c$ is sometimes referred to as the *control influence matrix*.

The corresponding state and output equations are

$$\begin{cases} \dot{\mathbf{z}} = \boldsymbol{A}\mathbf{z} + \boldsymbol{B}_c\mathbf{u}_c(t) + \boldsymbol{B}_e\mathbf{u}_e(t) \,, \\ \mathbf{y}(t) = \boldsymbol{C}\mathbf{z} \,, \end{cases} \qquad (6.2)$$

where the inputs are assumed not to directly influence the outputs through the direct link matrix $\boldsymbol{D}$. The input gain matrices are simply

---

[1]In control terminology the controlled system is usually referred to as *plant*. In the following sections the more specific term *structure* will also be used: No attempt to deal with control theory in general is intended.

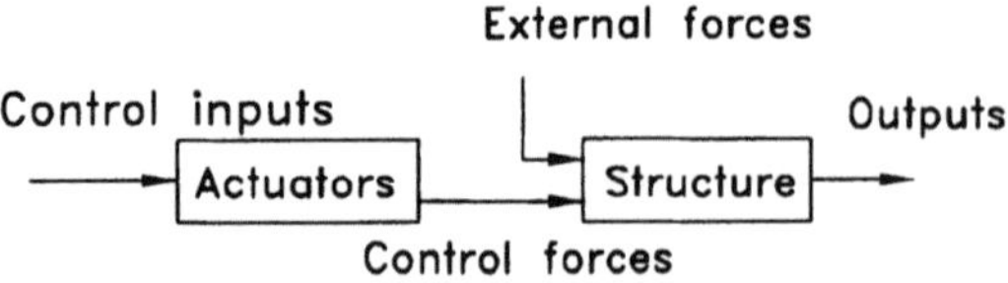

FIGURE 6.1. Block diagram of a structure on which both control and external forces are acting.

$$\boldsymbol{B}_c = \left[\begin{array}{c} \mathbf{M}^{-1}\mathbf{T}_c \\ 0 \end{array} \right], \qquad \boldsymbol{B}_e = \left[\begin{array}{c} \mathbf{M}^{-1}\mathbf{T}_e \\ 0 \end{array} \right].$$

The designer must deal with the whole system, made of the structure as well as the actuators, the control system that must provide the latter with the control inputs, and, in the case of active systems, the source of power for the actuators.

When designing an active system, the primary concern is often shifted from obtaining the required response to achieving stability. Or, better, stability becomes a prerequisite that must be fulfilled before thinking about performances.

This is new compared with the usual approach to structural dynamics: In most structures, stability was taken for granted, because the structure could only dissipate energy and free vibrations were bound to be extinguished sooner or later. The designer had to provide sufficient damping, but at least he was sure that the motion was stable. Notable exceptions were the cases in which the structure could absorb energy from its environment, as with aeroelastic structures or rotating machines. If even a small fraction of the energy from the aerodynamic field or kinetic energy could seep into the vibration, strong self-excited vibrations could take place. In these cases, stability has always been a primary concern.

For ideal and colocated active systems, a theorem that ensures marginal stability exists, but all real-world active systems are prone to instability: When the structure is acted on by actuators of any type, the control system must avoid supplying energy to any vibration motion, otherwise an unstable behaviour can occur.

The behaviour of any active structure is the result of the integration of the behaviour of the structural subsystem with that of the controller, the actuators, and the sensors; the only reasonable approach is to design the system as a whole. To add a controller to an aleready existing structure or one designed without taking into account the presence of the former may lead to performances far from the expected.

Structural dynamics, control engineering, transducer design, and electronics must merge from the beginning with that interdisciplinary approach often referred to as *mechatronics*.

6.2   Control systems

The science of control systems and the related technology saw enormous advances in recent decades, and it is impossible to summarize them satisfactorily in a text on structural dynamics. Only a few remarks on control systems will be reported here, limited to what can be useful in the context of structural control; the interested reader can find the relevant information in many textbooks on the subject. Classical control theory deals with linear, or at least linearized, control systems. The basic tools are those typical of linear system dynamics, namely, block diagrams, phase- or state-space equations, transfer functions, and eigenstructure analysis.

The control systems used for structural control can be based on transducers (sensors and/or actuators) of different types, such as mechanical, electrical, hydraulic, and pneumatic. Recently, however, electronic-based systems are becoming more common, both for all-electrical applications and in the form of electromechanical, electrohydraulic, and so on, applications. The electronic part can be based on analog or digital circuits; the first are preferred for simpler applications, where they are still cheaper than the latter. With the diffusion of microprocessor systems, digital techniques became more common, particularly for their flexibility and ability to perform very complex tasks.

Independently from the physical configuration, control systems can be divided into two categories: passive and active control systems. The first operate without any external energy supply, using the energy stored in the structure as potential or kinetic energy as a consequence of the dynamic response of the structure to be controlled to supply the control forces. Passive devices in many cases act as dampers. For example, piezoceramic materials can be used as both sensors and actuators: If a piezoelectric element is simply shunted by a resistor, a sort of electric damper is obtained. By also introducing an inductor into the circuit, the capacitance of the piezoelectric element forms, together with the other components of the circuit, a resonant inductor-resistor-capacitor system whose transfer function can be designed to obtain performance similar to that of a damped vibration absorber. Note that piezoelectric elements can also be used in active systems, as both sensors and actuators. Layers of piezoelectric materials can also perform as distributed sensory or actuating devices.

Active control systems are equipped with an external power supply system to provide the control forces. There are cases in which the amount of control energy is minimal and in this case the term semiactive systems is sometimes used. Active systems can be either manual or automatic, but only the latter are important in structural control, particularly if true dynamic control is required.

Both active and passive control systems can operate as open-loop or closed-loop systems. Open-loop control systems, sometimes referred to as *predetermined control systems*, react to the variation of the input parameters of the controlled system without actually measuring its output parameters in order to check whether the response of the system conforms with the required values. Using the example in the previous section, a system that changes the stiffness of the supports of a rotor depending on its angular velocity, is an open-loop system. In closed-loop or feedback systems, the control system monitors the outputs of the controlled system, compares their actual values with predetermined reference values, and uses the information so obtained to perform the control action. Closed- and open-loop techniques can be used simultaneously, as in the case of feedback systems in which the rapidity of the response of the controller is incremented by also monitoring the excitation and using this information to help control the system (feedforward technique).

An example of an active closed-loop control system taken outside the field of structural control is the driver of a vehicle. A blind driver who tries to drive home relying on his knowledge of the road would be an example of open-loop control.

An active magnetic bearing is an example of an active closed-loop control system. Most passive systems are closed-loop systems. Closed-loop systems are usually preferred when the control is required to react to unknown external or internal disturbances, but they are usually more complex and costly than open-loop systems. When the control is performed by mechanical means, as, for example, in the case of Watt's regulator, it is often quite arbitrary to state what is the controlled system and what is the controller.

Another common distinction for control systems is that between regulators and servomechanisms or tracking systems. A regulator is used to maintain the system in a predetermined condition, which can be an equilibrium position, a velocity, or an acceleration, in spite of external disturbance. The spring of the inverted pendulum in Figure 6.2 can be considered a regulator, even if it cannot achieve its goal with precision in the presence of a constant disturbance: If a constant force F acts on the pendulum (the controlled system), the spring cannot restore the position with $\theta = 0$. The input to the system controlled by a regulator is the condition to be maintained, and it is usually referred to as the *set point*. It is a constant, but it can be changed in many cases from one constant value to another. In the case of a servomechanism, however, the reference input changes in time and the control system tries to obtain an output of the controlled system that follows the reference. In this case the output can also be a position, a velocity, an acceleration, or any other relevant quantity.

A control system can be a single-input-single-output (SISO) or multiple-input-multiple-output (MIMO) or multivariable system. In the first case, the control system reacts to only one of the outputs of the controlled system (the output can be a single state or a combination of states) with just one control input u_c: They are clearly not restricted to systems with a single degree of freedom. MIMO control systems use a number of outputs of the plant to act on a number of control inputs. When each input separately controls a single output, the term *decentralized control* is used. Decentralized control is sometimes resorted to with the aim of weakening the coupling of a MIMO plant.

Example 6-1
Consider the pendulum shown in Figure 6.2; it is stabilized in its inverted position by a linear spring. In Figure 6.2a the pendulum-spring assembly is considered the system, while in Figure 6.2b the pendulum is considered the controlled system, and the spring is the controller.

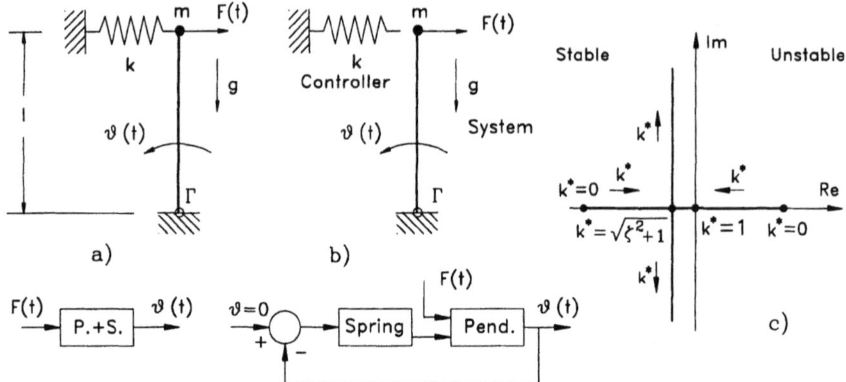

FIGURE 6.2. Inverted pendulum stabilized by a linear spring: (a) The whole assembly is considered the system; (b) the pendulum is considered the system and the spring is considered the controller; (c) root locus in nondimensional form.

If a force $F(t)$ acts as a disturbance on the pendulum and a viscous damping with coefficient Γ is provided by the hinge, the linearized equation for small displacements about the inverted upright position, obtained as dynamic equilibrium equation for rotations about the hinge point, is

$$ml^2\ddot{\theta} + \Gamma\dot{\theta} + (kl^2 - mgl)\theta = lF(t)\,.$$

The transfer function of the system of Figure 6.2a is then

$$G(s) = \frac{1}{ml^2 s^2 + \Gamma s + kl^2 - mgl}\,.$$

If the scheme of Figure 6.2b is considered, the equation of motion of the system is

$$ml\ddot{\theta} + \frac{\Gamma}{l}\dot{\theta} - mg\theta = F_c + F(t)\,,$$

where F_c is the control force and the equation of the control system is $F_c = -kl\theta$.
The characteristic equation of the controlled system, including the controller, is in nondimensional form

$$s^{*2} + 2\zeta s^* + k^* - 1 = 0\,,$$

where

$$s^* = s\sqrt{\frac{l}{g}}\,, \qquad \zeta = \frac{\Gamma}{2m\sqrt{gl^3}}\,, \qquad k^* = k\frac{l}{mg}\,.$$

The root locus is plotted in Figure 6.2c. The breakaway point occurs for $k^* = \sqrt{1 + \zeta^2}$, and the system is stable if $k^* > 1$, i.e., if $k > mg/l$. The last condition could be obtained directly from the equation of motion, stating that the total stiffness of the system must be positive to have stable behaviour.

6.3 Controlled linear systems

6.3.1 Controllability and observability

Consider the linear system whose behaviour is described by equations (6.2). The possibility of controlling it through the control inputs $\mathbf{u}_c$ and of knowing its state from the observation of its outputs $\mathbf{y}$ are defined as its controllability and observability.

If it is possible to determine a law for the inputs $\mathbf{u}_c(t)$ defined from time t_0 to time t_f, which allows the system to be driven from the initial state to a desired state and in particular to a state with all state variables equal to zero, the system is said to be *controllable*. If this can be done for any arbitrary initial time t_0 and initial state, then the system is completely controllable. To check whether the system is controllable, it is possible to write the controllability matrix

$$\mathbf{H} = \left[\begin{array}{ccccc} \boldsymbol{B}_c & \boldsymbol{A}\boldsymbol{B}_c & \boldsymbol{A}^2\boldsymbol{B}_c & \ldots & \boldsymbol{A}^{n-r}\boldsymbol{B}_c \end{array} \right], \qquad (6.3)$$

where r is the number of control inputs, i.e., the number of columns of matrix $\boldsymbol{B}_c$. If such a matrix, which has a total of n rows and $n \times r$ columns, has rank n, the system is completely controllable. The controllability matrix can also be written considering products $\boldsymbol{A}^i\boldsymbol{B}_c$ until $i = n - 1$ instead of $i = n - r$. In this way, the number of columns of the controllability matrix is $r \times n$ instead of n, but nothing is changed, because the added columns are linear combinations of the others.

In a similar way, a linear system is said to be *observable* if it is possible to determine the state at time t_0 from the laws of the inputs $\mathbf{u}(t)$ and the outputs $\mathbf{y}(t)$ defined from time t_0 to time t_f. If this can be done for any arbitrary initial time t_0 and initial state, then the system is completely observable. To check whether a linear system with fixed parameters is observable, it is possible to write the observability matrix

$$\mathbf{O} = \begin{bmatrix} \boldsymbol{C}^T & \boldsymbol{A}^T \boldsymbol{C}^T & (\boldsymbol{A}^T)^2 \boldsymbol{C}^T & \cdots & (\boldsymbol{A}^T)^{n-m} \boldsymbol{C}^T \end{bmatrix}, \qquad (6.4)$$

where m is the number of outputs, i.e., the number of rows of matrix $\boldsymbol{C}$.

If such a matrix, which has a total of n rows and $n \times m$ columns, has rank n, the system is completely observable.

It is easy to verify that in the case of a linear structural system with a single degree of freedom both conditions for observability and controllability are always verified. This is, however, not necessarily the case for systems with many degrees of freedom, where observability and controllability must be checked in each case.

The study of the controllability and observability matrices allows one to obtain a sort of on–off answer on the issue of whether a structure can be controlled by a given set of actuators or observed by a set of sensors but does not state how controllable or observable it is. Other criteria allowing one to obtain a measure of the controllability and observability of a plant, and then allowing to search for the best position of transducers, are described in the literature[2].

It is possible to define the controllability and observability starting from the equations written in modal coordinates; in this way it is possible to state which modes can be controlled and observed by a given set of transducers. The obvious general rule is that a mode cannot be either observed or controlled by a transducer located near one of the nodes of the relevant mode shape.

6.3.2 *Open-loop control*

Consider the structure (plant, in the usual control terminology) in Figure 6.1. The actuators added to provide suitable control forces $\mathbf{F}_c$ are driven by a controller to which a number of reference inputs $\mathbf{r}(t)$ is provided (block diagram in Figure 6.3a). If the controller is linear, the control input can thus be expressed as

$$\mathbf{u}_c = \mathbf{K}_r \mathbf{r}(t). \qquad (6.5)$$

A more complex example of an open-loop system is a system with input compensation, in which a device supplies a set of control inputs

---

[2]See for example J.L. Junkins, Y. Kim, *Introduction to Dynamics and Control of Flexible Structures*, AIAA, Washington D.C., 1993.

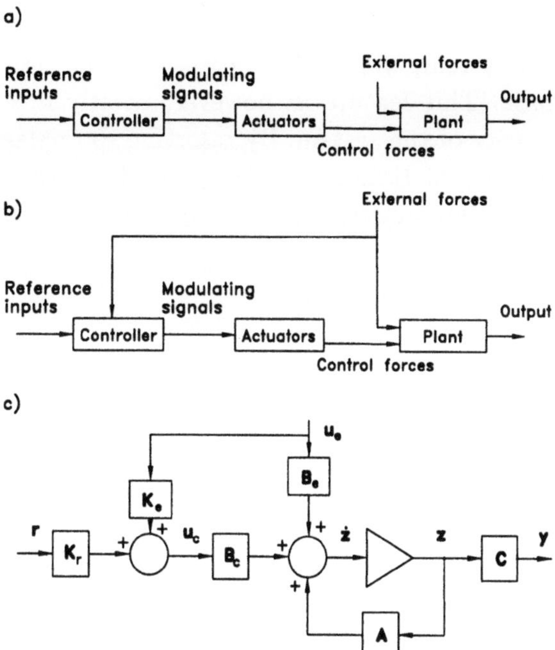

FIGURE 6.3. Open-loop control of the plant of Figure 6.1: (a) tracking system in which the control input tries to follow an external reference; (b) open loop control with external input compensation; (c) block diagram of the system shown in (b).

$\mathbf{u}_c$ that are functions not only of the reference inputs $\mathbf{r}(t)$ but also of the external forces $\mathbf{f}_e(t)$, or better, of the external inputs $\mathbf{u}_e(t)$, applied to the system (Figures 6.3b, c)

$$\mathbf{u}_c = \mathbf{K}_r \mathbf{r}(t) + \mathbf{K}_e \mathbf{u}_e(t) . \tag{6.6}$$

The matrices of the gains of the control system $\mathbf{K}_r$ and $\mathbf{K}_e$ have as many rows as the control inputs and as many columns as the reference and external inputs, respectively. The total state equation of the controlled system can be obtained by introducing equations (6.6) into equation (6.2)

$$\begin{cases} \dot{\mathbf{z}} = \boldsymbol{A}\mathbf{z} + (\boldsymbol{B}_c \mathbf{K}_e + \boldsymbol{B}_e)\,\mathbf{u}_e(t) + \boldsymbol{B}_c \mathbf{K}_r \mathbf{r}(t) , \\ \mathbf{y}(t) = \boldsymbol{C}\mathbf{z} . \end{cases} \tag{6.7}$$

In general, an open-loop system relies on the model of the plant to obtain a command input that, supplied to it, causes the output to follow a desired pattern. This strategy requires very good knowledge of the dynamics of the controlled system and is usually applied only as a feedforward component in conjunction with a feedback controller.

The free response of the system is then not affected by the presence of the control system, which plays a role only in determining the forced response. This feature is, however, strictly linked with the complete linearity of the system. By resorting to Laplace transform and assuming that at time $t = 0$ the value of all state variables is zero, the transfer function linking the outputs of the system with the external inputs can be easily computed

$$\mathbf{G}(s) = \boldsymbol{C}\left(s\mathbf{I} - \boldsymbol{A}\right)^{-1}\left(\boldsymbol{B}_c\mathbf{K}_e(s) + \boldsymbol{B}_e\right). \tag{6.8}$$

Note that the gains of the control system are considered functions of the Laplace variable s.

6.3.3 Closed-loop control

Consider now a structure controlled by a closed-loop system (Figure 6.4). The reference inputs $\mathbf{r}(t)$ interact with the outputs of the system $\mathbf{y}$ to supply suitable control inputs $\mathbf{u}_c$ to a set of actuators that produce the control forces. The actuators can be active systems, or passive elements, such as the spring in Figure 6.2b. The general block diagram of a feedback system is shown in Figure 6.2b. The open-loop transfer functions of the controlled system and control system are indicated as $G_{ol}(s)$ and $H(s)$, respectively.

Consider a SISO system. If no external disturbances are acting, i.e., $\mathbf{u}_e(t) = 0$, the output $y(t)$ is linked with the reference input $r(t)$ by the following relationship, written in the Laplace domain

$$y(s) = G_{ol}(s)\left[r(s) - H(s)y(s)\right]. \tag{6.9}$$

The closed-loop transfer function is then

$$G_{cl}(s) = \frac{y(s)}{r(s)} = \frac{G(s)}{1 + G(s)H(s)}. \tag{6.10}$$

In the case of MIMO systems, the input-output relationship is

$$\mathbf{y}(s) = \left(\mathbf{I} + \mathbf{G}_{ol}(s)\mathbf{H}(s)\right)^{-1}\mathbf{G}(s)\mathbf{r}(s). \tag{6.11}$$

The state equation of the controlled system is still equation (6.2), but now the control inputs $\mathbf{u}_c$ are determined by the outputs $\mathbf{y}(t)$ of the system and the reference inputs $\mathbf{r}(t)$

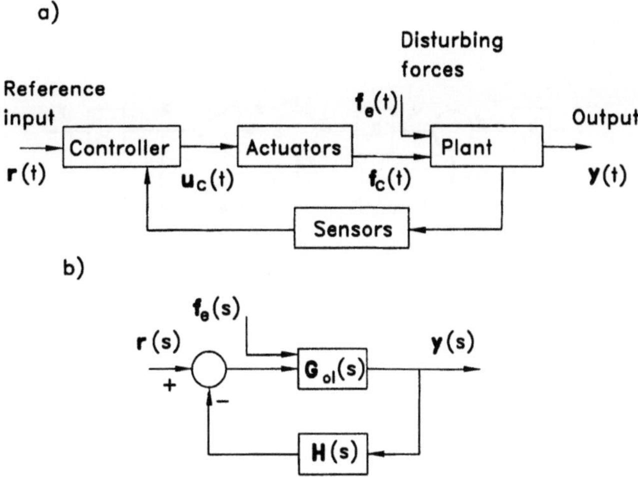

FIGURE 6.4. (a) Block diagram of the multi-degree-of-freedom structure in Figure 6.1 controlled by a feedback system; (b) general block diagram of a feedback system in the Laplace domain.

$$\mathbf{u}_c = \mathbf{K}_r \mathbf{r} - \mathbf{K}_y \mathbf{y}\,, \tag{6.12}$$

where the size of all matrices and vectors depends on the number of control and reference inputs and the number of outputs of the system. The state equation of the controlled system is then

$$\dot{\mathbf{z}} = \left(\boldsymbol{A} - \boldsymbol{B}_c \mathbf{K}_y \boldsymbol{C}\right)\mathbf{z} + \boldsymbol{B}_c \mathbf{K}_r \mathbf{r}(t) + \boldsymbol{B}_e \mathbf{u}_e(t)\,. \tag{6.13}$$

If the control system is a regulator, vector $\mathbf{r}$ contains just the constants that define the set point. If the aim of the control system is to maintain the structure in the static equilibrium position in spite of the presence of the perturbing inputs $\mathbf{u}_e(t)$, the reference inputs are equal to zero, and the equation of motion can be simplified. Note that the presence of the control loop affects the free behaviour of the system as well as the forced response. The very stability of the system can be affected and, while the control system can be used to increase the stability of the structure or to give an artificial stability to an unstable system, the behaviour of the system must be carefully studied to avoid the presence of the control system inducing unwanted instabilities in some modes.

The block diagram corresponding to equation (6.13) is shown in Figure 6.5. This type of feedback is usually referred to as *output feedback*, because the loop is closed using just the outputs of the

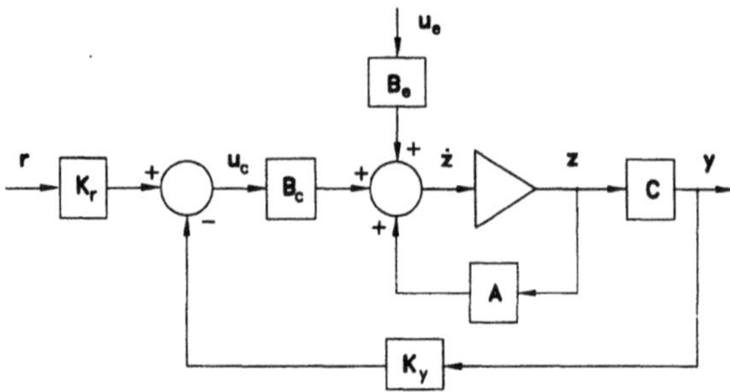

FIGURE 6.5. Block diagram of the multi-degree-of-freedom structure in Figure 6.4 controlled by an output feedback system.

system. The design of such a control system consists of determining the gain matrices $\mathbf{K}_y$ and $\mathbf{K}_r$, which enable the system to achieve the desired behaviour.

In the case of structural control, it is possible to write the equation of motion of the controlled system in the configuration space. Assume that the reference input is vanishingly small (i.e., the structure must be kept in the undeformed configuration); the output of the system can be written in the form

$$\mathbf{y}(t) = \begin{bmatrix} \boldsymbol{C}_1 & 0 \\ 0 & \boldsymbol{C}_2 \end{bmatrix} \left\{ \begin{array}{c} \dot{\mathbf{x}} \\ \mathbf{x} \end{array} \right\} \tag{6.14}$$

and the gain matrix $\mathbf{K}_y$ has the form

$$\mathbf{K}_e = \begin{bmatrix} \mathbf{G}_1 & 0 \\ 0 & \mathbf{G}_2 \end{bmatrix}. \tag{6.15}$$

These assumptions correspond to a sort of separation between the control inputs linked to the position and the velocity outputs of the system. They lead to control forces that can be expressed as

$$\mathbf{f}_c = -\mathbf{T}_c\mathbf{G}_1\boldsymbol{C}_1\dot{\mathbf{x}} - \mathbf{T}_c\mathbf{G}_2\boldsymbol{C}_2\mathbf{x}. \tag{6.16}$$

The equation of motion of the controlled system is then

$$\mathbf{M}\ddot{\mathbf{x}} + (\mathbf{C} + \mathbf{T}_c\mathbf{G}_1\boldsymbol{C}_1)\dot{\mathbf{x}} + (\mathbf{K} + \mathbf{T}_c\mathbf{G}_2\boldsymbol{C}_2)\mathbf{x} = \mathbf{f}_e. \tag{6.17}$$

Gains $\mathbf{G}_1$ and $\mathbf{G}_2$ then cause an increase of the damping and stiffness characteristics of the structure, respectively.

If the actuators and sensors are collocated, i.e., the actuators exert their control forces in the same location in which the sensors observe the displacements and the velocities of the structure, the output gains matrices are equal to the transpose of the control influence matrix: $\boldsymbol{C}_1 = \boldsymbol{C}_2 = \mathbf{T}_c^T$. If the gain matrices $\mathbf{G}_1$ and $\mathbf{G}_2$ are fully populated, symmetric positive defined matrices, then the damping and stiffness perturbations due to the control system $\mathbf{T}_c\mathbf{G}_1\boldsymbol{C}_1$ and $\mathbf{T}_c\mathbf{G}_2\boldsymbol{C}_2$, are themselves symmetric positive semidefinite matrices. This is enough to state that the controlled system is asymptotically stable.

The practical interest of this statement is, however, reduced by the considerations that in practice it can be very difficult to achieve a perfect sensor-actuator collocation and that in actual systems the behaviour of sensors and actuators differ from the ideal behaviour here assumed.

By resorting to Laplace transform and assuming that at time $t = 0$ the value of all state variables is zero, the closed-loop transfer function linking the outputs of the system $\mathbf{y}(s)$ with the external $\mathbf{u}_e(s)$ inputs when all reference inputs are equal to zero can be easily computed. For a system with output feedback, it follows that

$$\mathbf{G}(s) = \boldsymbol{C}\left(s\mathbf{I} - \boldsymbol{A} + \boldsymbol{B}_c\mathbf{K}_e(s)\boldsymbol{C}\right)^{-1}\boldsymbol{B}_e. \tag{6.18}$$

Also in this case, the gains of the control system are considered functions of the Laplace variable s. The poles of the closed-loop system are then the roots of the characteristic equation

$$\det\left(s\mathbf{I} - \boldsymbol{A} + \boldsymbol{B}_c\mathbf{K}_e(s)\boldsymbol{C}\right) = 0. \tag{6.19}$$

Similar equations hold for systems with state feedback and for the transfer functions linking the output with the reference input $\mathbf{r}(s)$.

Example 6-2
Consider the inverted pendulum in Figure 6.2. The system has a single degree of freedom and, hence, $n = 2$. The vectors and matrices included in the equation of motion of the controlled system are

$$\mathbf{z} = \left\{ \begin{array}{c} \dot{\theta} \\ \theta \end{array} \right\}, \qquad \boldsymbol{A} = \left[\begin{array}{cc} -\dfrac{\Gamma}{ml^2} & \dfrac{g}{l} \\ 1 & 0 \end{array} \right],$$

$$\boldsymbol{B}_e = \boldsymbol{B}_c = \left\{ \begin{array}{c} \dfrac{1}{ml} \\ 0 \end{array} \right\}, \qquad \mathbf{u}_e = F.$$

The output and the control parameter of the system coincide with the angular displacement θ, and the gain matrix $\mathbf{K}_y$ states that the dependence of the generalized force exerted by the spring as a function of the angular displacement: $\mathbf{y} = \theta$ $\mathbf{u} = \theta$, $\boldsymbol{C} = [0,1]$ and $\mathbf{K}_y = kl^2$. Because the set point is characterized by $\theta = 0$, it follows that $\mathbf{r} = 0$. Introducing the aforementioned values of all the relevant matrices and vectors into equation (6.13) the same equation of motion of the controlled system seen in Example 6-1 is obtained.

6.3.4 Basic control laws

In the preceding sections the controller was assumed to provide a set of control inputs $\mathbf{u}$ to the controlled system, which are determined just as linear combinations of the outputs of the system $\mathbf{y}$ and the control inputs $\mathbf{r}(t)$. In the case of output control, if the number of reference inputs r is equal to the number of outputs of the plant m and the two gain matrices $\mathbf{K}_y$ and $\mathbf{K}_r$ are equal, a control law of this type can be further simplified in the form

$$\mathbf{u} = \mathbf{K}_p \mathbf{e}_y = \mathbf{K}_p (\mathbf{r} - \mathbf{y}),$$

where the elements of vector $\mathbf{e}_y(t)$ are the errors of the output of the system, and matrix $\mathbf{K}_p$ contains the gains of the control system. This type of control is said to be *proportional control* (hence, the gain in this case can be called *proportional gain*) and provides a large corrective action when the instantaneous errors are large. It has the main advantage of being very simple but it has several disadvantages, such as a lack of precision in certain instances and the possibility of producing instability. In a certain sense, it stiffens the system.

A different choice, always within the frame of linear systems, is the so-called derivative control, which, in the case of SISO systems, can be summarized in the form

$$u = K_d \frac{de_y}{dt}.$$

This type of control reacts more to the increase of error than to the error itself and provides large corrective actions when the errors increase at a high rate. In a way, it provides a sort of damping to the system and enhances stability, but it is insensitive to constant errors and not very sensitive to errors that accumulate slowly. It is prone to cause a drift in the output of the system, but this disadvantage is not critical in structural control, where the control system must prevent vibrations, and its reaction to static or quasi-static forces is of little importance. In harmonic motion, its effectiveness increases with the frequency of the perturbation.

A third possibility is the so-called integral control, which can be summarized in the form

$$u = K_i \int_0^t e_y(t)dt \,.$$

It reacts to the accumulation of errors and causes a slow-reacting control action. Its disadvantages are mainly that it is insensitive to high frequencies and prone to cause instability.

Due to the different characteristics of the different control laws, often a law that combines the aforementioned control strategies, and possibly others based on higher-order derivatives or integrals, is used. This type of control is usually referred to as *proportional-integral-derivative* (PID) control. In the Laplace domain, the relationship between the control input to the system $u(s)$ and the error $e_y(s)$ for a SISO PID control system can be expressed in the form

$$u(s) = P(s)e_y(s) = \left(sK_d + K_p + K_i \frac{1}{s} \right) e_y(s) \,. \tag{6.20}$$

Equation (6.20) defines the gains of the control system as functions in the Laplace domain.

The outputs of the plant are a linear combination of the states and are also influenced by the velocities. In general, a purely proportional control has some of the features of derivative control and, in particular, a damping effect. It must be noted that the gains are usually functions of the Laplace variable s or, in the case of harmonic response, of the frequency λ, which depends on the actual physical configuration of the control system. A first effect, which is often unwanted, is due to the impossibility, for all the control system-actuator combinations, to react in an infinitely fast way to the inputs provided by the sensors. A simple way of modeling this delay is to assume that the control system acts as a first-order system, i.e., a

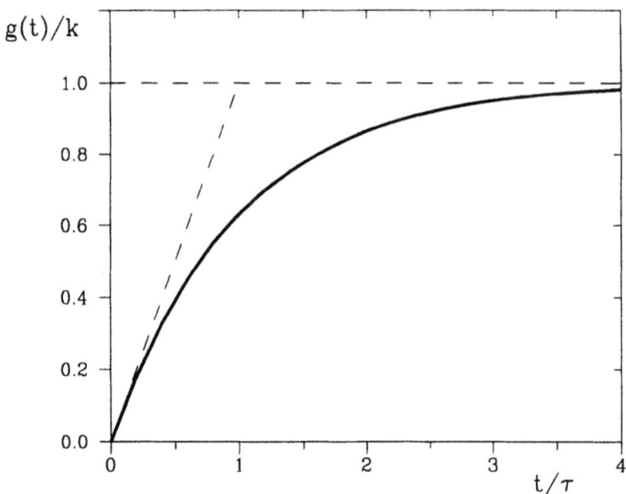

FIGURE 6.6. Nondimensional response of a first-order system to a unit step.

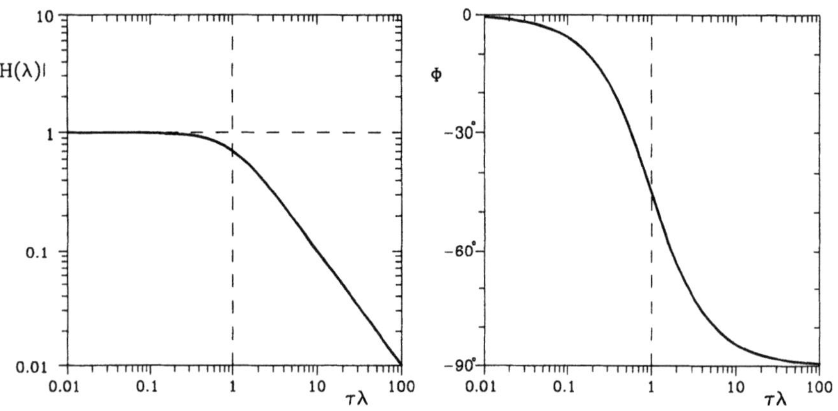

FIGURE 6.7. Bode diagram of the frequency response of a first-order system.

system governed by a first-order differential equation. For a SISO proportional controller it follows that

$$\tau \dot{u} + u = k e_y .$$

The transfer function of the control system is

$$\frac{e_y(s)}{u(s)} = k\frac{1}{1 + s\tau} . \tag{6.21}$$

The physical meaning of parameters k and τ is immediate. The first is the gain of the control system; the second is the time constant for a step input. The response $g(t)$ to a unit step input, as defined by equation (1.124), is reported in nondimensional form in Figure 6.6. It is simply

$$g(t) = k \left(1 - e^{-t/\tau}\right) . \tag{6.22}$$

The frequency response of a first-order system can be easily obtained from the transfer function (6.21) by substituting $i\lambda$ to s. Separating the real part from the imaginary part, its expression is

$$\Re[H(\lambda)] = \frac{k}{1 + \tau^2\lambda^2}, \qquad |H(\lambda)| = \frac{k}{\sqrt{1 + \tau^2\lambda^2}},$$
$$\Im[H(\lambda)] = -\frac{k\tau\lambda}{1 + \tau^2\lambda^2}, \qquad \Phi = \arctan(-\tau\lambda) . \tag{6.23}$$

The bode diagram of the frequency response is plotted in Figure 6.7.

The circular frequency $\lambda_g = 1/\tau$ is the so-called cut-off frequency. Note that the period corresponding to the cut-off frequency defined here does not coincide with the time constant but is $T_g = 2\pi\tau$. The first-order system acts as a low pass filter with an attenuation, at frequencies higher than the cut-off frequency, of about 6 dB/oct or 20 dB/dec. It also introduces a phase lag of the response that goes from zero at very low frequency to 90° at a frequency tending to infinity. At the cut-off frequency the phase lag is 45°. The effectiveness of the control system is quickly reduced at frequencies higher than the cut-off frequency.

The time constant of the control system–actuators combination is a most important parameter that can dictate the choice of a particular layout, especially when high-frequency operation is required. In most cases the control system can be modeled as a first-order system only as a very rough approximation. There are usually limits above which a linear model is no longer possible: All types of control systems can supply an output that is limited in magnitude. When the maximum response is attained, the saturation phenomenon occurs and the gain decreases as a function of the input, giving way to a nonlinear response.

Apart from the presence of a cut-off frequency, a dependence of the control law from the frequency can be purposely devised by adding a compensator with an appropriate law $K(s)$. For example, if the disturbance frequencies are well determined, an ideal control system would have a very small gain except in correspondence to the frequencies of the disturbances to be suppressed. A support that insulates a device from external disturbances should behave as a very stiff system at all frequencies except those that must be suppressed, where its stiffness should approach zero. The use of a compensator allows the frequency response of the system to be tailored to fulfill the design goals of the particular application, to both achieve the required performance and ensure adequate robustness. This can be done easily with electronic control systems, which can be designed to obtain almost arbitrary transfer functions, provided the required output does not exceed the maximum limits.

In many cases, some of the parameters of the system are not well known or are prone to change in time. There are many examples in the field of structural dynamics, such as the case of hysteretic damping, which is usually poorly known and is affected by many parameters in a way that is often impossible to control, or that of the wandering unbalance, which is shown by many rotors. One of the advantages of a feedback control system is that they can usually compensate for these unwanted effects, because they measure directly the outputs and act to keep them within stated limits. Feedback control systems can even compensate for the uncertainties and variations of their own parameters.

A system that is little affected by changes of operating conditions, by parameter variations, and by external disturbances is said to be robust and robustness is one of the basic requisites of control systems. Generally speaking, the sensitivity of the quantity q to the variations of parameter α is measured by the derivative $\partial q / \partial \alpha$. By computing the sensitivities of the relevant characteristics of the system (eigenvalues, frequency responses, etc.) to the variation of the critical parameters, it is possible to assess its robustness. The root sensitivity $S_{i,\alpha}$ of the ith root s_i of the transfer function to parameter α, for example, can be defined as

$$S_{i,\alpha} = \alpha \left(\frac{\partial s_i}{\partial \alpha} \right) = \frac{\partial s_i}{\partial \log(\alpha)}. \tag{6.24}$$

The derivative is, in many cases, computed numerically, by giving a small variation to the parameter under study and computing a new value of the relevant characteristic of the system. If the dynamic matrix $\boldsymbol{A}$ can be differentiated with respect to parameter α, the sensitivity of the ith eigenvalue s_i of the dynamic matrix $\partial q / \partial \alpha$ can be computed in closed form

$$\frac{\partial s_i}{\partial \alpha} = \mathbf{q}_{Li}^T \frac{\partial \mathbf{A}_i}{\partial \log(\alpha)} \mathbf{q}_{Ri} \,, \tag{6.25}$$

where $\mathbf{q}_{Li}$ and $\mathbf{q}_{Ri}$ are, respectively, the ith left and right eigenvectors. Similar but more complicated expressions can be found in the literature for the eigenvector sensitivity.

Example 6-3

Consider the linear mechanical system with a single degree of freedom shown in Figure 6.8a. The control force is supplied by an actuator acting on an auxiliary spring of stiffness k_1 and governed by an active control system.

With reference to the figure, the equation of motion of the system is

$$m\ddot{x} + c\dot{x} + kx = F(t) + F_c(t)\,.$$

The control force can be expressed as $F_c = -k_1(x - u)$, and the equation of motion of the controlled system is

$$m\ddot{x} + c\dot{x} + (k + k_1)x = k_1 u + F(t)\,.$$

Consider a closed-loop control system in which the displacement $u(t)$ is a linear function of the displacement x. The block diagram, in which the gain of the control system P is referred to as $G_c(s)$, is shown in Figure 6.8b. This type of control is a regulator, in the sense described in Section 6.2 with a set point corresponding to $x = 0$. The transfer function of the total system is

$$G(s) = \frac{1}{ms^2 + cs + k + k_1 + k_1 G_c(s)}\,.$$

In the case of proportional control, in which the displacement of the actuator is proportional to the displacement of the system $u = -G_c x$, the control system actually adds a spring with stiffness equal to k_1 multiplied by the gain. If a derivative control is applied, the control system is equivalent to a damper, with damping coefficient equal to k_1 multiplied by the gain. Note that in this case the gain is not a nondimensional quantity.

Other cases of interest are shown in Figures 6.8c and e, in which the actuator of the control system acts through a damper or on an auxiliary mass. The last layout is usually referred to as an *active vibration absorber*. The expressions of the control force are

$$F_c = -c_1(\dot{x} - \dot{u}) \quad \text{and} \quad F_c = -m_1(\ddot{x} - \ddot{u})\,.$$

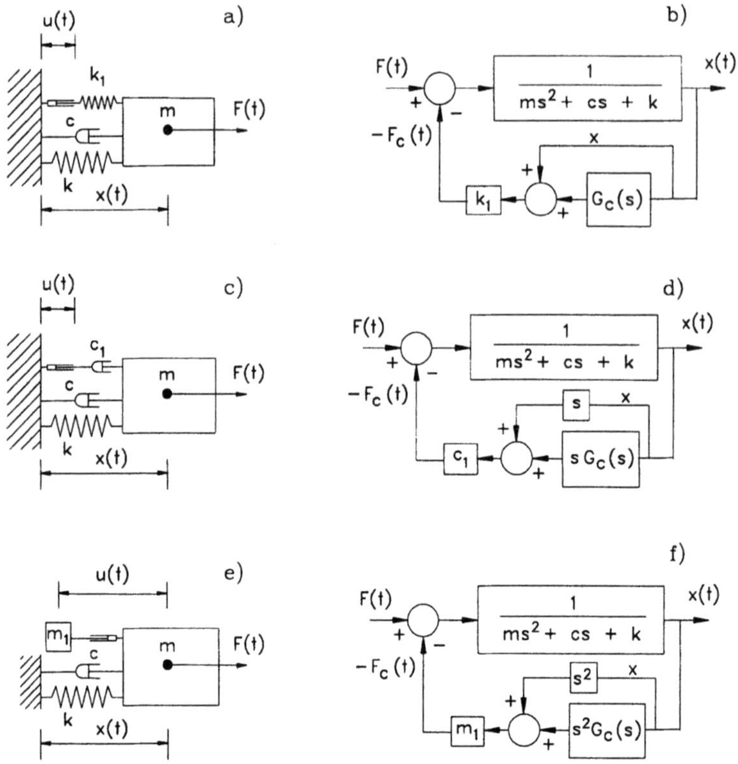

FIGURE 6.8. (a) Sketch and (b) block diagram of a system with a single degree of freedom controlled by an actuator through an auxiliary spring (c) and (d), an auxiliary damper, or (e) and (f) an auxiliary mass.

If the control law is of the proportional type ($u = -G_c x$), the transfer functions of the two systems are, respectively,

$$G(s) = \frac{1}{ms^2 + (c + c_1)s + k + sc_1 G_c(s)},$$

$$G(s) = \frac{1}{(m + m_1)s^2 + cs + k + s^2 m_1 G_c(s)}.$$

In the first case, a proportional control introduces an active damping into the system, an integral control law in which $u(t)$ is proportional to the integral of the displacement or in which $\dot{u}(t)$ is proportional to $x(t)$, introduces an active stiffness. If a derivative control is used, the control force is proportional to the acceleration, and the effect is similar to that of an added mass.

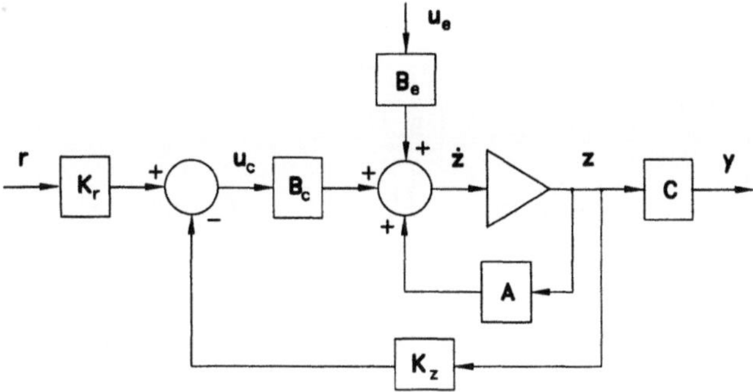

FIGURE 6.9. Block diagram of the multi-degree-of-freedom structure in Figure 6.4 controlled by a state feedback system.

6.3.5 State feedback and state observers

A more complete type of feedback is the state feedback shown in Figure 6.9, in which all atate variables are used to close the loop. The control inputs are obtained in this case by the equation

$$\mathbf{u}_c = \mathbf{K}_r\mathbf{r} - \mathbf{K}_z\mathbf{z},$$

and the product $\mathbf{K}_y\boldsymbol{C}$ in equation 6.13 must be substituted by matrix $\mathbf{K}_z$.

Usually state feedback is regarded as an ideal situation, but for practical reasons, linked with the impossibility of measuring many state variables, output feedback is applied.

The alternative is to use a device that estimates the state variables of the plant, performing a sort of simulation of its behaviour (Figure 6.10, the figure refers to a regulation problem in which the reference inputs $\mathbf{r}$ are equal to zero. It is usually referred to as a *state observer*. This is possible if the system is observable, in the sense defined in Section 6.3.1.

The behaviour of the observer is defined by matrices $\boldsymbol{A}_0$, $\boldsymbol{B}_0$, and $\boldsymbol{C}_0$. If they were equal to the matrices of the system, the observer would be an exact model of the plant. In general, this is impossible and the observer is only an approximated model, often a reduced order one. In structural control, if digital techniques are used, the observer can be a finite element model, perhaps reduced using Guyan reduction or similar procedures.

To ensure that the observer evolves in time like the actual plant, a feedback is introduced: The difference between the output of the observer $\hat{\mathbf{y}}$ and that of the system $\mathbf{y}$ is introduced, through a gain

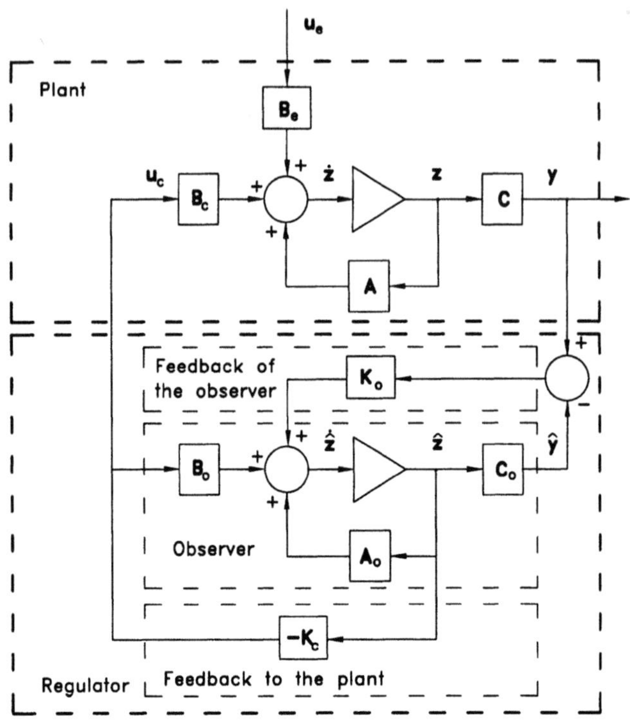

FIGURE 6.10. Block diagram of the multi-degree-of-freedom structure in Figure 6.4 controlled by a regulator with a state observer.

matrix $\mathbf{K}_0$, into the observer.

The states of the observer $\hat{\mathbf{z}}$ are readily available, and the states of the plant are not such (this is why the observer is used) and the control feedback is closed through the control gain matrix $\mathbf{K}_c$.

The equations allowing the study of the closed-loop behaviour of the system of Figure 6.10 are

$$\begin{cases} \dot{\mathbf{z}} = \boldsymbol{A}\mathbf{z} + \boldsymbol{B}_c\mathbf{u}_c + \boldsymbol{B}_e\mathbf{u}_e\,, \\ \mathbf{y} = \boldsymbol{C}\mathbf{z}\,, \\ \dot{\hat{\mathbf{z}}} = \boldsymbol{A}_0\hat{\mathbf{z}} + \boldsymbol{B}_0\mathbf{u}_c + \mathbf{K}_0(\mathbf{y} - \hat{\mathbf{y}})\,, \\ \dot{\hat{\mathbf{y}}} = \boldsymbol{C}_0\hat{\mathbf{z}}\,, \\ \mathbf{u}_c = -\mathbf{K}_c\hat{\mathbf{z}}\,. \end{cases} \qquad (6.26)$$

The observer with its feedback branches constitutes the reguator of the plant, which can be considered a system that has the output of the plant $\mathbf{y}$ as input and outputs the control inputs $\mathbf{u}_c$, which are fed back to the plant. The last three equations (6.26), which define the behaviour of the regulator, can be written in the form

$$\begin{cases} \dot{\hat{z}} = A_{reg}\hat{z} + B_{reg}y\,, \\ u_c = C_{reg}\hat{z}\,. \end{cases} \tag{6.27}$$

where $A_{reg} = A_0 + K_0C_0 - B_0K_c$, $B_{reg} = K_0$, and $C_{reg} = -K_c$.

If the observer has the same number of states as the plant, the error of the observer e_0 can be easily defined as the difference between the actual state of the system z and the state $\hat{z}$, which has been approximated by the observer: $e_0 = z - \hat{z}$. By subtracting the third equation (6.26) from the first, it follows that

$$\dot{e}_0 = \Big[A - A_0 - K_0C_0 - C\Big]z + \Big(A_0 - K_0C_0\Big)e_0 + \Big(B_c - B_0\Big)u_c + B_e u_e\,. \tag{6.28}$$

If $A_0 = A$, $C_0 = C$ (actually it is enough that $A_0 = A + K_0(C_0 - C)$ and $B_0 = B_c$, equation (6.28) reduces to

$$\dot{e}_0 = \Big(A - K_0C\Big)e_0 + B_e u_e\,. \tag{6.29}$$

Consider the free behaviour of the system described by equation (6.29). If the real part of all eigenvalues of matrix $A_0 = A - K_0C$ is negative, the error of the observer tends to zero for time tending to infinity and the observer is asymptotically stable. The observer can thus be designed by stating a set of eigenvalues of matrix A_0, whose real parts are negative and that minimize the time delay between the estimated and actual state vectors. It is a common suggestion[3] that the poles of the observer (i.e., the eigenvalues of matrix A_0) be placed on the Argand plane on the left of the poles of the controlled structure. However, a fast observer obtained in this way can be prone to amplify the disturbances that are always present in the signals from the sensors. The choice of the poles must be a trade-off between the requirements of quick response and disturbance rejection.

The computation of matrix K_0, which causes matrix A_0 to yield the required eigenvalues, can be performed using the pole-placement procedure, provided that the rank of the observability matrix O is n, i.e., the system is observable. This procedure is straightforward if the system has a single output, while for multivariable systems some arbitrary choices must be made.

---

[3] See, for example, H.H.E. Leipholz, M. Abdel Rohman, *Control of Structures*, Martinus Nijhoff, Dordrecht, 1986, 106

The state equation of the complete system shown in Figure 6.10 is

$$\left\{ \begin{array}{c} \dot{\mathbf{z}} \\ \dot{\mathbf{e}}_0 \end{array} \right\} = \left[\begin{array}{cc} \mathbf{A} - \mathbf{B}_c \mathbf{K}_c & \mathbf{B}_c \mathbf{K}_c \\ \mathbf{0} & \mathbf{A} - \mathbf{K}_0 \mathbf{C} \end{array} \right] \left\{ \begin{array}{c} \mathbf{z} \\ \mathbf{e}_0 \end{array} \right\} + \left\{ \begin{array}{c} \mathbf{B}_e \mathbf{u}_e \\ \mathbf{B}_e \mathbf{u}_e \end{array} \right\} .$$

The characteristic equation of the closed-loop system is

$$\det \left[\begin{array}{cc} s\mathbf{I} - \mathbf{A} + \mathbf{B}_c \mathbf{K}_c & \mathbf{B}_c \mathbf{K}_c \\ \mathbf{0} & s\mathbf{I} - \mathbf{A} + \mathbf{K}_0 \mathbf{C} \end{array} \right] = 0 , \tag{6.31}$$

i.e.,

$$\det \left(s\mathbf{I} - \mathbf{A} + \mathbf{B}_c \mathbf{K}_c \right) \det \left(s\mathbf{I} - \mathbf{A} + \mathbf{K}_0 \mathbf{C} \right) = 0 . \tag{6.32}$$

The roots of equation (6.32) are the eigenvalues of the controlled system plus the eigenvalues of the observer and the relevant parts of the characteristic equation can be solved separately.

To prescribe the free behaviour of the controlled system and of the observer in terms of natural frequencies and decay rates, the n eigenvalues $s_1, s_2, \ldots, s_n$ of the dynamic matrix $\mathbf{A} - \mathbf{B}_c \mathbf{K}_c$ (for the former) and of the dynamic matrix $\mathbf{A} - \mathbf{K}_0 \mathbf{C}$ (for the second) can be stated. Note that if they are complex, there must be a number of conjugate pairs in order to obtain real values of the gains. The characteristic equation of the controlled system (corresponding to the first part of equation (6.32)) is

$$\det(s\mathbf{I} - \mathbf{A} + \mathbf{B}_c \mathbf{K}_c) = (s - s_1)(s - s_2) \ldots (s - s_n) = 0 . \tag{6.33}$$

The n eigenvalues can be used to compute the n coefficients of the characteristic polynomial. By equating the expressions of the coefficients on the left-hand side of equation (6.33) with the coefficients so computed, n equations, which can be used to compute the n unknown elements of matrix $\mathbf{K}_c$, are obtained. The computation of the unknowns can be performed easily using Ackermann's formula

$$\mathbf{K}_c = \left[\begin{array}{ccccccc} 0 & 0 & 0 & \ldots & 0 & 1 \end{array} \right] \mathbf{H}^{-1} \mathbf{N} , \tag{6.34}$$

where $\mathbf{H}$ is the controllability matrix defined by equation (6.3), and matrix $\mathbf{N}$ is obtained from the coefficients $a_0, a_1, \ldots, a_{n-1}$ of the characteristic polynomial (subscripts refer to the power of the unknown) through the formula

$$\mathbf{N} = \boldsymbol{A}^n + a_{n-1}\boldsymbol{A}^{n-1} + a_{n-2}\boldsymbol{A}^{n-2} + \ldots + a_0\mathbf{I}. \qquad (6.35)$$

If the controllability matrix, which in the case of a single-input system is a square matrix, is not singular, there is no difficulty in computing the gains of the control system $\mathbf{K}_c$. However, this is essentially a theoretical statement because in practice the numerical evaluation of matrix $\mathbf{K}_c$ for systems whose order of the dynamic matrix is higher than a few units can be affected by large numerical errors.

Apart from the computational difficulties encountered when applied to MIMO systems, the pole assignment method allows one to state the poles of the controlled system but does not allow one to control its eigenvectors. This can result in badly conditioned eigenvector matrices, which is highly undesirable. It is much more convenient to resort to methods that are generally referred to as *eigenstructure assignment* procedures, which allow to state, within ample limits, both the eigenvalues and the eigenvectors of the controlled system. For a detailed presentation of such methods, the reader is again advised to refer to specialized textbooks on structural control.

Optimal control techniques, which are increasingly applied, are mostly based on the minimization of performance indices, which include the error of the control system but can take into account the costs or energy needed to perform the control action. A common definition of the performance indices is

$$J = \int_0^\infty \left[\mathbf{z}^T(t)\mathbf{Q}\mathbf{z}(t) + \mathbf{u}_c^T(t)\mathbf{R}\mathbf{u}_c(t) \right] dt. \qquad (6.36)$$

This formulation, to be applied to systems in which the desired condition is that with all state variables equal to zero, minimizes the integral in time of the deviation from this condition, i.e., of the control errors and control inputs, which are in some way linked with the energy needed for the control function. The various state variables and control inputs can have different importance, as indicated by the weight matrices $\mathbf{Q}$ and $\mathbf{R}$ introduced into the definition of the performance index.

Under wide assumptions, it is possible to demonstrate that the feedback gain matrix $\mathbf{K}_c$ for an optimal state feedback can be written in the form

$$\mathbf{K}_c = \mathbf{R}^{-1}\boldsymbol{B}_c^T\mathbf{P}\,, \qquad (6.37)$$

where $\mathbf{P}$ is the solution of the algebraic Riccati equation

$$\boldsymbol{A}^T\mathbf{P} + \mathbf{P}\boldsymbol{A} - \mathbf{P}\boldsymbol{B}_c\mathbf{R}^{-1}\boldsymbol{B}_c^T\mathbf{P} + \mathbf{Q} = \mathbf{0}\,. \qquad (6.38)$$

The regulator so obtained is referred to as an optimal linear quadratic regulator, where *linear* refers to the linearity of the controlled system and *quadratic* refers to the quadratic performance index. Similar considerations can be made for the observer, if $\mathbf{y}$ and $\hat{\mathbf{z}}$ are substituted to $\mathbf{z}$ and $\mathbf{u}_c$ into the performance index (6.36).

The detailed design of the control system and observer is, however, well beyond the scope of this text, and the reader is advised to refer to the many texts in which the optimal control theory and the solution of algebraic Riccati equations are dealt with.[4]

6.4 Modal approach to structural control

Structural models usually have many degrees of freedom, particularly if they have been obtained through the finite element method. They are then impractical for the design of the control system, which usually requires to deal with low-order models of the plant. Moreover, models obtained through the FEM have a large number of high-frequency modes with little physical meaning, because they are due more to the discretization procedure than to the structure itself.

A first way to circumvent this problem is the use of hybrid models, in which the structure is modeled as an elastic continuum and the controller is modeled as a discrete system. This approach suffers from the same drawbacks seen in Chapter 2 for the continuum models, namely, the difficulties encountered in modeling any shape except the simplest ones.

A very effective way to obtain a reduced order model for the design of the control system is that of using the Guyan reduction: The model of the plant can be condensed until a small number of master

---

[4]See, for example, the already-mentioned book J.L. Junkins, Y. Kim, *Introduction to Dynamics and Control of Flexible Structures*, AIAA, Washington D.C., 1993.

degrees of freedom are retained. The comparison of the open-loop natural frequencies and vibration modes of the reduced oirder model with the corresponding ones obtained from the full model allows verification of the validity of this approach in any relevant case.

Finally, the approach most frequently resorted to is that of using modal coordinates and retaining only a reduced number of modes.

The equation of motion of the controlled structure, i.e., of the structure on which the control forces $\mathbf{f}_c$ are acting, can be reduced in modal form by resorting to the eigenvectors of the undamped system

$$\overline{\mathbf{M}}\ddot{\boldsymbol{\eta}} + \overline{\mathbf{C}}\dot{\boldsymbol{\eta}} + \overline{\mathbf{K}}\boldsymbol{\eta} = \boldsymbol{\Phi}^T\mathbf{f}_c + \boldsymbol{\Phi}^T\mathbf{f}_e. \tag{6.39}$$

Note that equation (6.39) holds for both discrete and continuous systems, with the only difference being that in the latter case there is, theoretically, an infinity of modes and that the modal damping matrix $\overline{\mathbf{C}}$ is usually obtained by directly assessing the modal damping of the various modes and hence is usually diagonal. The eigenvectors will be assumed normalized in such a way that the modal masses have a unit value, i.e., the modal mass matrix is an identity matrix. The modal damping matrix is not, in general, a diagonal matrix, and the modes are not uncoupled. However, as structures are in most cases very little damped, the coupling of the modes is fairly weak and the modal damping matrix can be approximated by a diagonal matrix simply by neglecting its out-of-diagonal elements.

The modal equation in the phase space is then

$$\begin{aligned}\dot{\mathbf{z}} &= \boldsymbol{A}\mathbf{z} + \boldsymbol{B}_c\mathbf{u}_c(t) + \boldsymbol{B}_e\mathbf{u}_e(t)\,,\\ \mathbf{y} &= \boldsymbol{C}\mathbf{z}\,,\end{aligned} \tag{6.40}$$

where

$$\mathbf{z} = \left\{\begin{array}{c}\dot{\boldsymbol{\eta}}\\ \boldsymbol{\eta}\end{array}\right\}\,,\qquad \boldsymbol{A} = \left[\begin{array}{cc}-\overline{\mathbf{C}} & -\overline{\mathbf{K}}\\ \mathbf{I} & \mathbf{0}\end{array}\right]\,,$$

$$\boldsymbol{B}_c = \left[\begin{array}{c}\boldsymbol{\Phi}^T\mathbf{T}_c\\ \mathbf{0}\end{array}\right]\,,\qquad \boldsymbol{B}_e = \left[\begin{array}{c}\boldsymbol{\Phi}^T\mathbf{T}_e\\ \mathbf{0}\end{array}\right]\,, \tag{6.41}$$

and the output gain matrix links the outputs $\mathbf{y}$ of the system with the modal coordinates. Note that if the damping matrix is diagonal, the first equation (6.40) uncouples in a number of sets of two equations, each dealing with a single modal coordinate. This uncoupling, which is usually only approximate, does not hold once a closed-loop control system is included into the equation. The coupling due to the control system can be very strong, much stronger than that linked with damping which is often neglected. In a way, the effect of the action of the control system can often be assimilated to that of very high damping, and some of the modes of the controlled system can even be overdamped: No small-damping assumption holds anymore. Apart from the strong coupling due to the added damping, the presence of the control system changes the mode shapes; it is true that the new mode shapes can be expressed as combinations of the modes of the uncontrolled system, but this further increases modal coupling. Due to this coupling, the results obtained by considering only a limited number of modes in the design of the control system are only an approximation of the behaviour of the complete model.

In particular, the errors introduced by the fact that the outputs of the system $\mathbf{y}$ are also influenced by the modes that have been neglected are referred to as *observation spillover*. Those introduced by the fact that the control forces due to the control system do not only influence the modes considered but also the neglected ones, are referred to as *control spillover*. Spillover can cause the system to behave in ways different from the predicted one and in some cases can even cause instability. Note that active systems are prone to instability, because energy is supplied to the structure and, if this occurs in a wrong way, it can cause the excitation of some modes instead of damping them.

If an observer is used to approximate only a reduced number of modes, it is better to use an equation that considers the vectors $\mathbf{z}$ (of order n, now containing the modal coordinates and their derivatives) and $\hat{\mathbf{z}}$ (of order n_r, with $n_r < n$, containing the estimated values of the modal coordinates on which control is performed and their derivatives) state variables of the system instead of equation (6.30), which considers $\mathbf{z}$ and the errors $\mathbf{e}_0$

$$\left\{ \begin{array}{c} \dot{\mathbf{z}} \\ \dot{\hat{\mathbf{z}}} \end{array} \right\} = \left[\begin{array}{cc} \mathbf{A} & -\mathbf{B}_c\mathbf{K}_c \\ \mathbf{K}_0\mathbf{C} & \mathbf{A} - \mathbf{B}_0\mathbf{K}_c - \mathbf{K}_0\mathbf{C}_0 \end{array} \right] \left\{ \begin{array}{c} \mathbf{z} \\ \hat{\mathbf{z}} \end{array} \right\} + \left\{ \begin{array}{c} \mathbf{B}_e\mathbf{u}_e \\ 0 \end{array} \right\} .$$

(6.42)

The various matrices have different sizes, and they combine in such a way as to yield a closed-loop dynamic matrix that has $n + n_r$ rows and columns. Its eigenvalues are the closed-loop roots of the system and their study allows the effects of spillover to be predicted.

Although the dangers of spillover suggest the opportunity of avoiding the modal approach when designing the control system for a structure, modal control has been used successfully in many cases; it allows the solution of the problem of designing the control system in a simple and straightforward way, particularly when the number of modes to be considered is low. Even if modal control is widely accepted, the possible dangers due to spillover must always be kept in mind, and an evaluation of the effects of higher-order modes is in many cases advisable.

Note that equations (6.40) and (6.42) also hold in the case of continuous systems, once their modal characteristics have been obtained. They contain only a form of proportional control but, as in equation (6.42) the control forces are also linked to the modal velocities, the control system provides a form of damping as in the case of derivative control.

Example 6-4
Consider the prismatic homogeneous beam shown in Figure 6.11a. It is supported at one; at the other end it is supported by an actuator that can supply both the static force to balance the self-weight and dynamic forces to control vibration. The control system will be studied using modal analysis, taking into account the first three modes, one of which is a rigid-body motion.
Using the equations in Section 2.2.4 and the values reported in Table 2.2, the eigenvectors, normalized to yield unit values for the modal masses, the modal masses, and stiffnesses are as follows:
- Mode 0 (rigid-body mode):

$$q_0(\zeta) = \zeta\sqrt{\frac{3}{m}}\,, \qquad \overline{M} = 1\,, \qquad \overline{K} = 0,$$

- ith mode (any i):

$$q_i(\zeta) = \sqrt{\frac{2}{m}}\left[\sin(\beta_i\zeta) + \frac{\sin(\beta_i)}{\sinh(\beta_i)}\sinh(\beta_i\zeta)\right],$$

$$\overline{M} = 1\,, \qquad \overline{K} = \beta_i^4\frac{EI_y}{ml^3}\,,$$

where $m = \rho Al$ is the total mass of the beam, $\zeta = z/l$ is a nondimensional coordinate, and coefficients β_i are $\beta_i = (i+1/4)\pi$, i.e., $\sin(\beta_i) = -1^i/\sqrt{2}$.

The modal forces due to self-weight are $\overline{f}_0 = -g\sqrt{3m}/2$, $\overline{f}_i = 0$. The modal forces due to the control force f_c are $\overline{f}_{c_0} = f_c\sqrt{3/m}$, $\overline{f}_{c_i} = 2F_c-1^i/\sqrt{m}$. Comparing the expressions of the modal forces, it follows immediately that the static component of the control force is $\overline{f}_{c_{st}} = mg/2$. Note that all modal coordinates have a static value even if the modal forces due to self-weight vanish: In the equilibrium condition the static control force makes the modal coordinate of mode zero vanish while the others yield the static deflected shape

$$\eta_{i_{st}} = (-1)^i \frac{gl^3\sqrt{m^3}}{(i+1/4)^4\pi^4 EI_y} \, .$$

Due to the linearity of the problem, the dynamic study will be performed neglecting the static forces and deflections; this is possible only within the linear range, particularly if no saturation occurs in the actuator. By introducing the constant

$$\lambda_1 = \pi^2 \sqrt{\frac{EI_y}{ml^3}} \, ,$$

which is nothing other than the first natural frequency of the same beam simply supported at both ends, matrices $\boldsymbol{A}$ and $\boldsymbol{B}_c$ reduce to

$$\boldsymbol{A} = \begin{bmatrix} \boldsymbol{0} & -\lambda_1^2 \begin{bmatrix} 0 & 0 & 0 & \cdots \\ 0 & 1,25^4 & 0 & \cdots \\ 0 & 0 & 2,25^4 & \cdots \\ \cdots & \cdots & \cdots & \cdots \end{bmatrix} \\ \boldsymbol{I} & \boldsymbol{0} \end{bmatrix} ,$$

$$\boldsymbol{B}_c = \frac{1}{\sqrt{m}} \left\{ \begin{Bmatrix} \sqrt{3} \\ -2 \\ 2 \\ \cdots \\ 0 \end{Bmatrix} \right\} .$$

Consider as a first case an output-feedback control system with a single sensor at the free end of the beam yielding both displacement and velocity. The output vector $\mathbf{y}$ and matrix $\boldsymbol{C}$ are then

$$\mathbf{y} = \begin{Bmatrix} \dot{x}(l) \\ x(l) \end{Bmatrix} ,$$

$$\boldsymbol{C} = \frac{1}{\sqrt{m}} \begin{bmatrix} \sqrt{3} & -2 & 2 & \cdots & 0 & 0 & 0 & \cdots \\ 0 & 0 & 0 & \cdots & \sqrt{3} & -2 & 2 & \cdots \end{bmatrix} .$$

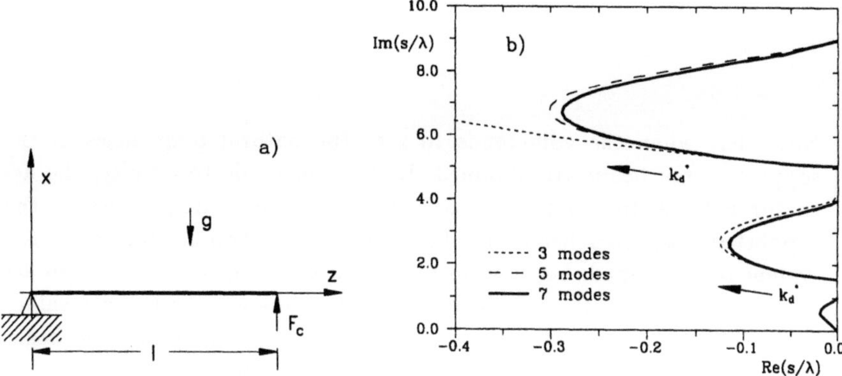

FIGURE 6.11. Beam supported at one end and actively controlled at the other: (a) sketch of the system; (b) root loci.

The control gain matrix $\mathbf{K}_y$ has one row, because it must supply the control input to a single actuator, and two columns, because there are two outputs of the system: $\mathbf{K}_y = \begin{bmatrix} k_v & k_d \end{bmatrix}$, where k_d is the proportionality constant between the force of the actuator and the displacement (dimensionally it is a stiffness), and k_v is the proportionality constant between the force and the velocity (dimensionally it is a damping coefficient). By introducing the nondimensional gains

$$k_v^* = \frac{k_v}{\lambda_1 m}, \qquad k_d^* = \frac{k_d}{\lambda_1^2 m} = k_d \frac{l^3}{EI_y \pi^4},$$

the closed-loop dynamic matrix is

$$\mathbf{\mathcal{A}}_{cl} = \begin{bmatrix} \lambda_1 k_v^* \mathbf{R} & -\lambda_1^2 \mathbf{S} \\ \mathbf{I} & \mathbf{0} \end{bmatrix},$$

where

$$\mathbf{R} = \begin{bmatrix} 3 & -2\sqrt{3} & 2\sqrt{3} & \dots \\ -2\sqrt{3} & 4 & -4 & \dots \\ 2\sqrt{3} & -4 & 4 & \dots \\ \dots & \dots & \dots & \dots \end{bmatrix},$$

$$\mathbf{S} = \begin{bmatrix} 3k_d^* & -2\sqrt{3}k_d^* & 2\sqrt{3}k_d^* & \dots \\ -2\sqrt{3}k_d^* & 1{,}25^4 + 4k_d^* & -4k_d^* & \dots \\ 2\sqrt{3}k_d^* & -4k_d^* & 2{,}25^4 + 4k_d^* & \dots \\ \dots & \dots & \dots & \dots \end{bmatrix}.$$

The nondimensional root loci for the first three modes with various values of the gain k_d^* are plotted in Figure 6.11b. To state a relationship between the two elements of the gain matrix, the gain k_v^* has been assumed to comply with the relationship $k_v^* = 0.1\sqrt{k_d^*/3}$, which corresponds to assume a damping ratio $\zeta = 0.1$ for the controlled first mode if mode coupling is neglected.

Note that when the gain tends to zero the natural frequencies of the supported-free beam are obtained. If the gain tends to infinity, the actuator reduces to a rigid support and the natural frequencies of the supported-supported beam are obtained. In these two limiting cases the system is undamped, because no structural damping was taken into account. For values of the gain included between them, intermediate values of the frequency are obtained but the roots are complex and the negative real part evidences the damping role of the control system. Another consideration regards spillover. The figure has been drawn by taking into account three, five, and seven modes. The response at the first mode is correct in all three cases; that at the second mode requires at least five modes to be computed correctly, but the errors due to spillover with three modes are not large. For the response at the third mode seven modes are required. Using five modes the results are still acceptable, but with three modes the errors due to spillover lead to unacceptable results.

Consider now a state feedback, obtained by resorting to a single displacement sensor at the free end and adding an observer to the system. Take into account only the first three modes. Matrices $\boldsymbol{A}$, $\boldsymbol{B}_c$, and $\boldsymbol{C}$ are

$$
\boldsymbol{A} = \begin{bmatrix} \boldsymbol{0} & -\lambda_1^2 \begin{bmatrix} 0 & 0 & 0 \\ 0 & 1.25^4 & 0 \\ 0 & 0 & 2.25^4 \end{bmatrix} \\ \mathbf{I} & \boldsymbol{0} \end{bmatrix},
$$

$$
\boldsymbol{B}_c = \frac{1}{\sqrt{m}} \left\{ \left\{ \begin{array}{c} \sqrt{3} \\ -2 \\ 2 \\ 0 \end{array} \right\} \right\},
$$

$$
\boldsymbol{C} = \frac{1}{\sqrt{m}} \begin{bmatrix} 0 & 0 & 0 & \sqrt{3} & -2 & 2 \end{bmatrix}.
$$

By assuming a unit value for λ_1, the controllability and observability matrices and their determinants are

$$
\mathbf{H} = \frac{1}{\sqrt{m}} \begin{bmatrix} \sqrt{3} & 0 & 0 & 0 & 0 & 0 \\ -2 & 0 & 4.883 & 0 & -11.92 & 0 \\ 2 & 0 & -51.26 & 0 & 1{,}314 & 0 \\ 0 & \sqrt{3} & 0 & 0 & 0 & 0 \\ 0 & -2 & 0 & 4.883 & 0 & -11.92 \\ 0 & 2 & 0 & -51.26 & 0 & 1{,}314 \end{bmatrix},
$$

$$\mathbf{O} = \frac{1}{\sqrt{m}} \begin{bmatrix} 0 & \sqrt{3} & 0 & 0 & 0 & 0 \\ 0 & -2 & 0 & 4.883 & 0 & -11.92 \\ 0 & 2 & 0 & -51.26 & 0 & 1{,}314 \\ \sqrt{3} & 0 & 0 & 0 & 0 & 0 \\ -2 & 0 & 4.883 & 0 & -11.92 & 0 \\ 2 & 0 & -51.26 & 0 & 1314 & 0 \end{bmatrix},$$

$$\det(\mathbf{H}) = -1.01 \times 10^8,, \qquad \det(\mathbf{O}) = -1.01 \times 10^8.$$

Assume that the six nondimensional eigenvalues of the controlled system are $-0.2 \pm 0.6i$, $-0.2 \pm 2i$, and $-0.5 \pm 5i$, and those of the observer are $-0.3 \pm 0.6i$, $-0.5 \pm 2i$, and $-0.9 \pm 5i$. By using the pole-placement procedure, the following gain matrices for the controller and observer are obtained

$$\mathbf{K}_c = \lambda_1\sqrt{m}\begin{bmatrix} 0.452 & -0.149 & 0.460 & 0.381\lambda_1 & -0.784\lambda_1 & 0.341\lambda_1 \end{bmatrix},$$

$$\mathbf{K}_0 = \lambda_1\sqrt{m}\begin{bmatrix} 0.456\lambda_1 & -1.011\lambda_1 & 1.555\lambda_1 & 0.746 & -0.268 & 0.786 \end{bmatrix}^T.$$

A simple check shows that the required roots of the system are obtained. To check the errors due to spillover, the eigenvalues of the systems were again computed, taking into account 10 modes, only the first three of which are controlled using the observer and the controller defined earlier. The following 26 nondimensional eigenvalues have been obtained: $-0.200 \pm 0.584i$, $-0.303 \pm 0.621i$, $-0.266 \pm 1.949i$, $-0.598 \pm 2.071i$, $-0.893 \pm 4.765i$, $-0.482 \pm 5.108i$, $0.0213 \pm 10.599i$, $0.0103 \pm 18.069i$, $0.0047 \pm 27.564i$, $0.0024 \pm 39.063i$, $0.0013 \pm 52.563i$, $0.0008 \pm 68.063i$, and $0.0005 \pm 85.563i$.

The first six are similar to the ones stated for the controller and the observer. The remaining ones are not very different from the corresponding natural frequencies of the uncontrolled system: 10.562, 18.063, 27.563, 39.063, 52.563, 68.063, and 85.563, but have a positive real part. Spillover causes a slight instability of the system, which, however, in the current case is due to the fact that no damping of the beam has been considered. The very small damping introduced by the beam itself should be sufficient to achieve stable behaviour in all modes.

Example 6-5
Consider the same system studied in Example 6-4. Study its response to a shock load of the type prescribed by MIL-STD 810 C, basic design (see Example 1-10) and compare the results obtained with those obtained for a simply supported beam with the same dimensions.

The total mass of the beam is $m = \rho Al = 3.124$ kg and the value of parameter λ_1 in the previous example is 147.74 rad/s. Assuming a unit value for the modal masses, the modal stiffnesses for the first eight modes (including the rigid-body mode) are: $\overline{K}_0 = 0$, $\overline{K}_1 = 53{,}290$, $\overline{K}_2 = 559{,}400$, $\overline{K}_3 = 2.435 \times 10^6$, $\overline{K}_4 = 7.121 \times 10^6$, $\overline{K}_5 = 1.658 \times 10^7$, $\overline{K}_6 = 3.331 \times 10^7$, and $\overline{K}_7 = 6.030 \times 10^7$. Because the uncontrolled system is very lightly damped, a modal damping approximation is accepted.

The modal damping for the various modes can be computed as $C_i = \eta \lambda_i$: $\overline{C}_0 = 0$, $\overline{C}_1 = 2.308$, $\overline{C}_2 = 7.479$, $\overline{C}_3 = 15.605$, $\overline{C}_4 = 26.683$, $\overline{C}_5 = 40.720$, $\overline{C}_6 = 57.710$, and $\overline{C}_7 = 77.655$.

By assuming the same eigenfrequencies of the three controlled modes as in Example 6-3, the following gain matrices for the controller and the observer are obtained:

$$\mathbf{K}_c = \begin{bmatrix} 118.1 & -38.85 & 120.0 & 14{,}705 & -30{,}239 & 13{,}152 \end{bmatrix},$$

$$\mathbf{K}_0 = \begin{bmatrix} 17{,}571 & -38{,}985 & 59{,}983 & 195.86 & -69.88 & 205.28 \end{bmatrix}^T.$$

As seen in Example 6-4, to check the errors due to spillover the eigenvalues of the systems were computed again, taking into account 10 modes. This computation is now performed taking into account damping in order to verify whether the damping of the system can to counteract the instabilizing effects of spillover. The following 26 eigenvalues, expressed in rad/s, have been obtained: $-29.73 \pm 86.09i$, $-44.60 \pm 92.20i$, $-43.95 \pm 290.45i$, $-84.68 \pm 302.97i$, $-129.03 \pm 700.17i$, $-77.97 \pm 758.20i$, $-4.59 \pm 1{,}565.7i$, $-11.81 \pm 2{,}669.2i$, $-19.67 \pm 4{,}072.2i$, $-28.51 \pm 5{,}771.0i$, $-38.63 \pm 7{,}765.4i$, $-50.13 \pm 10{,}055i$, and $-63.13 \pm 12{,}641i$.

The first six are very similar to the ones for the controller and the observer and are almost not changed (to working accuracy) by the presence of damping. The remaining ones, which are not very different from the corresponding natural frequencies of the uncontrolled system, now have a negative real part. The damping of the system counteracts the instability due to spillover, even if the very low absolute value of some decay rates (particularly for the seventh mode) suggests a behaviour with a small stability margin. The very low value assumed for structural damping accounts for this type of behaviour.

The shock, applied to the frame that supports the beam, is prescribed as a linearly increasing acceleration, reaching 20 g in 11 ms, and then decreasing abruptly to zero. The shock load is seen from the beam as a constant distributed load whose absolute value is given by the mass multiplied by the time-varying acceleration. Only the modal force related to the first mode (mode 0, rigid-body mode) is different from zero $\overline{F}_0 = -a\sqrt{3m}/2$.

The response has been computed using equation (6.39), taking into account 20 state variables for the beam (10 modes) and six state variables for the observer, by integrating the equation in time using the Euler method. The external input u_e is the time-varying acceleration $-a$, while vector $\boldsymbol{B}_e$ has all elements vanishingly small except the first, which is $B_{e_1} = \sqrt{3m}/2$. The response at the center of the beam has been reconstructed from the modal response and is plotted in Figure 6.12a. In Figure 6.12b the stress at the center of the beam, computed from the mode shapes and modal coordinates, has been plotted. In the same figures the values obtained applying the same shock to a similar beam simply supported at both ends are plotted for comparison.

From the figure it is clear that the active control system is very effective in damping out the free oscillations caused by the shock, while the simply supported beam gives way to a lower value of the maximum displacement. This is, however, to be expected, because the control system is assumed to be quite soft and then it allows a sort of large rigid-body motion under the effect of the shock.

The reduction of the stress peak is very large, since the rigid-body motion, which absorbs most of the energy of the shock, does not contribute to the stresses. The time history of the force applied by the actuator at the free end of the beam is shown in Figure 6.13. Note that this example has been reported only to show some features of structural control. No attempt to optimize the control system or verify its feasibility from an energy viewpoint has been performed. A purely proportional control law with no cut-off frequency has been used: no actual control systems can follow this law; the highest controlled frequency is, however, about 120 Hz and any system with a cut-off frequency higher than this value would do the job.

Example 6-6

Consider the same beam studied in Example 6-5. The beam is now a structural member moved by an actuator that exerts a torque at the hinged end (Figure 6.14a). The beam is controlled by an open-loop system that governs the actuator torque following a predetermined pattern.

Assuming that the torque is controlled following either a square-wave pattern (bang-bang control) (Figure 6.14b) or a more elaborate double-versine time history (Figure 6.14c) and neglecting weight, compute the maximum torque needed to achieve a rotation of 45° ($\pi/4$ rad) in 0.5 s and the time history of the tip of the beam during and after the maneuver.

Rigid-body dynamics. The equation of motion of the beam as a rigid body is simply $\ddot{\theta} = T(t)/J$, where $J = ml^2/3$ is the moment of inertia of the beam about the axis of rotation.

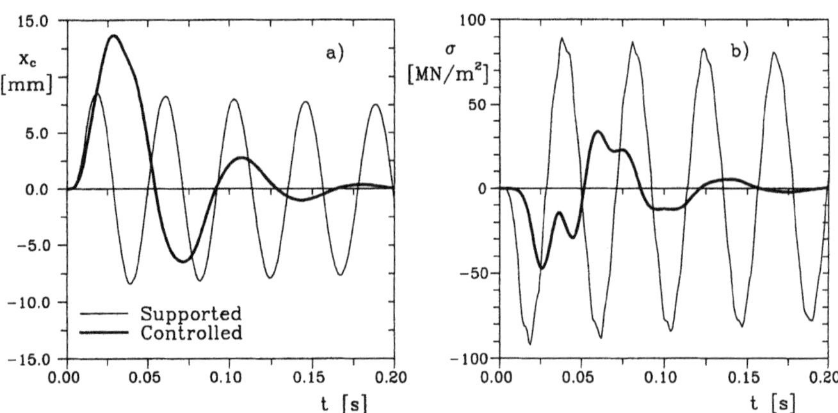

FIGURE 6.12. Time history of the displacement at the (a) center and of the stress at the (b) same point of the beam in Figure 6.11 when a shock load of the type prescribed by MIL-STD 810 C, basic design, is applied to it; comparison with a simply supported beam.

In the case of the bang-bang control the time history of the torque is $T = T_{max}$ for $0 \leq t \leq t_{max}/2$ and $T = -T_{max}$ for $t_{max}/2 < t \leq t_{max}$. Note that the square-wave law is that which allows the minimum travel time for a given value of the maximum driving torque. The control law is symmetrical in time and the speed of the beam at time t_{max} reduces to zero. By integrating the equation of motion, the time history of the displacement is

$$\begin{cases} \theta = \dfrac{3T_{max}}{2ml^2}t^2 & \text{for } 0 \leq t \leq \tfrac{t_{max}}{2}, \\ \theta = \dfrac{3T_{max}}{4ml^2}\left(-t_{max}^2 + 4t_{max}t - 2t^2\right) & \text{for } \tfrac{t_{max}}{2} \leq t \leq t_{max}. \end{cases}$$

The relationship linking the torque with the displacement θ_{max} and the time needed for the rotation is $T_{max} = 4ml^2\theta_{max}/3t_{max}^2$.

In the case of the double-versine control, the time history of the torque is

$$\begin{cases} T = \dfrac{T_{max}}{2}\left[1 - \cos\left(\dfrac{4\pi}{t_{max}}t\right)\right] & \text{for } 0 \leq t \leq \tfrac{t_{max}}{2}, \\ T = -\dfrac{T_{max}}{2}\left[1 - \cos\left(\dfrac{4\pi}{t_{max}}t\right)\right] & \text{for } \tfrac{t_{max}}{2} \leq t \leq t_{max}. \end{cases}$$

Also in this case the control law is symmetrical in time and the speed of the beam at time t_{max} reduces to zero. By integrating the equation of motion, the time history of the displacement is, for $0 \leq t \leq t_{max}/2$ and $t_{max}/2 < t \leq t_{max}$, respectively,

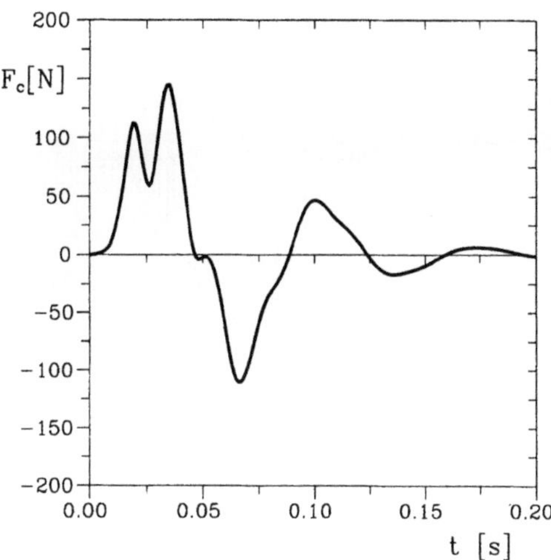

FIGURE 6.13. Time history of the control force of the same beam in Figure 6.12.

$$\theta = \frac{3T_{max}}{4ml^2}\left\{t^2 + \frac{t_{max}^2}{8\pi^2}\left[\cos\left(\frac{4\pi}{t_{max}}t\right) - 1\right]\right\},$$

$$\theta = \frac{3T_{max}}{8ml^2}\left\{-t_{max}^2 + 4t_{max}t - 2t^2 - \frac{t_{max}^2}{4\pi^2}\left[\cos\left(\frac{4\pi}{t_{max}}t\right) - 1\right]\right\}.$$

The relationship linking the torque with the displacement θ_{max} and the time needed for the rotation is $T_{max} = 8ml^2\theta_{max}/3t_{max}^2$. The values of the maximum torque are then $T_{max} = 13.09$ Nm for the square wave and $T_{max} = 26.17$ Nm for the double versine.

Dynamic behaviour of the beam. The dynamic behaviour of the beam will be studied in the xz-reference frame of Figure 6.14a. Such a frame is not inertial and, consequently, inertia forces due to its motion must be introduced into the equations. The position in an inertial (nonrotating) reference frame $x'z'$ of a general point of the beam, located at coordinate z, is

$$\left\{\begin{array}{c} x' \\ z' \end{array}\right\} = \left\{\begin{array}{c} z\sin(\theta) + u_x\cos(\theta) \\ z\cos(\theta) - u_x\sin(\theta) \end{array}\right\},$$

where $u_x(t)$ is the displacement due to the bending of the beam. By differentiating the position, the kinetic energy of a length dz of beam is readily obtained $dT = \frac{1}{2}\rho A dz(\dot{\theta}^2 z^2 + \dot{u}_x^2 + u_x^2\dot{\theta}^2 + 2z\dot{\theta}\dot{u}_x)$. By assuming that the law $\theta(t)$ is stated, performing the relevant derivatives with respect to u_z and $\dot{u}_z$, the inertial terms entering into the equation of motion for the bending behaviour of the beam are $\rho A dz(\ddot{u}_x + z\ddot{\theta}^2 + u_x\dot{\theta}^2)$.

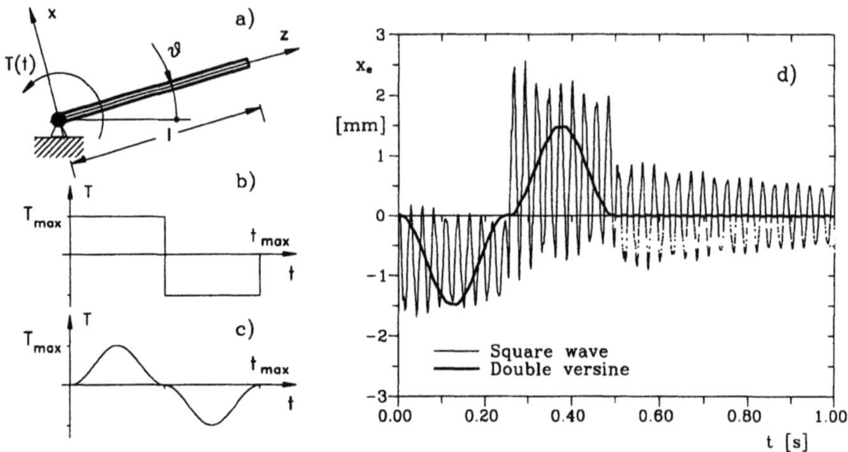

FIGURE 6.14. Beam rotating about one of its ends under the effect of the driving torque $T(t)$: (a) sketch of the system at time t; (b) and (c) time histories of the control torque during the maneuver: square-wave and double-versine patterns; (d) time history of the displacement of the free end: displacement from the rigid-body position, computed by numerically integrating the equations of the first five modes.

The first term is the usual term that appears in the equation of motion of the stationary beam. The second term is due to the acceleration of the beam and can easily be accounted for as an external excitation, because it does not contain the deformation of the beam. The third term is a component of the centrifugal force due to rotation acting in a direction perpendicular to the z-axis. The last effect can be neglected, as the stiffening effect due to the centrifugal field, if the angular velocity is small enough. Actually, in the small movement studied, the acceleration is high but the angular velocity maintains a low value. Moreover, the aim of the study is mainly to predict the free behaviour of the beam after the required position is reached and the two neglected effects stop acting. Clearly they are not noninfluent, because the behaviour after the beam has stopped depends on what happens during the motion, but the assumption that their effect is small can be accepted, at least in a first-approximation study.

The beam can be studied as a beam at standstill, under the effect of the moment $T(t)$ and the inertia force $-\rho A z \ddot{\theta}(t)$. The study will be performed using modal analysis. The mode shapes, modal masses, and stiffnesses have been computed in Example 6-4. The modal forces due to the inertia forces are $\overline{f}_{i_0} = -l\ddot{\theta}\sqrt{m/3}$ and $\overline{f}_{i_i} = 0$. The modal forces due to the driving torque $T(t)$ are

$$\overline{f}_{T_i} = T(t)\frac{dq_i(z)}{dz},$$

i.e.

$$\overline{f}_{T_0} = \frac{T(t)}{l}\sqrt{\frac{3}{m}}, \qquad \overline{f}_{T_i} = \frac{T(t)}{l}\sqrt{\frac{2}{m}}\beta_i\left[1 + \frac{\sin(\beta_i)}{\sinh(\beta_i)}\right].$$

Remembering that $\ddot{\theta} = 3T/ml^2$, the equation of motion for the rigid-body mode is simply

$$\ddot{\eta}_0 = \frac{T(t)}{l}\sqrt{\frac{3}{m}} - l\sqrt{\frac{3}{m}}\ddot{\theta} = 0,$$

i.e., because at time $t = 0$ both $\eta_0 = 0$ and $\dot{\eta}_0 = 0$, the first modal coordinate is always equal to zero. This result is obvious, because the driving torque has been computed by imposing the rigid-body motion of the beam. The equations of motion of the other modes are

$$\ddot{\eta}_i + \overline{C}_i\dot{\eta}_i + \overline{K}_i\eta_i = \frac{T(t)}{l}\sqrt{\frac{2}{m}}\beta_i\left[1 + \frac{\sin(\beta_i)}{\sinh(\beta_i)}\right].$$

The displacement at the free end of the beam, with respect to the position of the rigid-body motion is simply

$$x_e = \sum_{i=1}^{\infty}\eta_i q_i(t) = \frac{2}{\sqrt{m}}\sum_{i=1}^{\infty}(-1)^i\eta_i.$$

The modal equations of motion can be solved by numerical integration or using Duhamel's integral. A solution based on the numerical integration of the first five modes is shown in Figure 6.14d. From the figure, it is clear that the double-versine control succeeds in positioning the beam without causing long-lasting vibrations as in the case of the bang-bang control pattern. The latter, however, strongly excites the first mode, which is little damped. The results are linked with the particular application, because a control input of the versine type can also excite some modes; however, the fact that the square-wave control input is more prone to exciting vibrations than more smooth control laws is a general feature. In many cases, particular control laws that are much better than the square-wave and versine ones are used, and much theoretical and experimental work has been devoted to identifying optimal control laws. In a practical case, if high positioning precision is required, a closed-loop control must be associated with the open-loop control shown here. If the response of the control device is fast enough, deformation modes can also be controlled.

6.5 Dynamic study of rotors on magnetic bearings

Magnetic suspension has been proposed for many applications, from vehicles to models in wind tunnels. The substitution of the conventional mechanical suspension system of rotors, based on rolling elements or lubricated journal bearings with a magnetic suspension system, is very promising for many reasons. Magnetic bearings can drastically reduce bearing drag, while completely avoiding the presence of lubricant and wear. The field in which the advantages of magnetic bearings are most important is, however, that of the dynamic behaviour of the rotor: The stiffness and damping of the bearing system can be tailored for the application and can be adjusted following the operating conditions. In the case of active bearings, the control system can be used not only to maintain the rotor in the required position as a rigid body, minimizing the effects of unbalance and reducing the needs of strict balancing tolerances, but also to control the deformation modes. The main goal of magnetic suspension is to keep the suspended body in the proper position and avoid rigid-body modes: A complete suspension must then constrain the six modes of a rigid body in space. In the case of a vehicle or rotor, one of the rigid-body modes must be left unconstrained (rotation about the axis, for a rotor) and a five-axis suspension is needed. Electromagnetic forces can be exerted by a passive device, based on permanent magnets or by uncontrolled electromagnets, or by an active device outfitted by a suitable control system. Strictly speaking, the two types of magnetic suspension systems should be defined as uncontrolled or controlled, because a device based on an uncontrolled electromagnet is active in the sense that it receives energy from the outside. However, the use of the terms *active* and *passive* with this meaning is well established in magnetic-bearing technology and will be followed in this section.

The Earnshaw theorem states that a magnetic suspension based on a five-passive-axes layout is unstable except if diamagnetic materials are used. The choice is then between the use of a hybrid suspension system, in which at least one of the degrees of freedom of the rotor is constrained by a mechanical bearing and that of a device that is at least partially active. The solutions span from that of a one-active-axis suspension, in which two passive radial bearings control the four degrees of freedom linked with the lateral behaviour of the rotor and one axial active bearing restrains axial motion to a fully active five-axis suspension, with active radial and axial bearings.

In this section the equations of motion for the lateral behaviour of a linear axisymmetrical rotor running on a set of active radial bearings will be obtained. By resorting to the complex coordinates defined in Section 4.6.2 (equation (4.70)), the equation of motion of the rotor-stator system is

$$\mathbf{M}\ddot{\mathbf{q}} + (\mathbf{C}_n + \mathbf{C}_r - i\omega\mathbf{G})\dot{\mathbf{q}} + (\mathbf{K} - i\omega\mathbf{C}_r)\mathbf{q} = \qquad (6.43)$$
$$= \mathbf{f}_c + \mathbf{f}_n + \omega^2\mathbf{f}_r e^{i\omega t}.$$

where vector $\mathbf{f}_c$ includes the forces exerted by the magnetic bearings on the rotor and the stator, and the stiffness matrix $\mathbf{K}$ is singular with four vanishing eigenvalues because the four rigid-body motions of the rotor are unconstrained. The real part of vector $\mathbf{f}_c$ relates to the forces due to the magnetic bearings in the xz-plane, while its imaginary part is linked with the forces exerted in the yz-plane.

Assume that each actuator is made by four electromagnets as shown in Figure 6.15: Coils 1 and 3 are responsible for the force in the x-direction and coils 2 and 4 for the force in the y-direction. As a first approximation, the force exerted by a single coil depends on the current i and the air gap t between stator and rotor following the law

$$F = K\left(\frac{i}{t}\right)^2, \qquad (6.44)$$

where K is a constant that includes all design parameters of the actuator.

Usually, to linearize the overall behaviour of the bearing, the coils are fed with a constant current, referred to as *bias current*, on which the control current is superimposed (see Example 6-8). A so-called static compensation current i_0' is superimposed to the bias current i_0 of the two coils acting in the same direction to withstand any static load. The bias and compensation current in coils 1 and 3 acting in the x-direction can be written as $i_0 + i_0'$ and $i_0 - i_0'$. The total x-component of the force F_c exerted by the actuator is

$$F_x = K\left[\left(\frac{i_0 + i_0' + i_x}{c - d - u_x}\right)^2 - \left(\frac{i_0 - i_0' - i_x}{c + d + u_x}\right)^2\right], \qquad (6.45)$$

where i_x, c, d and u_x are, respectively, the control current, the radial clearance, the static offset, and the radial displacement, and the constants K of the two coils are assumed to be equal. Equation (6.45)

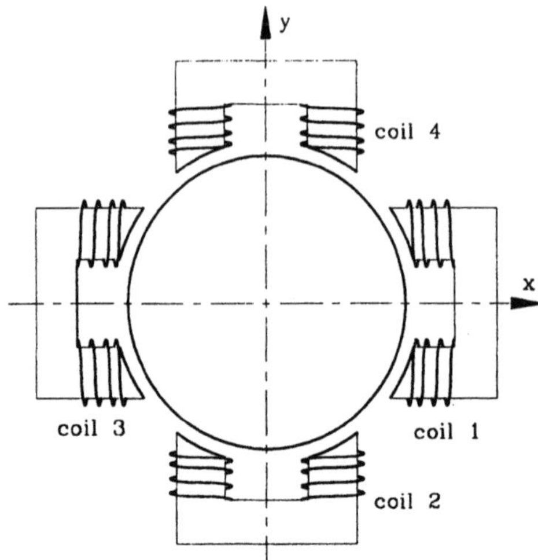

FIGURE 6.15. Sketch of an electromagnetic actuator of a magnetic bearing.

holds only if $i_0 - i_0' - i_x > 0$: In the opposite case, the controller should switch off one of the coils, and the system works in nonlinear conditions.

An actuator that works with a bias current high enough to allow all coils to be always energized is said to work as a "class A" actuator while a "class B" actuator works with only one of the two coils on, the one that carries the load. The intermediate condition, in which at least one coil is switched on and off can be referred to as "mixed-mode" working.

The force expressed by equation (6.45) can be linearized about the condition with $i_x = 0$, $u_x = 0$ as

$$F_x = F_0 + K_i i_x + K_u u_x \,, \tag{6.46}$$

where

$$F_0 = K \left[\left(\frac{i_0 + i_0'}{c - d} \right)^2 - \left(\frac{i_0 - i_0'}{c + d} \right)^2 \right] \,,$$

$$K_i = \left(\frac{\partial F_x}{\partial i_x} \right)_{\substack{i_x = 0 \\ u_x = 0}} = 2K \left[\frac{i_0 + i_0'}{(c - d)^2} + \frac{i_0 - i_0'}{(c + d)^2} \right] \,, \tag{6.47}$$

$$K_u = \left(\frac{\partial F_x}{\partial u_x} \right)_{\substack{i_x = 0 \\ u_x = 0}} = 2K \left[\frac{(i_0 + i_0')^2}{(c - d)^3} + \frac{(i_0 - i_0')^2}{(c + d)^3} \right] \,.$$

The mechanical static stiffness k of the controlled magnetic bearing depends on the gains K_i of the actuator defined earlier and on the gains K_c, K_a and K_t of the controller, the power amplifier and the transducer, and on K_u, which is the absolute value of the negative mechanical stiffness, expressing the instability of the bearing in open-loop conditions predicted by the Earnshaw theorem:

$$k = -K_u + K_d = -K_u + K_i K_c K_a K_t. \qquad (6.48)$$

If the actuator operates in a central position, i.e., if $d = 0$, equation (6.48) reduces to

$$F_0 = 4K\frac{i_0 i_0'}{c^2}, \qquad K_i = 4K\frac{i_0}{c^2}, \qquad K_u = 4K\frac{i_0^2 + i_0'^{\,2}}{c^3}. \qquad (6.49)$$

The force F_0 acting on the bearing determines the value of the current i_0' and hence affects the stiffness of the bearing. As a consequence, if the static forces acting in the x- and y-directions are not equal, as in the case of horizontal rotors, the behaviour is anisotropic even if the geometrical and electrical characteristics are equal in the two directions. This effect, which is particularly strong in the case of bearings operating with low bias currents, can be accounted for using different gains of the controller for the two directions but, generally speaking, complete compensation is not possible.

When this occurs, the rotor exhibits the usual behaviour of a rotor running on anisotropic bearings and, if damping is low enough, backward whirling occurs in the speed range spanning between the two rigid-body critical speeds in the two reference planes. The behaviour of a spindle running on a five-active-axes magnetic suspension with decentralized PID controller is summarized in Figure 6.16.[5] The figure has been obtained recording the amplitude and phase of the synchronous components of the displacements in the x- (horizontal) and y- (vertical) directions during a spin-down test. The greater stiffness in vertical direction is clear from the figure and occurs albeit the controller has a higher gain in horizontal direction to partially compensate for the anisotropy. A very simple PID decentralized controller has been used; if a more sophisticated control technique were

---

[5] S. Carabelli, C. Delprete, G. Genta, "Control strategies for decentralized control of active magnetic bearings", 4th International Symposium on Magnetic Bearings, Zurich, August 1994.

implemented, both anisotropy and amplitude at the critical speeds could have been controlled better.

Alternatively, a bias current larger than that required to withstand the static forces acting on the bearing can be used to obtain an almost isotropic behaviour.

Equation (6.43) can be written with reference to the state space as

$$\dot{\mathbf{z}} = \boldsymbol{A}\mathbf{z} + \boldsymbol{B}_c\mathbf{u}_c(t) + \boldsymbol{B}_n\mathbf{u}_n(t) + \boldsymbol{B}_r\mathbf{u}_r(t)e^{i\omega t}, \qquad (6.50)$$

where

- vector $\mathbf{z}$ contains the n complex state variables $\dot{\mathbf{q}}$ and $\mathbf{q}$,

- vector $\mathbf{u}_c(t)$ contains the r control input functions. It can be written in complex form, and the real and imaginary parts of its components are related to the forces in the two coordinate planes xz and yz. Vectors $\mathbf{u}_n(t)$ and $\mathbf{u}_r(t)$ are input vectors related to nonrotating and rotating forces, which, in the most general case, can be functions of time. They can be real vectors, in some cases simply scalar quantities, but it is possible to formulate the equations in such a way that they are complex.

- $\boldsymbol{A}$ is the complex dynamic matrix of the system,

$$\boldsymbol{A} = \begin{bmatrix} -\mathbf{M}^{-1}(\mathbf{C}_n + \mathbf{C}_r - i\omega\mathbf{G}) & -\mathbf{M}^{-1}(\mathbf{K} - i\omega\mathbf{C}_r) \\ \mathbf{I} & \mathbf{0} \end{bmatrix},$$
$$(6.51)$$

- matrices $\boldsymbol{B}_i$ are real or complex matrices that link the forces due to the control devices, nonrotating and rotating forces with the various inputs.

The equations linking the outputs of the system with the state vectors and control equations are the same as those already seen in Section 6.3 and need not be repeated here. The only difference between a controlled structure and a controlled rotor is the presence of the imaginary terms due to the gyroscopic matrix added to the damping matrix and those due to rotating damping added to the stiffness matrix. The gain matrices and vector of the state variables are complex.

Consider the rigid rotor sketched in Figure 6.17a suspended on a four-active-axes magnetic suspension. Assume that the study of the

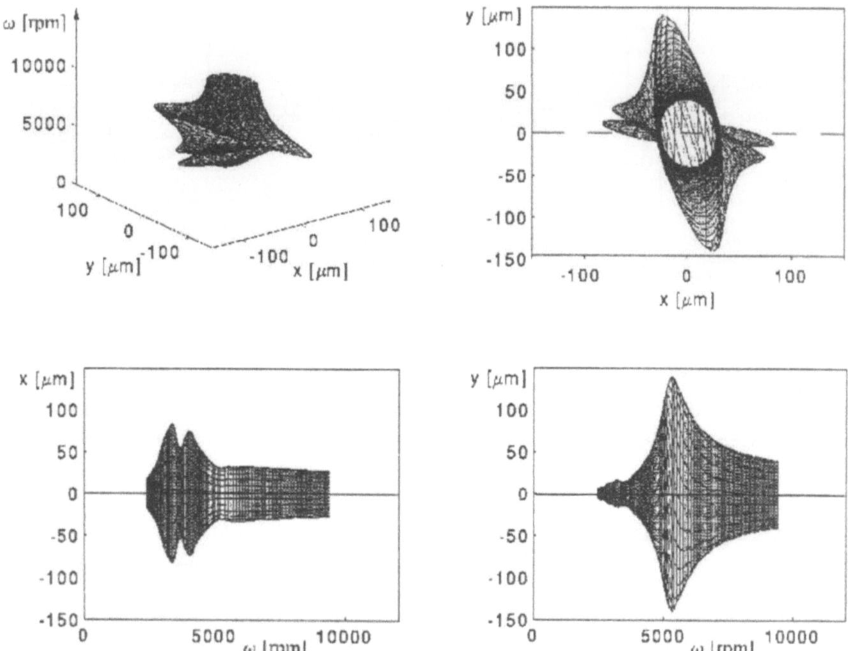

FIGURE 6.16. Dynamic behaviour of a spindle running on a five-active-axes magnetic suspension with decentralized PID controller; same type of representation of the orbits at different speeds as in Figure 4.25, except that in this case the experimental values of the synchronous components of the displacements measured during a spin-down test have been used.

lateral behaviour of the system can be performed separately from the axial behaviour and that the transducers are located at the same axial coordinate as the magnetic bearings. Note that the last assumption can be dropped when a model with many degrees of freedom is used and the displacements at the location of both sensors and actuators enter the equation of motion. The rotor is assumed to be affected by static and couple unbalance and a constant lateral force due to weight acts in the xz-plane. There are two complex degrees of freedom, and, using the complex-coordinate notation seen in Section 4.6.2, the dynamic matrix and the state vector are

$$\boldsymbol{A} = \begin{bmatrix} 0 & 0 & 0 & 0 \\ 0 & i\omega J_p/J_1 & 0 & 0 \\ 1 & 0 & 0 & 0 \\ 0 & 1 & 0 & 0 \end{bmatrix} , \qquad \mathbf{z} = \left\{ \begin{array}{c} \dot{x} + i\dot{y} \\ \dot{\phi}_y - i\dot{\phi}_x \\ x + iy \\ \phi_y - i\phi_x \end{array} \right\} . \qquad (6.52)$$

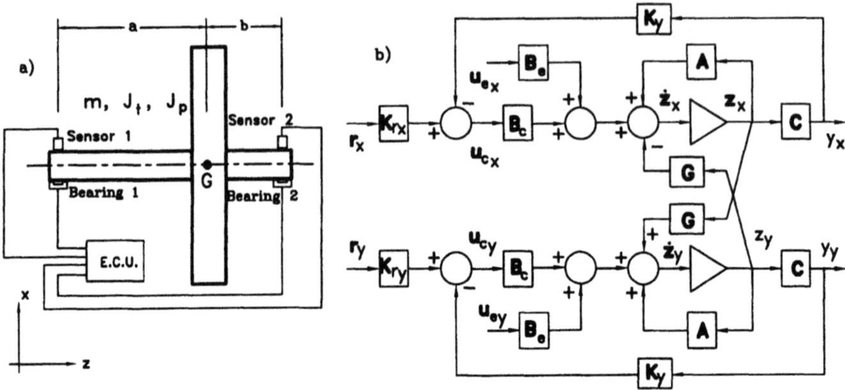

FIGURE 6.17. Rigid rotor on a four-axis active magnetic suspension: (a) sketch of the system; (b) block diagram, evidencing gyroscopic coupling between the xz- and yz-planes.

The nonrotating forces due to weight, the rotating forces due to unbalance, and the control forces can be summarized in the following input vectors:

$$\boldsymbol{B}_n\mathbf{u}_n = \left\{\begin{array}{c} -g \\ 0 \\ 0 \\ 0 \end{array}\right\}, \qquad \boldsymbol{B}_r\mathbf{u}_r e^{i\omega t} = \omega^2 \left\{\begin{array}{c} \epsilon e^{i\alpha} \\ \chi(1 - J_p/J_t) \\ 0 \\ 0 \end{array}\right\} e^{i\omega t},$$

$$(6.53)$$

$$\boldsymbol{B}_c\mathbf{u}_c = \left[\begin{array}{cc} 1/m & 1/m \\ -a/J_t & b/J_t \\ 0 & 0 \\ 0 & 0 \end{array}\right] \left\{\begin{array}{c} F_{c_{x_1}} + iF_{c_{y_1}} \\ F_{c_{x_2}} + iF_{c_{y_2}} \end{array}\right\}.$$

In a similar way, the output of the system, i.e., the complex displacements and velocities at the transducer location, are

$$\mathbf{y} = \left\{\begin{array}{c} \dot{x}_1 + i\dot{y}_1 \\ \dot{x}_2 + i\dot{y}_2 \\ x_1 + iy_1 \\ x_2 + iy_2 \end{array}\right\} = \left[\begin{array}{cccc} 1 & -a & 0 & 0 \\ 1 & b & 0 & 0 \\ 0 & 0 & 1 & -a \\ 0 & 0 & 1 & b \end{array}\right] \mathbf{z}. \qquad (6.54)$$

Instead of an actual multivariable control system, it is possible to independently control each bearing. This simplifies the control unit design, which reduces to a set of simpler units, each acquiring the information regarding the displacements and velocities in the two radial directions at the bearing location and supplying the control inputs to the actuators that provide the control forces in the corresponding directions. Furthermore, each axis of each bearing can be controlled independently, using a set of SISO control systems. To provide the required damping, the information regarding velocities is also required, possibly using two velocity sensors, but usually by differentiating in time the displacements. In the case of deformable rotors, however, a state feedback performed through a suitable observer seems to be more appropriate, particularly where the response to external disturbances is concerned.[6]

Assuming a complete axial symmetry, which is consistent only with the case of a bearing completely unloaded from static forces, using a simple proportional control such as the one summarized by equation (6.20), and assuming that the reference position is characterized by a zero value of the displacements, the control forces for the rotor of Figure 6.17a can be expressed as

$$\left\{ \begin{array}{c} F_{c_{x_1}} + iF_{c_{y_1}} \\ F_{c_{x_2}} + iF_{c_{y_2}} \end{array} \right\} = \left\{ \begin{array}{c} F_{c_{x_{1_0}}} + iF_{c_{y_{1_0}}} \\ F_{c_{x_{2_0}}} + iF_{c_{y_{2_0}}} \end{array} \right\} + \tag{6.55}$$

$$+ \left[\begin{array}{cccc} k_{v_1} & 0 & K_{d_1} - K_{u_1} 0 & 0 \\ 0 & k_{v_2} & 0 & K_{d_2} - K_{u_2} \end{array} \right] \mathbf{z} .$$

In equation (6.55) the two bearings are considered uncoupled and each is characterized by a stiffness $K_d - K_u$ (equation (6.48)) and a damping coefficient k_v.

If the system is nonaxisymmetrical, the use of complex coordinates implies a form of the relevant equations that include the conjugate of the state vector (see Section 4.8) The complex-coordinate approach is, however, not the only possible one and is best suited to cases in which the characteristics of the magnetic bearings are isotropic in the xy-plane. If this is not the case, due to anisotropies due to the bearing themselves or to the control system, the real and imaginary parts of equations from (6.50) to (6.55) can be separated in the same way seen in Chapter 4 for general rotor dynamics. Real equations with double size are then obtained. The dynamic matrix $\mathbf{A}$ is in this case

---

[6]P. Larocca, D. Fermental, E. Cusson, "Performance comparison between centralized and decentralized control of the Jeffcott rotor", 2nd Int. Symp. on Magnetic Bearings, Tokyo, July 1990.

$$\boldsymbol{A} = \begin{bmatrix} -\mathbf{M}^{-1}(\mathbf{C}_n + \mathbf{C}_r) & -\omega\mathbf{M}^{-1}\mathbf{G} & -\mathbf{M}^{-1}\mathbf{K} & -\omega\mathbf{M}^{-1}\mathbf{C}_r \\ \omega\mathbf{M}^{-1}\mathbf{G} & -\mathbf{M}^{-1}(\mathbf{C}_n + \mathbf{C}_r) & \omega\mathbf{M}^{-1}\mathbf{C}_r & -\mathbf{M}^{-1}\mathbf{K} \\ \mathbf{I} & \mathbf{0} & \mathbf{0} & \mathbf{0} \\ \mathbf{0} & \mathbf{I} & \mathbf{0} & \mathbf{0} \end{bmatrix},$$

$$(6.56)$$

where the various matrices are referred to flexural behaviour in the xz-plane. The relevant matrices for the rigid rotor of Figure 6.17, studied using real coordinates, are

$$\boldsymbol{A} = \begin{bmatrix} \begin{bmatrix} 0 & 0 & 0 & 0 \\ 0 & 0 & 0 & \omega J_p/J_t \\ 0 & 0 & 0 & 0 \\ 0 & -\omega J_p/J_t & 0 & 0 \end{bmatrix} & \mathbf{0} \\ \mathbf{I} & \mathbf{0} \end{bmatrix}, \qquad (6.57)$$

$$\boldsymbol{B}_n \mathbf{u}_n = \left\{ \begin{Bmatrix} -g \\ 0 \\ 0 \\ 0 \end{Bmatrix} \atop \mathbf{0} \right\},$$

$$\mathbf{z} = \begin{bmatrix} \dot{x} & \dot{\phi}_y & \dot{y} & -\dot{\phi}_x & x & \phi_y & y & \phi_x \end{bmatrix}^T,$$

$$\boldsymbol{B}_r \mathbf{u}_r = \left\{ \begin{Bmatrix} \epsilon\cos(\alpha + \omega t) \\ \chi(1 - J_p/J_t)\cos(\omega t) \\ \epsilon\sin(\alpha + \omega t) \\ \chi(1 - J_p/J_t)\sin(\omega t) \end{Bmatrix} \atop \mathbf{0} \right\}, \qquad (6.58)$$

$$\boldsymbol{B}_c \mathbf{u}_c = \left[\begin{bmatrix} 1/m & 0 & 1/m & 0 \\ -a/J_t & 0 & b/J_t & 0 \\ 0 & 1/m & 0 & 1/m \\ 0 & -a/J_t & 0 & b/J_t \end{bmatrix} \atop \mathbf{0} \right] \begin{Bmatrix} F_{c_{x_1}} \\ F_{c_{y_1}} \\ F_{c_{x_2}} \\ F_{c_{y_2}} \end{Bmatrix},$$

$$\mathbf{y} = \begin{Bmatrix} \dot{x}_1 \\ \dot{y}_1 \\ \dot{x}_2 \\ \dot{y}_2 \\ x_1 \\ y_1 \\ x_2 \\ y_2 \end{Bmatrix} = \begin{bmatrix} 1 & 0 & -a & 0 & 0 & 0 & 0 & 0 \\ 0 & 1 & 0 & -a & 0 & 0 & 0 & 0 \\ 1 & 0 & b & 0 & 0 & 0 & 0 & 0 \\ 0 & 1 & 0 & b & 0 & 0 & 0 & 0 \\ 0 & 0 & 0 & 0 & 1 & 0 & -a & 0 \\ 0 & 0 & 0 & 0 & 0 & 1 & 0 & -a \\ 0 & 0 & 0 & 0 & 1 & 0 & b & 0 \\ 0 & 0 & 0 & 0 & 0 & 1 & 0 & b \end{bmatrix} \mathbf{z} \,, \qquad (6.59)$$

$$\mathbf{K}_y = \begin{bmatrix} k_{v_{x_1}} & 0 & 0 & 0 & k_{d_{x_1}} & 0 & 0 & 0 \\ 0 & k_{v_{y_1}} & 0 & 0 & 0 & k_{d_{y_1}} & 0 & 0 \\ 0 & 0 & k_{v_{x_2}} & 0 & 0 & 0 & k_{d_{x_2}} & 0 \\ 0 & 0 & 0 & k_{v_{y_2}} & 0 & 0 & 0 & k_{d_{y_2}} \end{bmatrix} \,.$$

$$(6.60)$$

Note that the coupling between the xz- and yz-planes is due only to the gyroscopic terms. This is clearly shown in the block diagram of Figure 6.17b, where matrices $\boldsymbol{A}$ are those of the corresponding nongyroscopic system in each of the two planes xz and yz. The block $\mathbf{G}$ supplies the gyroscopic coupling between the two planes, and its gain is proportional to the spin speed ω. Note that in Figure 6.17 an output control acting separately in the xz- and yz-planes is used. On the contrary, no assumption on a control system that acts separately on the two bearings has been made.

A simple proportional control, or, better, a PID control, can be used but more complex control laws are common. As a first thing, actual controller-actuator systems have a cut-off frequency, related to their construction scheme and more influenced by the characteristics of the actuator than of the controller. A more complex dependence of the gain from the frequency can be used to obtain better performance: A very high static stiffness can be very useful in many applications, as machine-tool spindles, while low stiffness at higher frequency can allow the system to work in supercritical conditions, using self-centering to reduce balancing requirements. Compensators are usually introduced into the control system in order to perform these tasks.

Apart from the dependence of the gain from the frequency, it is also possible to introduce a dependence of the control law on the spin speed, producing a Campbell diagram in which the dependence of the various natural frequencies from the spin speed is different from that due to the gyroscopic effect alone. Also, the unbalance response can be strongly affected by the dependence of the gains of the control system on the speed and very smooth running can be achieved without the need of very strict balancing tolerances. While the dependence of the stiffness from the frequency can shift the critical speeds, its dependence on the speed can make some critical speeds disappear altogether.

The separate control of each bearing can be suitable if only rigid-body behaviour is to be controlled, but a true multivariable control system, based on an observer to perform state feedback, can be used, even if at the cost of added complexity. This is particularly important when the control system is aimed at controlling the deformation modes of the system. Note that the modal approach is still possible, but the modal gyroscopic matrix is, generally speaking, nondiagonal, as is the modal damping matrix. However, while the latter is usually very small and the errors due to neglecting the out-of-diagonal terms are negligible, the same does not hold for the first. The modal gyroscopic matrix can be reduced to a diagonal matrix to uncouple the modes even in this case, but errors due to spillover can be large. Their amount depends on the actual rotor configuration and must be checked in each case.

Using either the complex- or real-coordinates approach, it is possible to study the closed-loop dynamics of the system. The Campbell diagram can thus be plotted and the unbalance response can be computed, as for uncontrolled rotors. Although the results obtainable are not in principle dissimilar from those that are typical of rotors running on other types of bearings, the particular features of magnetic bearings can produce a very different behaviour. The control system can introduce a damping that is much higher than that achievable with mechanical means. This is very favourable in both fighting high-speed instability and reducing unbalance response when crossing critical speeds. The whole pattern of the Campbell diagram can be strongly affected: The usual consideration that damping affects the decay rate but has little influence on the natural frequencies is no longer valid at the high damping levels obtainable in controlled systems. Some modes can become, in the whole speed range or only a part of it, overdamped, and the relevant branch of the Campbell diagram can go to zero and then disappear. The cut-off frequency of the control system and actuators cannot be higher than the maximum rotational frequency of the rotor, reducing both the stiffness and the damping of the supports at the frequencies of synchronous whirling. This effect can be advantageous for the unbalance response, particularly in the case of reduction of the stiffness.

Many strategies aimed at obtaining a sort of self-balancing have been attempted and used. The simplest is that of using a notch filter centered on a frequency that follows the spin speed in such a way as to remain synchronous. Although this is effective in achieving a self-centered configuration, it has poor rejection characteristics for disturbances at the synchronous frequency and is not advisable. A feedforward strategy in which the synchronous disturbance due to unbalance is measured by averaging in time the outputs of the sensors at that frequency and then is fed to the controller, which forces the rotor to spin about its principal axis of inertia, yields very interesting results.

A last dynamic problem that has to be studied in the case of a rotor on magnetic bearings is that of the dynamic behaviour in the case of failure of the control system. Usually rotors running on magnetic bearings are provided by a set of emergency touch-down bearings, usually of the rolling-element type, that can perform the spin-down task in the case of high-speed failure of the control system. The dynamic study of the spin-down on the emergency bearings is very difficult, because it is mainly a nonlinear process, and it is performed by numerically integrating the equations of motion in time. The rotor first falls down until it is in contact with the emergency bearings, then a phase of rubbing is initiated, with possible rebounds, until the inner race of the emergency bearings gains the required speed and the rotor spins in a new configuration. This rigid-body behaviour is accompanied by deflections and vibrations, and it is very important that the structural integrity of the system is warranted and that undue contact between the rotor and the stator is avoided.

Example 6-7
Consider a rigid, isotropic rotor supported by two magnetic radial bearings and assume that the geometric configuration corresponds with that of Figure 6.17.

The main data are: $m = 3$ kg; $J_p = 0.02$ kg m^2; $J_t = 0.015$ kg m^2; $a = 200$ mm; $b = 100$ mm. Compute the gain matrix of the control system assuming that each bearing is controlled separately and that a complete axial symmetry is required. Plot the Campbell diagram and compute the control forces at 20,000 rpm if the rotor has a residual static unbalance following the ISO quality grade G 2.5. The control system should provide uncoupling between translational and rotational modes and locate the first critical speed in the vicinity of 2,500 rpm $= 261.8$ rad/s. Assume that the time constant of the control system is equal to 1 ms.

Using complex coordinates, the closed-loop dynamic matrix $\mathbf{A} - \mathbf{B}_c\mathbf{K}_y\mathbf{C}$ is

$$
\mathbf{A}_{cl} =
\begin{bmatrix}
-\dfrac{k_1}{m} & -\dfrac{k_2}{m} & -\dfrac{k_4}{m} & -\dfrac{k_5}{m} \\[2mm]
-\dfrac{k_2}{J_t} & -\dfrac{k_3}{J_t} & -\dfrac{k_5}{J_t} & -\dfrac{k_6}{J_t} \\[2mm]
1 & 0 & 0 & 0 \\[1mm]
0 & 1 & 0 & 0
\end{bmatrix} ,
$$

where $k_1 = k_{v_1} + k_{v_2}$, $k_2 = -ak_{v_1} + bk_{v_2}$, $k_3 = a^2 k_{v_1} + b^2 k_{v_2} - i\omega J_p$, $k_4 = k_{d_1} + k_{d_2}$, $k_5 = -ak_{d_1} + bk_{d_2}$, and $k_6 = a^2 k_{d_1} + b^2 k_{d_2}$.

Uncoupling between translational and rotational modes implies that $ak_{v_1} = bk_{v_2}$, $ak_{d_1} = bk_{d_2}$, and the condition on the critical speed of the corresponding undamped system yields $\sqrt{(k_{d_1} + k_{d_2})/m} = \omega_{cr} = 261.8$. The values of the stiffness gains thus obtained are $k_{d_1} = 6.85 \times 10^4$ Nm, $k_{d_2} = 1.37 \times 10^5$ Nm and yield the following terms to be included in the closed-loop gain matrix: $k_{d_t} = k_{d_1} + k_{d_2} = 2.056 \times 10^5$ N/m, and $k_{d_r} = a^2 k_{d_1} + b^2 k_{d_2} = 4{,}110$ Nm/rad.

The damping gains can be obtained assuming a given value for the damping ratio of a particular mode, say, the translational mode. Assuming a damping ratio of 1/3, the following values are obtained: $k_{v_1} = 349.1$ Ns/m, $k_{v_2} = 174.5$ Ns/m, $k_{v_t} = k_{v_1} + k_{v_2} = 523.6$ Ns/m, and $k_{v_r} = a^2 k_{v_1} + b^2 k_{v_2} = 10.47$ Nsm/rad.

The static deflection of the system is immediately obtained: $x_{st} = -mg/k_{d_t} = 0.14$ mm. Note that the control system can be set to compensate for the static deflection and to maintain the rotor in the center of the bearing notwithstanding the weight.

The stiffness gain so computed also includes the negative stiffness linked with the open-loop behaviour of the bearing; the time constant of the sensor-controller-actuator subsystems should be applied only to the part of the overall gain linked with K_d and not to that linked with K_u. As a first order approximation the contribution of the negative stiffness will be neglected.

The Campbell diagram can thus be easily computed. Because the controller is assumed to have a nonvanishing time constant, the gain matrix $\mathbf{K}_y$ is not constant:

$$\mathbf{K}_y = \frac{1}{1+s\tau} \begin{bmatrix} k_{v_1} & 0 & k_{d_1} & 0 \\ 0 & k_{v_2} & 0 & k_{d_2} \end{bmatrix}.$$

By transforming the equations of motion in the Laplace domain and remembering that the translational and rotational modes are uncoupled, the equations allowing the plotting of the Campbell diagram are

$$m\tau s^3 + ms^2 + sk_{v_t} + k_{d_t} = 0$$

for the translational mode and

$$J_t \tau s^3 + \left(J_t - i\omega\tau J_p\right)s^2 + \left(k_{v_r} - i\omega J_p\right)s + k_{d_r} = 0$$

for the rotational mode. The whirl frequency, i.e., the imaginary part of the Laplace variable s, and the decay rate, i.e., the real part of s, are plotted in Figure 6.18. Also, in the same figure the solution obtained for a vanishingly small time constant of the control system is reported. The roots locus is shown in Figure 6.19.

Due to uncoupling between translational and rotational rigid-body modes, the response to static unbalance can be plotted by resorting to the equation for translational motion alone

$$\left(-m\omega^2 + i\frac{\omega}{1+i\omega\tau}k_{v_t} + \frac{1}{1+i\omega\tau}k_{d_t}\right)(x+iy) = m\epsilon\omega^2 e^{i\omega t}.$$

The amplitude of the response to static unbalance is plotted in nondimensional form in Figure 6.20. Note that the system is very much damped, but the presence of a cut-off frequency, affecting the damping and the stiffness, causes an increase of the amplitude at the critical speed. The rotor self-centers very quickly in the supercritical range. Unbalance quality grade G 2.5 at 20,000 rpm = 2,094 rad/s corresponds to an eccentricity $\epsilon = 1.19\mu$m, i.e., with a mass of 3 kg, to a residual unbalance $m\epsilon = 3.58$ g mm. The response of the system at 20,000 rpm is $(x+iy)_0 = -1.236 - 0.012i\,\mu$m for $\tau = 1$ ms and $(x+iy)_0 = -1.203 - 0.102i\,\mu$m for $\tau = 0$

The total control force acting on the rotor is simply

$$F_c = \left(i\frac{\omega}{1+i\omega\tau}k_{v_t} + \frac{1}{1+i\omega\tau}k_{d_t}\right)(x+iy) =$$

$$= m\omega^2(x+iy) + m\epsilon\omega^2 e^{i\omega t}.$$

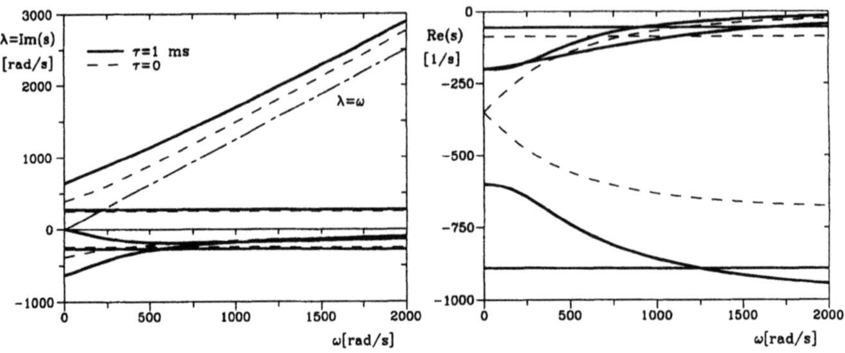

FIGURE 6.18. Campbell diagram of a rigid rotor on magnetic bearings: (a) imaginary part and (b) real part of the Laplace variable s.

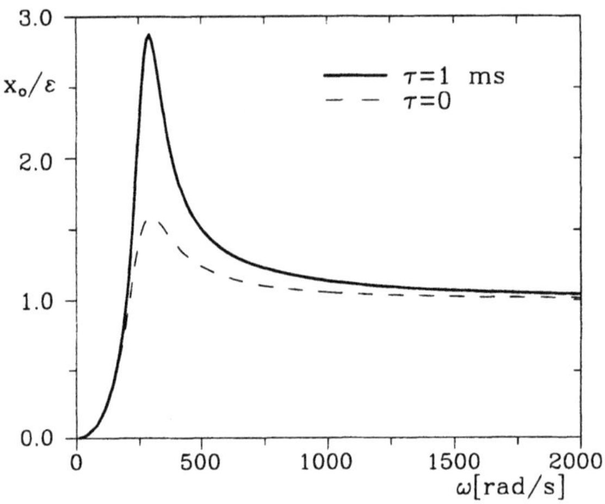

FIGURE 6.19. Roots locus of the same system in Figure 6.16.

The modulus of the control force is then $|F_c| = 0.594$ N for $\tau = 1$ ms and $|F_c| = 1.347$ N for $\tau = 0$, and is shared between the two bearings following their stiffnesses (2/3 on bearing 1 and 1/3 on bearing 2). Note that the control forces are very small, even if the balancing quality grade is not very high, particularly in the case of the less stiff control, with higher time constant. Actually, very smooth running is obtained with even less strict balancing tolerances.

Example 6-8
Consider a rotor running on a one-active-axis magnetic suspension. The radial bearings are of the passive type, based on permanent magnets.

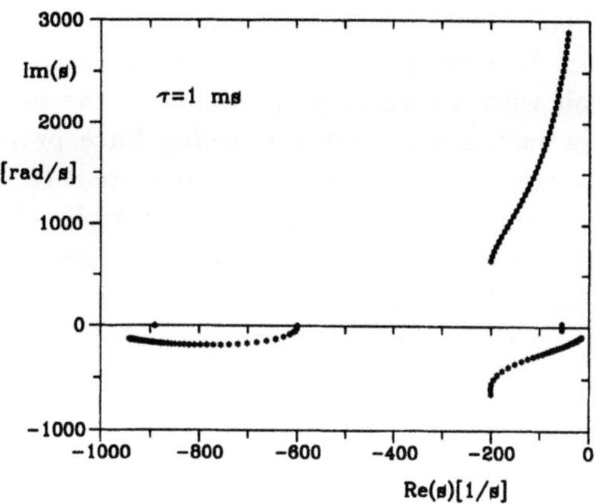

FIGURE 6.20. Unbalance response of a rigid rotor on magnetic bearings.

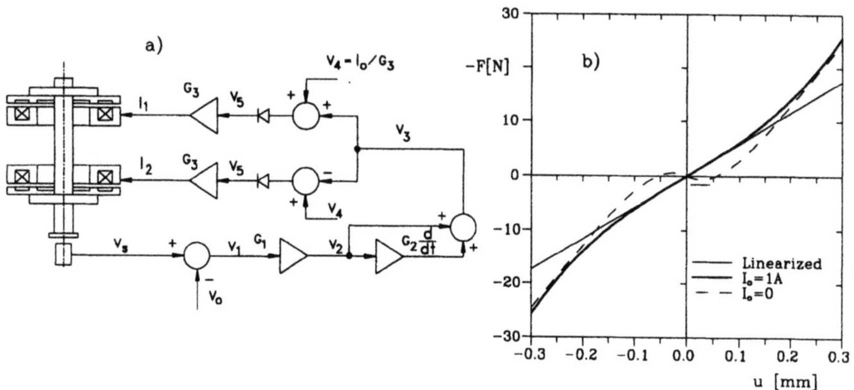

FIGURE 6.21. Active axial magnetic bearing. (a) Scheme of the control system; (b) force-displacement characteristics.

The axial suspension must be active, because the passive radial bearings have a negative (unstabilizing) axial stiffness. The mass of the rotor is $m = 0.8$ kg, and the linearized axial stiffness of the radial suspension is $k_z = -5 \times 10^4$ 104 N/m. The axial actuator is made by a couple of electromagnets whose force-current characteristic is $F_c = \mu_0 S N^2 I^2 / 4d^2$, where I is the current, $\mu_0 = 4\pi \times 10^{-7}$ is the vacuum permeability, $S = 1.2 \times 10^{-3} \text{m}^2$ is the surface of the pole pieces, $N = 160$ is the number of turns, and $d = 0.5$ mm is the nominal air gap.

The axial position is measured by a sensor that outputs a voltage proportional to the axial displacement u from the nominal position (air gap of 2.5 mm) through the law $V_s = V_0 + au$, where constants V_0 and a take the values 0.85 V and 300 V/m, respectively, when the input voltage to the sensor is 12 V.

The quadratic law linking the magnetic force to the current in the coil compels a choice between two alternatives: to use a control system that supplies the coil with a current proportional to the square root of the displacement in such a way that a restoring force proportional to the displacement is exerted or to add a constant current to the current proportional to the displacement supplied by the control system. In the latter case the coils exert two forces, which in the equilibrium position are equal and opposite; when the shaft is displaced axially the force due to one coil is reduced while the other increases and the restoring effect is obtained. A detailed scheme of the control system, following the second of the two approaches, is shown in Figure 6.21a.

Note that the voltage signal $V_2 = V_1 G_1$ is differentiated to supply an input to the current generators that is proportional to both the axial displacement and velocity. A damping effect is thus obtained. The voltage supplied to the current generators driving the coils is then

$$V_5 = V_4 \mp G_1 V_1 \mp G_1 G_2 \frac{dV}{dt} = V_4 \mp aG_1(u + G_2 \dot{u}),$$

where the upper signs hold for the coil nearer the end at which the sensor is located. The current supplied to the coils is then $I_i = G_3 V_5 = I_0 \pm aG_1 G_3(u + G_2 \dot{u})$, the upper sign being referred to coil 1. When both coils are operating, the force they exert (positive if directed upward, i.e., in a direction to cause an increase of the displacement u) is

$$F = \frac{\mu_0 S N^2}{4} I_0 \left[\frac{I_2^2}{(d-u)^2} - \frac{I_1^2}{(d+u)^2} \right].$$

The expression of the force can be linearized, obtaining

$$F = \frac{\mu_0 S N^2}{4} \left[aG_1 G_3(u + G_2 \dot{u}) - \frac{I_0}{d} u \right].$$

The total stiffness and damping of the axial suspension are, respectively,

$$K_z = \frac{\mu_0 S N^2}{4} I_0 \left[aG_1 G_3 - \frac{I_0}{d} \right] - k_z, \qquad C_z = \frac{\mu_0 S N^2}{4} I_0 aG_1 G_3 G_3.$$

The second term in brackets is the open-loop stiffness, which is negative. It can cause the axial instability of the bearing, which, if the current is not controlled, is unstable.

To achieve a stable behaviour in the axial direction, the stiffness must be positive. Assuming the following values for the current I_0 and the gains, $I_0 = 1$ A, $G_1 = 10$, $G_2 = 3 \times 10^{-4}$ s, $G_3 = 0.9$ A/V, the values of the axial stiffness and damping are $K_z = 58{,}100$ N/m and $C_z = 125$ Ns/m.

The system is then axially stable. The axial natural frequency and the damping ratio are $\lambda_n = 269$ rad/s and $\zeta = 0.29$. The force-displacement characteristic of the axial suspension is reported in Figure 6.21b. The curve obtained from the linearized equation is compared with those obtained from the nonlinearized equation and the curve obtained when no constant current is supplied ($I_0 = 0$).

6.6 Exercises

Exercise 6-1

Consider the damped pendulum in Figure 6.2. Plot the root locus with the following data. Compute the value of the stiffness of the spring in such a way that the natural frequency of the system is 50 Hz, the damping ratio and time history of the system following a shock due to the impact of a ball of mass of 1 kg traveling at 30 m/s on the bob of the pendulum. Assume a perfectly elastic shock. Data: length $l = 1$ m, $m = 2$ kg, $g = 9.806$ m/s^2, $\Gamma = 100$ Nm/rad.

Exercise 6-2

Substitute the spring in Example **??** with an actuator driven by a proportional-derivative control system and a transducer reading angle θ. Compute the gains in such a way that the frequency of the damped oscillations of the system is 50 Hz and the damping ratio is 0.2. Compute the response to the same impulse studied in Exercise 6-1.

Exercise 6-3

Consider the system with three degrees of freedom studied in Example 1-2 and add a control system that exerts a moment on disc 2. Consider as input to the system the rotation θ_A of the supporting point and as output the rotation of disc 2. Compute the matrices of the system and check whether it is controllable and observable. Give a step input to the system with a rotation $\theta_A = 10°$, and assume as new reference position the static equilibrium configuration rotated of angle θ_A. Compute the gains in such a way that the overshot in the first mode is not greater than 50%, assuming an output feedback. Compute the time histories of the response at all discs.

Exercise 6-4

Consider the system of the preceding exercise using state feedback through an observer controlling only the first mode. Design the observer and repeat the computations.

Exercise 6-5

Consider the system in Exercise 6-4. Compute the gains of the control system in such a way that the poles are $-0.3 \pm i$, $-0.8 \pm i4$, and $-1.5 \pm i6$, and the poles of the observer have the same imaginary part and a real part with absolute value 50% greater. Compute the time history of the response to the same step input of the preceding exercises.

Exercise 6-6

Consider the quarter-car model of Example 1-9. Substitute the shock absorber with a "skyhook damper," i.e., an active device that exerts on the sprung mass a force proportional to its absolute vertical velocity. Assume that the feedback loop is based on the vertical acceleration of the sprung mass. Compute the frequency response of the system; plot the maximum value of the acceleration of the sprung mass as a function of the gain of the control system and choose the parameters that allow the minimum value of the peak acceleration to be obtained. Compute the variable component of the vertical force exerted by the tire on the road and the value of the gain that minimizes its peak value.

Exercise 6-7

Compute the power spectral density of the response of the active suspension obtained in Exercise 6-6 (with the value of the gain minimizing the vertical acceleration) when traveling on a normal road following ISO standards (Figure 1.32d). Compare the results with those obtained in Example 1-9 for a conventional suspension.

Exercise 6-8

Consider the turbomolecular pump of Example 4-2. Design a magnetic suspension system to achieve the same goals with a shaft stiff enough to allow the rotor to be considered rigid in the whole working range. Compute the response to a residual static unbalance corresponding to a quality grade G 6.3. Assume a time constant for the control system equal to 1 ms.

Exercise 6-9

Repeat the computations in the previous exercise without neglecting the compliance of the shaft. Assume that the shaft has a diameter of 8 mm.

Exercise 6-10

Compute the Campbell diagram of the rotor Example 4-3 operating on two magnetic bearings. Assume a bearing stiffness of 10^7 N/m and use a simple control in which each bearing is controlled independently, without attempts to specifically control deformation modes.

Appendix A
Solution Methods

A.1 General considerations

The computations needed to perform the dynamic analysis of systems with many degrees of freedom are, in many cases, very involving. If the number of degrees of freedom is high, only the use of digital computers allows the obtaining of the natural frequencies, the mode shapes, and the forced response of the system, or better, of its discretized model. Many numerical methods have been developed to efficiently perform the mentioned tasks; their formulation and implementation have been and still are the object of intensive research work. The detailed study of these methods is well beyond the scope of this text, but an engineer using structural analysis codes needs to have at least an approximate knowledge of the relevant solution methods, particularly when he has to choose among various possibilities offered by computer programs. He does not need to be an applied mathematician, but it is important that he has at least an idea of how the machine "crunches the numbers."

Four mathematical problems will be dealt with here, namely, the solution of a set of linear equations, the solution of eigenproblems, the solution of sets of nonlinear equations, and the numerical integration of sets of differential equations, both linear and nonlinear.

Many books[1] and a countless number of papers have been devoted to these four problems; the short outline presented here has the aim of supplying general information to the structural analyst who has to use the relevant computer codes. No detailed formulas or practical details on coding are included because many subroutines are available, and the author does not advise the preparation and use of home-brewed codes, particularly in this case.

A.2   Solution of linear sets of equations

The basic mathematical problem encountered in the static analysis of linear systems with many degrees of freedom is the solution of a set of linear equations, whose matrix of the coefficients, the stiffness matrix, is generally real, positive defined, and symmetrical, and often has a band structure with a bandwidth much smaller than the matrix. The response of the system to a harmonic excitation can be computed by solving equation (1.71) where the dynamic stiffness matrix, while being symmetrical and usually retaining a band structure, can be nonpositive definite. If the model includes damping, the dynamic stiffness matrix is complex.

The set of linear equations

$$\mathbf{Ax} = \mathbf{b}\,, \tag{A.1}$$

where both $\mathbf{A}$ and $\mathbf{b}$ are complex, can be transformed into the set of real equations, at the expense of doubling the size of the problem

$$\begin{bmatrix} \Re(\mathbf{A}) & -\Im(\mathbf{A}) \\ \Im(\mathbf{A}) & \Re(\mathbf{A}) \end{bmatrix} \left\{ \begin{array}{c} \Re(\mathbf{x}) \\ \Im(\mathbf{x}) \end{array} \right\} = \left\{ \begin{array}{c} \Re(\mathbf{b}) \\ \Im(\mathbf{b}) \end{array} \right\}\,, \tag{A.2}$$

Note that the matrix of the coefficients of equation (A.2) is nonsymmetrical even if that of equation (A.1) is. Also, its band structure can be far less prominent than that of the original matrix.

When the conjugates $\overline{\mathbf{x}}$ of the unknowns $\mathbf{x}$ explicitly enter the equations

---

[1]See, for example, F.G. Curtis, *Applied numerical analysis*, Addison Wesley, Reading, 1978; J.H. Wilkinson, C. Reinsh, *Linear algebra: handbook for automatic computing*, Springer, New York, 1971; J. Stoer, R. Burlish, *Introduction to numerical analysis*, Springer, New York, 1980; W.H. Press, B.P. Flannery, S.A. Teukolsky, and W.T. Vetterling, *Numerical recipes, the art of scientific computing*, Cambridge Univ. Press, Cambridge, 1986.

$$\mathbf{Ax} + \mathbf{B}\overline{\mathbf{x}} = \mathbf{c}, \tag{A.3}$$

the corresponding real set of equations is

$$\begin{bmatrix} \Re(\mathbf{A}) + \Re(\mathbf{A}) & -\Im(\mathbf{A}) + \Im(\mathbf{B}) \\ \Im(\mathbf{A}) + \Im(\mathbf{B}) & \Re(\mathbf{A}) - \Re(\mathbf{B}) \end{bmatrix} \left\{ \begin{array}{c} \Re(\mathbf{x}) \\ \Im(\mathbf{x}) \end{array} \right\} = \left\{ \begin{array}{c} \Re(\mathbf{c}) \\ \Im(\mathbf{c}) \end{array} \right\}. \tag{A.4}$$

The solution of a linear set of equations is usually indicated by writing the inverse of the matrix of the coefficients. The inversion of the matrix of coefficients is, however, a most inefficient way to solve linear sets of equations, and notations involving it must not be considered an indication of how to perform the computations.

When the solution to equation (A.1) is actually sought, two different types of techniques can be used, namely, direct and iterative algorithms. While in the early developments of the FEM, iterative algorithms were widely used, there is now a general agreement on the application of direct techniques. They are all more or less related to the Gauss elimination technique, which was devised more than a century ago. It is based on the transformation of the set of equations in order to eliminate the unknowns one by one until a single equation with one unknown is obtained. The unknowns can thus be computed one by one. The procedures of this type can be subdivided into two steps: the triangularization of the matrix of the coefficients, in which the equation yielding one of the unknowns is obtained, and the subsequent backsubstitution in which all other unknowns are subsequently found. The first part is by far the longest, where the computer time is concerned. The total number of elementary operations to be performed is of the order of $n^3/3$.

The Gauss method can be interpreted as a series of $n-1$ transformations of the matrix of coefficients $\mathbf{A}$ and vector $\mathbf{b}$ of equation (A.1), which yields an equation in which the matrix of coefficients is an upper triangular matrix. Such a transformation can be summarized as the multiplication of both sides of equation (A.1) by a nonsingular matrix $\mathbf{G}$, such that matrix $\mathbf{U} = \mathbf{GA}$ is an upper triangular matrix. The first of the two steps of the Gauss method is then the construction of matrix $\mathbf{G}$ and the computation of the products $\mathbf{U} = \mathbf{GA}$ and $\mathbf{b}^* = \mathbf{Gb}$, while the second is the solution of the equation $\mathbf{Ux} = \mathbf{b}^*$. It is then clear that the products of matrix $\mathbf{A}$ and vector $\mathbf{b}$ by the transformation matrix $\mathbf{G}$ can be performed separately and, in particular, when several sets of equations with the same matrix of coefficients but different vectors $\mathbf{b}$ have to be performed, matrix $\mathbf{U}$ can be computed only once, and the parts of the computation to be performed several times are only those related to products $\mathbf{Gb}$ and the backsubstitution, which are less costly from a computational viewpoint.

When no exchange of lines is required, matrix $\mathbf{A}$ can be decomposed in the form $\mathbf{A} = \mathbf{LU}$, where $\mathbf{L}$ is a lower triangular matrix and $\mathbf{U}$ is the aforementioned upper triangular matrix. Such decomposition is often referred to as LU factorization or, in two forms slightly different from each other, Doolittle and Crout factorizations.

When matrix $\mathbf{A}$ is symmetrical and positive definite the two triangular matrices $\mathbf{L}$ and $\mathbf{U}$ are the transposes of each other and the transformation takes the form $\mathbf{A} = \mathbf{LL}^T$. This form is referred to as Choleski factorization. Because the solution of the set of equations through Choleski factorization is faster than the use of the regular Gauss method, involving only about $n^3/6$ operations, most finite element codes use, for the static solution, this algorithm. The presence of a band structure can further simplify the computation and many algorithms that take into account this feature have been developed.

When several sets of equations with the same matrix of coefficients have to be solved, it is customary to write the equation in the form

$$\mathbf{AX} = \mathbf{B}, \tag{A.5}$$

where matrix $\mathbf{B}$ is a rectangular matrix whose columns are the various vectors $\mathbf{b}$ of the different sets of equations and, similarly, the columns of matrix $\mathbf{X}$ are the unknowns vectors $\mathbf{x}$. Note that equation (A.5) is not only a notational shortcut, $\mathbf{X}$ is an unknown matrix that when multiplied by matrix $\mathbf{A}$ yields matrix $\mathbf{B}$. If the number of sets of equations is m, equation (A.5) is actually a set of $m \times n$ equations yielding the $m \times n$ unknown elements of matrix $\mathbf{X}$.

The precision obtainable with the aforementioned techniques depends on the structure of the matrix of the coefficients. If it is well conditioned, the result is usually very good; however, it is possible to apply iterative procedures that allow the refining of the results obtained through direct techniques. Consider the set of equations $\mathbf{Ax} = \mathbf{b}$ and the approximate solution $\mathbf{x}^{(1)}$. The exact solution $\mathbf{x}$ can be written in the form $\mathbf{x} = \mathbf{x}^{(1)} + \delta\mathbf{x}^{(1)}$, where the last term expresses the errors introduced by the approximate solution technique.

Equation (A.1) can be written in the form

$$\mathbf{A}\delta\mathbf{x}^{(1)} = \mathbf{r}^{(1)} \quad \text{where} \quad \mathbf{r}^{(1)} = \mathbf{b} - \mathbf{A}\mathbf{x}^{(1)}, \tag{A.6}$$

which can be used to compute the error $\delta\mathbf{x}^{(1)}$. Note that the solution of equation (A.6) is straightforward, because it requires only the factorization of matrix $\mathbf{A}$, which has already been performed. The procedure can be repeated several times, each time getting nearer the correct solution. In most cases, however, the precision of the result directly obtained is sufficient, and no iterative refinement is required.

If the matrix of coefficients were diagonal, the solution would have been straightforward because each equation would have directly yielded one unknown. In any case, it is possible to take the elements of the main diagonal of the matrix of coefficients and to separate them from the other elements

$$\mathbf{A} = \mathbf{A}^* + \mathbf{A}^{**}, \tag{A.7}$$

where $\mathbf{A}^*$ is diagonal and $\mathbf{A}^{**}$ is a matrix with zero element on the main diagonal. Equations (A.1) can be written in the form

$$\mathbf{A}^*\mathbf{x} = \mathbf{b} - \mathbf{A}^{**}\mathbf{x}, \tag{A.8}$$

which can easily be solved iteratively. A trial vector $\mathbf{x}^{(0)}$, usually with all elements equal to 0, is introduced on the right-hand side and a new value $\mathbf{x}^{(1)}$ is computed by solving a set of uncoupled equations. The procedure is then repeated until convergence is obtained. This iterative solution scheme is referred to as the Jacobi method. When solving the ith equation, the new values of the first $(i-1)$ unknowns have already been obtained and the new values can be used directly. The latter scheme is known as the Gauss-Seidel method.

A condition that is sufficient, although not necessary, for ensuring convergence of the Jacobi method is that matrix $\mathbf{A}$ be diagonally dominant, i.e., in each row the element on the diagonal is greater than the sum of the other elements.

It is possible to demonstrate that when the matrix of the coefficients is symmetrical and positive defined the Gauss-Seidel method converges. When the Gauss-Seidel and Jacobi methods both converge, the first is faster than the latter. To maximize the chances of obtaining convergence and to make it faster, the order of the equations should be rearranged in such a way that the largest elements lie on the main diagonal.

In some cases it is necessary to explicitly obtain the inverse of a matrix, as when performing matrix condensation. Remembering equation (A.5), if matrix $\mathbf{B}$ is the identity matrix $\mathbf{I}$, the unknown matrix $\mathbf{X}$ is nothing other than the inverse $\mathbf{A}^{-1}$. A simple way to compute the inverse of a matrix is by factorizing it and then obtaining the various columns by solving n sets of equations in which the various vectors $\mathbf{b}$ have all terms equal to zero except the term corresponding to the number of the column to be found, which has a unit value.

If matrix $\mathbf{A}$ is complex, the real and imaginary parts of its inverse $\mathbf{A}^{-1}$ can be computed from the following real equation:

$$\begin{bmatrix} \Re(\mathbf{A}) & -\Im(\mathbf{A}) \\ \Im(\mathbf{A}) & \Re(\mathbf{A}) \end{bmatrix} \begin{Bmatrix} \Re(\mathbf{A}^{-1}) \\ \Im(\mathbf{A}^{-1}) \end{Bmatrix} = \begin{Bmatrix} \mathbf{I} \\ \mathbf{0} \end{Bmatrix}. \tag{A.9}$$

Note that when matrix $\mathbf{A}$ has a band structure its inverse usually does not have the same type of structure. If the matrix is stored in the memory of the computer in such a way as to take advantage of its band structure, the memory required for storage of the inverse can be much greater than that needed for the original matrix.

A.3   Computation of eigenfrequencies

A.3.1   General considerations

The first and most important step for the study of the dynamic behaviour of a linear system is the evaluation of its eigenfrequencies and mode shapes. When using discretized models, this basic step reduces to the mathematical problem of finding the eigenvalues and eigenvectors of the dynamic matrix of the system. Even if the size of the matrices can be reduced by applying the condensation and substructuring techniques seen in Chapter 2, the solution of an eigenproblem whose size is only a few hundred can still require long and costly computations.

The complexity of the problem depends not only on its size but also on the characteristics of the relevant matrices and the requirements of the particular problem. First, the user can be interested in obtaining only the eigenvalues or both eigenvalues and eigenvectors.

Generally speaking, the problem can then be attacked at three different levels, namely, it can reduce to the search of a single, usually the lowest, eigenfrequency or of a selected number of eigenfrequencies, usually the lowest or those included in a given range, or of all eigenfrequencies.

The first alternative was very popular when automatic computation was not available or very costly. The first natural frequency could be rapidly evaluated with limited costs, but there was no chance to perform any modal analysis. Nowadays this approach is only used in the first steps of the design procedures, in order to be sure that no natural frequency is lower than a given value, usually stated in the design specifications. Approximate techniques, which yield a value lower than the correct one, are sufficient in this case, and more detailed computations can be postponed to a subsequent stage of the analysis, when the design is better defined.

Usually the dynamic analysis of models with many degrees of freedom follows the second approach. The knowledge of a certain number of eigenvalues and eigenvectors allows the performance of an approximate modal analysis and the computation of all the required dynamic characteristics of the system. In particular, the FEM yields a large number of vibration modes, due to the large number of degrees of freedom of the mathematical model, but many of them, usually those with the highest eigenfrequencies, have little physical relevance and are strongly influenced by the discretization scheme used. If they are discarded, no relevant information on the dynamic behaviour of the system is lost.

The last approach is so demanding where the complexity of the computations is concerned that it is used only when dealing with systems with a very small number of degrees of freedom, perhaps obtained through a large-scale condensation of a more complex model. However, the algorithms that search all the eigenvalues are more efficient than the selective ones, for a given number of eigenvalues found and is a common opinion that, when more than about 20% of the eigenvalues are required, it is more convenient to find all of them. Note that the algorithms that search all eigenvalues do not usually find them in any prescribed order; as a consequence it is not possible to start the search and stop the algorithm after a given number of solutions has been found, because there is always the possibility that a solution lying within the field of interest has been lost.

As already stated for the solution of linear sets of equations, but to a greater extent, the choice of the most convenient method depends on the structure of the relevant matrices and the aims of the search, and it is not possible to state which is the best method, in general. No attempt to deal in detail with the various mathematical aspects of the problem will be done here because the aim of this section is only to supply some general information on the more common algorithms. The user can find more details in any good textbook of numerical analysis, and any computer center has a large number of subroutines, ready to be linked in any custom-made program.

If the matrix whose eigenvalues and eigenvectors are required is real, the eigenanalysis can yield either real or complex-conjugate results. If, on the contrary, the starting matrix is complex, the complex results are not conjugate. Consider the general eigenproblem

$$(\mathbf{A} - \lambda\mathbf{I})\mathbf{x}_0 = \mathbf{0}, \tag{A.10}$$

already written in standard form, where matrix $\mathbf{A}$ is complex. It can be transformed into the real eigenproblem

$$\left(\begin{bmatrix} \Re(\mathbf{A}) & -\Im(\mathbf{A}) \\ \Im(\mathbf{A}) & \Re(\mathbf{A}) \end{bmatrix} - \lambda\mathbf{I} \right) \left\{ \begin{matrix} \Re(\mathbf{x}_0) + i\Im(\mathbf{x}_0) \\ \Re(\mathbf{x}_0) - i\Im(\mathbf{x}_0) \end{matrix} \right\} = \mathbf{0}, \tag{A.11}$$

whose size is double that of the original problem. Equation (A.11) can be easily solved using the standard algorithms for nonsymmetrical matrices and yields $2n$ solutions: the n eigenvalues and eigenvectors of equation (A.10) and their conjugates. In all those cases in which the sign of the imaginary part of the eigenvalues is important, a procedure that can distinguish between the actual eigenvalues of the original problem and those added when doubling the size of the matrices must be devised. This can easily be done by checking the structure of the eigenvector corresponding to each eigenvalue: If the real part of the first n elements is equal to the imaginary part of the remaining ones, a solution of the original problem has been found; otherwise the solution is discarded.

The Routh-Hurwitz criterion, which allows the assessing of whether some of the eigenvalues have a positive real part is sometimes used as an alternative to the actual solution of the eigenproblem for the study of the stability of the system as, when the time history for the free motion is assumed to be of the type $\mathbf{x} = \mathbf{x}_0 e^{st}$, instability is strictly linked with the presence of eigenvalues with positive real part. The Routh-Hurwitz criterion is based on computations that are much simpler than those required to actually solve the eigenproblem; however, when the size of the matrix is not very small, the application of the criterion itself can lead to long computations and, now that the direct solution of the eigenproblem is made possible by the use of automatic computation, it can be questionable whether it is worthwhile to resort to an approach that, at any rate, yields results of lesser interest. The main disadvantage of the Routh-Hurwitz criterion is actually that it shows whether a system is stable but not how stable it is. This can be circumvented by performing an eigenvalue shift, i.e., by substituting s with $(s_1 - \sigma)$, modifying accordingly the eigenproblem, and repeating the stability study. It is possible to state whether the real part of some of the eigenvalues is greater than $-\sigma$, i.e., whether in the complex plane some eigenvalues lie on the right of the line of equation $\Re(s) = -\sigma$. Note that the computation of the new eigenproblem can involve long computations, particularly if n is high.

Due to the mentioned drawbacks, the Routh-Hurwitz criterion will not be dealt with more here; the relevant equations can be found in many texts of applied mathematics and dynamics, as in, for example, A.F. D'Souza, *Design of control systems*, Prentice-Hall, Englewood Cliffs, 1988, 199.

A.3.2 The Rayleigh quotient

Assume an arbitrary n-dimensional vector $\mathbf{x}$. The ratio

$$R = \frac{\mathbf{x}^T \mathbf{K} \mathbf{x}}{\mathbf{x}^T \mathbf{M} \mathbf{x}} \tag{A.12}$$

is a number that lies between the smallest and largest eigenvalues. If $\mathbf{x}$ is an eigenvector, the Rayleigh quotient expressed by equation (A.12) coincides with the corresponding eigenvalue. Moreover, if the arbitrary vector $\mathbf{x}$ is a linear combination of a reduced set of eigenvectors, the Rayleigh quotient is included in a field spanning the minimum and maximum of the eigenvalues corresponding to the given eigenvectors. If vector $\mathbf{x}$ is close to a generic eigenvector with an error ϵ, the Rayleigh quotient is close to the corresponding eigenvalue with an error of the order of the square of ϵ. This means that if the Rayleigh quotient is considered a function of vector $\mathbf{x}$, it is stationary in the neighbourhood of any eigenvector.

A.3.3 The Dunkerley formula

The so-called Dunkerley formula has been a very widespread tool for the computation of an approximate value of the lowest eigenfrequency of undamped systems and is still reported in many handbooks, even if often with other names and in modified forms. Its usefulness lies mostly in the feature of supplying an approximation of the lowest natural frequency that is surely lower than the exact value. It can be used with confidence when it is necessary to verify that the first natural frequency is higher than a given value.

It is based on the compliance formulation (first equation (1.35)), in which the highest eigenvalue corresponds to the lowest eigenfrequency. It is well known that the sum of the eigenvalues of a matrix is equal to the sum of the elements on its main diagonal. If, as is usually the case, the first natural frequency is much lower than the others, the square of its reciprocal is very close to the sum of the eigenvalues and then to the trace of the dynamic matrix in the compliance formulation.

Because the Dunkerley formula is used in general for the lumped-parameters models, the mass matrix is a diagonal matrix and the elements on the main diagonal of the dynamic matrix $\mathbf{D} = \mathbf{K}^{-1}\mathbf{M}$ can be computed simply as $d_{ii} = \beta_{ii}m_{ii}$. It then follows that

$$\frac{1}{\lambda_1^2} \approx \sum_{\forall i} \frac{1}{\lambda_i^2} = \sum_{\forall i} \beta_{ii}m_{ii}. \tag{A.13}$$

The use of equation (A.13) requires computation of the compliance matrix; it can therefore be used very simply when the elastic behaviour of the system is expressed in terms of coefficients of influence. In such a case it is not even necessary to compute all coefficients, because it is sufficient to know those that are on the main diagonal. If, however, the stiffness approach is followed, as when using the FEM, the compliance matrix must be computed by inverting the stiffness matrix and the use of the Dunkerley formula may be inconvenient. An exception is the case in which the stiffness matrix has already been factorized, e.g., for the solution of the static problem.

A.3.4   Vector iteration method

The lowest natural frequency can be easily computed using an iterative procedure, known in general as the *vector iteration method*, and, sometimes, in structural dynamics, as the *Stodola method*. Also, in this case the procedure allows the computation of the highest eigenvalue of the dynamic matrix, and then, when used to obtain the lowest eigenfrequency, the compliance formulation must be used.

By introducing the dynamic matrix $\mathbf{D} = \mathbf{K}^{-1}\mathbf{M}$ into the first equation (1.35), it can be rewritten as

$$\mathbf{D}\mathbf{x}_0 = \frac{1}{\lambda^2}\mathbf{x}_0 \, . \tag{A.14}$$

If vector $\mathbf{x}_0$ coincides with one of the eigenvectors, the result obtained by premultiplying it by the dynamic matrix is a vector proportional to $\mathbf{x}_0$. The constant of proportionality is the relevant eigenvalue, i.e., the reciprocal of the square of the corresponding eigenfrequency λ_i.

A fast converging iterative procedure can be devised. Assume a trial vector $\mathbf{x}'$: Because any vector in the space of the configurations of the system can be expressed as a linear combination of the eigenvectors through the modal coordinates, it can be written in the form

$$\mathbf{x}' = \mathbf{\Phi}\eta' \, . \tag{A.15}$$

Equation (A.15) is, at this stage of the computation, just a formal statement, because the matrix of the eigenvectors $\mathbf{\Phi}$ is still unknown and the modal coordinates corresponding to $\mathbf{x}'$ cannot be computed.

Premultiplying vector $\mathbf{x}'$ by the dynamic matrix, a second vector $\mathbf{x}''$ is readily obtained:

$$\mathbf{x}'' = \mathbf{D}\mathbf{x}' = \mathbf{D}\mathbf{\Phi}\eta' = \sum_{\forall i} \mathbf{D}\mathbf{q}_i\eta_i' \, . \tag{A.16}$$

Remembering equation (A.16), it follows that

$$\mathbf{x}'' = \mathbf{D}\mathbf{\Phi}\eta' = \sum_{\forall i} \frac{1}{\lambda_i^2}\mathbf{D}\mathbf{q}_i\eta_i' \, . \tag{A.17}$$

Because the first vector is not an eigenvector, $\mathbf{x}''$ is not proportional to $\mathbf{x}'$. It can, however, be expressed as a linear combination of the eigenvectors of the system: $\mathbf{x}'' = \mathbf{\Phi}\eta''$. The modal coordinates of $\mathbf{x}''$ and of $\mathbf{x}'$ are then linked by the relationship

$$\eta_i'' = \frac{1}{\lambda_i^2}\eta_i'. \tag{A.18}$$

Equation (A.18) states that by premultiplying vector $\mathbf{x}'$, corresponding to the modal coordinates η', by the dynamic matrix, a second vector $\mathbf{x}''$ is obtained whose modal coordinates η'' can be obtained from those of the former simply by multiplying them by $1/\lambda_i^2$. Because ratio $1/\lambda_i^2$ for the first mode is greater (usually much greater, but here it is not strictly needed) than the same ratio for the other modes, it is clear that the first modal coordinate of vector $\mathbf{x}''$ is greater, in relative terms, than that of vector $\mathbf{x}'$. This physically means that the shape of vector $\mathbf{x}''$ is more similar to the first mode shape than that of vector $\mathbf{x}'$.

By repeating the procedure, this similarity increases, iteration after iteration, because the modal coordinates of the nth iterate $\mathbf{x}^{(n)}$ are

$$\eta_i^{(n)} = \frac{1}{\lambda_i^{2n}}\eta_i'. \tag{A.19}$$

After a certain number of iterations, it is possible to obtain a vector that coincides, apart from a small error that can be arbitrarily small, with the first eigenvector. The first eigenfrequency is then computed through equation (A.14). Practically, the starting vector can be chosen arbitrarily: It can be coincident with the static deflected shape but this is not really important. The choice of a vector that is not too different from the first eigenvector allows convergence in a smaller number of iterations to be obtained, but the method converges so fast that the number of iterations is usually very low, even if the starting vector is chosen randomly. At each iteration, the vector is normalized and premultiplied by the dynamic matrix, until the normalization factor at the ith iteration is different from that at the $(i-1)$-th iteration by a quantity smaller than a given tolerance. The last normalization factor so obtained is the reciprocal of the square of the lowest natural frequency.

If the starting vector $\mathbf{x}'$ has a first modal coordinate η_1' that is exactly 0, the procedure should theoretically converge to the second eigenvector. Actually, it is sufficient that the modal coordinate η_1', although very small, is not exactly zero, as happens as a consequence of computational approximation, and that convergence to the first mode is obtained at any rate.

The vector iteration method is similar in aim to the Dunkerley formula, with important differences. In the former case, the result can be refined to obtain an error that is arbitrarily small, independent of how much smaller the first natural frequency is with respect to the others (if the first two eigenfrequencies are very close, convergence can be slow but is, at any rate, ensured). In the second case, the error cannot be corrected and depends on the relative magnitude of the eigenvalues. While the former also allows computation of the mode shape, the second yields only the value of the frequency.

To obtain the second eigenfrequency, i.e., to make the vector iteration method converge to the second eigenvector, it is necessary to use a starting vector whose first modal coordinate is exactly equal to zero and to verify at each iteration that this feature is maintained. The last statement means that at each iteration the vector obtained must be modified to remove the small component of the first mode that creeps in due to computational errors. If this component is not removed, it would, in subsequent iterations, outgrow all other components.

Consider a generic vector $\mathbf{x}'$, whose modal coordinates are η'. It is possible to demonstrate that if the first modal coordinate is equal to zero then

$$\mathbf{x}'^T\mathbf{M}\mathbf{q}_1 = 0 . \tag{A.20}$$

Equation (A.20) can be written in the form

$$\sum_{i=1}^{n} \eta_i'\mathbf{q}_i^T\mathbf{M}\mathbf{q}_1 = 0. \tag{A.21}$$

All terms of the sum on the right hand side are equal to zero, the first because the first modal coordinate is equal to zero, all other terms owing to the fact that the eigenvectors are m-orthogonal. To ensure that the first modal coordinate of vector x' is vanishingly small, it is then enough to verify that it satisfies equation (A.20).

Because, in general, equation (A.20) is not satisfied, it is possible to use it to modify one of the elements of vector $\mathbf{x}'$ in order to transform it into a new vector $\mathbf{x}'^*$ with the required characteristics. Equation (A.20) can be written in the form

$$\sum_{i=1}^{n} \left(x_i' \sum_{j=1}^{n} m_{ij} q_{j1} \right) = 0, \tag{A.22}$$

which can readily be solved in the first element x_1':

$$x_1' = \frac{\sum_{i=2}^{n} \left(x_i' \sum_{i=1}^{n} m_{ij} q_{j1} \right)}{\sum_{j=1}^{n} m_{1j} q_{j1}}. \tag{A.23}$$

This transformation can be implemented by premultiplying vector $\mathbf{x}'$ by a matrix $\mathbf{S}$, which is usually referred to as the sweeping matrix

$$\mathbf{x}'^* = \mathbf{S}\mathbf{x}', \tag{A.24}$$

where

$$\mathbf{S} = \begin{bmatrix} 0 & \alpha_1 & \alpha_2 & \alpha_3 & \dots & \alpha_n \\ 0 & 1 & 0 & 0 & \dots & 0 \\ 0 & 0 & 1 & 0 & \dots & 0 \\ \dots & \dots & \dots & \dots & \dots & \dots \\ 0 & 0 & 0 & 0 & \dots & 1 \end{bmatrix} \tag{A.25}$$

and

$$\alpha_i = -\frac{\sum_{j=1}^{n} m_{(i+1)j} q_{j1}}{\sum_{j=1}^{n} m_{1j} q_{j1}}.$$

Instead of premultiplying the vector obtained at each iteration by the sweeping matrix, it is computationally more efficient to postmultiply the dynamic matrix by the sweeping matrix and to perform the iterative computation using the modified dynamic matrix

$$\mathbf{D}^{(2)} = \mathbf{D}\mathbf{S}. \tag{A.26}$$

Once the second eigenvector has also been computed, the computation can proceed by computing a new sweeping matrix, which also ensures that the second modal coordinate of the relevant vector vanishes, postmultiplying the original dynamic matrix by the sweeping matrix, and repeating the iterative computation. Generally speaking, to obtain the sweeping matrix for the computation of the $(m+1)$-th eigenvector, a set of m coupled linear equations must be solved. The computation of the sweeping matrix gets more complex while increasing the order of the eigenvector to be computed.

Alternatively, instead of using the sweeping matrix it is possible to resort to the so-called *deflated dynamic matrices*. The deflated matrix for the computation of the second eigenvector can be computed using the formula

$$\mathbf{D}^{(2)} = \mathbf{D} - \frac{1}{\lambda_1^2}\mathbf{q}_1\mathbf{q}_1^T\mathbf{M}\,, \qquad (A.27)$$

where the first eigenvector has been normalized in such a way that the first modal mass has a unit value.

Equation (A.27) can be proved by simply noting that when a vector $\mathbf{x}'$ is premultiplied by matrix $\mathbf{D}^{(2)}$, it follows that

$$\mathbf{D}^{(2)}\mathbf{x}' = \mathbf{D}\mathbf{x}' - \frac{1}{\lambda_1^2}\mathbf{q}_1\mathbf{q}_1^T\mathbf{M}\mathbf{x}'\,, \qquad (A.28)$$

i.e., writing the equation in terms of modal coordinates,

$$\mathbf{D}^{(2)}\mathbf{x}' = \sum_{i=1}^{n}\eta_i\mathbf{D}\mathbf{q}_i - \frac{1}{\lambda_1^2}\sum_{i=1}^{n}\eta_i\mathbf{q}_1\mathbf{q}_1^T\mathbf{M}\mathbf{q}_i\,. \qquad (A.29)$$

Remembering that the eigenvectors are m-orthogonal, only one of the terms of the last sum in equation (A.29) is different from zero; the term in which the first modal mass, which has a unit value, is present. Equation (A.29) reduces to

$$\mathbf{D}^{(2)}\mathbf{x}' = \eta_1\left[\mathbf{D}\mathbf{q}_1 - \frac{1}{\lambda_1^2}\mathbf{q}_1\right] + \sum_{2=1}^{n}\eta_i\mathbf{D}\mathbf{q}_i = \sum_{i=2}^{n}\eta_i\mathbf{D}\mathbf{q}_i\,. \qquad (A.30)$$

The right-hand side of equation (A.30) does not contain the first eigenvector, and the deflated matrix expressed by equation (A.27) allows performance of the iterations without the danger that the computation again converges on the first mode. In a similar way, it is possible to show that the deflated matrix for the computation of the third eigenfrequency can be obtained from $\mathbf{D}^{(2)}$ using the formula

$$\mathbf{D}^{(3)} = \mathbf{D}^{(2)} - \frac{1}{\lambda_2^2}\mathbf{q}_2\mathbf{q}_2^T\mathbf{M}\,. \qquad (A.31)$$

Similar formulas hold for all subsequent modes. Note that in this case the computation of each mode is not more difficult than the computation of the previous ones. However, the approximation with which the results are obtained gets worse and the use of the vector iteration method, using both sweeping matrices and deflated matrices is advisable only when a very small number of eigenfrequencies are to be obtained.

A.3.5 *Transformation of the matrices of the eigenproblem*

Many techniques aimed at solving the eigenproblem take advantage of various transformations of the relevant matrices, which, while leaving unmodified the eigenvalues and eigenvectors or modifying them in a predetermined way, allow the solution to be obtained in a much simpler way. The first transformation is usually referred to as *eigenvalue shifting*. If the stiffness matrix is substituted by

$$\mathbf{K}^* = \mathbf{K} - a\mathbf{M}\,, \tag{A.32}$$

the eigenvalues of the modified problem $(\mathbf{K}^* - \lambda^{*2}\mathbf{M}^*)\mathbf{x} = 0$ are related to those of the original eigenproblem by the simple relationship

$$\lambda^{*2} = \lambda^2 - a\,. \tag{A.33}$$

Transformation (A.32) can be very useful when the original stiffness matrix is singular as, with an appropriate choice of the eigenvalue shift a, it is possible to obtain a matrix $\mathbf{K}^*$ that is positive definite. Another use of the eigenvalue shifting is that of hastening the convergence of iterative techniques: Because the speed of convergence depends on the ratio between the second eigenvalue and the first one, an appropriate shift that increases this ratio can allow faster computations.

Consider a transformation of the type

$$\mathbf{K}^* = \mathbf{Q}\mathbf{K}\mathbf{Q}^T\,; \qquad \mathbf{M}^* = \mathbf{Q}\mathbf{M}\mathbf{Q}^T\,. \tag{A.34}$$

Under wide assumptions on the transformation matrix $\mathbf{Q}$, the eigenvalues of the transformed problem are the same as those of the original one. If the transformed matrices are diagonal, the eigenproblem is immediately solved.

Many methods have been developed with the aim of determining a transformation matrix that can diagonalize the mass and stiffness matrices, usually working in subsequent steps. A particular case is that of Jacobi method, devised to deal with the case in which the eigenproblem, reduced to standard form, has a symmetrical dynamic matrix, i.e., matrix $\mathbf{M}$ is an identity matrix, possibly multiplied by a

constant, and then the eigenvectors are orthogonal. Because matrix
M is already diagonal, the transformation matrix must be orthogonal
and can be assumed to be a rotation matrix. The transformation of
the stiffness matrix can thus be thought of as a sequence of rotations
of the dynamic matrix, until a diagonal matrix is obtained. Note
that an infinity of rotations is theoretically needed to obtain exactly
a diagonal matrix, but, in practice, with a finite number of steps a
matrix that is diagonal within the required accuracy is obtained.

A set of n^2 rotations applied to all combinations of rows and
columns, or better, of $(n^2 - n)/2$ rotations, is referred to as a *Jacobi
sweep*. The number of sweeps needed to achieve the required precision is in most cases between 6 and 10, if particular strategies are
followed in the procedure. The total number of matrix rotations is
then between $3n^2$ and $6n^2$. Many computer programs based on the
Jacobi procedure, with different modifications to hasten convergence
and extend it to cases that cannot be reduced to a symmetric dynamic matrix, are in common use and have been included in dynamic
analysis codes.

Other methods use similar iterative sequences of matrix transformations, such as the LR algorithm and the QR algorithm. The latter
is often considered the most efficient general-purpose algorithm to
find all eigenvalues and eigenvectors, real or complex, of a general
matrix. A method that is now considered very efficient for the computation of a reduced set of eigenvalues is the Lanczos method. It is
based on the transformation of the relevant matrices into tridiagonal
matrices and in the subsequent computation of selected eigenvalues.
The various factorization techniques mentioned in Section A.2, such
as LU or Choleski factorization, are often used before starting the
eigenvalue solution procedure. Also balancing procedures, aimed at
avoiding large differences between the elements of the matrices can
be very useful. An eigensolution code is actually a sequence of many
procedures that transform the relevant matrices, find the eigenvalues
and the eigenvectors, and backtransform the results.

A.3.6 Subspace iteration technique

The subspace iteration method is one of the most popular approaches
to the computation of the first m eigenvalues and eigenvectors, where
$m < n$. The method starts by stating that the m eigenvectors of interest are a linear combination of p (with $p > m$) vectors **r** arbitrarily

chosen

$$\mathbf{q}_i = \mathbf{Q}\mathbf{a}_i , \qquad (A.35)$$

where

$$\mathbf{Q} = [\mathbf{r}_1\mathbf{r}_2 \ldots \mathbf{r}_p] \qquad \text{and} \qquad i = 1,2,\ldots,m.$$

Vector $\mathbf{a}_i$ contains the p coefficients of the linear combination yielding the ith eigenvector. The size of the vectors and matrices is n for $\mathbf{q}_i$ and $\mathbf{r}$, $n \times p$ for $\mathbf{Q}$, and p for $\mathbf{a}_i$. This procedure has an immediate geometrical meaning: It states that the m eigenvectors that are sought lie in a p-dimensional subspace of the space of the configurations, which is identified by vectors $\mathbf{r}_i$.

The Rayleigh quotient

$$R = \frac{\mathbf{q}_i^T \mathbf{K}\mathbf{q}_i}{\mathbf{q}_i^T \mathbf{M}\mathbf{q}_i} = \frac{\mathbf{a}_i^T \mathbf{K}^*\mathbf{a}_i}{\mathbf{a}_i^T \mathbf{M}^*\mathbf{a}_i}, \qquad (A.36)$$

where matrices $\mathbf{M}^*$ and $\mathbf{K}^*$ are obtained through transformation (A.34), coincides with the ith eigenvalue if $\mathbf{q}_i$ coincides exactly with the eigenvector. Moreover, it is possible to state that the linear combination coefficients $\mathbf{a}_i$ leading to an eigenvector can be obtained by imposing a stationarity condition of the Rayleigh quotient (A.36). This stationarity condition can be expressed by the equation

$$(\mathbf{K}^* - R\mathbf{M}^*)\mathbf{a} = 0. \qquad (A.37)$$

Equation (A.37) defines an eigenproblem yielding the Rayleigh quotients, i.e., the eigenvalues and the corresponding vectors $\mathbf{a}_i$, which allow the eigenvectors of the original problem to be found. Obviously, because the size of the eigenproblem is p, only p eigenvalues can be found. Due to the reduced size of the eigenproblem, any standard technique, such as the Jacobi method can be used without further problems. The eigenvectors so obtained can be transformed back to the original n-dimensional space by premultiplying them by matrix $\mathbf{Q}$. If the first p eigenvectors of the original problem lie exactly in the subspace identified by matrix $\mathbf{Q}$, the solution so obtained would be exact. Because vectors $\mathbf{r}_i$ have been chosen more or less arbitrarily, the solution is only an approximation. Ritz vectors, as defined in Section 1.10, can be used and the approach outlined earlier is usually referred to as the *Ritz method* for the computation of eigenvalues and eigenvectors.

An iterative technique aimed at refining the result obtained through the Ritz method is the essence of the subspace iteration technique. The computation starts by choosing a set of p initial trial vectors, where the number p of dimensions of the subspace is greater than the number m of eigenvalues to be computed. A rule generally followed is to choose p as the minimum value between $2m$ and $m+8$. From the p trial vectors $\mathbf{x}_i$ a set of Ritz vectors $\mathbf{r}_i^*$ is computed through the equation

$$\mathbf{KQ} = \mathbf{MX}, \qquad (A.38)$$

where matrix $\mathbf{X}$ contains vectors $\mathbf{x}_i$. The matrices are then transformed to the subspace defined by the Ritz vectors and the eigenproblem is solved using the Jacobi method, as outlined earlier.

A convergence test, aimed at verifying whether the first m eigenvalues obtained are close enough to the true eigenvalues, is performed. If this is not the case, the eigenvectors so obtained are assumed to be new trial vectors $\mathbf{x}_i$, and the procedure is repeated. The procedure converges to the first m eigenvalues unless one of the trial vectors is m-orthogonal to one of the eigenvectors to be found. It is possible to devise a procedure to verify this occurrence and modify the initial choice accordingly. The first trial vector can be assumed arbitrarily, for example, a vector with all unit elements, the second as a vector with all zero terms except the one in the position in which the original matrices have the smallest ratio k_{ii}/m_{ii}, which is assumed to have unit value. The subsequent vectors are similar to the second one, with the second, third, and so on smallest value of k_{ii}/m_{ii}. Because at each iteration a set of equations whose coefficient matrix is $\mathbf{K}$ has to be solved, the factorization of this matrix can be performed only once at the beginning of the computation and need not be repeated anymore.

A.4   Solution of nonlinear sets of equations

The solution of nonlinear sets of equations is still a difficult problem for which a completely satisfactory general solution does not exist. If the set can be reduced to a single nonlinear equation, the bisection method, although usually not very efficient, ensures that all real solutions can be found. Many other methods are applicable to this case. In the case of a set with many equations, two approaches, both iterative, are usually possible.

The simplest one is the use of a Jacobi or Gauss-Seidel iterative procedure, already seen for linear sets of equations. If the set of nonlinear equations is written separating the diagonal part of the matrix of the coefficients of the linear part from the out-of-diagonal part as in equation (A.7), the equation allowing the obtaining of $\mathbf{x}^{(i+1)}$, at the $(i+1)$-th iteration from vector $\mathbf{x}^{(i)}$ at the ith iteration is

$$\mathbf{A}^{*}\mathbf{x}^{(i+1)} = \mathbf{b} - \mathbf{A}^{**}\mathbf{x}^{(i)} + \mathbf{g}^{(i)}, \qquad (A.39)$$

where $\mathbf{g}(x)$ is the nonlinear part of the set of equations.

The convergence of the procedure is, in general, not sure, depending also on the choice of the trial vector used to start the computation and, if multiple solutions exist, their domains of attraction can have complex shapes.

The Newton-Raphson algorithm is often regarded as the best choice for the solution of sets of nonlinear equations. It is based on an iterative solution of linear equations obtained through a series expansion of the nonlinear original equations truncated after the first term. It is performed by writing the equations to be solved in the form

$$\mathbf{p}(x) = 0 \qquad (A.40)$$

and expanding the nonlinear functions $\mathbf{p}(x)$ in the neighbourhood of the solution $\mathbf{x}^{(o)}$ in the form

$$\mathbf{p}(x) = \mathbf{p}(x^{(o)}) + \mathbf{S}(x^{(o)}) \left(\mathbf{x} - \mathbf{x}^{(o)}\right), \qquad (A.41)$$

where the elements of the Jacobian matrix $\mathbf{S}$ are

$$S_{ij} = \frac{\partial p_i(x)}{\partial x_j}. \qquad (A.42)$$

The equation allowing the obtaining of $\mathbf{x}^{(i+1)}$ at the $(i+1)$-th iteration from vector $\mathbf{x}^{(i)}$ at the ith iteration is

$$\mathbf{x}^{(i+1)} = -h\mathbf{S}^{-1}\mathbf{x}^{(i+1)}\{p(x^{(i)})\}, \qquad (A.43)$$

where h is a relaxation factor that can be used to hasten convergence but is usually taken equal to unity.

Usually the method converges to one of the solutions of the equation, but the convergence characteristics are strongly influenced by the initial assumption of the trial vector $\mathbf{x}^{(0)}$. There are cases in which, with selected values of $\mathbf{x}^{(0)}$, the iterative procedure does not lead to convergence but locks itself in a cycle in which a number of vectors $\mathbf{x}$ are cyclically found. When multiple solutions exist, the domains of attraction of the various solutions can have very complicated shapes, with fractal geometries often found. Moreover, solutions that are physically unstable also have their own domain of attraction: solutions that are physically impossible can be found when starting from selected initial values. The numerical stability of the solution obtained through the Newton-Raphson method then has nothing to do with the physical stability of the same solution. Recently, much research work has been devoted to the Newton-Raphson method, and the bibliography on the subject is rapidly growing.[2].

A.5   Numerical integration in time of the equation of motion

An increasingly popular approach to the computation of the time history of the response from the time history of the excitation is the numerical integration of the equation of motion. It must be expressly stated that while Duhamel's integral can be applied only to linear systems, the numerical integration of the equation of motion can also be performed for nonlinear systems. However, any solution obtained through this numerical approach must be considered the result of a numerical experiment and usually gives little general insight on the relevant phenomena. The numerical approach does not substitute other analytical methods but rather provides a very powerful tool to investigate cases that cannot be dealt with in other ways.

There are many different methods that can be used to perform the integration of the equation of motion. All of them operate following the same guidelines: The state of the system at time $t + \Delta t$ is computed from the known conditions that characterize the state of the system at time t. The finite time interval Δt must be small

---

[2]The books by H.O. Peitgen: *Newton's method and dynamical systems*, Kluwer Academic Publishers, Dordrecht, 1988, and *The beauty of fractals*, Springer, New York, 1986, can be very useful.

enough to allow the use of simplified expressions of the equation of motion without incurring errors that are too large. The mathematical simulation of the motion of the system is thus performed step by step, increasing the independent variable t with subsequent finite increments Δt.

The various methods use different simplified forms of the equation of motion and, consequently, the precision with which the conditions at the end of each step are obtained depends on the particular method used. All the simplified forms must tend to the exact equation of motion when the time interval Δt tends to zero. It is therefore obvious that the higher the precision obtainable at each step from a given method is, the longer the time interval that allows to obtaining the required overall precision. The choice of the method is therefore a trade-off between the simplicity of the expression used to perform the computation in each step, which influences the computation time required, and the total number of steps needed to follow the motion of the system for the required time.

The simplest methods are those based on the substitution of the differentials in the equation of motion with the corresponding finite differences and, among them, the simplest is the so-called Euler method, which operates in the phase space, i.e., requires the transformation of the second-order differential equations of motion into a set of first-order equations. In the case of linear systems, it deals with equation (1.11). The ratios between the finite differences $(\mathbf{z}_2 - \mathbf{z}_1)/\Delta t$ computed between instants t_2 and t_1, separated by the time interval Δt, are substituted to the derivative $\dot{\mathbf{z}}$, and the approximate average values $(\mathbf{z}_2 + \mathbf{z}_1)/2$ are substituted to the instantaneous values of the same variables and the equation that allows the computation of the state variables at time t_2 is

$$(2\mathbf{I} - \Delta t \boldsymbol{A})\,\mathbf{z}_2 = (2\mathbf{I} + \Delta t \boldsymbol{A})\,\mathbf{z}_1 + \Delta t \boldsymbol{B}\left[\mathbf{u}(t_2) + \mathbf{u}(t_1)\right] . \qquad (\text{A.44})$$

The matrix of the coefficients of this set of linear equations is constant if the step of integration is not changed during the simulations and needs to be factorized only once.

In the case of systems with parameters that are variable with time, this does not occur, and the relevant matrix must be factorized at each step. In the case of nonlinear systems, the nonlinear part of the equation can be introduced into functions $\mathbf{u}$, which become $\mathbf{u}(z_i, t)$, and the discretization in time is usually performed by replacing the differentials with the finite differences and using the values of the state variables at time t_1

$$z_2 = z_1 + \Delta t \left(\boldsymbol{A} z_1 + \boldsymbol{B} u(z_1, t_1) \right). \tag{A.45}$$

Another method that is based on the direct substitution of the finite differences to the differentials but operates directly with the second-order equations is the central differences method. The first and second derivatives of the displacement in the generic instant t_i can be expressed as functions of the positions assumed by the system in the same instant and in those that precede and follow it by the time interval Δt

$$\begin{aligned}
\dot{\mathbf{x}}_i &= (\mathbf{x}_{i+1} - \mathbf{x}_{i-1}) \frac{1}{2\Delta t}, \\
\ddot{\mathbf{x}}_i &= (\mathbf{x}_{i+1} - 2\mathbf{x}_i + \mathbf{x}_{i-1}) \frac{1}{(\Delta t)^2}.
\end{aligned} \tag{A.46}$$

By writing the dynamic equilibrium equation of the system at time t_i using the expressions (A.46) for the derivatives and solving it in $\mathbf{x}_{i+1}$, it follows that

$$\begin{aligned}
\left(2\mathbf{M} + \Delta t\mathbf{C} \right) \mathbf{x}_{i+1} &= \left(4\mathbf{M} - 2(\Delta t)^2\mathbf{K} \right) \mathbf{x}_i + \\
&+ \left(-2\mathbf{M} + \Delta t\mathbf{C} \right) \mathbf{x}_{i-1} + 2(\Delta t)^2 \mathbf{f}(t_i).
\end{aligned} \tag{A.47}$$

Equation (A.47) allows computation of the position at time t_{i+1} once the positions at times t_{i-1} and t_i are known. Note that if the time increment Δt is kept constant, the factorization of the matrix of coefficients of equation (A.47) can be performed only once at the beginning of the computation.

The computation must start from the initial conditions, which are usually expressed in terms of positions and velocities at time t_o. The central differences method, however, requires knowledge of the positions at two instants that precede the relevant one and not of positions and velocities at a single instant. The positions at the instant that precede the initial time can be easily extrapolated

$$\mathbf{x}_{-1} = \mathbf{x}_0 - \mathbf{v}_0 \Delta t + \ddot{\mathbf{x}}_0 \frac{(\Delta t)^2}{2}, \tag{A.48}$$

where the acceleration $\ddot{\mathbf{x}}_0$ at time t_o is easily computed from the equation of motion. The iterative computation can thus start.

The central differences method is said to be *explicit*, because the dynamic equilibrium equation is written at time t_i, at which the position is known. On the contrary, the dynamic equilibrium equation can be written at time t_{i+1}, at which the position is unknown. In this case, the method is said to be *implicit*.

One of the most common implicit methods is the so-called Newmark method. It is based on the extrapolation of the positions and velocities at time t_{i+1} as functions of the accelerations at the same time, which is assumed to be an unknown. Their expressions are

$$\dot{\mathbf{x}}_{i+1} = \dot{\mathbf{x}}_i + \Delta t \left[(1 + \gamma)\ddot{\mathbf{x}}_i + \gamma\ddot{\mathbf{x}}_{i+1} \right],$$

$$\mathbf{x}_{i+1} = \mathbf{x}_i + \Delta t\,\dot{\mathbf{x}}_i + (\Delta t)^2 \left[\left(\tfrac{1}{2} - \beta \right) \ddot{\mathbf{x}}_i + \beta\ddot{\mathbf{x}}_{i+1} \right]. \tag{A.49}$$

Parameter β is linked to the assumed time history of the acceleration in the time interval from t_i to t_{i+1}. If, for example, the acceleration is assumed to be a constant, equal to its average value in the mentioned interval, it follows that $\beta = 1/4$. A linearly variable acceleration leads to $\beta = 1/6$. Parameter γ, which is usually stated at the value $1/2$, controls the stability of the method.

The equation of motion written at time t_{i+1} can be used to obtain the displacements $\mathbf{x}_{i+1}$ at the end of the time step. It is

$$\left(2\mathbf{M} + \Delta t\mathbf{C} + 2\beta(\Delta t)^2\mathbf{K} \right)\mathbf{x}_{i+1} = \left(4\mathbf{M} - 2(1 - 2\beta)(\Delta t)^2\mathbf{K} \right)\mathbf{x}_i +$$

$$+2\beta(\Delta t)^2 \left[\mathbf{f}(t_{i+1}) + \left(\tfrac{1}{\beta} - 2 \right) \mathbf{f}(t_i) - \mathbf{f}(t_{i-1}) \right] +$$

$$-\left(2\mathbf{M} + \Delta t\mathbf{C} + 2\beta(\Delta t)^2\mathbf{K} \right)\mathbf{x}_{i-1}. \tag{A.50}$$

Also in this case the matrix of the coefficients of equation (A.50) does not change from step to step and can be factorized only once.

Another implicit method is the so-called Houbolt method, in which the response of the system is approximated by a third-power law spanning for three time intervals of amplitude Δt. The acceleration and velocity at time $t + \Delta t$ are expressed by the following functions of the position at time $t - 2\Delta t$, $t - \Delta t$, t, and $t + \Delta t$

$$\ddot{\mathbf{x}}_{i+1} = (2\mathbf{x}_{i+1} - 5\mathbf{x}_i + 4\mathbf{x}_{i-1} - \mathbf{x}_{i-2})\frac{1}{(\Delta t)^2}\,,$$

$$\dot{\mathbf{x}}_{i+1} = (11\mathbf{x}_{i+1} - 18\mathbf{x}_i + 9\mathbf{x}_{i-1} - 2\mathbf{x}_{i-2})\frac{1}{6\Delta t}\,. \tag{A.51}$$

By introducing the values of the velocity and acceleration computed earlier into the equation of motion written at time $t + \Delta t$, the following expression, which can be solved in the unknown values of the displacements, is readily obtained:

$$\left(2\mathbf{M} + \tfrac{11}{6}\Delta t\mathbf{C} + (\Delta t)^2\mathbf{K}\right)\mathbf{x}_{i+1} = \left(5\mathbf{M} + 3\Delta t\mathbf{C} + (\Delta t)^2\mathbf{K}\right)\mathbf{x}_i +$$

$$- \left(4\mathbf{M} + \tfrac{3}{2}\Delta t\mathbf{C}\right)\mathbf{x}_{i-1} + \left(\mathbf{M} + \tfrac{1}{3}\Delta t\mathbf{C}\right)\mathbf{x}_{i-2} + (\Delta t)^2\mathbf{f}(t_{i+1})\,. \tag{A.52}$$

To start the computation, the positions at times 0, $-\Delta t$, and $-2\Delta t$ must be known. Because this step of the computation is not critical, a rough approximation can be obtained by simply assuming constant speed before time $t = 0$. A better approximation can be obtained using the second equation (A.46) for the acceleration at time 0 (which can be computed directly from the equation of motion in which the initial conditions for the position and speed have been introduced), the following equation expressing the velocity

$$\dot{\mathbf{x}}_0 = (2\mathbf{x}_1 + 3\mathbf{x}_0 - 6\mathbf{x}_{-1} + \mathbf{x}_{-2})\frac{1}{6\Delta t} \tag{A.53}$$

and equation (A.52) with $i = 0$. They are three linear equations that yield the unknown positions at times Δt, $-\Delta t$ and $-2\Delta t$.

Other methods, such as the Wilson method, are based on the incremental formulation of the equation of motion. The dynamic equilibrium equation at time t_{i+1} can be written in the form

$$\mathbf{M}(\ddot{\mathbf{x}}_i + \Delta\ddot{\mathbf{x}}) + \mathbf{C}(\dot{\mathbf{x}}_i + \Delta\dot{\mathbf{x}}) + \mathbf{K}(\mathbf{x}_i + \Delta\mathbf{x}) = \mathbf{f}(t_i) + \Delta\mathbf{f}\,. \tag{A.54}$$

which can be solved in $\Delta\ddot{\mathbf{x}}$ and $\Delta\dot{\mathbf{x}}$, yielding

$$\Delta\ddot{\mathbf{x}} = \frac{6}{(\Delta t)^2}\Delta\mathbf{x} - \frac{6}{\Delta t}\dot{\mathbf{x}}_i - 3\ddot{\mathbf{x}}_i\,,$$

$$\Delta\dot{\mathbf{x}} = \frac{3}{\Delta t}\Delta\mathbf{x} - 3\dot{\mathbf{x}}_i - \frac{\Delta t}{2}\ddot{\mathbf{x}}_i\,. \tag{A.55}$$

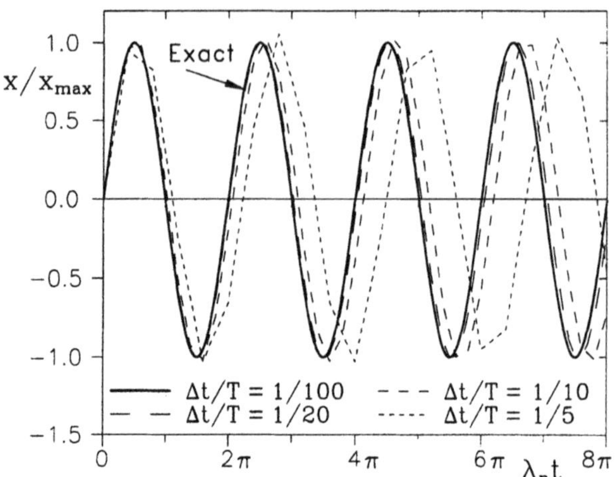

FIGURE A.1. Numerical integration of the equation of motion of a system with a single degree of freedom and comparison with the analytical solution; Newmark method with different values of the time step.

Equations (A.55) can be introduced into equation (A.54), obtaining

$$\left(6\mathbf{M} + 3\Delta t\mathbf{C} + (\Delta t)^2\mathbf{K}\right)\Delta\mathbf{x} = \mathbf{M}\left(6\Delta t\dot{\mathbf{x}}_i + 3(\Delta t)^2\ddot{\mathbf{x}}_i\right) +$$
$$+ \mathbf{C}\left(3(\Delta t)^2\dot{\mathbf{x}}_i - \frac{(\Delta t)^3}{2}\ddot{\mathbf{x}}_i\right) + (\Delta t)^2\Delta\mathbf{f}. \tag{A.56}$$

Equation (A.56) thus allows computation of the position of the system at time t_{i+1}.

Step-by-step integration methods can be unconditionally stable, when the error between the approximate and exact solutions remains bounded at increasing values of the time interval Δt. If the method is stable only for small enough values of Δt but becomes unstable if the larger values of Δt are assumed, the method is said to be conditionally stable.

The central differences method, for example, is only conditionally stable and, in the case of systems with a single degree of freedom, a value of Δt smaller than the period of the free oscillations T divided by π has to be chosen

$$\Delta t < \frac{T}{\pi} = 2\sqrt{\frac{m}{k}}. \tag{A.57}$$

The aforementioned formulation of the Wilson and Houbolt methods is unconditionally stable while the Newmark method is unconditionally stable only if $\gamma = 1/2, \beta >= 1/4$. To obtain accurate results, it is, at any rate, necessary to use values of the time interval that are smaller than those needed to achieve stability. In the case of systems with a single degree of freedom, it is, consequently, not very important to use unconditionally stable methods.

When systems with many degrees of freedom are studied, however, the choice can become quite important: The higher-order modes, although physically of little importance, can drive the solution to instability and compel the use of a value of the time increment much smaller than that which would be necessary to achieve the required precision. In this case there are two viable choices: To use unconditionally stable methods or to resort to the modal approach, which allows high-frequency modes to be discarded and has the added advantage of accepting different values of the step of integration for the various modes.

To evaluate the errors introduced by the numerical integration, it is possible to simulate the motion of a system with a single degree of freedom excited by a harmonic forcing function. In this case, two types of errors are readily seen: a decrease of the amplitude in time, which is greater than that due to the actual damping of the system, and an increase of the period. The first effect is equivalent to the introduction of a numerical damping which, in some cases, can be exploited to reduce stability problems. The second effect is well shown by the curves reported in Figure A.1, obtained integrating the equation of motion of an undamped system with a single degree of freedom using the Newmark algorithm. No forcing function has been introduced, and the initial conditions corresponding to an impulse excitation have been used.

In any case, a correct choice of the time step Δt ensures that both effects are very small and that the required accuracy is obtained.

The step-by-step time integration of the equation of motion can also be applied to the study of nonlinear systems. All methods seen earlier can be used with small modifications, provided the nonlinear forces are included in the forcing vector $\mathbf{f}$. In the case of explicit methods there is no difficulty, because the equilibrium equations are written at time t_i in which the displacements are known and the nonlinear functions can be easily computed. In the case of the central differences method, the equation yielding the values of the unknowns at time x_{i+1} is

$$
\begin{aligned}
\left(2\mathbf{M} + \Delta t\mathbf{C}\right)\mathbf{x}_{i+1} &= \left(4\mathbf{M} - 2(\Delta t)^2\mathbf{K}\right)\mathbf{x}_i + \\
&+ \left(-2\mathbf{M} + \Delta t\mathbf{C}\right)\mathbf{x}_{i-1} - 2(\Delta t)^2\mathbf{g}_i + 2(\Delta t)^2\mathbf{f}(t_i) ,
\end{aligned}
\tag{A.58}
$$

where vector function g expresses the nonlinear part of the model. The matrix of the coefficients is also, in this case, a constant and can be factorized only once at the beginning of the computation.

Other methods, particularly implicit methods, include the unknown to be found in each step into the nonlinear part of the equation of motion, requiring the solution of a set of nonlinear equations at each step, or, more often, the inclusion of the derivatives of the nonlinear functions in the matrix of the coefficients of the set of linear equations. This has the advantage of obtaining greater precision, given the time step Δt, at the cost of performing the factorization of the matrix of the coefficients at each step.

In the Newmark method, for example, if the nonlinear function g is included in the forcing functions, equation (A.50) cannot be solved directly. An iterative scheme can, however, be easily devised. When starting the computation at a new integration step, a trial value of the acceleration at the end of the step (time t_{i+1}) is assumed: $\ddot{\mathbf{x}}_{i+1}^*$. It can be chosen in an arbitrary way, but if a value not too different from the actual one is chosen, the convergence is faster. At the first trial, it is possible to assume that the acceleration at the end of the step is the same as at the beginning or, better, to extrapolate the value from the previous ones.

Once the acceleration at time t_{i+1} has been assumed, it is possible to compute the velocities and position at the same time using equation (A.49). These values can be introduced into the equation of motion written at time t_{i+1}, as the nonlinear functions can now be computed, obtaining a set of equations in the unknowns $\ddot{\mathbf{x}}_{i+1}$, which is, generally speaking, linear because the equations of motion are usually linear in the accelerations.

The values $\ddot{\mathbf{x}}_{i+1}$ obtained by solving this set of equations can be compared with the trial values $\ddot{\mathbf{x}}_{i+1}^*$. If the difference between them is small enough, accelerations $\ddot{\mathbf{x}}_{i+1}$ can be assumed to be correct and the results obtained for the state of the system at the end of the step are retained. The computation proceeds to the following step.

If, on the contrary, they are not close enough, a new set of accelerations $\ddot{\mathbf{x}}^{**}_{i+1}$ is assumed and the computation is repeated. The choice of $\ddot{\mathbf{x}}^{**}_{i+1}$ is a delicate one because the number of iterations needed to achieve convergence is strongly influenced by it. If the values computed at the end of the step are directly assumed for the new trial, the correction is generally too large, and oscillations about the correct value, which can be quite strong, can be caused.

To avoid these mathematical oscillations, some sort of numerical damping can be introduced, assuming that

$$\ddot{\mathbf{x}}^{**}_{i+1} = \ddot{\mathbf{x}}^{*}_{i+1} + \frac{1}{\beta'}(\ddot{\mathbf{x}}_{i+1} - \ddot{\mathbf{x}}^{*}_{i+1}). \tag{A.59}$$

If $\beta' = 1$, no damping is introduced. With growing β' the oscillations are reduced, until, for a "critical value" of β', the maximum speed of convergence is reached. If β' is further increased, the number of iterations needed increases again. The optimum value of β' must be obtained by trials for each case.

The Newmark method is often used for the numerical simulation of nonlinear systems, because it allows us to achieve a good precision with relatively long steps of integration. It is, however, only one of the many possible choices. Among the others, various formulations of the Runge-Kutta method, of different order, and with or without adaptive time steps, various predictor-corrector methods, and Nystrom and Bulirsh-Stoer methods can be listed.[3].

---

[3]They are described in many books, such as J. Stoer, R. Burlish, *Introduction to numerical analysis*, Springer-Verlag, New York, 1980, and W.H. Press, B.P. Flannery, S.A. Teukolsky, W.T. Vetterling, *Numerical recipes, the art of scientific computing*, Cambridge Univ. Press, Cambridge, England, 1986.

Bibliography

So many papers and books have been published on mechanical vibrations that any attempt to supply a complete or even representative bibliography is bound to fail. On only the subject of rotor dynamics, for example, Loewi and Piarulli (1961) list 554 titles. Every year, there are many conferences on rotor dynamics, and tens of papers are presented at each. Also, many papers and books that deal with other subjects contain parts related to mechanics of vibration. For example, books on springs, gears, bearings, and so on, contain chapters on the dynamic problems of these machine elements, which are often of interest to those who study the general problem of mechanical vibrations.

The author chose to give a list of books he considers oriented toward the subject of this book. It is not aimed at being comprehensive; the author is sure that many books that should have been included, were left out. Proceedings of conferences or collections of papers are also not included, or the list would have been too long. Where translations in different languages exist and an English edition was found, the latter was listed.

The bibliography is subdivided into four parts, and the titles are listed in each part by order of the year of publication. Books published in the same year are listed in alphabetical order by last name of the first author. The first part includes books on structural dynamics in general. There are also some titles on the more general topic of system dynamics, which the author thought were relevant to mechanical vibrations. No intention to list books on system dynamics in general was considered.

As a personal suggestion of the author to all who start the study of the subject of mechanical vibrations, the short book by Bishop, *Vibration*, (Cambridge Univ. Press, 1979), is well worth reading for its nonmathematical approach, which makes the reading easy, pleasant, and involving. For a more mathematical approach to the subject, the classical book of Lord Rayleigh (1877) is still very much worthwhile reading. Its title is somewhat misleading, because it deals with mechanical vibrations in general.

General mechanical vibrations

Lord Rayleigh, *The Theory of Sound*, McMillan, London, 1877 (Last edition 1944).

O. Foppl, *Grundzüge der Technischen Schwingungslehere*, Springer, Berlin, 1923.

E.A. Wedemeyer, *Automobilschwingungslehere*, Friedr. Vieweg & Sohn, Braunschweig, 1930.

G. Krall, *Meccanica Tecnica delle Vibrazioni*, Zanichelli, Bologna, 1940.

M.F. Gardner, J.L. Barnes, *Transients in Linear Systems*, Wiley, New York, 1942.

S. Timoshenko, D.H. Young, *Advanced Dynamics*, McGraw-Hill, New York, 1948.

A.A. Andronov, C.E. Chaikin, *Theory of Oscillations*, Princeton, 1948.

Y. Rocard, *Dynamique Générale des Vibrations*, Masson, Paris, 1948.

C.E. Crede, *Vibration and Shock Isolation*, Wiley, New York, 1951.

H.M. Hausen and P.F. Chenea, *Mechanics of Vibration*, Wiley, New York, 1952.

A. Tenot, *Mesure des Vibrations et Isolation des Assises de Machines*, Dunod, Paris, 1953.

F.R. Erskine Crossley, *Dynamics in Machines*, Ronald Press Co., New York, 1954.

A.R. Holowenko, *Dynamics of Machines*, Wiley, New York, 1955.

R. Mazet, *Mécanique Vibratoire*, Librairie Polytechnique Béranger, Paris, 1955.

R.E.D. Bishop, D.C. Johnson, *Vibration Analysis Tables*, Cambridge Univ. Press, Cambridge, 1956.

S.H. Crandall, *Engineering Analysis*, McGraw-Hill, New York, 1956.

J.P. Den Hartog, *Mechanical Vibrations*, McGraw-Hill, New York, 1956.

J.B. Hartman, *Dynamics of Machinery*, McGraw-Hill, New York, 1956.

N.O. Myklestad, *Fundamentals of Vibration Analysis*, McGraw-Hill, New York, 1956.

A.H. Church, *Mechanical Vibrations*, Wiley, New York, 1957.

G.W. Van Santen, *Vibrations Méchaniques*, Bibliothèque Technique Philips, Eindhoven, 1957.

R. Burton, *Vibration and Impact*, Dover, New York, 1958.

S. H. Crandall, *Random Vibration*, MIT Press, Cambridge, Mass., 1958.

C.H. Norris, R.J. Hansen, M.J. Holley, J.M. Biggs, S. Namyet, J.K. Minami, *Structural Design for Dynamic loads*, Mc Graw-Hill, New York, 1959.

V. Rocard, *General Dynamics of Vibrations*, Ungar, New York, 1960.

W.G. Bickley, A. Talbot, *An Introduction to the Theory of Vibrating Systems*, Clarendon Press, Oxford, 1961.

C.M. Harris, C.E. Crede, *Shock and Vibration Handbook*, McGraw-Hill, New York, 1961.

D.S. Jones, *Electrical and Mechanical Oscillations*, Routledge & Kegan, London, 1961.

O. Danek, L. Spacek, *Selbsterregte Schwingungen an Werkzengmaschinen*, V.E.B., Berlin, 1962.

J. Kozesnik, *Dynamics of Machines*, SNTL, Praga, 1962.

S.H. Crandall, *Random Vibration*, vol. 2, MIT Press, Cambridge, Mass., 1963.

S.H. Crandall, W.D. Mark, *Random Vibrations in Mechanical Systems*, Academic Press, New York, 1963.

R. Mathey, *Physique des Vibrations Mécaniques*, Dunod, Parigi, 1963.

J.R. Barker, *Mechanical and Electrical Vibrations*, Wiley, New York, 1964.

G. Buzdugan, *La Mesure des Vibrations mécaniques*, Eyeralles, Parigi, 1964.

J.M. Biggs, *Introduction to Structural Dynamics*, McGraw-Hill, New York, 1964.

W.C. Hurty, M.F. Rubinstein, *Dynamics of Structures*, Prentice Hall, Englewood Cliffs, 1964.

W.W. Seto, *Theory and Problems of Mechanical Vibrations*, Schaum, New York, 1964.

R.E.D. Bishop, G.M.L. Gladwell, *The Matrix Analysis of Vibration*, Cambridge Univ. Press, Cambridge, 1965.

W.T. Thompson, *Vibration Theory and Applications*, Prentice Hall, Englewood Cliffs, N.J., 1965.

S.A. Tobias, *Machine-tool Vibrations*, Blackie, Glasgow, 1965.

R.H. Cannon, *Dynamics of Physical Systems*, McGraw-Hill, New York, 1967.

L. Meirovitch, *Analytical Methods in Vibrations*, Macmillan, New York, 1967.

B.J. Lazan, *Damping of Materials and Members in Structural Mechanics*, Pergamon Press, Oxford, 1968.

R.H. Scxanlan, R. Rosenbaum, *Aircraft Vibration and Flutter*, Dover, New York, 1968.

J.C. Snowdon, *Vibration and Shock in Damped Mechanical Systems*, Wiley, New York, 1968.

J.M. Prentis, *Dynamics of Mechanical Systems*, Longman, London, 1970.

R.H. Wallace, *Understanding and Measuring Vibrations*, Springer, New York, 1970.

R. Baldacci, C.Ceradini, E. Giangreco, *Dinamica e Stabilità*, CISIA, Milano, 1971.

J.S. Bendat, A.G. Piersol, *Random Data: Analysis and Measurement Procedures*, Wiley, New York, 1971.

L.L. Beranek, *Noise and Vibration Control*, McGraw-Hill, New York, 1971.

T.V. Duggan, *Power Transmission and Vibration Considerations in Design*, Iliffe Books, London, 1971.

J.D. Robrun, C.J.Dodds, D.B. Macvean, V.R. Paling, *Random Vibration*, Springer, Vienna, 1971.

E. Sevin, W.D. Pilken, *Optimum Shock and Vibration Isolation*, The Shock and Vibration Information Center, Washington, D.C., 1971.

J.L. Shearer, A.T. Murphy, H.H. Richardson, *Introduction to System Dynamics*, Addison Wesley, Reading, Mass., 1971.

J.T. Broch, *Mechanical Vibration and Shock Measurements*, Brüel & Kjaer, Naerum, 1972.

L. Fryba, *Vibration of Solids and Structures Under Moving Loads*, Noordhoff, Groningen, 1972.

B. Parker, R. Crossley, *Modal Control, Theory and Applications*, Taylor & Francis, London, 1972.

S. H. Crandall, W,D. Mark, *Random Vibrations in Mechanical Systems*, Academic Press, New York, 1973.

D.G. Fertis, *Dynamics and Vibration of Structures*, Wiley, New York, 1973.

S. Timoshenko, D.H. Younger, W. Weaver, *Vibration Problems in Engineering*, Wiley, New York, 1974.

R.W. Clough, J. Penzien, *Dynamics of Structures*, Mc Graw-Hill, New York, 1975.

L. Meirovitch, *Elements of Vibration Analysis*, Mc Graw-Hill, New York, 1975.

D. E. Newland, *Random Vibrations and Spectral Analysis*, Longman, London, 1975.

R.D. Blevins, *Flow Induced Vibration*, Van Nostrand, New York, 1977.

A.B. Pippard, *The Physics of Vibration*, Cambridge Univ. Press, Cambridge, 1978.

F.S. Tse, I.E. Morse, R.T. Hinkle, *Mechanical Vibrations*, Allyn & Bacon, Boston, 1978.

R.E.D. Bishop, *Vibration*, Cambridge Univ. Press, Cambridge, 1979.

R.D. Blevins, *Formulas for Natural Frequency and Mode Shape*, Van Nostrand, New York, 1979.

J.B. Hunt, *Dynamic Vibration Absorbers*, Mech. Eng. Publications, London, 1979.

J.S. Bendat, A.G. Piersol, *Engineering Applications of Correlation and Spectral Analysis*, Wiley, New York, 1980.

I. Cochin, *Analysis and Design of Dynamic Systems*, Harper & Row, New York, 1980.

J. Donéa (Editor), *Advanced Structural Dynamics*, Applied Science Publ., London, 1980.

V.A. Svetlickij, *Vibrations Aleatoires des Systèmes Mécaniques*, Technique et Documentation, Parigi, 1980.

R.R. Craig, *Structural Dynamics*, Wiley, New York, 1981.

W. Soedel, *Vibrations of Shells and Plates*, Dekker, New York, 1981.

D.J. Gorman, *Free Vibration Analysis of Rectangular Plates*, Elsevier, New York, 1982.

F. Cesari, *Metodi di Calcolo nella Dinamica delle Strutture*, Pitagora, Bologna, 1983.

W. Gough, J.P.G. Richards, R.P. Williams, *Vibrations and Waves*, Wiley, New York, 1983.

M.Lalanne, P. Berthier, J. Der Hagopian, *Mechanical Vibrations for Engineers*, Wiley, New York, 1983.

V. Migulin, *Basic Theory of Oscillations*, Mir, Moscow, Russia, 1983.

H.J. Pain, *The physics of Vibrations and Waves*, Wiley, New York, 1983.

V.V. Bolotin, *Random Vibration of Elastic Systems*, Martinus Nijoff Publ., The Hague, 1984.

D.E. Newland, *An Introduction to Random Vibration and Spectral Analysis*, II ed., Longman, London, 1984.

J.S. Rao, K. Gupta, *Theory and Practice of Mechanical Vibrations*, Wiley Eastern, Delhi, 1984.

K. Zaveri, M. Phil, *Modal Analysis of Large Structures*, Brüel & Kjaer, Naerum, 1984.

R. Buckley, *Oscillations and Waves*, Adam Hilger, Bristol, 1985.

R.A. Ibrahim, *Parametric Random Vibration*, R.S.P., Wiley, New York, 1985.

L. Meirovitch, *Introduction to Dynamics and Control*, Wiley, New York, 1985.

P.C. Müller, W.O. Schiehlen, *Linear Vibrations*, Martinus Nijoff Publ., Dordrecht, 1985.

A.D. Nashif, D.I.G. Jones, J.P. Henderson, *Vibration Damping*, Wiley, New York, 1985.

J.S. Bendat, A.G. Piersol, *Random Data*, Wiley, New York, 1986.

W.K. Blake, *Mechanics of Flow-Induced Sound and Vibration*, Academic Press, New York, 1986.

G. Buzdugan, E. Mihailescu, M. Rades, *Vibration Measurement*, Martinus Nijoff Publ., Dordrecht, 1986.

D.J. Ewins, *Machinery Noise and Diagnostics*, Butterworth, London, 1986.

G.M.L. Gladwell, *Inverse Problems in Vibration*, Martinus Nijoff Publ., Dordrecht, 1986.

R. Gutowski, V.A. Swietlicki, *Dynamika Idragania Ukladow Mechanicznych*, *Panstwowa Wydawnictwo Naulowe*, Warsaw, 1986.

L. Meirovitch, *Elements of Vibrations Analysis*, McGraw-Hill, New York, 1986.

K. Piszczek, J. Niziot, *Random Vibration of Mechanical Systems*, Ellis Horwood, Chichester, 1986.

S.S. Rao, *Mechanical Vibrations*, Addison Wesley, Reading, Mass., 1986.

C.Y. Young, *Random Vibration of Structures*, Wiley, New York, 1986.

J.S. Anderson, M. Bratos Anderson, *Solving Problems in Vibrations*, Longman, Singapore, 1987.

A.A. Andronov, A.A. Vitt, S.E. Khaikin, *Theory of Oscillators*, Dover, New York, 1987.

M.T. De Almeida, *Vibrações Mechânicas para Engenheiros*, Blücher, San Paulo, 1987.

F. Küçükay, *Dynamic der Zahnradgetriebe*, Springer, Berlin, 1987.

R.H. Lyon, *Modal Analysis of Large Structures*, Brï & Kjaer, Naerum, 1987.

M. Roseau, *Vibrations in Mechanical Systems*, Springer, Berlin, 1987.

N.K. Bajaj, *The Physics of Waves and Oscillations*, McGraw-Hill, New Delhi, 1988.

C.F. Beards, *Vibrations and Control Systems*, Ellis Horwood, Chichester, 1988.

A.R. Guido, S. Della Valle, *Meccanica delle Vibrazioni, Dinamica dei Sistemi a Molti Gradi di Libertà*, Cooperativa Universitaria Editrice Napoletana, Napoli, 1988.

G.V. Berg, *Elements of Structural Dynamics*, Prentice Hall, Englewood Cliffs, 1989.

M. Del Pedro, P. Pahud, *Vibration Mechanics*, Kluver Academic Publishers, Dordrecht, 1989.

D.E. Newland, *Mechanical Vibration Analysis and Computations*, Longman, Singapore, 1989.

M.P. Norton, *Fundamentals of Noise and Vibration Analysis for Engineers*, Cambridge Univ. Press, Cambridge, 1989.

C. Carmignani, *Fondamenti di Dinamica Strutturale*, ETS, Pisa, 1990.

L. Meirovitch, *Dynamics and Control of Structures*, Wiley, New York, 1990.

M. Petyt, *Introduction to Finite Element Vibration Analysis*, Cambridge Univ. Press, Cambridge, 1990.

V. Prodonoff, *Vibrações Mechânicas, Simulaçao e Análise*, Maity Comunicaçao, Rio de Janeiro, 1990.

D. Schiff, *Dynamic Analysis and Failure Modes of Simple Structures*, Wiley, New York, 1990.

M. Del Pedro, P. Pahud, *Vibration Mechanics, Linear Discrete Systems*, Kluver Academic Publishers, Dordrecht, 1991.

J.F. Doyle, *Static and Dynamic Analysis of Structures*, Kluver Academic Publishers, Dordrecht, 1991.

G.I. Schuëller (Editor), *Structural Dynamics*, Springer, Berlin, 1991.

A.A. Shabana, *Theory of Vibration. Vol. I: An Introduction; Vol. II Discrete and Continuous Systems*, Springer, New York, 1991.

G.B. Warburton, *Reduction of Vibrations*, Wiley, New York, 1991.

V. Wowk, *Machinery Vibration, Measurement and Analysis*, McGraw Hill, New York, 1991.

A.D. Dimarogonas, S. Haddad, *Vibration for Engineers*, Prentice Hall, Englewood Cliffs, 1992.

D. Guicking, *Active Noise and Vibration Control, Reference Bibliography*, University of Göttingen, 1992.

S. Kaliski (Editor), *Vibrations and Waves*, Elsevier, Amsterdam, 1992.

J.L. Junkins, Y.Kim, *Introduction to Dynamics and Control of Flexible Structures*, AIAA, Washington, 1993.

S.G. Kelly, *Fundamental of Mechanical Vibrations*, McGraw-Hill, New York, 1993.

B.G. Koronev, L.M. Reznikov, *Dynamic Vibration Absorbers, Theory and Technical Applications*, Wiley, Chichester, 1993.

D.K. Miu, *Mechatronics, Electromechanics and Contromechanics*, Springer, New York, 1993.

D.E. Newland, *An Introduction to Random Vibrations, Spectral and Wavelet Analysis*, Longman, Singapore, 1993.

W.T. Thomson, *Theory of Vibration with Applications*, Chapman & Hall, London, 1993.

A. Bolton, *Structural Dynamics in Practice, a Guide for Professional Engineers*, McGraw-Hill, New York, 1994.

D.J. Inman, *Engineering Vibration*, Prentice Hall, Englewood Cliffs, 1994.

C. F. Beards, *Engineering Vibration Analysis with Applications to Control Systems*, Arnold, London, 1995.

K.G. McConnel, *Vibration Testing, Theory and Practice*, Wiley, New York, 1995.

S.S. Rao, *Mechanical Vibrations*, Addison Wesley, Reading, 1995.

C. F. Beards, *Structural Vibration, Analysis and Damping*, Arnold, London, 1996.

J.F. Doyle, *Wave Propagation in Structures*, Springer, New York, 1997.

M. Géradin, D. Rixen, *Mechanical Vibrations, Theory and Application to Structural Dynamics*, Wiley, New York, 1997.

Rotordynamics

F.M. Dimentberg, *Flexural Vibrations of Rotating Shafts*, Butterworth, London, 1961.

A. Tondl, *Some Problems of Rotor Dynamics*, Chapman & Hall, London, 1965.

R.G. Loewi, V.J. Piarulli, *Dynamics of Rotating Shafts*, The Shock and Vibration Information Center, Naval Res. Lab., Washington, D.C., 1969.

L. Buzzi, *Equilibratura*, CEMB, Mandello del Lario, 1971.

G. Schweitzer, *Critical Speeds of Gyroscopes*, Springer, Vienna, 1972.

H. Schneider, *Balancing Technology*, Schenck, Darmstadt, 1977.

A.D. Dimarogonas, S.A. Paipetis, *Analytical Methods in Rotor Dynamics*, Applied Science Publishers, London, 1983.

O. Marenholtz, *Dynamics of Rotors*, Springer, Vienna, 1984.

J. Rao, *Rotor Dynamics*, Wiley Eastern, Delhi, 1985.

N.F. Rieger, *Balancing of Rigid and Flexible Rotors*, The Shock and Vibration Information Center, U.S. DoD, Washington, D.C., 1986.

J.M. Vance, *Rotordynamics of Turbomachinery*, Wiley, New York, 1988.

M.S. Darlow, *Balancing of High Speed Machinery*, Springer, New York, 1989.

M.J. Goodwin, *Dynamics of Rotor-Bearing Systems*, Unwin Hyman, London, 1989.

M. Lalanne, G. Ferraris, *Rotordynamics Predictions in Engineering*, Wiley, New York, 1990.

D. Childs, *Turbomachinery Rotordynamics* , Wiley, New York, 1993.

E. Krämer, *Dynamics of Rotors and Foundations*, Springer, Berlin, 1993.

C.W. Lee, *Vibration Analysis of Rotors*, Kluver Academic Publishers, Dordrecht, 1993.

M. Lalanne, G. Ferraris, *Rotordynamics Predictions in Engineering, 2nd ed.*, Wiley, New York, 1998.

Torsional vibrations

W.A. Tuplin, *Torsional Vibration, Elementary Theory and Design Calculations*, Chapman & Hall, London, 1934.

T. O'Callaghan, *Berechnung von Torsionsschwingungen an Hand der Theorie der Effektiven Massen*, Fachverlag Schiele & schön, Berlin, 1958.

E.J. Nestorides, *A Handbook on Torsional Vibrations*, Cambridge Univ. Press, Cambridge, 1958.

W. Ker Wilson, *Torsional Vibration Problems*, Chapman & Hall, London, 1963.

Nonlinear and chaotic vibrations

J.J. Stoker, *Nonlinear Vibrations*, Interscience, New York, 1950. C. Hayashi, *Forced Oscillations in Non-linear Systems*, Nippon Printing and Publ. Comp., Osaka, 1953.

H. Kauderer, *Nichtlineare Mechanik*, Springer, Berlin, 1958.

N. Bogoliubov, I. Mitropolski, *Les Méthodes Asymptotiques en Théorie des Oscillations Non Linéaires*, Mir, Mosca, 1962.

M. Minorsky, *Nonlinear Oscillations*, Van Nostrand, Princeton, 1962.

C. Hayashi, *Nonlinear Oscillations in Physical Systems*, McGraw-Hill, New York, 1964.

N.V. Butenin, *Elements of the Theory of Nonlinear Oscillations*, Blaisdell, New York, 1965.

M. Roseau, *Vibrations Non Linéaires et Théorie de la Stabilité*, Springer, Berlin, 1966.

M. Urabe, *Nonlinear Autonomous Oscillations*, Academic Press, New York, 1967.

F. Dinca, C. Teodosiu, *Nonlinear and Random Vibrations*, Editura Academiei Rep. Soc. Romania, Bucarest, 1973.

A.H. Nayfeh, D.T. Mook, *Nonlinear Oscillations*, Wiley, New York, 1979.

V.M. Starzhinskii, *Applied Methods in the Theory of Nonlinear Oscillations*, Mir, Moscow, 1980.

P. Hagedorn, *Nonlinear Oscillations*, Clarendon Press, Oxford, 1981.

C. Sparrow, *The Lorentz Equations: Bifurcations, Chaos, and Strange Attractors*, Springer, New York, 1982.

J. Guckenheimer, P Holmes, *Nonlinear Oscillations, Dynamical Systems, and Bifurcations of Vector Fields*, Springer, New York, 1983.

C. Hayashi, *Nonlinear Oscillations in Physical Systems*, Princeton University Press, Princeton, 1985.

G. Schmidt, A. Tondl, *Nonlinear Vibration*, Cambridge Univ. Press, Cambridge, 1986.

J.M.T. Thompson, H.B. Stewart, *Nonlinear Dynamics and Chaos*, Wiley, New York, 1986.

F.C. Moon, *Chaotic Vibrations*, Wiley, New York, 1987.

T. Kapitaniak, *Chaos in Systems with Noise*, World Scientific, Singapore, 1988.

W. Szemplinska-Stupnicka, *The behaviour of Nonlinear Vibrating Systems*, Kluwer Academic Publishers, Dordrecht, 1990.

R.R. Mohler, *Nonlinear Systems*, Prentice Hall, Englewood Cliffs, 1991.

A. Tondl, *Quenching of Self-Excited Vibrations*, Academia, Prague, Czechoslovakia, 1991.

S.R. Bishop, *Nonlinearity and Chaos in Engineering Dynamics*, Wiley, Chichester, 1994.

Index

MIX
Papier aus verantwortungsvollen Quellen
Paper from responsible sources
FSC® C105338
FSC
www.fsc.org

If you have any concerns about our products,
you can contact us on
ProductSafety@springernature.com

In case Publisher is established outside the EU,
the EU authorized representative is:
Springer Nature Customer Service Center GmbH
Europaplatz 3, 69115 Heidelberg, Germany

Printed by Libri Plureos GmbH
in Hamburg, Germany